COMPLETE EDITION

TECHNOLOGY IN ACTION

ALAN EVANS

KENDALL MARTIN

MARY ANNE POATSY

Library of Congress Cataloging-in-Publication Data

Evans, Alan.
   Technology in action / Alan Evans, Kendall Martin, Mary Anne Poatsy.
   Complete ed.
     p. cm.
     ISBN 0-13-142393-2
  1. Microcomputers.   I. Martin, Kendall. II. Poatsy, Mary Anne. III. Title.
   QA76.5.E9195 2004
   004.16—dc22

                                                                                                         2003022578

Vice President/Publisher: Natalie E. Anderson
Executive Acquisitions Editor: Jodi McPherson
Acquisitions Editor: Melissa Sabella
Developmental Editor: Shannon Leuma
Editorial Assistant: Alana Meyers
Senior Media Project Manager: Cathleen Profitko
Senior Marketing Manager: Emily Williams Knight
Marketing Assistant: Nicole Beaudry
Senior Managing Editor: Gail Steier de Acevedo
Senior Project Manager, Production: April Montana
Production Assistant: Jessica Rivchin
Manufacturing Buyer: April Montana
Design Manager: Maria Lange
Art Director, Interior Design: Kevin Kall
Art Director, Cover and Illustration: April Montana
Cover and Interior Illustration: Pre-Press Company, Inc.
Cover Design: Quorum Creative Services
Composition/Full-Service Project Management: Pre-Press Company, Inc.
Cover Printer: Lehigh Press, Inc.
Printer/Binder: VonHoffmann Press

Credits and acknowledgments borrowed from other sources and reproduced, with permission, in this textbook appear on appropriate page within text (or on pages 617–624).

Microsoft® and Windows® are registered trademarks of the Microsoft Corporation in the U.S.A. and other countries. Screen shots and icons reprinted with permission from the Microsoft Corporation. This book is not sponsored or endorsed by or affiliated with the Microsoft Corporation.

Copyright © 2005 by Pearson Education, Inc., Upper Saddle River, New Jersey, 07458. Pearson Prentice Hall. All rights reserved. Printed in the United States of America. This publication is protected by Copyright and permission should be obtained from the publisher prior to any prohibited reproduction, storage in a retrieval system, or transmission in any form or by any means, electronic, mechanical, photocopying, recording, or likewise. For information regarding permission(s), write to: Rights and Permissions Department.

Pearson Prentice Hall™ is a trademark of Pearson Education, Inc.
Pearson® is a registered trademark of Pearson plc
Prentice Hall® is a registered trademark of Pearson Education, Inc.

Pearson Education LTD.   Pearson Education Australia PTY, Limited
Pearson Education Singapore, Pte. Ltd   Pearson Education North Asia Ltd
Pearson Education, Canada, Ltd   Pearson Educación de Mexico, S.A. de C.V.
Pearson Education–Japan   Pearson Education Malaysia, Pte. Ltd

10 9 8 7 6 5 4 3 2
ISBN 0-13-142393-2

DEDICATION

For my wife Patricia, whose patience, understanding,
and support made this work possible.

—Alan Evans

For all the teachers, mentors, and gurus
who have popped in and out of my life.

—Kendall Martin

For my husband, Ted, who unselfishly took on more than
his fair share and supported me throughout; and for my children,
Laura, Carolyn, and Teddy, who sacrificed having a mom
for a while to make this happen.

—Mary Anne Poatsy

ABOUT THE AUTHORS

Alan D. Evans, M.S., CPA
Alan is currently the Director of Computer Science for Montgomery County Community College. Alan's parents instilled him with a love of education at an early age. After a successful career in business, Alan finally realized his true calling was education. He has been teaching at the collegiate level for the past 4 years.

Alan makes presentations at technical conferences and meets regularly with computer science faculty and administrators from other colleges to discuss curriculum development. Currently, he is conducting student assessments to determine competency levels of incoming students in computer literacy and MS Office applications. This benchmarking will be used to develop baseline literacy standards for all incoming freshman at the college.

Kendall E. Martin, Ph.D.
Kendall has been teaching since 1988 at institutions including Villanova University, DeSalles University, Arcadia University, Ursinus College, Community College of Morris, and Montgomery County Community College at both the undergraduate and master's degree level.

Kendall's education includes a B.S. in Electrical Engineering from the University of Rochester and an M.S. and Ph.D. in Engineering from the University of Pennsylvania. She has industrial experience in research and development environments (AT&T Bell Laboratories) as well as experience from several start-up technology firms.

At Ursinus College, Kendall developed a successful faculty training program for distance education instructors, and she makes conference presentations during the year.

Mary Anne Poatsy, MBA, CFP
Mary Anne is an adjunct faculty member at Montgomery County Community College, teaching various computer application and concepts courses in face-to-face and online environments. She has been teaching introductory computer courses since 1997.

Mary Anne holds a B.A. in Psychology and Elementary Education from Mount Holyoke College and an MBA in Finance from Northwestern University's Kellogg Graduate School of Management. Mary Anne has more than 8 years of educational experience ranging from elementary and secondary education to Montgomery County Community College, Muhlenberg College, and Bucks County Community College, as well as training in the professional environment.

ACKNOWLEDGEMENTS

First and foremost, we would like to thank our students. We constantly learn from them while teaching, and they are a continual source of inspiration and new ideas.

We could not have written this book without the loving support of our families. Our spouses and children made sacrifices (mostly in time not spent with us) to permit us to make this dream into a reality.

Our heartfelt thanks go to Shannon Leuma, our developmental editor. Shannon took three novice authors and helped them become a cohesive team. She continually went above and beyond the requirements of her job in a quest to make this book special.

Although working with the entire team at Prentice Hall was a truly enjoyable experience, a few individuals deserve special mention. Jodi McPherson, our executive editor, was inspirational from the very start. She helped us grow the ideas that we had into an actual book. Jodi's support and encouragement, along with that of our acquisitions editor, Melissa Sabella, made this book live! As Senior Media Project Manager, Cathi Profitko worked tirelessly to ensure that the media accompanying the text was professionally produced and delivered in a timely fashion. Despite the inevitable problems that always crop up when producing multimedia, she handled all challenges with a smile. She is absolutely one of the hardest working individuals we have ever known. Emily Knight, who is in charge of marketing our text, has been a great supporter of this project since the beginning. Her enthusiasm is contagious as she displays a remarkable work ethic. And we can't forget Natalie Anderson, Vice President of Information Technology Business Publishing, who is our publisher. Natalie has a wonderful sense of humor, which helps smooth over the inevitable bumps in the road encountered on a project of this magnitude. Our heartfelt appreciation goes to April Montana, Senior Project Manager of Production, who worked tirelessly to ensure our book was published on time and looked fabulous. She had a very difficult task as the timeline was short, the art was complex, and there were many people with whom she had to coordinate tasks. Her dedication and hard work helped make this book a reality.

There were many people we did not meet in person at Prentice Hall, and elsewhere, who made significant contributions by designing the book, illustrating, composing the pages, producing multimedia, and securing permissions. We thank them all, particularly, supplement authors, Cindy Buell and Mike Mitri.

Also deserving of thanks are the many experienced textbook authors who provided invaluable advice and encouragement to three neophytes at the start of this project.

Many of our colleagues at Montgomery County Community College made suggestions and provided advice during this project. We appreciate all the help everyone provided, but we would particularly like to thank John Mack, Jerri Williams, Diane Coyle, and Dianne Meskauskas who worked directly on the supplements to the book.

And finally, we would like to thank the following reviewers and the many others who contributed their time, ideas, and talents to this project. We appreciate the time and energy that you put into your comments as they helped us turn out a better product.

ACKNOWLEDGEMENTS

REVIEWERS

Wilma Andrews	*Virginia Commonwealth University*
Linda Belton	*Springfield Technical Community College*
Julie Boyles	
Gerald U. Brown Jr.	*Tarrant County College*
Judy Cestaro	*California State University—San Bernardino*
Debra Chapman	*The University of South Alabama*
Françoise Corey	*CSU, Long Beach*
Thad Crews	*Western Kentucky University*
John Cusaac	*Fullerton College*
Susan N. Dozier	*Tidewater Community College*
Annette Duvall	*Albuquerque Technical Vocational Institute*
Catherine L Ferguson	*University of Oklahoma*
Beverly Fite	*Amarillo College*
Richard A. Flores	*Citrus College*
Sherry Green	*Purdue—Calumet*
Debra Gross	*The Ohio State University*
Judy Irvine	*Seneca College*
Kathy Johnson	*DeVry Chicago*
Stephanie Jones	*South Plains College*
Robert R. Kendi	*Lehigh University*
Jackie Lamoureux	*Albuquerque Technical Vocational Institute*
Judith Limkilde	*Seneca College—King Campus*
Richard Linge	*Arizona Western College*
Joelene Mack	*Golden West College*
Dana McCann	*Central Michigan University*
Lee McClain	*West Washington University*
Daniela Marghitu	*Auburn University*
Laura Melella	*Fullerton College*
Josephine G. Mendoza	*California State University, San Bernardino*
Rebecca A. Mundy	*University of Southern California*
Linda Mushet	*Golden West College*
Omar Nooraldeen	*Cape Fear Community College*
Woody Pekoske	*North Carolina State University*
Paul Quan	*Albuquerque Technical Vocational Institute*
Kriss Stauber	*El Camino College*
Neal Stenlund	*Northern Virginia Community College*
Song Su	*East Los Angeles College*
Goran Trajkovski	*Towson University*
Linda Turpen	*Albuquerque Technical Vocational Institute*
Bill VanderClock	*Bentley Business University*
Mary Anne Zlotow	*College of DuPage*

FOR THE INSTRUCTOR

Why We Wrote This Book

Our 15 years of teaching computer concepts have coincided with sweeping innovations in computing technology that have affected every facet of society. From ATM's to the World Wide Web, computers are more than ever a fixture of our daily lives—and the lives of our students. But although today's students have a greater comfort level with their digital environment than previous generations, their textbooks haven't caught up. Even the best books spend too much time introducing hardware students already know, and not enough time explaining all the fun and productive things that same hardware can do.

We wrote *Technology in Action* to address this problem by focusing on today's student. Instead of a history lesson on the microchip, we focused on what tasks students can accomplish with their PC, and skills they can apply immediately in the workplace and at home. The result is a book that sparks student interest by focusing on the material they want to learn (such as how to set up a home network), while teaching the material they need to learn (such as how networks work). The sequence of topics is carefully set up to mirror the typical student learning experience.

As they read through this text, your students will progress through stages of increasing difficulty:

1. Examining why it's important to be computer fluent and how computers impact our society
2. Examining the basic components of the computer
3. Connecting to the Internet
4. Exploring software
5. Learning the operating system and personalizing the computer
6. Evaluating and upgrading the PC
7. Exploring home networking and keeping the computer safe from hackers
8. Going mobile with cell phones, PDAs, Tablet PCs, and laptops

FOR THE INSTRUCTOR

The focus of these early chapters is on practical uses for the computer, with real-world examples to help the learner place computing in a familiar context. Exciting SOUND BYTE multimedia—fully integrated with the text—accelerates student mastery of complex topics.

For students ready to go beyond the basics, special BEHIND THE SCENES chapters venture deeper into the realm of computing through in-depth explanations of how elements of the system unit (CPU, motherboard, RAM) work. They are specifically designed to keep more experienced students engaged, and challenge them with interesting research assignments.

In addition, we have included DIG DEEPER sections throughout the book to encourage further study by taking an in-depth look at a topic of student interest, such as how a firewall works or how speech recognition software works, that they can explore at their own pace.

Now that the computer has become a ubiquitous tool in our lives, a new approach to computer concepts is warranted. This book is designed to reach the students of the 21st century.

What's Inside

- **Questions Students Ask** These section headings reflect the natural progression of a typical PC learning experience by integrating questions students ask consistently every semester.
- **Sound Bytes** These dynamic multimedia tutorials help explain complex concepts.
- **Dig Deeper** These sections cover technical topics in depth to challenge more advanced students.
- **Bits and Bytes** In these sections, students learn how to maintain their computers, how to develop good habits for safe computing, as well as other interesting facts about computers.
- **Trends in IT** These sections explore new and emerging technologies, careers in computing, and ethical considerations involved in computing.

FOR THE INSTRUCTOR

About the Multimedia—Designed by the Authors

Technology in Action features a unique and innovative student resource—interactive multimedia labs written specifically to the text—that demystify even the most complex topics. These **Sound Bytes** are indicated by an icon; at least two appear in each chapter.

The Sound Byte labs all include a multimedia lesson with a video or animation clip to illustrate a concept. After viewing the lesson students can take a multiple-choice quiz to assess their retention of the material and then immediately apply their knowledge with fun and interesting activities specifically tailored to each Sound Byte.

For example, your student reads about random access memory in Chapter 6—seeing the Sound Byte icon, she views the accompanying multimedia lab for Installing RAM. She takes the quiz, reinforcing the objective information, and then does the activity which sends her to the Crucial Technology Web site to find information on matching the correct RAM type to her system and gathering the latest prices.

For students very new to computers there are lessons in the features of PDA devices, laptop tours and more; for those ready to go beyond the basics, lessons include creating a database in Access and writing a simple Web page in Word.

For your convenience, the Sound Bytes are included on the Instructor's Resource CD and are also available to students on a CD or on the companion Web site.

FOR THE INSTRUCTOR

The Sound Bytes available are:

- A Day in the Life of a Network Technician
- Best Utilities for Your Computer
- Binary Numbers
- CD/DVD Reading & Writing
- Computer Architecture
- Connecting to the Internet
- Constructing a Simple Web Page
- CPU—Running Programs
- Creating a Web-Based E-mail Account
- Creating an Access Database
- Customizing Windows XP
- Enhancing Photos with Image-Editing Software
- File Compression
- File Management
- Finding Information on the Web
- Hard Disk Anatomy/Technology & Disk Defragmentation
- Healthy Computing
- Improving an Access Database
- Installing a CDRW Drive
- Installing a Network
- Installing a Personal Firewall
- Installing RAM
- Letting your Computer Clean up after Itself
- Looping Around in the IDE
- Memory Hierarchy
- PDAs on the Road and at Home
- Port Tour
- Programming for End Users: Macros
- Protecting Your Computer
- Questions to Ask Before you Buy a Computer
- Securing Wireless Networks
- Surge Protectors and UPSs
- Tablet and Laptop Tour
- Using Speech Recognition
- Using Windows XP to Evaluate CPU Performance
- Virtual Computer Tour
- Welcome to the Web
- What's My IP Address? (and Other Interesting Facts about Networks)
- Where does Binary Show Up?

FOR THE INSTRUCTOR

Instructor Resources

The new and improved Prentice Hall Instructor's Resource Center CD-ROM includes the tools you expect from a Prentice Hall Computer Concepts text like:

- The Instructor's Manual in Word and PDF formats
- Solutions to questions and exercises from the book and Web site
- Multiple, customizable PowerPoint Slide Presentations for each chapter
- Computer Concepts Animations
- TechTV Videos
- Image Library of IT and computer related images
- Sound Bytes
- Test Bank with TestGen & Quizmaster Software

This CD-ROM is an interactive library of assets and links. This CD writes custom "index" pages that can be used as the foundation of a class presentation or online lecture. By navigating through this CD, you can collect the materials that are most relevant to your interests, edit them to create powerful class lectures, copy them to your own computer's hard drive, and/or upload them to an online course management system.

TechTV

TechTV is the San Francisco-based cable network that showcases the smart, edgy and unexpected side of technology. By telling stories through the prism of technology, TechTV provides programming that celebrates its viewers' passion, creativity, and lifestyle.

TechTV's programming falls into three categories:

1. **Help and Information,** with shows like *The Screen Savers*, TechTV's daily live variety show featuring everything from guest interviews and celebrities to product advice and demos; Tech Live, featuring the latest news on the industry's most important people, companies, products and issues; and *Call for Help*, a live help and how-to show providing computing tips and live viewer questions.

2. **Cool Docs,** with shows like *The Tech Of . . .* , a series that goes behind the scenes of modern life and shows you the technology that makes things tick;

FOR THE INSTRUCTOR

Performance, an investigation into how technology and science are molding the perfect athlete; and *Future Fighting Machines*, a fascinating look at the technology and tactics of warfare.

3. **Outrageous Fun,** with shows like *X-Play*, exploring the latest and greatest in videogaming, and *Unscrewed with Martin Sargent*, a new late-night series showcasing the darker, funnier world of technology.

For more information, log onto **www.techtv.com** or contact your local cable or satellite provider to get TechTV in your area.

TestGen Software

TestGen Software: TestGen is a test generator program that lets you view and easily edit test bank questions, transfer them to tests, and print in a variety of formats suitable to your teaching situation. The program also offers many options for organizing and displaying test banks and tests. Powerful search and sort functions let you easily locate questions and arrange them in the order you prefer.

QuizMaster, also included in this package, allows students to take tests created with TestGen on a local area network. The QuizMaster Utility built into TestGen lets instructors view student records and print a variety of reports. Building tests is easy with Test Gen and exams can be easily uploaded into WebCT, BlackBoard, and Course Compass.

Tools for Online Learning

Training and Assessment
www2.phgenit.com/support

Prentice Hall offers Performance Based Training and Assessment in one product—Train & Assess IT. The training component offers computer-based training that a student can use to preview, learn, and review Computer Concepts, Microsoft Office applications, and other software-related skills. Built-in prescriptive testing suggests a study path based not only on student test results but also on the specific textbook chosen for the course.

The Assessment component offers computer-based testing that shares the same user interface as Train IT and is used to evaluate a student's knowledge about specific topics in Word, Excel, Access, PowerPoint, Outlook, the Internet, and Computing Concepts. It does this in a task-oriented environment to demonstrate pro-

FOR THE INSTRUCTOR

ficiency as well as comprehension of the topics by the students. Assess IT offers more administrative features for the instructor and additional questions for the student.

Assess IT also allows professors to test students out of a course, place students in appropriate courses, and evaluate skill sets.

Companion Web site
WWW.PRENHALL.COM/TECHINACTION

This text is accompanied by a Companion Web site at **www.prenhall.com/techinaction**. Features of this new site include an interactive study guide, downloadable supplements, online end-of-chapter materials, additional internet exercises, Tech TV videos, Web resource links such as Careers in IT and crossword puzzles plus Technology Updates and Bonus Chapters on the latest trends and hottest topics in information technology. All links to Internet exercises will be constantly updated to ensure accuracy for students.

Unique to the *Technology in Action* Web site are a series of innovative multimedia labs written specifically to the text that demystify even the most complex topics. These **Sound Bytes** are indicated by an icon with at least two appearing per chapter and are also available on the CD that accompanies the text. (For a list of available Sound Bytes, please see page xii.)

CourseCompass
www.coursecompass.com

CourseCompass is a dynamic, interactive online course-management tool powered exclusively for Pearson Education by Blackboard. This exciting product allows you to teach market-leading Pearson Education content in an easy-to-use, customizable format.

Blackboard
www.prenhall.com/blackboard

Prentice Hall's abundant online content, combined with Blackboard's popular tools and interface, result in robust Web-based courses that are easy to implement, manage, and use—taking your courses to new heights in student interaction and learning.

FOR THE INSTRUCTOR

WebCT
www.prenhall.com/webct

Course management tools within WebCT include page tracking, progress tracking, class and student management, a grade book, communication tools, a calendar, reporting tools, and more. GOLD LEVEL CUSTOMER SUPPORT, available exclusively to adopters of Prentice Hall courses, is provided free of charge upon adoption and provides you with priority assistance, training discounts, and dedicated technical support.

OneKey

OneKey lets you in to the best teaching and learning resources all in one place. OneKey for *Technology in Action* is all your students need for out-of-class work conveniently organized by chapter to reinforce and apply what they've learned in class and from the text. OneKey is all you need to plan and administer your course. All your instructor resources are in one place to maximize your effectiveness and minimize your time and effort. OneKey for convenience, simplicity, and success.

FOR THE STUDENT

Using This Book

This book is arranged in the order you would most likely follow when exploring the computer. You'll start in the second chapter by looking at your computer as you would if you were assembling it for the first time, exploring each piece and its function within your system. Next, because so many people use the Internet to communicate, conduct research, and even shop, Chapter 3 explores the Internet and its many features. Even if you're an experienced Internet user, this chapter will help you use the Internet more effectively and more safely.

In Chapter 4, you'll look more closely at the application software that you'll most likely encounter in your daily life, both at work and at home. Then, in Chapter 5, you'll explore your computer's operating system. In doing so, you'll learn about the different system software programs you can use to keep your computer in top shape, as well as ways in which you can keep your files and folders organized.

Once you understand the pieces of your system and the three elements that make it a useful tool (the Internet, application software, and the operating system), Chapter 6 will help you evaluate your computer system to see if it is meeting your needs. By exploring and evaluating your system's parts, you'll learn whether you need to upgrade your computer and how to go about doing so.

It's likely that you have more than one computer in your home, workplace, or school. To share common resources between computers, you need to know about networking. Thus, in Chapter 7, you'll learn about home computer networks, as well as how to protect yourself from hackers and viruses. Then, in Chapter 8, you'll explore mobile computing devices such as cell phones, PDAs, laptops, and tablet PCs.

Finally, in Chapter 9, you'll find out just how your computer's hardware really works. You'll learn more about your CPU and the types of RAM available on the market today, and how these components affect the performance of your computer.

Along the way, Technology in Focus features will teach you more about digital technology, protecting your computer and the data on it, as well as the history of the personal computer.

No matter how much you use the computer, you probably still have a lot of questions about how to use it *best*. Throughout this book, you'll find references to multimedia components called **Sound Bytes** that *show* you the answers to some frequently asked questions, such as how to set up a firewall and how to use anti-virus software effectively.

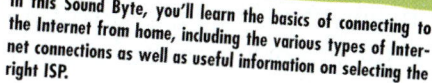

SOUND BYTE
CONNECTING TO THE INTERNET
In this Sound Byte, you'll learn the basics of connecting to the Internet from home, including the various types of Internet connections as well as useful information on selecting the right ISP.

FOR THE STUDENT

In each chapter, you'll also find **Bits and Bytes**, boxes that contain interesting facts and helpful tips on how to maintain and better use your computer.

Also scattered throughout the book are **Trends in IT** features that examine computer-related ethical issues, careers in technology, as well as emerging technologies.

Finally, throughout the book you'll find **Dig Deeper** features. These features take an in-depth look at various computer concepts, such as how a hard disk drive or a computer firewall works.

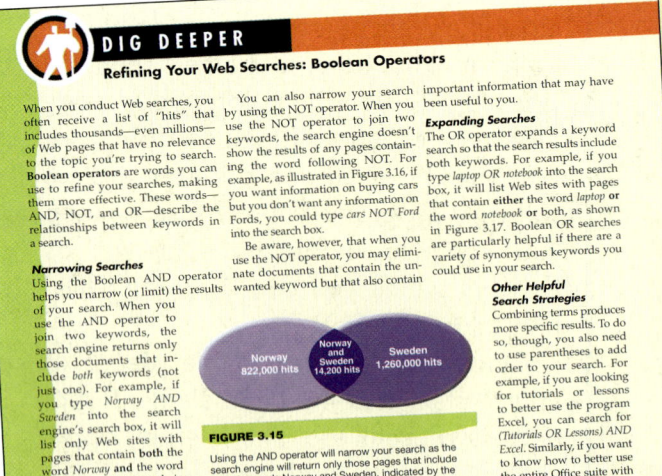

The wonderful thing about computers is that there is something new to learn about them every day. Developing an understanding of how computers can make your life easier is critical to your future success. So, let's get started on our exploration of computers.

Please visit the Web site at **www.prenhall.com/techinaction** where you will find bonus chapters on the latest trends in information technology, an interactive study guide, online end-of-chapter materials, Careers in IT, TechTV Videos, Technology Updates, crossword puzzles, and more.

CONTENTS AT A GLANCE

CHAPTER 1	Why Computers Matter to You: Becoming Computer Fluent 3
CHAPTER 2	Looking at Computers: Understanding the Parts 31
CHAPTER 3	Introducing the Internet: Making the Most of the Web's Resources 65
CHAPTER 4	Application Software: Programs that Let You Work and Play 107
TECHNOLOGY IN FOCUS	**DIGITAL ENTERTAINMENT** 150
CHAPTER 5	Using System Software: The Operating System, Utility Programs, and File Management 167
CHAPTER 6	Evaluating Your System: Understanding and Assessing Hardware 211
CHAPTER 7	Networking and Security: Connecting Computers and Keeping Them Safe from Hackers and Viruses 251
TECHNOLOGY IN FOCUS	**PROTECTING YOUR COMPUTER AND BACKING UP YOUR DATA** 294
CHAPTER 8	Mobile Computing: Keeping Your Data on Hand 309
CHAPTER 9	Behind the Scenes: Inside the System Unit 347
TECHNOLOGY IN FOCUS	**THE HISTORY OF THE PC** 376
CHAPTER 10	Behind the Scenes: Software Programming 387
CHAPTER 11	Behind the Scenes: Databases and Information Systems 429
CHAPTER 12	Behind the Scenes: Networking and Security 473
CHAPTER 13	Behind the Scenes: The Internet: How It Works 515
TECHNOLOGY IN FOCUS	**CAREERS IN IT** 554

Glossary 569
Index 585
Credits 617

TABLE OF CONTENTS

CHAPTER 1   Why Computers Matter to You: Becoming Computer Fluent   3

Technology in Action: Becoming Computer Fluent   3

Becoming a Savvy Computer User and Consumer   4

Being Prepared for Your Career   6
Computers in Retail: "Let Me Look That Up For You"   6
Computers in the Arts: "Shall We Dance?"   7
Computers in the Medical Field:
 "Bring the Virtual Patient in, Doctor"   8
Computers in the Detective Fields:
 "Put Down That Mouse—You're Under Arrest!"   9
Computers in the Legal Fields:
 "Welcome to the Virtual Courtroom"   9
Computers in the Classroom:
 "And on the Left, You See the Mona Lisa"   10
Computers at Home:
 "Just Program It and Forget It"   11

Getting Ready for the Technology of Tomorrow   13
Nanoscience:
 The Next "Big" Thing   13
Biomedical Chip Implants:
 Combining Humans With Machines?   14
Artificial Intelligence:
 Will Computers Become Human?   15

Understanding the Challenges Facing a Digital Society   15

Becoming Computer Fluent   16
Computers Are Data Processing Devices   16
Bits and Bytes:
 The Language of Computers   17
Computer Hardware   17
Computer Software   19
Computer Platforms: PCs and Macs   20
Specialty Computers   21

Using This Book   22

TABLE OF CONTENTS

CHAPTER 2 — Looking at Computers: Understanding the Parts 31

Technology in Action:
   Setting Up the System 31

Your Computer's Hardware 32

Input Devices 32
Keyboards 32
Mice and Other Pointing Devices 35
Inputting Sound 37

Output Devices 38
Monitors 38
Printers 40
Outputting Sound 42

The System Unit 44
On the Front Panel 44
On the Back: Ports 46
Inside the System Unit 48

Setting It All Up: Ergonomics 50

CHAPTER 3 — Introducing the Internet: Making the Most of the Web's Resources 65

Technology in Action:
   Interacting with the Internet 65

Internet Basics 66
A Brief History of the Internet 66
The Web vs. the Internet 66
The Internet's Clients and Servers 66

Connecting to the Internet 68
Dial-Up Connections 68
Broadband Connections 69
Choosing the Right Internet Connection Option 73

Finding an Internet Service Provider 73
Choosing an ISP 75

Navigating the Web: Web Browsers 75
Browser Features 76

**Getting Around the Web:
   URLs, Hyperlinks, and Other Tools** 76
URLs 76
Hyperlinks and Beyond 78
Favorites and Bookmarks 78

TABLE OF CONTENTS

Searching the Web:
Search Engines and Subject Directories 79
Search Engines 79
Subject Directories 82

Communicating Through the Internet:
E-Mail and Other Technologies 83
E-Mail 83
Chat Rooms 87
Instant Messaging 88
Newsgroups 90

Conducting Business Over the Internet:
E-Commerce 90
E-Commerce Safeguards 91
Cookies 92

Multimedia.com 94

CHAPTER 4 Application Software:
Programs That Let You Work and Play 107

Technology in Action: Using Application Software 107

The Nuts and Bolts of Software 108

Productivity Software 108
Word Processing Software 108
Spreadsheet Software 110
Presentation Software 113
Database Software 114
Productivity Software Tools 115
Integrated Software Applications vs. Software Suites 116

Financial and Business-Related Software 120
Personal Financial Software 121
General Business Software 122
Specialized Business Software 123

Graphics and Multimedia Software 123
Desktop Publishing Software 124
Image-Editing Software 124
Drawing Software 126
Video-Editing Software 127
Computer-Aided Design Software 128
Web Page Authoring Software 128

Educational and Reference Software 129
Educational Software 129
Reference Software 130

Entertainment Software 131

Communications Software 132

TABLE OF CONTENTS

Getting Help with Software 133

Buying Software 134
Standard Software 134
Discounted Software 134
Freeware and Shareware 135
Software Versions and System Requirements 136

Installing/Uninstalling and Opening Software 136

TECHNOLOGY IN FOCUS — **DIGITAL ENTERTAINMENT** 150

CHAPTER 5 — Using System Software: The Operating System, Utility Programs, and File Management 167

Technology in Action:
Working with System Software 167

System Software Basics 168

Operating System Categories 168
Real-Time Operating Systems 169
Single-User Operating Systems 169
Multiuser Operating Systems 170

Desktop Operating Systems 170
Microsoft Windows® 170
Mac OS® 172
UNIX® 173
Linux 173

What the Operating System Does 175
The User Interface 176
Processor Management 176
Memory and Storage Management 178
Hardware and Peripheral Device Management 179
Software Application Coordination 179

The Boot Process: Starting Your Computer 180
Step 1: Activating BIOS 181
Step 2: Performing the Power-On Self-Test 181
Step 3: Loading the Operating System 182
Step 4: Checking Further Configurations and Customizations 182
Handling Errors in the Boot Process 182

The Desktop and Windows Features 184

Organizing Your Computer: File Management 186
Organizing Your Files 187

TABLE OF CONTENTS

    Viewing and Sorting Files and Folders 188
    Naming Files 190
    Working with Files 191

Utility Programs 192
    Display Utilities 192
    The Add or Remove Programs Utility 194
    File Compression Utilities 194
    System Maintenance Utilities 195
    System Restore and Backup Utilities 198
    The Task Scheduler Utility 199
    Accessibility Utilities 199

CHAPTER 6 — Evaluating Your System: Understanding and Assessing Hardware 211

Technology in Action: A Case of PC Envy 211

To Buy or To Upgrade: That Is the Question 212

What Is Your Ideal Computer? 213

Assessing Your Hardware: Evaluating Your System 214

Evaluating the CPU Subsystem 214
    Upgrading Your CPU 216

Evaluating RAM: The Memory Subsystem 217
    Virtual Memory 219
    Adding RAM 220

Evaluating the Storage Subsystem 220
    The Hard Disk Drive 221
    Portable Storage Options:
        The Floppy and Beyond 223
    Upgrading Your Storage Subsystem 226

Evaluating the Video Subsystem 228
    Video Cards 228
    Monitors 230

Evaluating the Audio Subsystem 231
    Speakers 231
    Sound Cards 232

Evaluating Port Connectivity 233
    Adding Ports:
        Expansion Cards and Hubs 235

Evaluating System Reliability 237

Making the Final Decision 239

TABLE OF CONTENTS

CHAPTER 7   Networking and Security: Connecting Computers and Keeping Them Safe from Hackers and Viruses   251

Technology in Action: The Problems of Sharing   251

Networking Fundamentals   252

Network Architectures   253
Describing Networks Based on Network Control   253
Describing Networks Based on Distance   254

Network Components   255
Transmission Media   255
Network Adapters   256
Network Navigation Devices   256
Networking Software   256

Types of Peer-to-Peer Networks   256
Power Line Networks   256
Phone Line Networks   258
Ethernet Networks   259
Wireless Networks   262

Choosing a Peer-to-Peer Network   266

Configuring Software for Your Home Network   268

Keeping Your Home Computer Safe   269

Computer Threats: Hackers   270
What Hackers Steal   272
Trojan Horses   272
Denial of Service Attacks   273
How Hackers Gain Access   274

Computer Safeguards: Firewalls   275
Types of Firewalls   275
Is Your Computer Secure?   275

Computer Threats: Computer Viruses   278
Types of Viruses   279
Virus Classifications   280

Computer Safeguards: Antivirus Software   281
Other Security Measures   282

TECHNOLOGY IN FOCUS   **PROTECTING YOUR COMPUTER AND BACKING UP YOUR DATA**   294

TABLE OF CONTENTS

CHAPTER 8 — Mobile Computing:
Keeping Your Data on Hand 309

**Technology in Action:
Using Mobile Computing Devices** 309

**Mobile Computing:
Is It Right for You?** 310
Mobile Device Limitations 311

Mobile Computing Devices 312

Paging Devices 312

Cellular Phones 313
Cell Phone Hardware 313
How Cell Phones Work 315
Cell Phone Features: Text Messaging 316
Cell Phone Internet Connectivity 316

MP3 Players 317
MP3 Hardware 317
MP3 Flash Memory and File Transfer 317
MP3 Ethical Issues: Napster and Beyond 319

Personal Digital Assistants (PDAs) 321
PDA Hardware 321
PDA Operating Systems 322
PDA Memory and Storage 324
PDA File Transfer and Synchronization 325
PDA Internet Connectivity 326
PDA Software and Accessories 326
PDA or Cell Phone? 327

Tablet PCs 329
Tablet PC Hardware 329
Tablet Software 331
Tablet or PDA? 331

Laptops 332
Laptop Hardware 332
Laptop Operating Systems and Ports 333
Laptop Batteries and Accessories 334
Laptop or Desktop? 335

CHAPTER 9 — Behind the Scenes:
Inside the System Unit 347

**Technology in Action:
Taking a Closer Look** 347

TABLE OF CONTENTS

Digital Data: Switches and Bits   348
Electronic Switches   348
The Binary Number System   349

The CPU:
Processing Digital Information   353
The CPU Machine Cycle   355
Stage 1: The Fetch Stage   356
Stage 2: The Decode Stage   357
Stage 3: The Execute Stage   359
Stage 4: The Store Stage   359

RAM:
The Next Level of Temporary Storage   359
Types of RAM   360

Buses: The CPU's Data Highway   361

Making Computers Even Faster:
Advanced CPU Designs   363
Pipelining   363
Specialized Multimedia Instructions   365
Multiple Processing Efforts   365

TECHNOLOGY IN FOCUS   **THE HISTORY OF THE PC**   376

CHAPTER 10   Behind the Scenes:
Software Programming   387

Technology in Action:
Understanding Software Programming   387

The Life Cycle of an Information System   388
System Development Life Cycle   388

The Life Cycle of a Program   391

Describing the Problem:
The Problem Statement   392

Making a Plan:
Algorithm Development   393
Developing an Algorithm: Decision Making and Design   394
Top-Down Design and Object-Oriented Analysis   397

Coding:
Speaking the Language of the Computer   399
Categories of Programming Languages   399
Creating Code: Writing the Program   401
Compilation   404
Coding Tools: Integrated Development Environments   406

xxviii

TABLE OF CONTENTS

Debugging:
  Getting Rid of Errors 408

Finishing the Project:
  Testing and Documentation 409

Programming Languages:
  Many Languages for Many Projects 409
  Selecting the Right Language 410
  Windows Applications 410
  Web Applications: HTML and Beyond 413
  The Next Great Language 416

CHAPTER 11 Behind the Scenes: Databases and Information Systems 429

Technology in Action:
  Using Databases 429

Life Without Databases 430

Database Building Blocks 432
  Advantages of Using Databases 432
  Database Terminology 433
  Records and Tables 435
  Primary Keys and Foreign Keys 435

Database Types 436
  Relational Databases 436
  Object-Oriented Databases 436
  Object-Relational Databases 436

Database Management Systems:
  Basic Operations 437
  Creating Databases and Entering Data 437
  Input Forms 439
  Data Validation 439
  Viewing and Sorting Data 442
  Extracting or Querying Data 443
  Outputting Data 448

Relational Database Operations 448
  Normalization of Data 449

Data Storage 452
  Data Warehouses 452
  Populating Data Warehouses 454
  Data Staging 455
  Data Marts 456

Managing Data: Information Systems 457
  Office Support Systems 457
  Transaction Processing Systems 458
  Management Information Systems 459

xxix

Decision Support Systems 460
Internal and External Data Sources 460
Model Management Systems 460
Knowledge-Based Systems 461

Data Mining 461

CHAPTER 12 Behind the Scenes: Networking and Security 473

**Technology in Action:
Understanding How Networks Work 473**

Networking Advantages 474

Client/Server Networks 474

**Classifications of Client/Server Networks:
LANs, WANs, and MANs 475**

Constructing Client/Server Networks 478

Servers 478
Print Servers 478
Application Servers 479
Database Servers 480
E-Mail Servers 480
Communications Servers 480
Web Servers 481

Network Topologies 471
Bus Topology 481
Ring Topology 483
Star Topology 484
Comparing Topologies 486

Transmission Media 486
Wired Transmission Media 486
Wireless Media Options 488
Comparing Transmission Media 489

Network Operating Systems 491

Network Adapters 491

Network Navigation Devices 494
MAC Addresses 495
Repeaters and Hubs 496
Switches and Bridges 496
Routers 497

Network Security 498
Authentication 498
Access Privileges 499
Physical Protection Measures 500
Firewalls 501

CHAPTER 13 Behind the Scenes:
The Internet: How It Works 515

**Technology in Action:
Knowing How the Internet Works 515**

The Management of the Internet 516

Internet Networking 516
Connecting ISPs 517
T Lines 518
Network Access Points 519
Points of Presence 519
The Network Model of the Internet 520

Data Transmission and Protocols 522
Packet Switching 522
TCP/IP 523

IP Addresses and Domain Names 524
Domain Names 525
Domain Name Servers 528

Other Protocols: FTP and Telnet 529

HTTP, HTML, and Beyond 530
HTTP and SSL 531
HTML 531
The Common Gateway Interface 532
Client-Side Applications 533
XML 534

Communications Via the Internet 536
E-Mail 536
E-Mail Security: Encryption and Specialized Software 537
Instant Messaging 539

The Future of the Internet 541
The Large Scale Networking Program and Internet2 541
An Interplanetary Internet? 541
The Expanding Features of the Internet 542

TECHNOLOGY IN FOCUS **CAREERS IN IT** 554

Glossary 569
Index 585
Credits 617

xxxi

OBJECTIVES

After reading this chapter, you should be able to answer the following questions:

- What does it mean to be "computer fluent"? (p. 3)

- How does being computer fluent make you a savvy computer user and consumer? (pp. 4–5)

- How can becoming computer fluent help you in a career? (pp. 6–12)

- How can becoming computer fluent help you understand and take advantage of future technologies? (pp. 13–15)

- What kinds of challenges do computers bring to a digital society and how does becoming computer fluent help you deal with these challenges? (pp. 15–16)

- What exactly is a computer and what are its four main functions? (p. 16)

- What is the difference between data and information? (pp. 16–17)

- What are bits and bytes and how are they measured? (p. 17)

- What hardware does a computer use to perform its functions? (pp. 17–18)

- What are the two main types of software you find in a computer? (p. 19)

- What different kinds of computers are there? (pp. 20–21)

CHAPTER 1

Why Computers Matter to You:
Becoming Computer Fluent

TECHNOLOGY IN ACTION: BECOMING COMPUTER FLUENT

It's safe to say that computers are nearly everywhere in our society. You find them in schools, cars, airports, shopping centers, toys, medical devices, homes, and in many people's pockets. If you're like most Americans, you interact with computers almost every day, sometimes without even knowing it. Whenever you buy something with a credit card, you interact with a computer. Simply turning on your car engine requires that you use a computer. And, of course, most of us can't imagine our lives without e-mail. If you don't yet have a home computer and don't feel comfortable using one, you still can't have escaped the impact of technology: countless ads for computers, cell phones, digital cameras, and an assortment of Web sites surround us all each day. We're constantly reminded of the ways in which computers, the Internet, and technology are integral parts of our lives.

So, just by being a member of our society you know quite a bit about computers. But why is it important to learn more about computers, becoming what is called **computer fluent**? Being computer fluent means being familiar enough with computers that you understand their capabilities and limitations and know how to use them. But being computer fluent means more than just knowing about the parts of your computer. The following are some other benefits:

- Becoming computer fluent will help you use your computer more wisely and be a more knowledgeable consumer.
- Becoming computer fluent will help you in your career.
- Becoming computer fluent will help you better understand and take advantage of future technologies that will come your way.

In addition, understanding computers and their ethical, legal, and societal implications will make you a more aware participant in today's society.

Anyone can become computer fluent—no matter what your degree of technical expertise. Being computer fluent doesn't mean you need to know enough to program a computer or build one yourself. Just like with a car, if you drive one, you should know enough about it to take care of it and to use it effectively, but that doesn't mean you have to know how to build one. You should try to achieve the same familiarity with computers. In this chapter, we'll look at the ways in which computers can affect your life, now and in the future. We'll also look at just what a computer does as well as which parts help it perform its tasks.

Becoming a Savvy Computer User and Consumer

One of the benefits of becoming computer fluent is being a savvy computer user and consumer. What does this mean? The following are just a few examples of what it may mean to you:

- **Avoiding hackers and viruses.** Do you know what hackers and viruses are? Both can pose threats to computer security. Being aware of how hackers and viruses operate and knowing the damage they can do to your computer can help you avoid falling prey to them.

- **Protecting your privacy.** You've probably heard of identity theft—you see and hear news stories all the time of people whose "identities" are stolen and whose credit ratings are ruined by "identity thieves." But do you know how to protect yourself from identity theft when you're online?

- **Understanding the *real* risks.** Part of being computer fluent means being able to separate the *real* privacy and security risks from things you don't have to worry about. For example, do you know what a cookie is? Do you know whether it poses a privacy risk for you when you're on the Internet? What about a firewall? Do you know what one is? Do you really need one to protect your computer?

- **Using the Internet wisely.** Anyone who has ever searched the Web can attest that finding information and finding *good* information are two very different things. People who are computer fluent make the Internet a powerful tool and know how to effectively find the information they want. How familiar with the Web are you and how effective are your searches?

- **Avoiding Internet headaches.** If you have an e-mail account, chances are you have received the electronic form of junk mail called spam. What do you do when your e-mail inbox is crowded with spam? How can you keep from being overwhelmed by it? Computer fluency means avoiding spam and other Internet headaches.

- **Being able to maintain, upgrade, and troubleshoot your computer.** Computers are complex devices that can and do break down. Learning how to care for and maintain your computer (Figure 1.1) and knowing how to diagnose and fix problems can save you a lot of time and hassle. Do you know how to upgrade your computer if you want more memory, for example? Do you know which software and computer settings can help you keep your computer in top shape?

- **Making good purchasing decisions.** Everywhere you go you see ads like the one in Figure 1.2 for computers and other devices: laptops, printers, monitors, cell phones, digital cameras, and personal digital assistants (PDAs). Do you know what all the words in the ads mean? What is RAM? What is a CPU? What are MB, GB, GHz, and cache? How fast do you need your computer to be and how much memory should you have? Understanding computer "buzz words" and keeping up to date with technology will help you better determine which computers and devices match your needs.

FIGURE 1.1

Do you know how to maintain, troubleshoot, and upgrade your computer?

FIGURE 1.2

Do you know what all the words in a computer ad mean?

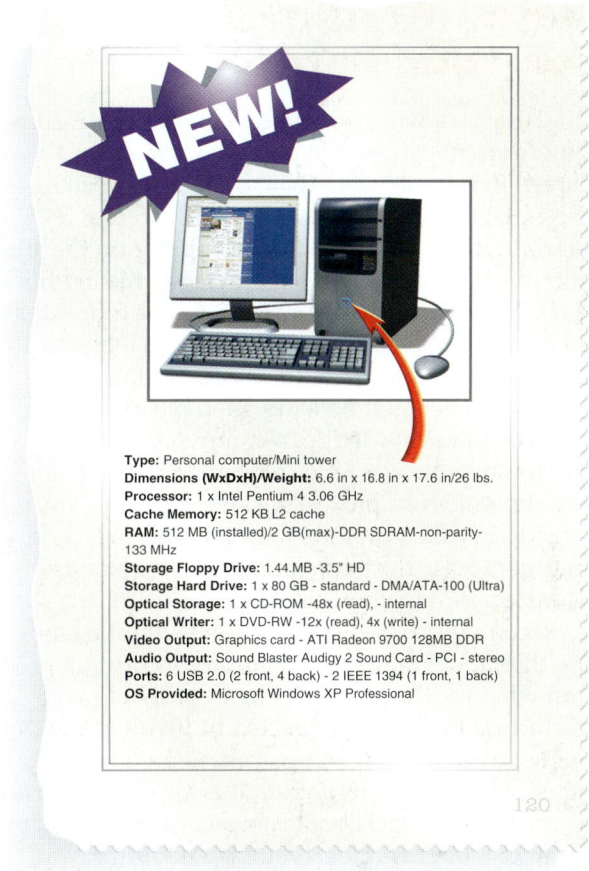

- **Knowing how to integrate the latest technology with your equipment.** Finally, becoming computer fluent means knowing how to be a better computer user in the future. It means knowing what technologies are on the horizon (some of which we'll discuss in this chapter) and how to integrate these technologies into your home setup when possible (Figure 1.3). Can you connect your laptop to a wireless network? What is "Bluetooth" and does your computer "have" it? Can a device with a USB 2.0 connector be plugged into an old USB 1.0 port? (For that matter, what is a USB port?) How much memory should your cell phone have? Knowing the answers to these and other questions will help you make better decisions as you consider whether investing in new technologies is right for you.

FIGURE 1.3

Becoming computer fluent means knowing how to be a better consumer and computer user in the future. Will you know whether your computer and devices work well with each other and with future technologies?

Being Prepared for Your Career

Regardless of what profession you pursue, if computers are not already in everyday use in that career, they most likely will be soon. As of 2001, the U.S. Department of Labor indicated that 53.5 percent of U.S. workers use a computer on the job, and this percentage will only increase. Meanwhile, the U.S. Department of Agriculture has found that employees who use a computer on the job earn 10 to 11 percent more than those who don't.

In fact, even getting a job often means having a rudimentary knowledge of computers. Drafting a resume and cover letter requires a basic understanding of word processing software, and you need to be at least a little familiar with the Internet to use the job search Web sites (such as Monster.com) at your disposal.

Becoming truly computer fluent—understanding the capabilities and limitations of computers and what you can do with them—will help you perform your job more effectively. If you're pursuing an engineering or business degree, you already know how important it is for you to learn as much as you can about computers. But if you're pursuing another career, you may not think you need computer skills. In this section, we'll discuss interesting ways in which computers are used in all sorts of careers.

COMPUTERS IN RETAIL: "LET ME LOOK THAT UP FOR YOU"

Many students have part-time jobs in retail stores. These jobs are often far from glamorous, but they do provide employees with experience using a host of computers, from the mundane to the complex. Most retail employees use simple computers when they process a transaction. These "point-of-sale" terminals (formerly known as cash registers) are in turn often connected to complex inventory and sales computer systems that provide immediate data to retail store managers and customers (see Figure 1.4). Sales clerks can perform searches for customers to determine what stores might have an item in stock that the customer couldn't find. And by analyzing their constantly updated records, managers can evaluate how their business is doing.

Before computers, if managers wanted to know how well a certain style of shoes was selling, for example, they would have to physically take inventory (count the remaining shoes). However, when a point-of-sale terminal records sales as they are made, it's easy for managers to query a sales database and determine how products are performing. Understanding what data can be captured by computers and how it can be analyzed to spot trends (in a process called "data mining") is key to running a successful retail business.

How do stores use the data they "mine"? Large retailers often study the data gathered from the terminals and determine which products are selling on a given day and in a specific location. This helps managers gauge how much merchandise they need to order from suppliers to replace stock that is sold. Managers can also use mined data to determine that for a certain product to sell well, they must lower its price—especially if they cut the price at one store and sales increased, for example. Data mining thus allows retailers to respond to consumer buying patterns.

FIGURE 1.4

Point-of-sale terminals not only update sales and inventory databases, but retail clerks can use them to search databases based on customers' inquiries.

COMPUTERS IN THE ARTS: "SHALL WE DANCE?"

Many art students think that because they're studying art, there is no sense in them studying computers. However, unless you plan on being a starving artist, you'll want to sell your work. To do so, you'll need to advertise your work to the public and/or contact art galleries to convince them to purchase or display your work. Wouldn't it be helpful if you knew how to construct a simple Web site like the one shown in Figure 1.5? Might you not need to send e-mails or write letters to your customers informing them of exhibits containing your work? Certainly, you could write letters without a computer, but it would be much less efficient and rob you of valuable time that you could spend creating your art.

But using computers in the arts goes way beyond using word processing software and the Internet. For example, the Atlanta Ballet, in conjunction with the Georgia Institute of Technology, is using computers to create "virtual" dancers and new performances for audiences. As shown in Figure 1.6, live dancers are wired with sensors that are connected to a computer that captures the dancer's movements. Based on the data it

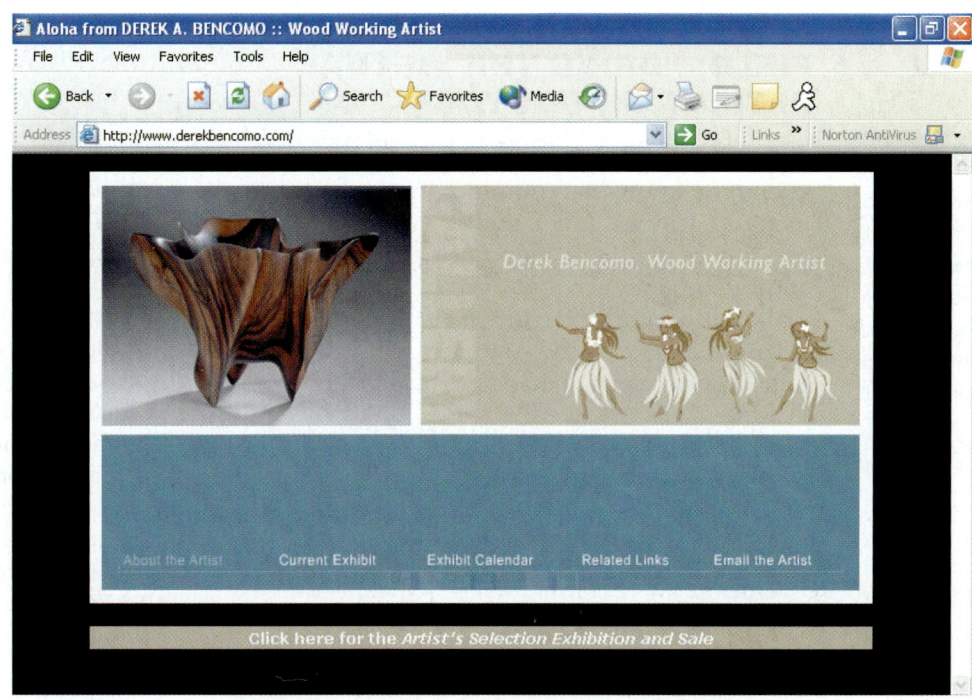

FIGURE 1.5

Derek Bencomo, a Hawaiian woodworking artist, uses his Web site (**www.derekbencomo.com**) to familiarize prospective customers with his work.

collects, the computer generates a virtual dancer on a screen. The computer operator can easily manipulate this virtual dancer as well as change the dancer's costume with a click of a mouse. This allows the ballet company to create new experiences for the audience by pairing virtual dancers with live dancers.

FIGURE 1.6

(a) While live dancers are performing center stage, virtual dancers are projected on either side to provide a new sensory experience for the audience.
(b) A technician monitors a wired dancer while the dancer's motions are captured by a computer.

TechTV

For more information on how surgeons use 3-D gaming technology, see Hot Topic: New Operation for 3-D Graphics—a TechTV clip found at www.prenhall.com/techinaction

COMPUTERS IN THE MEDICAL FIELD: "BRING THE VIRTUAL PATIENT IN, DOCTOR"

If you watch *ER* or have been to a doctor's office lately, you know that computers are a staple at medical facilities. Among other things, doctors and nurses use them to update patient records and to keep in touch with their patients via e-mail. But computer use in the medical field extends beyond simple computers. Imagine the following conversation at a medical facility:

Doctor: He's going into cardiac arrest. Adrenaline heart needle, stat!
Head Nurse: Blood pressure is 70 over 40 and falling doctor!
Doctor: Darn it, where's that needle!
OR Nurse: Adrenaline doctor.
Head Nurse: 60 over 30, doctor.
Doctor: Another 50 cc's of adrenaline.
Head Nurse: I've lost the pulse!
Doctor: We need to defibrillate . . . charging paddles . . . prepare to shock him . . . CLEAR!
Head Nurse: No response doctor. We've lost him.
Doctor: Darn! Well, let's boot him up and try again.

Training for physicians and nurses can be difficult at best. Often, the best way for medical students to learn is to experience a real emergency situation. The problem is that students are then confined to watching as the emergency unfolds while trained personnel actually care for the patient. Students rarely get to train in real-life situations, and when they do, a certain level of risk is involved. Medical students are now getting access to better training opportunities thanks to a computer technology called a **patient simulator** (shown in Figure 1.7). Patient simulators are life-sized mannequins that can speak, breathe, and blink (their eyes respond to external stimuli). They have a pulse and a heartbeat and respond just like humans to procedures such as the administration of intravenous drugs.

Medical students can train on patient simulators and experience firsthand how a human would react to their treatments without any risk to a live patient. The best thing about these "patients" is that if they "die," students can restart the computer simulation and try again. Teaching hospitals, universities, and medical schools are currently deploying patient simulators. In addition, the U.S. military is using patient stimulators to train doctors, nurses, and medics to respond to terrorist attacks that employ chemical and biological agents.

Doctors are also using specialized software to turn their handheld computers (PDAs) into powerful reference tools, enabling them to look up the symptoms of illnesses and the drugs used to treat them. When writing prescriptions, doctors can use their PDAs to verify dosages

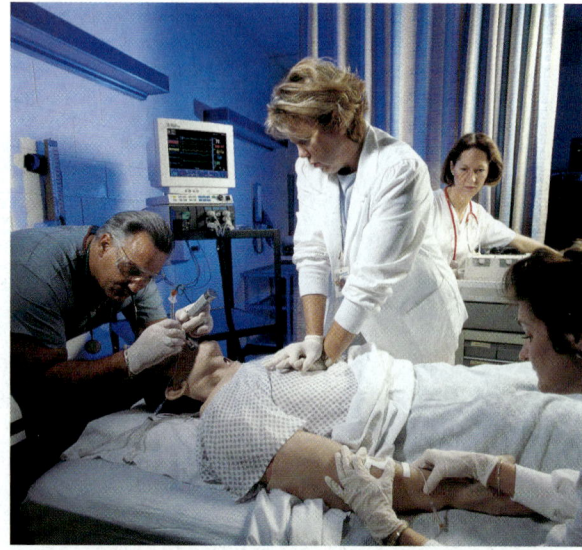
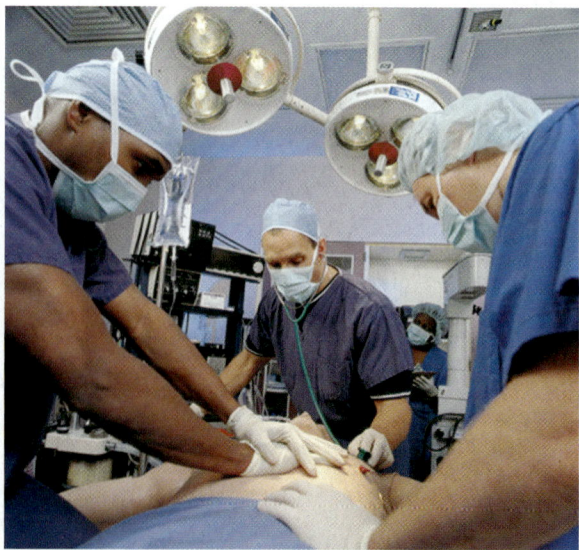

FIGURE 1.7

Patient simulators (made by Medical Education Technologies, Inc.) allow healthcare students to practice medical procedures without risk of injury or death to the patient.

and check the safety of prescribing certain drugs in combination. And using wireless technology, doctors can download a patient's medical history to their PDAs in a matter of seconds—no need to pull medical records from filing cabinets buried in the dark recesses of the hospital basement. Doctors can make notations to a patient's records and place orders for medication and tests electronically through the PDA, and technicians and nurses no longer have to struggle to decipher a doctor's handwriting.

COMPUTERS IN THE DETECTIVE FIELDS: "PUT DOWN THAT MOUSE—YOU'RE UNDER ARREST!"

When you think of a private investigator, what do you think of? Images of undercover agents sitting in a car at night with binoculars trained on a window hoping to catch a glimpse of an unfaithful spouse? Or countless hours spent trailing suspects on foot? Today, wearing out shoe leather to solve crimes is far from the only method available to investigators. Just like on the TV show *CSI*, computers are being used in police cars and crime labs to solve an increasing number of crimes. As an investigator today, you can save a great deal of time if you're just the least bit computer savvy.

One technique modern detectives are using to solve crimes is to employ computers to search the vast number of databases on the Internet. Proprietary law-enforcement databases such as the FBI's National Crime Information Center database enable police detectives to track down a wealth of information about individuals and businesses to help them solve crimes. Detectives are also using their knowledge of wireless networking to intercept and read suspects' e-mail or chat sessions while they are online, all from the comfort of a car parked outside the suspect's home (where legally permissible, that is).

As detective work goes more high tech, so, too, does crime. To fight such high-tech crime, a law-enforcement specialty called **computer forensics** (Figure 1.8) is growing. Already being used to send criminals behind bars, computer forensics is the application of computer systems and techniques to gather potential legal evidence. The ability to recover and read deleted or damaged files from a criminal's computer could provide

FIGURE 1.8

Computer forensics is the application of computer systems and techniques to gather potential legal evidence.

evidence for a trial. The tried and true techniques of personally interviewing and observing suspects are still crucial to investigative work, but using computers can make the entire process more efficient.

COMPUTERS IN THE LEGAL FIELDS: "WELCOME TO THE VIRTUAL COURTROOM"

In courtrooms today, video records of crimes in progress (often captured by cameras at convenience stores or gas stations) are sometimes displayed to the jury to help them understand how the crime unfolded. But what happens if there is no surveillance camera to record the crime? Paper diagrams, models, and still photos of the crime scene used to be the only choice for attorneys to illustrate their case. Now there is a much more exciting and lively alternative: computer forensics animations.

Computer forensics animations are extremely detailed (and often lifelike) recreations that have been generated with computers based on forensic evidence, depositions of witnesses, and the opinions of experts. Using sophisticated animation programs, similar to the ones Pixar Studios uses to create movies like *Toy Story* and *Monsters, Inc.*, forensic animators can depict one side's version of how events occurred, allowing the jury to

FIGURE 1.9

(a) This still excerpt from an animation depicts the collision of a tractor trailer and a passenger van. (b) This still excerpt from an animation depicts a shooting. The green and blue lines indicate bullet trajectories.

watch it unfold. Hearing witnesses describe an auto accident can't compare to watching it, and testimony of a person being shot is much more compelling when accompanied by a video recreation of the event (see Figure 1.9).

Before an animation can be shown to a jury, it must pass careful examinations by the opposing attorney and the judge to ensure it meets the standards set for admissibility to the courtroom. Certain judges contend that forensic animations will prejudice the jury and so prohibit them from being admitted into evidence. But as more courts accept these animations as evidence, forensic animation will be as commonplace in the courtroom as a judge's black robe. Chances are if you are entering the legal profession, you will encounter this type of evidence frequently.

COMPUTERS IN THE CLASSROOM: "AND ON THE LEFT, YOU SEE THE MONA LISA"

When teaching today, you need to be at least as computer fluent as your students. Computers are part of most schools, even preschools. And in many colleges, students are required to purchase their own computers. So teachers must have a working knowledge of computers to effectively integrate computer technology into the classroom.

The Internet has obvious advantages in the classroom as a research tool for students. And effective use of the Internet allows teachers to expose students to places they otherwise could not. Many museums and institutions have "virtual tours" on their Web sites that allow students to examine objects in their collection. Often, these virtual tours include three-dimensional photos that can be viewed from all angles. So even if you teach in Topeka, Kansas, you can take your students on a virtual tour of the National Air and Space Museum in Washington, D.C., for example (see Figure 1.10). There are no field trip expenses or permission slips needed.

Many teachers today are extending the reach of their courses beyond the classroom by using course management software. Course management software (such as Blackboard or WebCT) is a computer program that enables instructors and students to post messages (such as schedules) and other files (such as syllabi) to a course Web site they can access anytime. Although originally designed for online classes, course management software over the past few years has become a popular tool for traditional face-to-face classes as well.

BEING PREPARED FOR YOUR CAREER   **11**

Computers in the classroom will become more prevalent as prices continue to fall and parents demand their children be provided with the necessary computer skills they need to be successful in the workplace. Therefore, as an educator, you'll need to be computer fluent to plan constructive computerized lessons for your students and to use technology to interact with them.

COMPUTERS AT HOME: "JUST PROGRAM IT AND FORGET IT"

Running a household and raising children are time-intensive activities rivaling the commitment needed in any other profession. You may be familiar with many computerized entertainment products for the home (such as Sony Playstation and Nintendo GameCube), but these won't help you get your home maintenance work done any sooner. However, knowing how to use computers—being computer fluent—will help you be aware of the vast array of labor- and time-saving technological products being marketed to consumers.

Sick of cleaning your home or mowing the lawn? Robots are not just found in science labs and industrial settings anymore. Robotic "maintenance workers" are now emerging for the home market. Why should you have to waste time

FIGURE 1.10

Can't fit in a trip to Washington, D.C. this semester? As a teacher you can still conduct a virtual museum tour for your students using the Internet. This picture of the Wright Flyer from the National Air and Space Museum would be a natural conversation starter for a discussion about powered flight.

mowing the lawn, cleaning the pool, or vacuuming the floors when robots can do these tasks for you? Figure 1.11 shows a few of the robots you could be using to help make your career as a stay-at-home parent a little less stressful.

Meanwhile, so-called "smart devices"—devices such as temperature controls, lights, and security devices that contain computer chips—are becoming widely available for the home. Smart devices can control functions in the home without human intervention. And while attempts have been made to launch Internet-connected appliances (such as refrigerators that order food via the Internet when you're running low), these have not yet met with widespread success. Will your oven one day prepare your food without your supervision? Only time will tell. You can keep tabs on the latest products available to make your home easier to manage at sites like **www.smarthome.com**.

FIGURE 1.11

(a) The Aquabot from Aqua Products and (b) the RoboMower from Friendly Robotics help alleviate the tedium of outside chores. They roam your lawn or pool and perform their tasks with no human supervision necessary. (c) The Roomba robot vacuum manufactured by iRobot relieves you of the dreariness of vacuuming your floors. Roomba uses complex algorithms to ensure that the entire floor of a room is cleaned.

Getting Ready for the Technology of Tomorrow

If you are computer fluent you will be a smarter computer user and you will be prepared for virtually any career. But what are the other benefits of computer fluency? By understanding computers and how they work today, you'll be better able to take advantage of and understand the technologies of tomorrow. Let's take a look at some emerging technologies and how they may affect your life.

NANOSCIENCE: THE NEXT "BIG" THING

Have you ever heard of *nanoscience*? Developments in computing based on the principles of nanoscience are being touted as the next big wave in computing. Ironically, this realm of science focuses on very small objects. In fact, **nanoscience** involves the study of molecules and structures (called *nanostructures*) whose size ranges from 1 to 100 nanometers.

How big is this? The prefix *nano* stands for one-billionth. Therefore, a nanometer is one-billionth of a meter. To put this in perspective, a human hair is approximately 50,000 nanometers wide. Put side by side, 10 hydrogen atoms (the simplest atom) would measure approximately 1 nanometer. Anything smaller than a nanometer is just a stray atom or particle floating around in space. Therefore, nanostructures represent the smallest human-made structures that can be built.

Nanotechnology is the science revolving around the use of nanostructures to build devices on an extremely small scale. You may wonder what nanoscience and nanotechnology have to do with you and your life. Right now, not much, but someday scientists hope to use nanostructures to build computing devices too small to be seen by the naked eye. Nanowires, such as the one shown in Figure 1.12, could be used to create extremely small pathways in computer chips. Developments such as this could lead to computers the size of a pencil eraser that are far more powerful than today's desktop computers. Meanwhile, nanomachines (mechanical devices built on the nanoscale) could be designed to float through the human bloodstream and clear arterial blockages that lead to heart attacks and strokes.

If you watch Star Trek, you know that "nanoprobes" (tiny machines that can be injected into the bloodstream) have already been envisioned on the TV series' 24th century world. But nanotechnology is a relatively new field (less than 15 years old). Although we can create a carbon nanotube (Figure 1.13), we are still a long way from developing nanoscale machines. However, universities and government laboratories are investing billions of dollars in nanotechnology research every year. If you have an interest in science and engineering, this is the time to pursue an education in nanoscience, as you could be on the forefront of the next big technological breakthrough.

FIGURE 1.13

Here is a picture of a carbon nanotube developed in a laboratory. On the Star Trek TV series they have nanoprobes (nano-sized robots) that can do fantastic things, such as conditioning and repairing the human body to keep it in perfect working order. Although we have not yet been able to fashion nanotubes into robots, this could be the ultimate destination of current nanotechnology research.

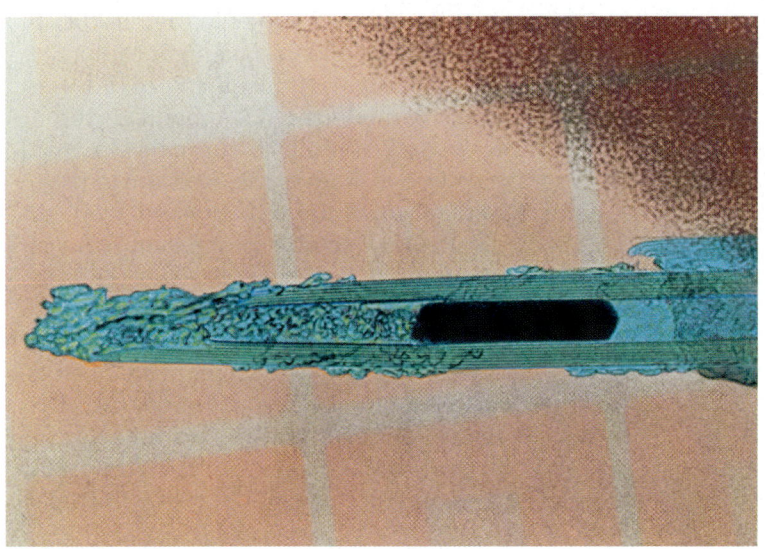

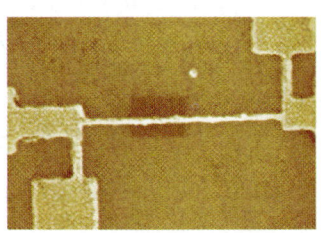

FIGURE 1.12

Comprising several types of atomic particles layered together, this nanowire could eventually be used to conduct electrical signals within a nanoscale computer—a computer smaller than a pencil eraser but more powerful than today's desktop computers.

BIOMEDICAL CHIP IMPLANTS: COMBINING HUMANS WITH MACHINES?

Mention implanting technology into the human body and some people conjure up images of the Borg, human-based life forms on *Star Trek* that have replaced parts of their bodies with cybernetic implants. The Borg are aliens that capture humans and install technology in their bodies to ensure obedience and enslavement to Borg society (the collective). Unlike the Borg, however, the goal of modern-day biomedical chip research is not to "assimilate" humans into a Borg-like collective. Rather, the goals are to provide technological solutions to physical problems (see Figure 1.14) and to provide a means for positively identifying individuals.

One potential application of biomedical chip implants is to provide sight to the blind. Macular degeneration and retinitis pigmentosa are two diseases that account for the majority of blindness in developing nations. Both diseases result in damage to the photoreceptors contained in the retina of the eye. (Photoreceptors convert light energy into electrical energy that is transmitted to the brain, allowing us to see.) Researchers are experimenting with chips that contain microscopic solar cells and are implanted in the damaged retina of patients. The idea is to have the chip take over for the damaged photoreceptors and transmit electrical images to the brain. Although these chips have been tested in patients, they have not yet restored anyone's sight. But uses of biomedical chips such as these are illustrative of the type of medical devices you may "see" in the future.

One form of biomedical chip already entering the market is a technology that can be used to verify a person's identity. Called the VeriChip, this "personal ID chip" is being marketed by a company called Applied Digital Solutions. VeriChips, about the size of a grain of rice (see Figure 1.15), are implanted underneath the skin. When exposed to radio waves from a scanning device, the chip emits a signal that transmits its unique serial number to the scanner. The scanner then connects to a database that contains the name, address, and serious medical conditions of the person in whom the chip has been implanted.

The company envisions the VeriChip speeding up airport security and being used together with other devices (such as electronic ID cards) to provide tamper-proof security measures. If someone stole your credit card, they couldn't use it if a salesclerk verified your identity via a chip before authorizing a transaction. Chips could eventually be developed so that they contain a vast wealth of information about the person in whom they are implanted. However, it remains to be seen whether the general public will accept having personal data implanted into their bodies.

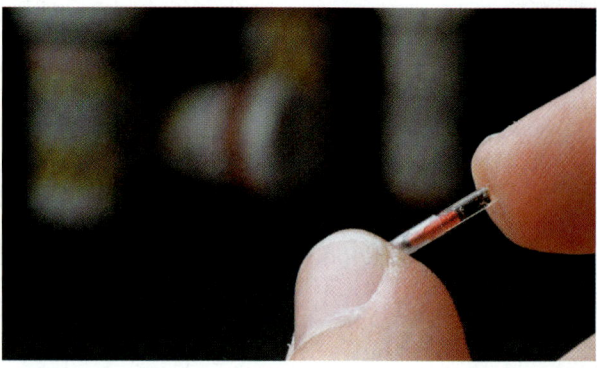

FIGURE 1.15

The VeriChip is a small device implanted directly under the skin. A special scanner can read the information stored on it.

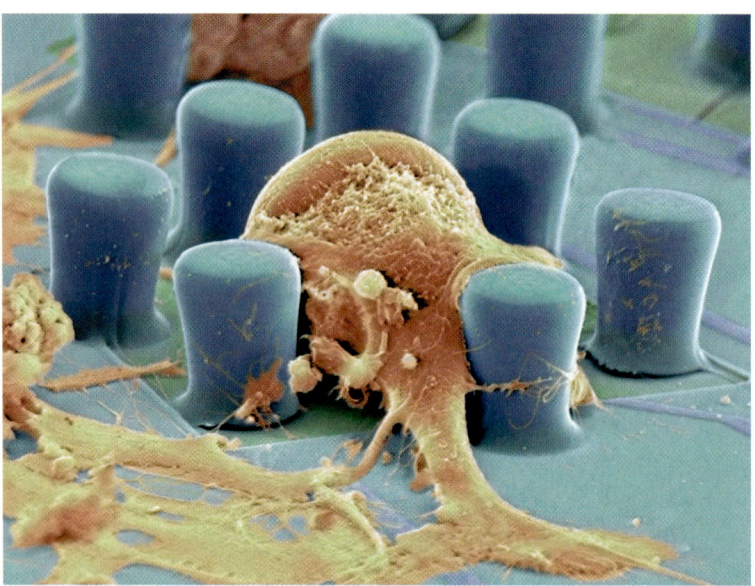

FIGURE 1.14

Researchers are currently experimenting with implantable chips such as the one shown above. Here we see a nerve cell on a silicon chip. The cell was cultured on the chip until it formed a network with nearby cells. The chip contains a transitor which stimulates the cell above it which in turn passes the signal to neighboring neurons.

ARTIFICIAL INTELLIGENCE: WILL COMPUTERS BECOME HUMAN?

Science fiction shows and movies such as *Star Wars* have always been populated with robots that emulate humans, seemingly effortlessly. So when will we have C-3PO or R2-D2 helping us in our home or office?

The answer is not anytime soon, although advances are being made. **Artificial intelligence (AI)** is the science that attempts to produce machines that display the same type of intelligence that humans do. Current computers can perform both calculations and the tasks that are programmed into them much faster than humans. As mentioned above, there are already robots on the market that can be programmed to do tasks such as lawn mowing. In addition, industrial robots, like the one shown in Figure 1.16, are used to perform tasks such as welding, painting, and heavy lifting in factories. However, the performance of these tasks is based on preprogrammed algorithms, not independent thought.

There are currently no computers that can emulate the human thought process. The human brain is superior to computers in a very important way: it is capable of processing almost a limitless number of tasks at the same time. Just to write a letter on your computer, your brain coordinates thousands of nerve impulses that control your hands, eyes, and thought processes. Although computers can multitask, not even the most powerful computers can handle the multitasking load of a human brain.

The reason computers can't emulate human thought yet is because no one fully understands how the human brain works. Scientific studies have confirmed that electrical activity between neurons somehow coalesces into thoughts, but scientists aren't quite sure how that happens. Until the human brain is fully understood, progress in the area of artificial intelligence is likely to proceed quite slowly, but you'll no doubt experience artificial intelligence breakthroughs in your lifetime.

Understanding the Challenges Facing a Digital Society

Part of becoming computer fluent is also being able to understand and form knowledgeable opinions on the challenges facing a digital society. While computers offer us a world of opportunities, they also pose ethical, legal, and moral challenges and questions. For example, how do you feel about the following:

In 2003, a man named Jay Walker (who founded the Web site Priceline.com) suggested a program called "US HomeGuard" to the federal government. The objective of the program is to install high-tech webcams (cameras connected to the Internet) in places that need to be protected from terrorist activities, such as airports, nuclear power plants, and reservoirs (see Figure 1.17). A central agency would be responsible for the webcam monitoring and would employ up to one million stay-at-home citizens as "spotters" to monitor the webcams for suspicious activity.

The webcams would take photos every 45 seconds or so and computers would compare the photos to the previous ones, noting any changes. If there was a change, such as a person throwing an object over a reservoir fence, the photo would be transmitted to at least three "spotters." If the

FIGURE 1.16

Industrial robots such as this one can be programmed to lift heavy objects and to perform repetitive tasks such as painting and welding. However, robots such as these can not think yet. They are just performing preprogrammed tasks.

spotters saw something odd, the system would capture more photos from the camera in question (and surrounding cameras) and send the images to another 10 spotters. If the second group confirmed the alert, security forces at the facility being attacked would swing into action.

Opponents of this technology say that it harkens back to George Orwell's book *1984*, in which citizens of a futuristic society are constantly under surveillance by a government entity known as Big Brother. Advocates of the plan say that only restricted areas would be under surveillance, and that the security benefits outweigh the privacy risks. The government is considering funding a pilot program to the tune of $40 million to see if the system is feasible. What do you think of this technology? Should the government be allowed to install such webcams? (See Figure 1.17.)

This is just one example of the kind of question active participants in today's digital society need to be able to think about, discuss, and, at times, take action on. Being computer fluent enables you to form *educated* opinions on these issues, and to take stands based on accurate information rather than media hype and misinformation. Here are a few other questions you, as a member of our digital society, may be expected to think about and discuss:

- What privacy risks do biomedical chips like the VeriChip pose? Do the privacy risks of such chips outweigh the potential benefits?
- Should companies be allowed to collect personal data from Web site visitors without their permission?
- Should spam be illegal? If so, what penalties should be levied on people who send spam?
- Is it ethical to download music off the Web without paying for it? What about copying a friend's software onto your computer?
- What are the risks involved with humans attempting to create computers that can learn and become more human?
- Should we rely solely on computers to provide security for sensitive areas such as nuclear power plants?

As a computer user, you will need to consider these and other questions to define the boundaries of the digital society in which you live.

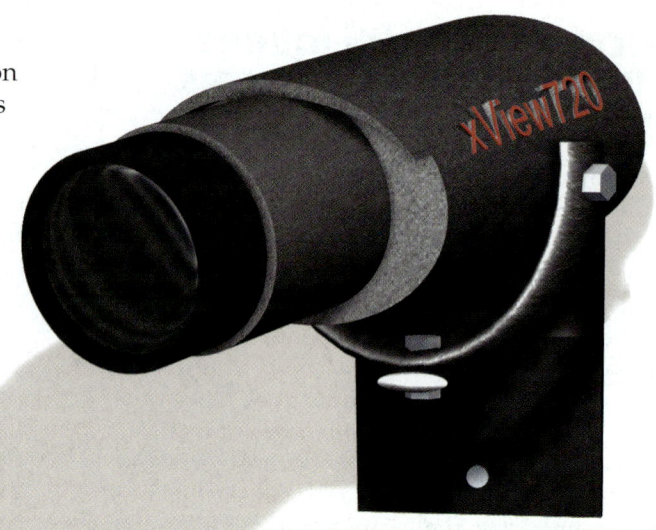

FIGURE 1.17

Should the government be allowed to install cameras like the one shown above, or do you think such technology should be outlawed?

Becoming Computer Fluent

By now you can see why becoming computer fluent is so important. But where do you start? As mentioned in the introduction, you glean some knowledge about computers just from being a member of our society. However, although you certainly know what a computer *is*, do you really understand how it works, what all its parts are, and what these parts do? In this section, we'll discuss what a computer does that makes it such a useful machine.

COMPUTERS ARE DATA PROCESSING DEVICES

Strictly defined, a **computer** is a data processing device that performs four major functions:

1. It gathers data (or allows users to input data).
2. It processes that data into information.
3. It outputs data or information.
4. It stores data and information.

To understand these four functions, you first need to understand the distinction between the terms *data* and *information*. People often use these terms interchangeably. Although in a simple conversation they may mean the same thing, when discussing computers, the distinction between data and information is an important one.

BECOMING COMPUTER FLUENT 17

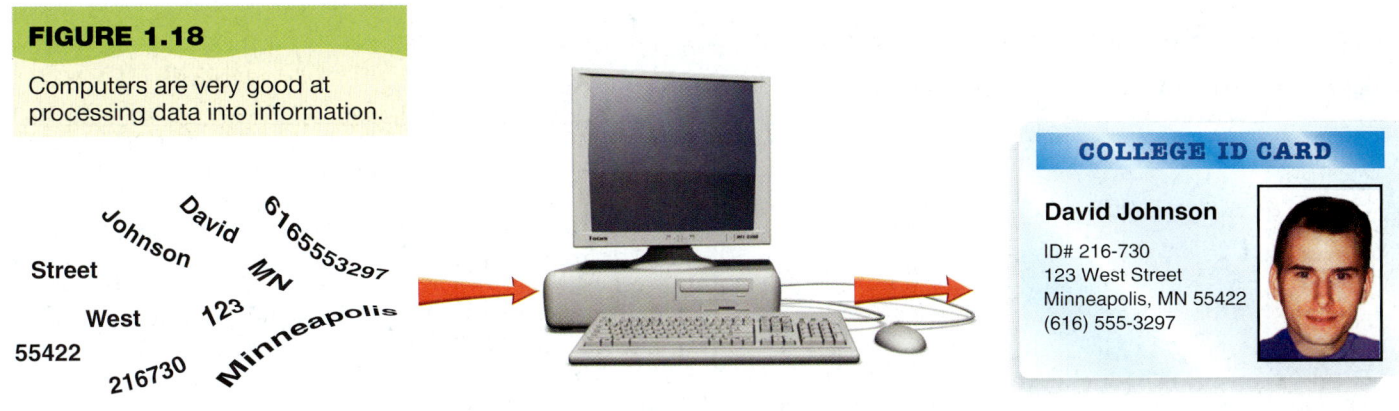

FIGURE 1.18
Computers are very good at processing data into information.

Input Data

Output Information

In computer terms, **data** is a representation of a fact or idea. Data can be a number, a word, a picture, or even a recording of sound. For example, the number "6165553297" and the names "David" and "Johnson" are pieces of data. But how useful are these chunks of data to you? **Information** is data that has been organized or presented in a meaningful fashion. When your computer provides you with a contact listing that indicates David Johnson can be reached by telephone at (616) 555-3297, the data mentioned above suddenly becomes useful—that is, it is *information*.

Computers are very good at processing data into information. When you first arrived on campus, you probably were directed to a place where you could get an ID card. You most likely provided a clerk with personal data (such as your name and address) that was entered into a computer. The clerk then took your picture with a digital camera (collecting more data). This information was then processed appropriately so that it could be printed on your ID card (see Figure 1.18). This organized output of data on your ID card is useful information. Finally, the information was probably stored as digital data on the computer for later use.

BITS AND BYTES: THE LANGUAGE OF COMPUTERS

How do computers process data into information? Unlike humans, computers work exclusively with numbers (not words). In order to process data into information, computers need to work in a language they understand. This language, called **binary language**, consists of just two numbers: 0 and 1. Everything a computer does (such as process data or print a report) is broken down into a series of 0s and 1s. Each 0 and 1 is a **binary digit**, or **bit** for short. Eight binary digits (or bits) combine to create 1 **byte**. In computers, each letter of the alphabet, each number, and each special character (such as the @ sign) consists of a *unique* combination of 8 bits, or a string of eight 0s and 1s. So, for example, in binary (computer) language, the letter K might be represented as 01001011. This equals 8 bits, or 1 byte. (We'll discuss binary language in more detail in Chapter 9.)

You've probably heard the terms kilobyte (KB) and megabyte (MB) before. But how do these fit into the bit and byte discussion here? Not only are bits and bytes used as the language that tells the computer what to do, they are also what the computer uses to represent the data and information it inputs and outputs. Word processing files, digital pictures, and even software programs are all represented inside a computer as a series of bits and bytes. These files and applications can be quite large, containing many millions of bytes. To make it easier to measure the size of these files, we need larger units of measure than a byte. Kilobytes, megabytes, and gigabytes are therefore simply amounts of bytes. As shown in Figure 1.19, a **kilobyte (KB)** is approximately 1,000 bytes, a **megabyte (MB)** is about a million bytes, and a **gigabyte (GB)** is about a billion bytes. As our information processing needs have grown, so too have our storage needs. Today, some computers can handle up to a petabyte of data—that's 1 quadrillion bytes!

COMPUTER HARDWARE

You've probably heard the term *hardware* used in reference to computers. An anonymous person once said that any part of a computer that

FIGURE 1.19 How Much is a Byte?

NAME	ABBREVIATION	NUMBER OF BYTES	RELATIVE SIZE
Byte	B	1 byte	Can hold one character of data.
Kilobyte	KB	2^{10} or 1,024 bytes	Can hold 1,024 characters or about half of a typewritten page double-spaced.
Megabyte	MB	2^{20} or 1,048,576 bytes	A floppy diskette holds approximately 1.4 MB of data, or approximately 768 pages of typed text.
Gigabyte	GB	2^{30} or 1,073,741,824 bytes	Approximately 786,432 pages of text. Since 500 sheets of paper is approximately 2 inches, this represents a stack of paper 262 feet high.
Terabyte	TB	2^{40} or 1,099,511,627,776 bytes	This represents a stack of typewritten pages almost 51 miles high.
Petabyte	PB	2^{50} or 1,125,899,906,842,624 bytes	The stack of pages is now 52,000 miles high, or about one-fourth the distance from the Earth to the moon.

you can kick with your foot when it doesn't work properly is classified as **hardware**. A more formal definition of hardware is any part of the computer you can physically touch. All hardware on the computer helps the computer to perform its various tasks (see Figure 1.20).

Most computer systems have hardware devices that you use to enter, or input, data (text, images, and sounds) into your computer. These devices, such as a keyboard and a mouse, are called **input devices**. In addition to the keyboard and the mouse, other computer input devices include scanners (which input text and photos), microphones (which input sounds), and digital cameras (which input photos and video).

You also use input devices to provide the steps and tasks the computer needs to process data into usable information. These steps and tasks are called **instructions**. Instructions may be in the form of a user response to a question posed while working in a software application or in the form of a command in which you instruct the computer what to do (such as clicking on an icon with your mouse).

As noted earlier, once data is entered into a computer, the computer processes that data. Those components that process data are located inside the **system unit**. The system unit is the metal or plastic case that holds all the physical parts of the computer together. The part of the system unit that is responsible for the processing (or the "brains" of the computer) is called the **central processing unit**, or **CPU**.

Another component inside the system unit that helps process data into information is **memory**. Memory chips hold (or store) the instructions or data that the CPU processes. The most common type of memory that a computer uses for processing data is random access memory, or RAM.

The CPU and memory are located on a special circuit board in the system unit called the **motherboard**. Once this data has been processed, it is classified as information.

In addition to input devices and the system unit, a computer includes devices that let you see your processed information. These devices, called **output devices**, include monitors and printers. Because they output sound, speakers are also considered output devices.

Finally, when your data has been input, processed, and output, you may want to store the data or information so that you can access and use it again. Specialized **storage devices** such as hard disk drives, floppy disk drives, and CD drives allow you to store your data and information.

We'll discuss all the hardware you find on a computer in much more detail in Chapter 2.

SOUND BYTE
VIRTUAL COMPUTER TOUR

In this Sound Byte, you'll take a video tour of the inside of your system unit. From opening the cover to locating the power supply, CPU, and memory, you'll learn to be more familiar with what's inside your computer.

BECOMING COMPUTER FLUENT

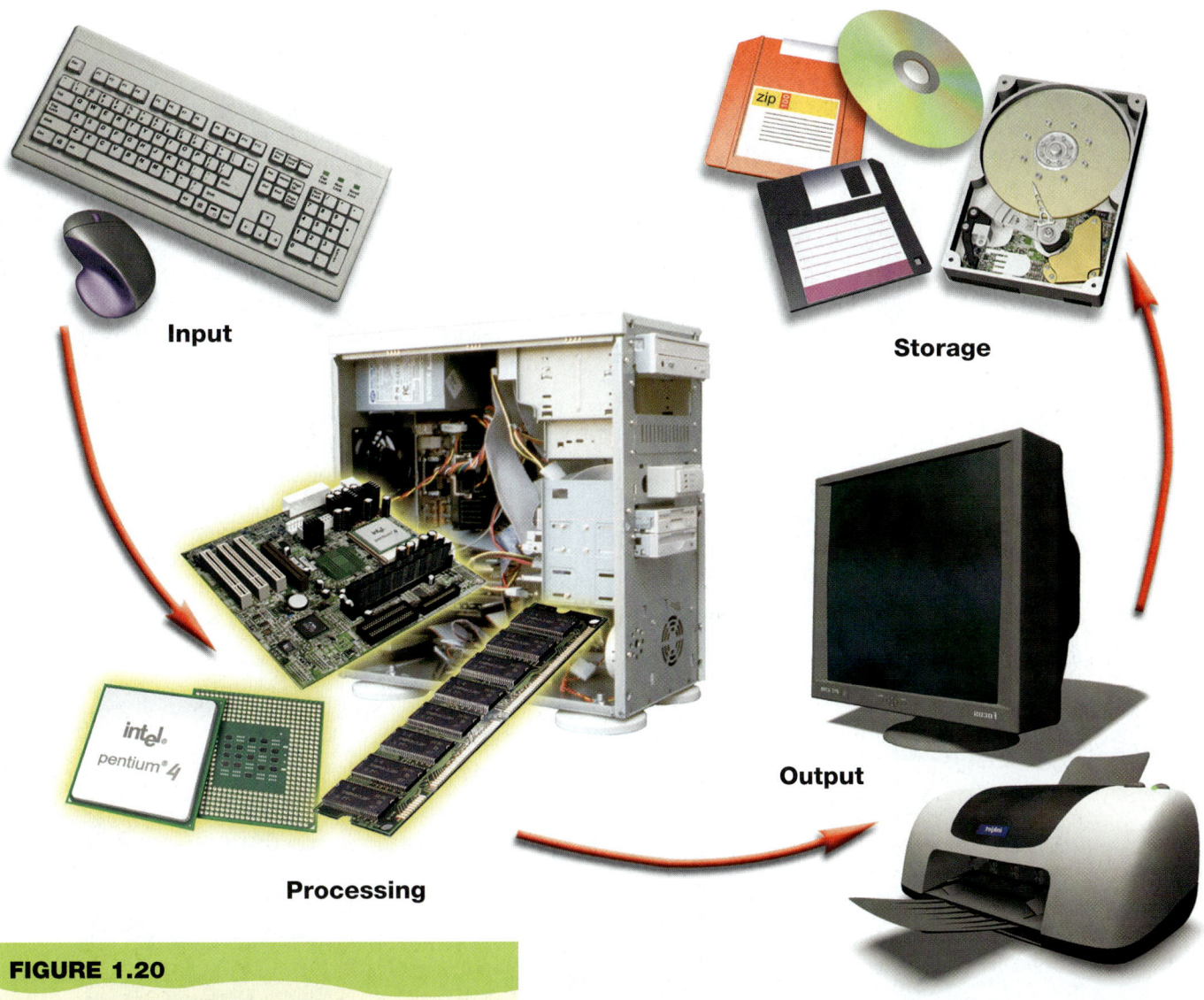

FIGURE 1.20
Each part of the computer serves a special function.

COMPUTER SOFTWARE

A computer needs more than just hardware to work: it also needs some form of software. Think of a book without words or a blank CD without music. Without words or music, these two common items are just shells that hold nothing. Similarly, a computer without software is a shell full of hardware components that can't do anything. **Software** is the set of computer programs that enables it to perform different tasks. There are two broad categories of software: application software and system software.

When you think of software, you are most likely thinking of application software. **Application software** is the set of programs you use on a computer to help you carry out tasks. If you've ever typed a document, created a spreadsheet, or edited a digital photo, for example, you've used a form of application software (See Figure 1.21).

System software is the set of programs that enables your computer's hardware devices and application software to work together. The most common type of system software is the **operating system (OS),** the program that controls the way in which your computer system functions. It manages the hardware of the computer system, including the CPU, memory, and storage devices, as well as input and output devices such as the mouse, keyboard, and printer. The operating system also provides a means by which users can interact with the computer. We'll cover software in greater depth in Chapters 4 and 5.

FIGURE 1.21

Software is the set of computer programs that enables the computer to perform different tasks.

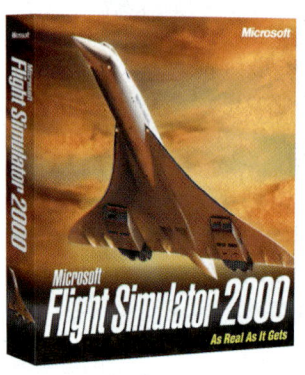

COMPUTER PLATFORMS: PCS AND MACS

The kind of operating system software you have depends on your computer's **platform**. The two most common platform types are the PC (short for personal computer, named after the original IBM personal computer) and the Apple Macintosh (or Mac). Macs and PCs use different CPUs, and therefore, each system processes information very differently. Correspondingly, their operating systems are also different. Macintosh computers use the Macintosh operating system (or Mac OS), whereas PCs generally run on Microsoft's Windows operating system. But that's where their differences end. You can still use both PCs and Macs to perform the same types of tasks (word processing and so on). The PC is not necessarily better than the Mac (or vice versa), but PCs have by far the larger market share (see Figure 1.22).

FIGURE 1.22

Ever since the early 1980s, PCs and Macs have been gunning for each other's market share. Although PCs have a much larger market share, Macs are still the platform of choice for many professions, such as graphic design.

SPECIALTY COMPUTERS

PCs and Macs are obviously not the only computers you may encounter. There are smaller, more mobile computers that you can carry around, such as laptops and PDAs. In addition, there are computers you may never come in direct contact with but that are important to our society nonetheless:

- **Servers** are computers that provide resources to other computers connected in a network. When you connect to the Internet, for example, your computer is communicating with a server at your Internet service provider (ISP). This server provides your computer with services that allow it to access the Internet. One server can provide services to many computers.
- **Mainframes** are large, expensive computers that support hundreds or thousands of users simultaneously. Mainframes excel at executing many different computer programs at the same time. Although many large companies still use mainframes, smaller and cheaper PC-based servers have replaced mainframes in many companies.
- **Supercomputers** are specially designed computers that can perform complex calculations extremely rapidly. They are used in situations where complex models requiring intensive mathematical calculations are needed (such as weather forecasting or atomic energy research). Supercomputers are the fastest and most expensive computers. The main difference between a supercomputer and a mainframe is that supercomputers are designed to execute a few programs as quickly as possible while mainframes are designed to handle many programs running at the same time (but at a slower pace).

No matter what the size, computers are an integral part of our lives, and they are constantly changing. Whether you're a novice or an experienced computer user or fall somewhere in between, you need to learn as much as you can about computers so that you can use them wisely.

SOUND BYTE
QUESTIONS TO ASK WHEN YOU BUY A COMPUTER

This Sound Byte will help you consider some important questions you need to ask when you buy a computer, such as whether you should get a laptop or a desktop, or whether you should purchase a new computer or a used or refurbished one.

Using This Book

This book is arranged in the order you would most likely proceed when exploring the computer. You'll start in the next chapter by looking at your computer as you would if you were assembling it for the first time, exploring each piece and its function within your system. Next, because so many people use the Internet to communicate, conduct research, and even shop, Chapter 3 explores the Internet and its many features. Even if you're an experienced Internet user, this chapter will help you use the Internet more effectively and more safely.

In Chapter 4, you'll look more closely at the application software that you'll most likely encounter in your daily life, both at work and at home. Then, in Chapter 5, you'll explore your computer's operating system. In doing so, you'll learn about the different system software programs you can use to keep your computer in top shape, as well as ways in which you can keep your files and folders organized.

Once you understand the pieces of your system and the three elements that make it a useful tool (the Internet, application software, and the operating system), Chapter 6 will help you evaluate your computer system to see if it is meeting your needs. By exploring and evaluating your system's parts, you'll learn whether you need to upgrade your computer and how to go about doing so.

It's likely that you have more than one computer in your home, workplace, or school. To share common resources among computers, you need to know about networking. Thus, in Chapter 7, you'll learn about home computer networks, as well as how to protect yourself from hackers and viruses. Then, in Chapter 8, you'll explore mobile computing devices such as cell phones, PDAs, laptops, and tablet PCs.

Finally, in Chapter 9, you'll find out just how your computer's hardware really works. You'll learn more about your CPU and the types of RAM available on the market today, and how these components affect your computer's performance.

Along the way, **Technology In Focus** features will teach you more about digital technology, protecting your computer and the data on it, as well as the history of the personal computer. You'll also find more material on the book's companion Web site, at **www.prenhall.com/techinaction**.

No matter how much you use the computer, you probably still have a lot of questions about how to use it *best*. Throughout this book, you'll find references to multimedia components called Sound Bytes (like the one shown in Figure 1.23) that *show* you the answers to some frequently asked questions, such as how to set up a firewall and how to use antivirus software effectively. You'll find these Sound Bytes in the book's CD as well as on the companion Web site (**www.prenhall.com/techinaction**).

In each chapter, you'll also find Bits and Bytes (like the one shown in Figure 1.24), boxes that contain interesting facts and helpful tips on how to maintain and better use your computer.

Also scattered throughout the book are Trends in IT features (like the one shown in Figure 1.25) that examine computer-related ethical issues, careers in technology, as well as emerging technologies.

Finally, throughout the book you'll find Dig Deeper features (like the one shown in Figure 1.26). These features take an in-depth look at various computer concepts, such as how a hard disk drive or a computer firewall works.

The wonderful thing about computers is that there is something new to learn about them every day. Developing an understanding of how computers can make your life easier is critical to your future success. So, let's get started on our exploration of computers. In the next chapter, you'll learn all about the parts of your computer and how these parts make computers such useful tools.

SOUND BYTE
CONNECTING TO THE INTERNET

In this Sound Byte, you'll learn the basics of connecting to the Internet from home, including the various types of Internet connections as well as useful information on selecting the right ISP.

FIGURE 1.23

Coordinating multimedia components are referenced in Sound Bytes

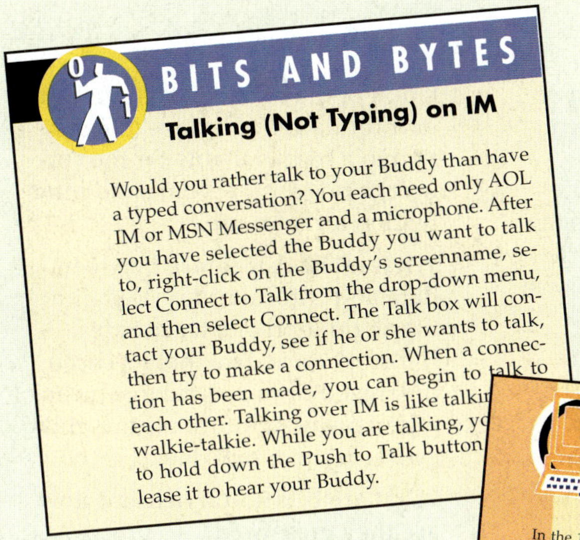
BITS AND BYTES

Talking (Not Typing) on IM

Would you rather talk to your Buddy than have a typed conversation? You each need only AOL IM or MSN Messenger and a microphone. After you have selected the Buddy you want to talk to, right-click on the Buddy's screenname, select Connect to Talk from the drop-down menu, and then select Connect. The Talk box will contact your Buddy, see if he or she wants to talk, then try to make a connection. When a connection has been made, you can begin to talk to each other. Talking over IM is like talki walkie-talkie. While you are talking, y to hold down the Push to Talk button lease it to hear your Buddy.

FIGURE 1.24

Bits and Bytes contain interesting facts and helpful computer maintenance and usage tips.

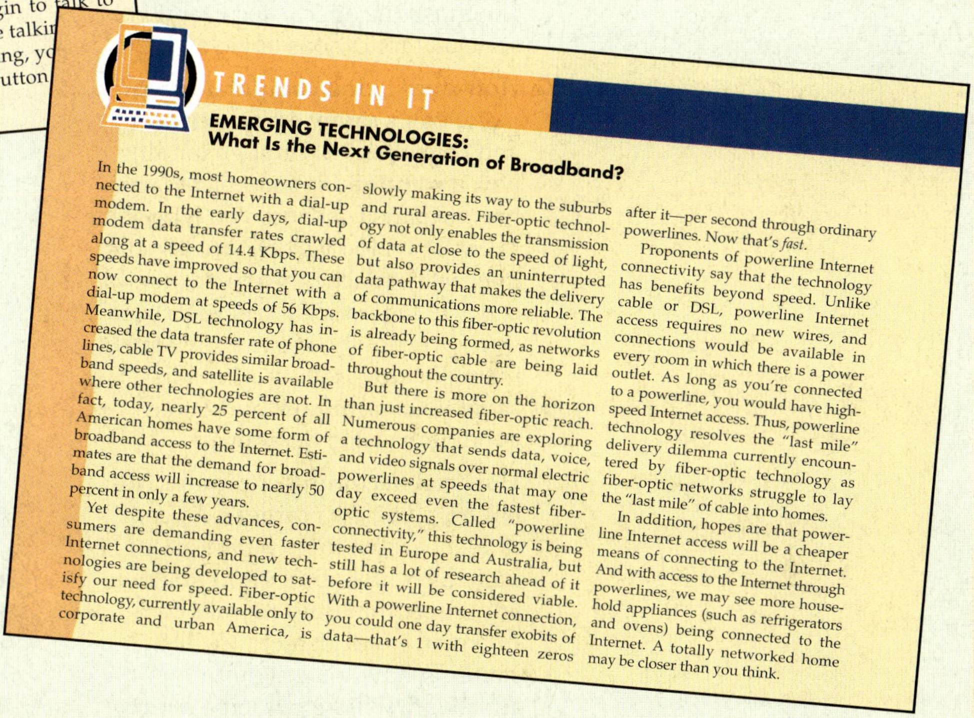
TRENDS IN IT

EMERGING TECHNOLOGIES: What Is the Next Generation of Broadband?

In the 1990s, most homeowners connected to the Internet with a dial-up modem. In the early days, dial-up modem data transfer rates crawled along at a speed of 14.4 Kbps. These speeds have improved so that you can now connect to the Internet with a dial-up modem at speeds of 56 Kbps. Meanwhile, DSL technology has increased the data transfer rate of phone lines, cable TV provides similar broadband speeds, and satellite is available where other technologies are not. In fact, today, nearly 25 percent of all American homes have some form of broadband access to the Internet. Estimates are that the demand for broadband access will increase to nearly 50 percent in only a few years.

Yet despite these advances, consumers are demanding even faster Internet connections, and new technologies are being developed to satisfy our need for speed. Fiber-optic technology, currently available only to corporate and urban America, is slowly making its way to the suburbs and rural areas. Fiber-optic technology not only enables the transmission of data at close to the speed of light, but also provides an uninterrupted data pathway that makes the delivery of communications more reliable. The backbone to this fiber-optic revolution is already being formed, as networks of fiber-optic cable are being laid throughout the country.

But there is more on the horizon than just increased fiber-optic reach. Numerous companies are exploring a technology that sends data, voice, and video signals over normal electric powerlines at speeds that may one day exceed even the fastest fiber-optic systems. Called "powerline connectivity," this technology is being tested in Europe and Australia, but still has a lot of research ahead of it before it will be considered viable. With a powerline Internet connection, you could one day transfer exobits of data—that's 1 with eighteen zeros after it—per second through ordinary powerlines. Now that's *fast*.

Proponents of powerline Internet connectivity say that the technology has benefits beyond speed. Unlike cable or DSL, powerline Internet access requires no new wires, and connections would be available in every room in which there is a power outlet. As long as you're connected to a powerline, you would have high-speed Internet access. Thus, powerline technology resolves the "last mile" delivery dilemma currently encountered by fiber-optic technology as fiber-optic networks struggle to lay the "last mile" of cable into homes.

In addition, hopes are that powerline Internet access will be a cheaper means of connecting to the Internet. And with access to the Internet through powerlines, we may see more household appliances (such as refrigerators and ovens) being connected to the Internet. A totally networked home may be closer than you think.

FIGURE 1.25

Trends in IT features examine computer-related ethical issues, careers in technology, and emerging technologies.

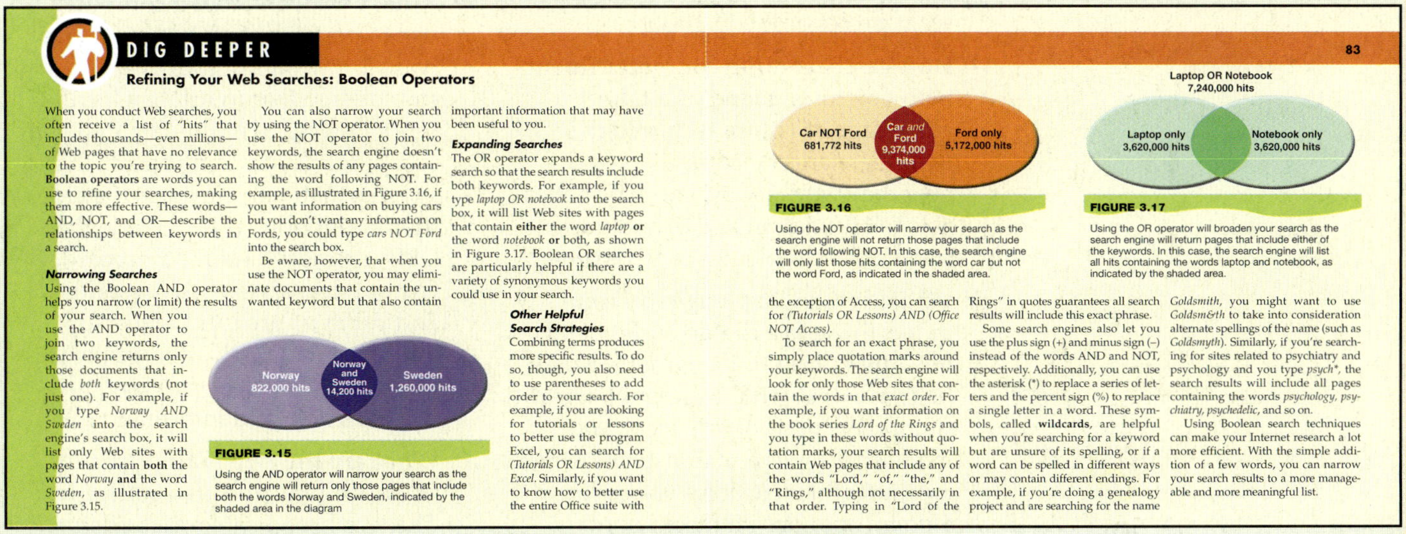
DIG DEEPER

Refining Your Web Searches: Boolean Operators

83

When you conduct Web searches, you often receive a list of "hits" that includes thousands—even millions—of Web pages that have no relevance to the topic you're trying to search. **Boolean operators** are words you can use to refine your searches, making them more effective. These words—AND, NOT, and OR—describe the relationships between keywords in a search.

Narrowing Searches

Using the Boolean AND operator helps you narrow (or limit) the results of your search. When you use the AND operator to join two keywords, the search engine returns only those documents that include *both* keywords (not just one). For example, if you type *Norway AND Sweden* into the search engine's search box, it will list only Web sites with pages that contain **both** the word *Norway* **and** the word *Sweden*, as illustrated in Figure 3.15.

Norway 822,000 hits | Norway and Sweden 14,200 hits | Sweden 1,260,000 hits

FIGURE 3.15

Using the AND operator will narrow your search as the search engine will return only those pages that include both the words Norway and Sweden, indicated by the shaded area in the diagram

You can also narrow your search by using the NOT operator. When you use the NOT operator to join two keywords, the search engine doesn't show the results of any pages containing the word following NOT. For example, as illustrated in Figure 3.16, if you want information on buying cars but you don't want any information on Fords, you could type *cars NOT Ford* into the search box.

Be aware, however, that when you use the NOT operator, you may eliminate documents that contain the unwanted keyword but that also contain important information that may have been useful to you.

Expanding Searches

The OR operator expands a keyword search so that the search results include both keywords. For example, if you type *laptop OR notebook* into the search box, it will list Web sites with pages that contain **either** the word *laptop* **or** the word *notebook* **or** both, as shown in Figure 3.17. Boolean OR searches are particularly helpful if there are a variety of synonymous keywords you could use in your search.

Other Helpful Search Strategies

Combining terms produces more specific results. To do so, though, you also need to use parentheses to add order to your search. For example, if you are looking for tutorials or lessons to better use the program Excel, you can search for *(Tutorials OR Lessons) AND Excel*. Similarly, if you want to know how to better use the entire Office suite with the exception of Access, you can search for *(Tutorials OR Lessons) AND (Office NOT Access)*.

Car NOT Ford 681,772 hits | Car *and* Ford 9,374,000 hits | Ford only 5,172,000 hits

FIGURE 3.16

Using the NOT operator will narrow your search as the search engine will not return those pages that include the word following NOT. In this case, the search engine will only list those hits containing the word car but not the word Ford, as indicated in the shaded area.

Laptop OR Notebook 7,240,000 hits

Laptop only 3,620,000 hits | Notebook only 3,620,000 hits

FIGURE 3.17

Using the OR operator will broaden your search as the search engine will return pages that include either of the keywords. In this case, the search engine will list all hits containing the words laptop and notebook, as indicated by the shaded area.

To search for an exact phrase, you simply place quotation marks around your keywords. The search engine will look for only those Web sites that contain the words in that *exact order*. For example, if you want information on the book series *Lord of the Rings* and you type in these words without quotation marks, your search results will contain Web pages that include any of the words "Lord," "of," "the," and "Rings," although not necessarily in that order. Typing in "Lord of the Rings" in quotes guarantees all search results will include this exact phrase.

Some search engines also let you use the plus sign (+) and minus sign (–) instead of the words AND and NOT, respectively. Additionally, you can use the asterisk (*) to replace a series of letters and the percent sign (%) to replace a single letter in a word. These symbols, called **wildcards**, are helpful when you're searching for a keyword but are unsure of its spelling, or if a word can be spelled in different ways or may contain different endings. For example, if you're doing a genealogy project and are searching for the name *Goldsmith*, you might want to use *Goldsm&th* to take into consideration alternate spellings of the name (such as *Goldsmyth*). Similarly, if you're searching for sites related to psychiatry and psychology and you type *psych**, the search results will include all pages containing the words *psychology, psychiatry, psychedelic,* and so on.

Using Boolean search techniques can make your Internet research a lot more efficient. With the simple addition of a few words, you can narrow your search results to a more manageable and more meaningful list.

FIGURE 1.26

Dig Deeper features add an in-depth look at various computer concepts.

Summary

1. **What does it mean to be "computer fluent"?** If you are computer fluent, you understand the capabilities and limitations of computers and know how to use them wisely. Being computer fluent also enables you to make informed purchasing decisions; use computers in your career; take advantage of future technologies; and grasp the ethical, legal, and societal implications of technolog.

2. **How does being computer fluent make you a savvy computer user and consumer?** By understanding how a computer is constructed and how its parts function, you'll be able to better avoid hackers, viruses, and Internet headaches; protect your privacy; separate the real risks from those you don't have to worry about; maintain, upgrade, and troubleshoot your computer; and make good purchasing decisions.

3. **How can becoming computer fluent help you in a career?** As computers become more a part of our daily lives, it is difficult to imagine any career that does not use computers in some fashion. Understanding how to use computers effectively will help you be a more productive employee, no matter what profession you choose.

4. **How can becoming computer fluent help you understand and take advantage of future technologies?** By understanding computers and how they work today, you'll be better able to take advantage of and understand the technologies of tomorrow.

5. **What kinds of challenges do computers bring to a digital society and how does becoming computer fluent help you deal with these challenges?** Computers pose ethical, legal, and moral challenges and questions. Being computer fluent enables you to form *educated* opinions on these issues.

6. **What exactly is a computer and what are its four main functions?** Computers are data processing devices. The computer (1) gathers data (or allows users to input data); (2) processes that data (performs calculations or some other manipulation of the data); (3) outputs data or information (displays information in a form suitable for the end user); and (4) stores data and information for later use.

7. **What is the difference between data and information?** Data is a representation of a fact or idea. Information is data that has been organized or presented in a meaningful fashion. Information is more powerful than raw data.

8. **What are bits and bytes and how are they measured?** To process data into information, computers need to work in a language they understand. This language, called binary language, consists of two numbers: 0 and 1. Each 0 and 1 is a binary digit, or bit; 8 bits create 1 byte. In computers, each letter of the alphabet, each number, and each special character consists of a unique combination of 8 bits (1 byte), or a string of eight 0s and 1s. For describing large amounts of storage capacity, the terms *kilobyte* (approximately 1,000 bytes) *megabyte* (approximately 1 million bytes) and *gigabyte* (approximately 1 billion bytes) are used.

9. **What hardware does a computer use to perform its functions?** Hardware is any part of the computer you can touch. Most computers have input devices to input data into your computer. Those components that process that data are located inside the system unit. They include the CPU, memory, and motherboard. Output devices (such as monitors and printers) are used to display information. Finally, storage devices (such as hard drives) are used to preserve data and information.

10. **What are the two main types of software you find in a computer?** The two broad categories of software are application software and system software.

11. **What different kinds of computers are there?** Aside from PCs and Macs, you'll find mobile computing devices, servers, mainframes, and supercomputers.

Key Terms

application software	p. 19	mainframes	p. 21
artificial intelligence (AI)	p. 15	megabyte (MB)	p. 17
binary digit (bit)	p. 17	memory	p. 18
binary language	p. 17	motherboard	p. 18
byte	p. 17	nanoscience	p. 13
central processing unit (CPU)	p. 18	nanotechnology	p. 13
computer	p. 16	operating system (OS)	p. 19
computer fluent	p. 3	output devices	p. 18
computer forensics	p. 9	patient simulator	p. 8
data	p. 17	platform	p. 20
gigabyte (GB)	p. 17	servers	p. 21
hardware	p. 18	software	p. 19
information	p. 17	storage devices	p. 18
input devices	p. 18	supercomputers	p. 21
instructions	p. 18	system software	p. 19
kilobyte (KB)	p. 17	system unit	p. 18

Buzz Words

Word Bank

- input
- CPU
- data
- platform
- biomedical chips
- application software
- mainframe
- output
- nanotechnology
- byte
- supercomputer
- information
- server
- storage
- processing
- computer forensics
- system software
- bit
- megabyte
- gigabyte

Instructions: Fill in the blanks using the words from the Word Bank above.

Due to the integration of computers into business and society, many fields of study are available now that were unheard of a few years ago. (1)_____, the study of very small computing devices built at the molecular level, will provide major advances in the miniaturization of computing. (2)_____ is already taking criminologists beyond what they could accomplish with conventional investigation techniques. And as (3)_____ become widespread, individuals will benefit from having computing devices implanted right in their bodies.

At the lowest level, computers manipulate data in units called (4)_____s. Since these units are too small to define data on their own (say standing for the letter Q), they are grouped together to form (5)_____s. (6)_____ represents raw facts or ideas. (7)_____ represents facts or ideas that have been organized or processed in some fashion to make them more meaningful. When storing data, large quantities of space are needed. The capacity of most hard drives today is measured in (8)_____s, which represents over 1 billion bytes of information.

For data to be acted upon by a computer, various components of the computer must interact with the data. Mice and keyboards are examples of (9)_____ devices that are used to enter data into the computer. The CPU is an example of a (10)_____ device that helps turn data into information. (11)_____ devices, such as monitors and printers, allow computers to provide information in a usable format. To save information and data for later use, (12)_____ devices such as hard drives and CD drives are used.

Critical Thinking Questions

Instructions: Albert Einstein used "Gedanken experiments," or critical thinking questions, to develop his theory of relativity. Some ideas are best understood by experimenting with them in our own minds. The following critical thinking questions are designed to demand your full attention but require only a comfortable chair—no technology.

1. This chapter listed a number of ways in which knowing about computers (or becoming computer fluent) will help you. How much do you know about computers? What else would you like to know? How do you think learning more about computers will help you in the future?

2. How are computers used in the profession you are in or plan to enter? Can you think of any careers in which people do not use computers? Can you imagine computers being used in these careers in the future? How?

3. This chapter briefly discussed data mining, a technique retailers use to study sales data and gather information from it. Have you heard of data mining before? How might a company like Wal-Mart or Target use data mining to better run their business? Can you think of any privacy risks data mining might pose?

4. As you learned in the chapter, nanotechnology is the science revolving around the use of nanostructures to build devices on an extremely small scale. What applications of tiny computers can you think of? How might nanotechnology impact your life?

5. This chapter discussed various uses of biomedical chips. Many biomedical chip implants that will be developed in the future will most likely be aimed at correcting vision loss, hearing loss, or other physical impediments. But chips could also be developed to improve physical or mental capabilities of healthy individuals. For example, chips could be implanted in athletes to make their muscles work better together, thereby allowing them to run faster. Or, your memory could be enhanced by providing additional storage capacity for your brain.

    a. Should biomedical implant devices that increase athletic performance be permitted in the Olympics? What about devices that repair a problem (such as blindness in one eye) but then increase the level of visual acuity in the affected eye so that it is better than normal vision?
    b. Would you be willing to have a chip implanted in your brain to improve your memory?
    c. Would you be willing to have a VeriChip implanted under your skin? What privacy risks do devices such as the VeriChip pose?

6. Artificial intelligence is the science that attempts to produce machines that display the same type of intelligence that humans do. Do you think humans will ever create a machine that can think? In your opinion, what are the ethical and moral implications associated with artificial intelligence?

7. IT careers are suited to a wide range of people at different points in their lives.

    a. Would an IT career have advantages for a single parent? How?
    b. Would an IT career be able to help someone pursue a later career in a non-technical field?
    c. How might an IT career assist someone in completing a college degree?

Team Time

Problem:

People are often overwhelmed by the relentless march of technology. Accessibility to information is changing the way we work, play, and interact with our friends, family, and coworkers. In this Team Time, we'll consider the future and reflect on how the advent of new technologies will affect our daily lives 10 years in the future.

Task:

Your group has just returned from a trip in a time machine 10 years into the future. Amazing changes have taken place in just a short time. To a large extent, consumer acceptance of technology makes or breaks a new technology. Your mission is to create a creative marketing strategy to promote the technological changes you observed in the future and accelerate their acceptance.

Process:

Divide the class into three or more teams.

1. With the other members of your team, use the Internet to research up-and-coming technologies (**www.howstuffworks.com** is a good starting point). Prepare a list of future innovations that you believe will occur in the next 10 years. Determine how they will be integrated into society and the effect they will have on our culture.
2. Present your group's findings to the class for debate and discussion. Note specifically how the rest of the class reacts to your reports on the innovations. Are they excited? Skeptical? Incredulous? Did they laugh off your ideas or become wildly enthusiastic?
3. Write a marketing strategy paper detailing how you would promote the technological changes that you envision for the future. Also note some barriers for acceptance the technology may have to overcome, as well as any legal or ethical challenges or questions you see the new technology posing.

Conclusion:

She future path of technology is determined by dreamers. If not for innovators such as Edison, Bell, and Einstein, we would not be as advanced a society as we are today. Innovators come from all walks of life and our creative energies must be exercised to keep them in tune. Don't be afraid to suggest technological advancements that seem outrageous today. In 1966, when the original Star Trek series was on television, hand-held communicators seemed astounding and beyond our reach. Yet the dreamers who created those communication devices for a science-fiction series spawned a multi-billion-dollar cell phone industry in the 21st century. The next technological wave may be started in your imagination!

MATERIALS ON THE WEB 29

Materials on the Web

In addition to the review materials presented here, you'll find extra materials on the book's companion Web site (**www.prenhall.com/techinaction**) that will help reinforce your understanding of the chapter content. These materials include:

Sound Byte Lab Guides

For each Sound Byte mentioned in the chapter, there is a corresponding lab guide located on the book's companion Web site. These guides review the material presented in the Sound Byte and direct you to various Web resources that examine the material. The Sound Byte Lab Guides for this chapter include:

- Virtual PC Tour
- Questions to Ask When You Buy a Computer

True/False and Multiple Choice Quizzes

The book's Web site includes a True/False and a Multiple Choice quiz for this chapter. You can take these quizzes, automatically check the results, and e-mail the results to your instructor.

Web Research Projects

The book's Web site also includes a number of Web research projects for this chapter. These projects ask you to search the Web for information on computer-related careers, milestones in computer history, important people and companies, emerging technologies, and the applications and implications of different technologies.

OBJECTIVES

After reading this chapter, you should be able to answer the following questions:

- What input devices do you use to get data into the computer? (pp. 32–38)

- What output devices do you use to get information out of the computer? (pp. 38–43)

- What's on the front of your system unit? (pp. 44–46)

- What's on the back of your system unit? (pp. 46–48)

- What's inside your system unit? (pp. 48–49)

- How do you set up your computer to avoid strain and injury? (pp. 50–51)

CHAPTER 2

Looking at Computers:
Understanding the Parts

TECHNOLOGY IN ACTION: SETTING UP THE SYSTEM

Jillian has just bought a new computer and is setting it up. Because she lives in a dorm, she spent more than she had planned to buy a flat-panel monitor, which she places on her small desk. It takes up far less room than her old monitor, which was big and bulky. Next she pulls out her system unit, which she knows is the component she'll connect all the other pieces of her system to. Although she wanted to buy the most powerful computer on the market, she bought one that was slightly less expensive. Still, it came with a CD/DVD player, floppy and Zip drives, a 100GB hard drive, and what the computer salesperson said was enough memory and power to do almost anything. She sets it on the floor next to her desk and attaches the monitor to it.

Next she pulls out her keyboard. She looked into buying a wireless keyboard, but since her budget was tight, she bought a standard keyboard instead. The box tells her it is a "USB" keyboard, so she finds what looks to be the right port on the back of her system unit and plugs it in. Her mouse also needs a USB port. Finding another USB port, she attaches the mouse there.

She sets up her speakers next. Although the salesperson told her she'd probably want to upgrade them, she decided to wait until she can afford it. She arranges them on her desk, then finds two round ports on her system unit that match their plugs. She sees the ports are labeled "speaker out" and "mic in," so she inserts the speakers into the "speaker out" port.

Last is her printer. She debated over which type of printer to buy, but decided to buy an inkjet because she prints a lot of color copies. She finds the right port on her system unit and connects it. She then plugs the monitor, speakers, printer, and system unit into the surge protector, which the salesperson told her would protect her devices from power surges. All that's left is to make sure her setup is comfortable, and she's ready to go.

What kind of monitor, keyboard, mouse, printer, and system unit do you have? Do you know all the options available and what the different components of your system do? In this chapter, we'll take a look at your computer's basic parts. You'll learn about input devices (such as the mouse and keyboard), output devices (such as monitors and printers), storage devices (such as the hard drive), as well as components inside the computer that help it to function. Finally, you'll learn how to set up your computer so that it's comfortable to work on.

32 CHAPTER 2 LOOKING AT COMPUTERS: UNDERSTANDING THE PARTS

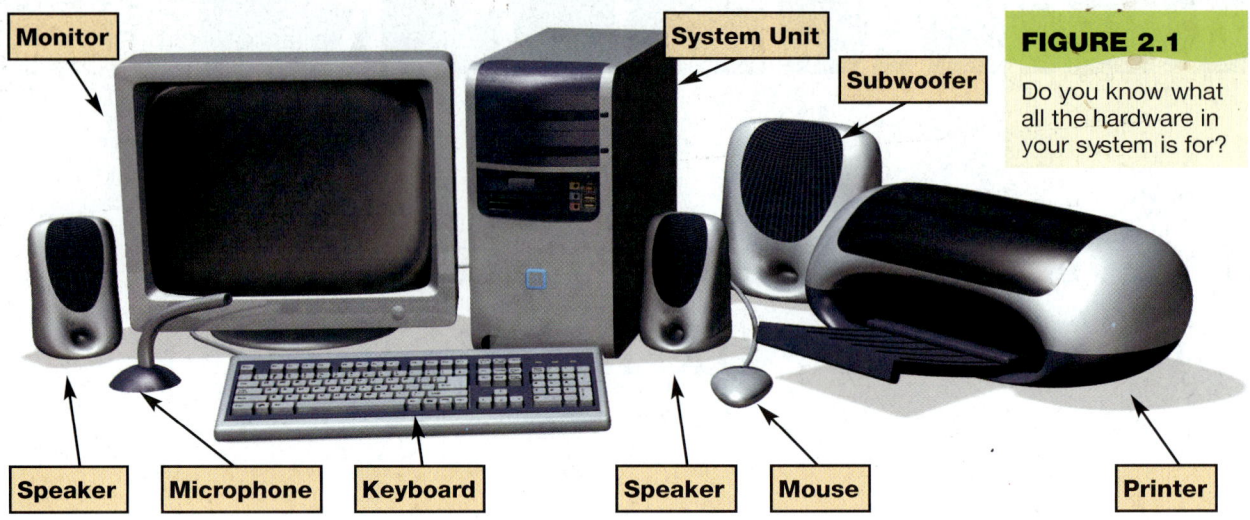

FIGURE 2.1
Do you know what all the hardware in your system is for?

Your Computer's Hardware

For the amount of amazing things computers can do, they are really quite simple machines. You learned in Chapter 1 that a basic computer system is made up of software and hardware. In this chapter, we'll look at your computer's **hardware**, the parts you can actually touch (see Figure 2.1). As you learned in Chapter 1, hardware components perform four main functions: they enable the computer to input data, process that data, and output and store the data and information. We'll begin our exploration of hardware by taking a look at your computer's input devices.

Input Devices

As you learned in Chapter 1, an **input device** enables you to enter data (text, images, and sounds) and instructions (user responses and commands) into the computer. The most common input devices are the **keyboard** and the **mouse**. You use keyboards to enter typed data and commands, whereas you use the mouse to enter user responses and commands. There are other input devices as well: microphones input sounds, while scanners and digital cameras input text and images.

KEYBOARDS

Aren't all keyboards the same?
Most desktop computers come with a standard keyboard, which uses the **QWERTY keyboard** layout, as shown in Figure 2.2a. This layout gets its name from the first six letters on the top-left row of alphabetic keys on the keyboard. However, over the years, there has been some debate over what is the best layout for keyboards. The QWERTY layout was originally designed for typewriters, not computers, and was meant to slow typists to prevent typewriter keys from jamming. The QWERTY layout is

FIGURE 2.2
(a) The first six keys on the top-left row give the QWERTY keyboard its name. (b) With a Dvorak keyboard, you can type most of the more commonly used words in the English language with the letters found on "home keys," the keys in the middle row of the keyboard.

BITS AND BYTES

Keeping Your Keyboard Clean

To keep your computer running at its best, it's important that you occasionally clean your keyboard. To do so, follow these steps:

1. Turn your computer off.
2. Disconnect the keyboard from your system.
3. Turn the keyboard upside down and *gently* shake out any loose debris. You may want to spray hard-to-reach places with compressed air (found in any computer store) or use a vacuum device made especially for computers. Don't use your home vacuum, as the suction is too strong and may damage your keyboard.
4. Wipe the keys with a cloth or cotton swab dampened with a mild cleaning solution. Don't spray or pour cleaning solution directly onto the keyboard, as even the slightest amount of liquid can destroy a keyboard. Make sure you hold the keyboard upside down or at an angle to prevent drips from running into the circuitry.

---

therefore considered inefficient as it slows typing speeds. Now that technology can keep up with faster typing, other keyboards are being considered.

The **Dvorak keyboard** is the leading alternative keyboard. The Dvorak keyboard puts the most commonly used letters in the English language on "home keys," the keys in the middle row of the keyboard, as shown in Figure 2.2b. The Dvorak keyboard's design reduces the distance your fingers travel for most keystrokes, increasing typing speed.

What do the other keys on the keyboard do? All keyboards have the standard set of alpha and numeric keys that you regularly use when typing. As shown in Figure 2.3, there are also other keys on a keyboard that have special functions:

- The **numeric keypad** allows you to enter numbers quickly.
- **Function keys** act as shortcut keys you press to perform special tasks. They are sometimes referred to as the "F" keys because they start with the letter F followed by a number. Each software application has its own set of tasks assigned to the function keys, although some are more universal. For example, the F1 key is usually the Help key in software applications. However, the F4 key performs a different shortcut in Microsoft Word than it does in Microsoft Excel.

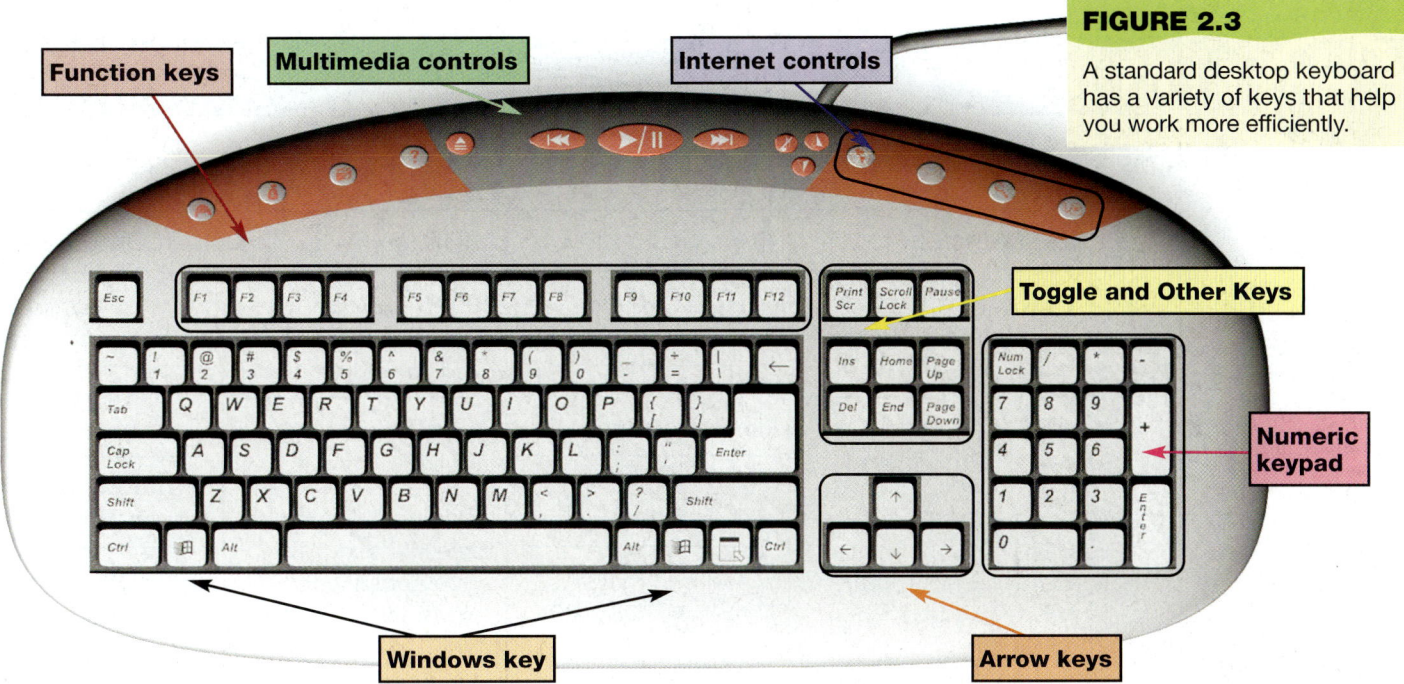

FIGURE 2.3
A standard desktop keyboard has a variety of keys that help you work more efficiently.

- The **Control key** is used in combination with other keys to perform shortcuts and special tasks. For example, holding down the Control key while pressing the letter B adds bold formatting to selected text. Similarly, you use the **Alt key** with other keys for additional shortcuts and special tasks. (On Macintosh computers, the Control key is the Apple or Command key while the Alt key is the Option key.)
- The **Windows key** is specific to the Windows operating system. Used alone, it brings up the Start menu; however, it's used most often in combination with other keys as shortcuts. For example, pressing the Windows key plus the letter E starts Windows Explorer.

Some newer keyboards also include keys or buttons that enable you to perform Internet commands (such as opening a Web browser or viewing e-mail), access Help features, or control your CD/DVD player.

What do the arrow keys do? Another set of controls on standard keyboards is the directional keys that move your **cursor**, the flashing I symbol that indicates where the next character will be inserted. The arrow keys move the cursor one space at a time, either up, down, left, or right.

Above the arrow keys, you'll usually find keys that move the cursor up or down one full page or to the beginning (Home) or end (End) of a line. The Delete (Del) key allows you to delete characters, whereas the Insert key allows you to insert or overwrite characters within a document. The Insert key is a **toggle key** because its function changes each time you press it: when toggled on, the Insert key inserts new text within a line of existing text. When toggled off, the Insert key *replaces* (or overwrites) existing characters with new characters as you type. Other toggle keys include the Num Lock key and the Caps Lock key, which toggle between an on/off state.

Are keyboards different on laptops? Because laptops are compact, they have fewer keys than standard keyboards. Still, a lot of the laptop keys have alternate functions so that you get the same capabilities from the limited keys as you do from the special keys on standard keyboards. You can also hook up traditional keyboards to most laptops, or you can use a specially designed keyboard, shown in Figure 2.4b, that fits on top of the laptop.

What about keyboards for PDAs? Generally, you enter data and commands into a personal digital assistant (PDA) via a **stylus**, a pen-shaped device that you use by tapping or writing on the PDA's touch-sensitive screen, as shown in Figure 2.5. However, some PDAs have built-in keyboards that allow you to type in text just as you would with a normal keyboard. If your PDA doesn't include a built-in keyboard, you can buy keyboards that attach to the PDA. We'll discuss PDA keyboards in more detail in Chapter 8.

BITS AND BYTES

Keystroke Shortcuts

You may know that you can combine certain keystrokes to take shortcuts within the Windows operating system. The following are a few of the most helpful shortcuts:

TEXT FORMATTING	FILE MANAGEMENT	CUT/COPY/PASTE	WINDOWS CONTROLS
CTRL+B Applies (or removes) **bold** formatting to selected text	**CTRL+O** Opens the Open dialog box	**CTRL+X** Cuts (removes) selected text from document	**Alt+F4** Closes the current window
CTRL+I Applies (or removes) *italic* formatting to selected text	**CTRL+N** Opens a new document	**CTRL+C** Copies selected text	**Ctrl+Esc** Opens the Start menu
CTRL+U Applies (or removes) underlining to selected text	**CTRL+S** Saves a document	**CTRL+V** Pastes selected text (previously cut or copied)	**Windows Key+F1** Opens Windows Help
	CTRL+P Opens the Print dialog box		**Windows Key+F** Opens the Search (Find Files) dialog box

FIGURE 2.4

(a) Laptop keyboards are more compact than traditional desktop keyboards, and usually don't include extra keys like the numeric keypad. (b) This full-size keyboard is designed to work with a laptop. Using the special folding legs, it sits right on top of your laptop keyboard, allowing you to have all the functionality of a standard keyboard.

Are there wireless keyboards? As its name indicates, a wireless keyboard doesn't use cables to connect to your computer. Rather, it's powered by batteries and sends data to the computer using a form of wireless technology. Infrared wireless keyboards communicate with the computer using infrared light waves (similar to how a remote control communicates with a TV). The computer receives the infrared light signals through a special infrared port. The disadvantage to infrared keyboards is that you need to point the keyboard directly at the infrared port on the computer in order for it to work.

The best wireless keyboards send data to the computer using radio frequency (RF). These keyboards contain a radio transmitter that sends out radio wave signals. These signals are received by a small receiver device that sits on your desk and is plugged into the back of the computer where the keyboard would normally plug in. Unlike infrared technology, RF technology doesn't require that you point the keyboard at the receiver for it to work. RF keyboards used on home computers can be placed as far as 6 to 30 feet from the computer, depending on their quality. RF keyboards used in business conference rooms or auditoriums can be placed as far as 100 feet away from the computer, but they are far more expensive than most home users can afford.

MICE AND OTHER POINTING DEVICES

What kinds of mice are there? The mouse you're probably most familiar with, like the one shown in Figure 2.6a, has a rollerball on the bottom, which moves when you drag the mouse across a mousepad. The movement of the rollerball controls the movement of your **pointer**, the I-beam or arrow that appears on the screen. Mice also have two or three buttons that enable you to execute commands and open shortcut menus. (Mice for Macintoshes usually have only one button.) Some newer mice have additional programmable buttons and wheels that let you scroll through documents or Web pages.

FIGURE 2.5

The stylus is the PDA's primary input device. You use it by tapping or writing on the PDA's touch-sensitive screen.

FIGURE 2.6

(a) A traditional mouse has a rollerball on the bottom, which moves when you drag the mouse across a mousepad. (b) A trackball mouse turns the traditional mouse on its back, allowing you to control the rollerball with your fingers. (c) An optical mouse has an optical laser (or sensor) on the bottom that detects its movement.

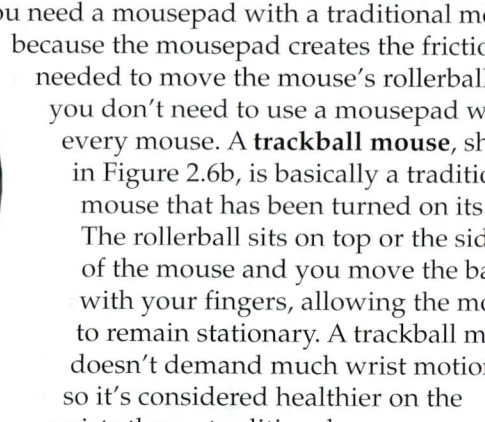
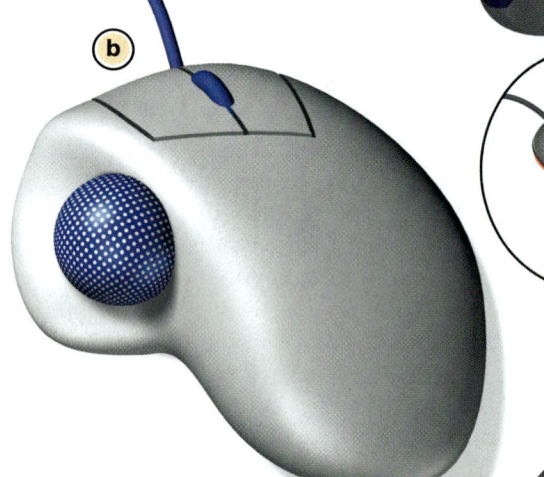

Why do I need to use a mousepad? You need a mousepad with a traditional mouse because the mousepad creates the friction needed to move the mouse's rollerball. But you don't need to use a mousepad with every mouse. A **trackball mouse**, shown in Figure 2.6b, is basically a traditional mouse that has been turned on its back. The rollerball sits on top or the side of the mouse and you move the ball with your fingers, allowing the mouse to remain stationary. A trackball mouse doesn't demand much wrist motion, so it's considered healthier on the wrists than a traditional mouse.

Another mouse that doesn't use a mousepad is the **optical mouse**, shown in Figure 2.6c. An optical mouse also doesn't contain a rollerball, but rather uses an internal sensor or laser to control the mouse's movement. The sensor sends signals to the computer, telling it where to move the pointer on the screen. Because they have no moving parts, optical mice have small advantages over traditional mice: there is less chance of parts breaking down and no way for dirt to interfere with the mechanisms. However, optical mice are often a bit more expensive than traditional mice.

Are there wireless mice? Just as there are wireless keyboards, there are wireless mice, both standard and optical. Wireless mice are similar to wireless keyboards in that they use batteries and send data to the computer via radio or light waves. If you also have an RF wireless keyboard, your RF wireless mouse and keyboard can share the same RF receiver.

What about mice for laptops? Because space is limited on laptops, the mouse is built into the keyboard area. Some laptops incorporate a trackball-like mechanism as a mouse. Others incorporate a **trackpoint**, a small, joystick-like nub that allows you to move the cursor with the tip of your finger. Other laptops have a **touchpad**, a small, touch-sensitive screen at the base of the keyboard. To use the touchpad, you simply move your finger across the pad. Some touchpads are also sensitive to taps, interpreting them as mouse-button clicks. Figure 2.7 shows some of the mouse options you'll find in laptops. Of course, if you prefer, you can always hook up a traditional mouse to your laptop as well.

BITS AND BYTES

Keeping Your Mouse Clean

It's amazing what kind of dirt and grime your mouse can pick up, even in the cleanest environment. You know your mouse needs cleaning when it becomes sluggish, nonresponsive, or jerky in its motion. To clean your mouse, follow these steps:

1. Turn your mouse over and remove the rollerball by turning the surrounding disk.
2. Take a dry cotton swab and remove any loose dust that has accumulated in the ball area.
3. Run a fresh cotton swab dampened in alcohol over the inside of the ball cavity, concentrating on the rollers.
4. Clean the mouse ball with alcohol to remove any oil and grime.
5. Let the components dry before putting them back together.

FIGURE 2.7

(a) This keyboard incorporates a trackpoint device that takes the place of a mouse.
(b) With a touchpad like the one on this keyboard, you control the pointer by moving your finger across the pad. The buttons at the bottom of the keyboard are used for right and left clicks of the mouse.

Are game controls considered mice? Game controls (such as joysticks and steering wheels) are not mice per se, but are considered input devices because they send data to the computer. Force-feedback joysticks and steering wheels deliver data in both directions: they translate your movements to the computer and translate its responses as forces on your hands, creating a richer simulated experience. If you like to move around a lot while you play games, you can purchase wireless game controllers at most computer stores.

Can I use a mouse with a PDA? As we mentioned earlier, the input device you use with PDAs is a *stylus*. You don't use a traditional mouse. However, some PDAs with built-in keyboards do include touchpads and track-point controls similar to those found in laptops. While you can't hook up a traditional desktop mouse to a PDA, there are micelike devices made especially for PDAs that eliminate the need for a stylus.

INPUTTING SOUND

How are microphones used? A microphone allows you to capture sound waves (such as your voice) and transfer them to digital format on your computer. Many computers include microphones as part of their basic equipment. If a microphone didn't come with your computer, you may want to buy one like the ones shown in Figure 2.8 if you plan to record your own audio files.

FIGURE 2.8

Headset and desktop microphones offer convenient, hands-free voice input.

You may also want to invest in a microphone if you plan to use **videoconferencing**. Videoconferencing technology allows a person sitting at a computer equipped with a personal video camera and a microphone to transmit video and audio across the Internet (or other communications medium). All computers participating in a videoconference need to have a microphone and speakers installed so that participants can hear one another.

Another use for microphones is **speech-recognition systems**, in which you operate your computer through a microphone, telling it to perform specific commands (such as to open a file) or to translate your spoken words into data input. Speech recognition has yet to truly catch on, but its popularity is growing. In fact, it's included in the software applications found in Office 2003. We'll discuss speech-recognition software in more detail in Chapter 4.

What's the best microphone to have? Microphone quality varies widely. For personal use, an inexpensive microphone is probably sufficient. However, if you plan to create professional products and sell them to others, you'll most likely need a professional recording studio microphone that could cost upwards of $250. Speech-recognition software requires a high degree of voice clarity, so it's best to have a headset-style microphone for this application.

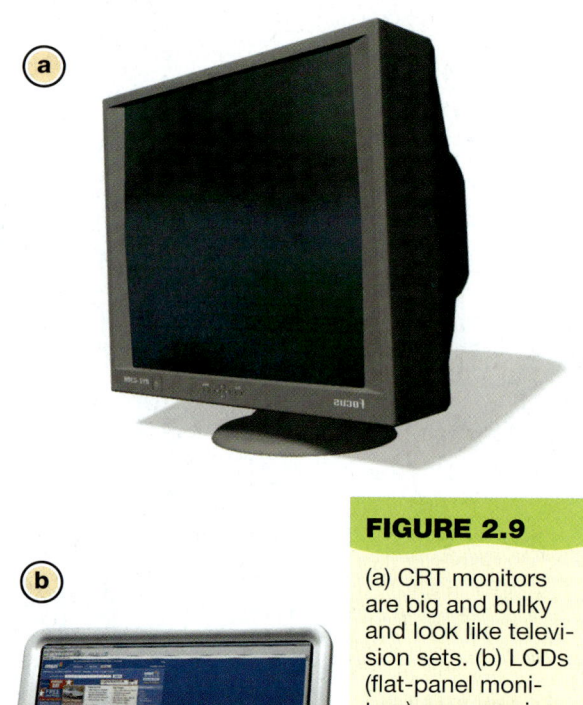

FIGURE 2.9
(a) CRT monitors are big and bulky and look like television sets. (b) LCDs (flat-panel monitors) save precious desktop space and weigh considerably less than CRT monitors.

Output Devices

As you learned in Chapter 1, **output devices** enable you to send processed data out of your computer. This can take the form of text, pictures (graphics), sounds (audio), and video. One common output device is a **monitor** (sometimes refered to as a **display screen**), which displays text, graphics, and video as "soft copies" (copies you can only see onscreen). Another common output device is the **printer**, which creates tangible or hard copies (copies you can touch) of text and graphics. Speakers are obviously the output devices for sound.

MONITORS

What different kinds of monitors are there? There are two basic types of monitors: CRTs and LCDs. If your monitor looks like a television set, it has a picture tube device called a **CRT (cathode ray tube)** like the one shown in Figure 2.9a. If your monitor is flat, such as those found in laptops, it's using **LCD (liquid crystal display)** technology, similar to that used in digital watches (see Figure 2.9b). LCD monitors (also called *flat-panel monitors*) are lighter and more energy efficient than CRT monitors, making them perfect for portable computers such as laptops.

CRT Monitors

How does a CRT monitor work? A CRT screen is a grid made up of millions of **pixels**, or tiny dots. Simply put, illuminated pixels are what create the images you see on your monitor. The pixels are illuminated by an electron beam that passes back and forth across the back of the screen very quickly—60 to 75 times a second—so that the pixels appear to glow continuously. Figure 2.10 shows in more detail how a CRT monitor works.

What factors affect the quality of a CRT monitor? A couple of factors affect the quality of a CRT monitor. One is the monitor's refresh rate. **Refresh rate** (sometimes referred to as *vertical refresh rate*) is the number of times per second the electron beam scans the monitor and recharges the illumination of each pixel. Current monitors have a maximum refresh rate of 75 Hertz (Hz). Hertz is a unit of frequency indicating cycles per second, in this case meaning the electron beam scans the monitor 75 times each second. The faster the refresh rate, the less the

OUTPUT DEVICES 39

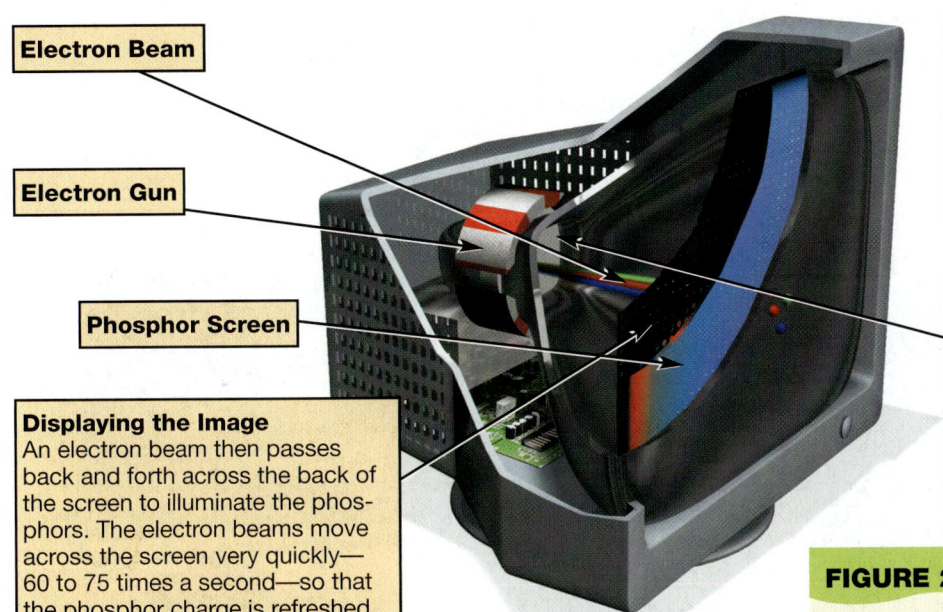

Electron Beam

Electron Gun

Phosphor Screen

Displaying the Image
An electron beam then passes back and forth across the back of the screen to illuminate the phosphors. The electron beams move across the screen very quickly—60 to 75 times a second—so that the phosphor charge is refreshed and appears to glow continuously.

The CRT screen is a grid made up of millions of **pixels**, or tiny dots. Each pixel is composed of three smaller elements called *phosphors* (each pixel contains a red, a green, and a blue phosphor). A phosphor is a substance that emits light when it is charged with electrons.

Creating the Colors
CRT monitors have an electron gun that shoots red, green, and blue electrons at the phosphors on the screen. The various colors of the image are created by mixing different combinations of red, green, and blue.

FIGURE 2.10

How a CRT Monitor Works

screen will flicker, the clearer the image will be, and the less eyestrain you'll experience.

The clearness or sharpness of the image—its **resolution**—is controlled by the number of pixels displayed on the screen. The higher the resolution, the sharper and clearer the image. Monitor resolution is listed as a number of pixels. A high-end monitor may have a maximum resolution of 1,600 x 1,200, meaning it contains 1,600 vertical columns with 1,200 pixels in each column. Note that most monitors allow you to adjust their resolution within a certain range. (We'll discuss setting your monitor's resolution in more detail in Chapter 6.)

Dot pitch is another factor that affects monitor quality. **Dot pitch** is the diagonal distance, measured in millimeters, between pixels of the same color on the screen. A smaller dot pitch means that there is less blank space between pixels, and thus a sharper, clearer image. A good CRT monitor has a dot pitch of .28 mm or less.

So if you're buying a new CRT monitor, choose the one with the highest refresh rate, the highest maximum resolution, and the smallest dot pitch.

LCD Monitors

How does an LCD monitor work? Like a CRT screen, an LCD screen is composed of a grid of pixels. However, instead of including a cathode ray tube, LCD monitors are made of two sheets of material filled with a liquid crystal solution. A fluorescent panel at the back of the LCD monitor generates light waves. When electric current passes through the liquid crystal solution, the crystals move around, either blocking the fluorescent light or letting the light shine through. This blocking or passing of light by the crystals causes images to be formed on the screen.

Are all LCD monitors the same? You'll generally find two types of LCD monitors on the market: **passive-matrix displays** and **active-matrix displays**. Less expensive LCD monitors use passive-matrix displays, while more expensive monitors use active-matrix displays. With passive-matrix technology, electrical current passes through the liquid crystal solution and charges groups of pixels, either in a row or a column. This causes the screen to brighten with

---

BITS AND BYTES

Cleaning Your Monitor

Have you ever noticed how quickly your monitor attracts dust? It's important to keep your monitor clean since dust build-up can act like insulation, keeping heat in and causing the electronic components to wear out much faster. To clean your monitor, follow these steps:

1. Shut off the monitor.
2. Wipe the monitor's surface using a sheet of fabric softener or a soft cloth dampened with window cleaner or water. Never spray anything directly onto the monitor. (Check your monitor's user manual to see if there are cleaning products you should avoid using.)
3. In addition to the screen, wipe away the dust from around the case.

Finally, don't place anything on top of the monitor as the items may block air from cooling it.

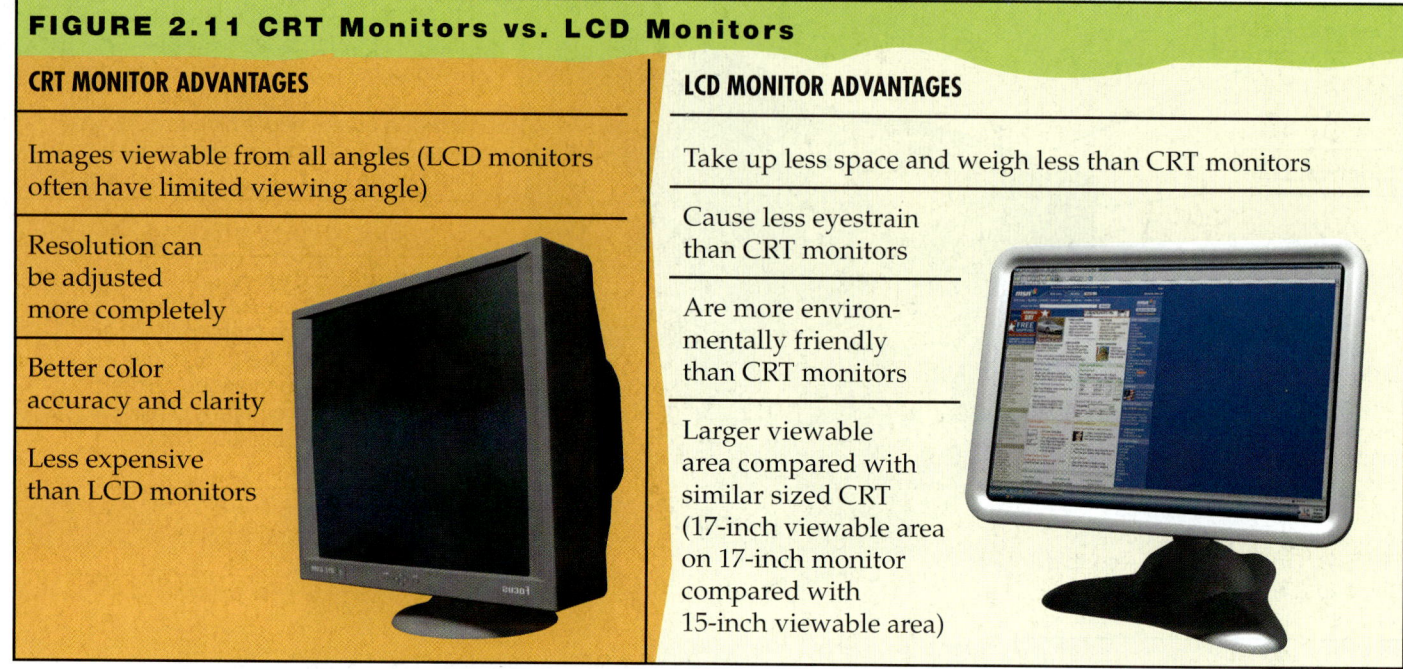

FIGURE 2.11 CRT Monitors vs. LCD Monitors

CRT MONITOR ADVANTAGES	LCD MONITOR ADVANTAGES
Images viewable from all angles (LCD monitors often have limited viewing angle)	Take up less space and weigh less than CRT monitors
Resolution can be adjusted more completely	Cause less eyestrain than CRT monitors
Better color accuracy and clarity	Are more environmentally friendly than CRT monitors
Less expensive than LCD monitors	Larger viewable area compared with similar sized CRT (17-inch viewable area on 17-inch monitor compared with 15-inch viewable area)

each pass of electrical current and subsequently fade. With active-matrix displays, each pixel is charged individually, as needed. The result is that an active-matrix display produces a clearer, brighter image.

LCD vs. CRT

Are LCD monitors better than CRT monitors? LCD monitors have a number of advantages over CRT monitors, as shown in Figure 2.11. Certainly size is an advantage, as LCD monitors take up far less space on a desktop than a CRT monitor. Additionally, LCD monitors weigh less, making them the obvious choice for mobile devices. LCD technology also causes less eyestrain than the refreshed pixel technology of a CRT.

Also, because of their different technologies, you can see more of an LCD screen than you can with the same size CRT monitor. For example, there are 17 inches of viewable area on a 17-inch LCD monitor but only 15 inches of viewable area on a 17-inch CRT monitor. LCDs are also more environmentally friendly, emitting less than half the electromagnetic radiation and using less power than their CRT counterparts.

One of the few disadvantages of LCD monitors is their limited viewing angle. Even with the most expensive models, you may have a hard time seeing the screen image clearly from an angle. Additionally, the resolution on a LCD screen is fixed and cannot be modified to the same degree as that of a CRT. CRT monitors also have better color accuracy than LCD monitors. Finally, LCD monitors are often more expensive than CRTs.

Printers

What are the different types of printers? There are two primary categories of printers: *impact* and *nonimpact*. **Impact printers** have tiny hammerlike keys that strike the paper through an inked ribbon, thus making a mark on the paper. The most common impact printer is the dot-matrix printer.

In contrast, **nonimpact printers** don't have mechanisms that strike the paper. Instead, they spray ink or use laser beams to make marks on the paper. The most common nonimpact printers are inkjet printers and laser printers. Such nonimpact printers have replaced dot-matrix printers almost entirely. They tend to be less expensive, quieter, faster, and offer better print quality than dot-matrix printers.

What were dot-matrix printers used for? You may remember the printers that used paper with perforated edges and track-feed holes down the side. Those were **dot-matrix printers**, the first computer printers (see Figure 2.12). They were revolutionary at the time because they allowed users to print a copy of the information displayed on their computer screens—something that hadn't been possible before. In addition to being slow and noisy, the output of the first dot matrix-printers wasn't nearly as good as that produced by typewriters.

Do people still use dot-matrix printers? Although inkjet and laser printers have replaced dot-matrix printers for everyday use, dot-matrix printers are still used for printing multipart forms such as invoices or contracts. This is because the pressure of the hammerlike keys can penetrate the multiple layers of paper.

OUTPUT DEVICES

BITS AND BYTES

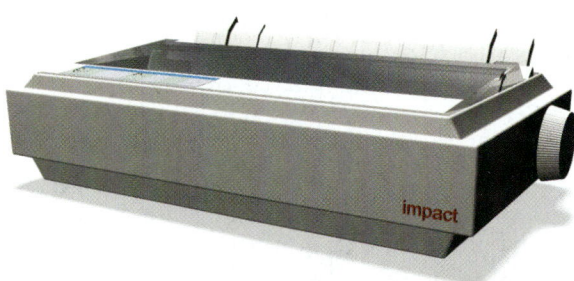

FIGURE 2.12
Dot-matrix printers have tiny hammer-like keys that strike the paper through an inked ribbon.

Does It Matter What Paper I Print On?

The quality of your printer is only part of what controls the quality of a printed image. The paper you use is equally important. If you're printing text-only documents for personal use, using low-cost paper is fine. However, if you're printing text-only documents for more formal use, such as resumes, you may want to choose a higher-quality paper. To do so, consider the paper's weight, whiteness (blue-white to creamy white), brightness (the brighter the paper, the easier it is to read), and opacity (especially important if you're printing on both sides of the paper).

If you're printing photos, paper quality can have a big impact on the results. Photo paper is more expensive than regular paper and comes in a variety of surface textures ranging from matte to high gloss. For a photo-lab look, high-gloss paper is the best choice. Semigloss (often referred to as satin) is good for portraits, while a matte surface is often used for black-and-white printing.

What are the advantages of inkjet printers? Compared with dot-matrix printers, **inkjet printers** are quieter, faster, and offer higher-quality printouts (see Figure 2.13). In addition, even good quality inkjet printers are affordable. Inkjet printers work by spraying tiny drops of ink onto paper. The first inkjet printers suffered from clogged inkjets, but over time, that problem was resolved. Their initial advantage, which continues today, is that inkjets print color images. In fact, when using the right paper, higher-end inkjet printers print images that look like professional-quality photos. Because of their high quality and low price, inkjet printers are the most popular printer for color printing.

Why would I want a laser printer? **Laser printers** are often preferred for their quick and quiet production and high-quality printouts (see Figure 2.14). Because they print quickly, laser printers are often used in schools and offices where multiple computers share one printer. Although more expensive to buy than inkjet printers, over the long run, laser printers are more economical than inkjets (they cost less per printed black-and-white page) when you include the price of ink and special paper in the overall cost.

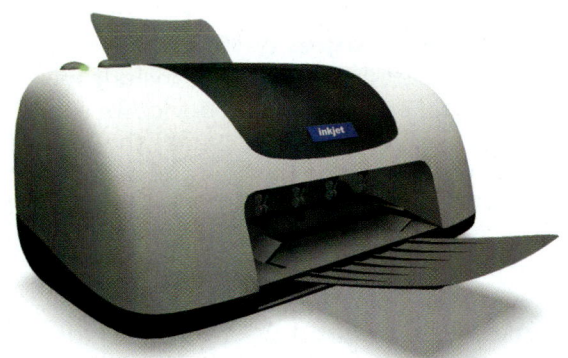

FIGURE 2.13
Inkjet printers are popular for home users, especially for color printing.

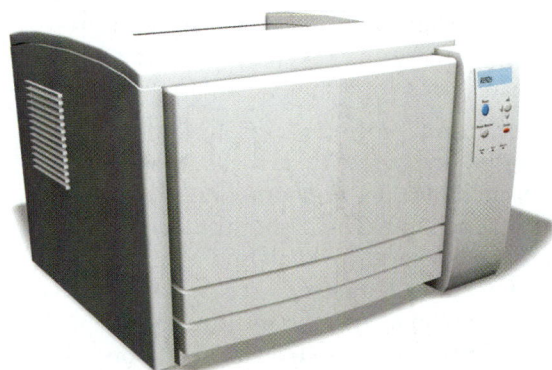

FIGURE 2.14
Laser printers print quickly and offer high-quality printouts.

TechTV For information on printer assisted crime see Counterfeit Computing—a TechTV clip found at www.prenhall.com/techinaction

Are there any other types of printers? Plotters are large printers used to produce oversize pictures that require precise continuous lines to be drawn, such as maps or architectural plans. Plotters use a computer-controlled pen that provides a greater level of precision than the series of dots that laser or inkjet printers are capable of making.

Thermal printers, such as the one shown in Figure 2.15, are another kind of specialty printer. These printers work by either melting wax-based ink onto ordinary paper (in a process called *thermal wax transfer printing*) or by burning dots onto specially coated paper (in a process called *direct thermal printing*). They are used in stores to print receipts and in airports for electronic ticketing, among other places. Thermal printers are also emerging as a popular technology for mobile and portable printing, for example in conjunction with PDAs.

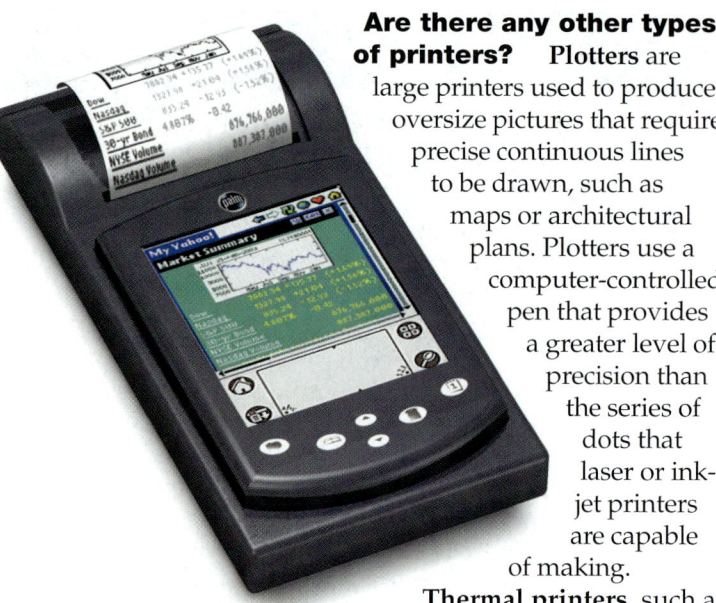

FIGURE 2.15
Thermal printers are often used with mobile computers.

Choosing a Printer

How do I select the best printer? There are several factors to consider when choosing a printer:

- **Speed.** A printer's speed determines how many pages it can print per minute (called pages per minute, or ppm). A reasonable printing speed for an inkjet printer is 2 to 6 ppm for text and slightly less for graphics. Laser printers print more quickly than inkjet printers (up to 30 or so ppm) but are more expensive.

- **Resolution.** A printer's resolution (or printed image clarity) is measured in dots per inch (dpi), or the number of dots of ink in a 1-inch line. The higher the dpi, the greater the level of detail and quality of the image. You'll sometimes see dpi represented as a horizontal number multiplied by a vertical number, such as 600 x 600, but you may also see the same resolution simply stated as 600 dpi. For general-purpose printing, 300 dpi is sufficient. If you're going to print photos, 1,200 dpi is better. The dpi for professional photo–quality printers is twice that.

- **Color Output.** If you're using an inkjet printer to print color images, buy a four-color (cyan, magenta, yellow, and black) or six-color printer (four-color plus light cyan and light magenta) for the highest-quality output. Some printers come with one ink cartridge for all colors, while others have two ink cartridges for black and color. The best setup is to have individual ink cartridges for each color so you can replace only the specific color cartridge that is empty. Color laser printers have separate toner cartridges for each color.

- **Memory.** Printers need memory in order to print. Inkjet printers run slowly if they don't have enough memory. If you plan to print small text-only documents on an inkjet printer, 1 to 2 megabytes (MB) of memory should be enough. You need about 4 MB of memory if you expect to print large text-only documents and 8 MB if you print graphic-intense files. Unlike inkjet printers, laser printers won't print at all without sufficient memory. To ensure your laser printer meets your printing needs, buy one with 16 MB of memory. Some printers allow you to add more memory later.

- **Use and Cost.** If you will be printing mostly black-and-white, text-based documents or will be sharing your printer with others, a laser printer is best because of its printing speed and overall economies for volume printing. If you're planning to print color photos and graphics, an inkjet is the better, more economical choice.

OUTPUTTING SOUND

What is the output device for sound? As noted earlier, most computers include inexpensive speakers as an output device for sound. These speakers are sufficient to play the standard audio clips you find on the Web and usually enable you to participate in teleconferencing. However, if you plan to watch a lot of DVDs or are particular about how your music sounds, you may want to upgrade to a more sophisticated speaker system, such as one that includes subwoofers (special speakers that produce only low bass sounds) and surround-sound (speaker systems set up in such a way that they "surround" you with sound). We'll discuss how to evaluate and upgrade your speaker system in more detail in Chapter 6.

DIG DEEPER

How Inkjet and Laser Printers Work

Ever wonder how a printer knows what to print, and how it puts ink in just the right places?

Most inkjet printers use *drop-on-demand* technology in which the ink is "demanded" and then "dropped" onto the paper. Two separate processes use drop-on-demand technology: Thermal bubble, used by Hewlett-Packard and Canon, and piezoelectric, used by Epson. The difference between the two processes is how the ink is heated within the print cartridge reservoir (the chamber inside the printer that holds the ink).

In the Thermal bubble process, the ink is heated in such a way that it expands (like a bubble) and leaves the cartridge reservoir through a small opening, or nozzle. Figure 2.16 shows the general process for Thermal bubble. In the piezoElectric process, each ink nozzle contains a crystal at the back of the ink reservoir that receives an electrical charge, causing the ink to vibrate and drop out of the nozzle.

Laser printers use a completely different process. Inside a laser printer is a big metal cylinder (or drum) that is charged with static electricity. When you ask the printer to print something, it sends signals to the laser in the laser printer, telling it to "uncharge" selected spots on the charged cylinder, corresponding to the document you wish to print. Toner, a fine powder that is used in place of liquid ink, is attracted to only those areas on the drum that are not charged. (These uncharged areas are the characters and images you want to print.) The toner is then transferred to the paper as it feeds through the printer. Finally, the toner is melted onto the paper. All unused toner is swept away before the next job starts the process all over again.

FIGURE 2.16

How an Inkjet Printer Works

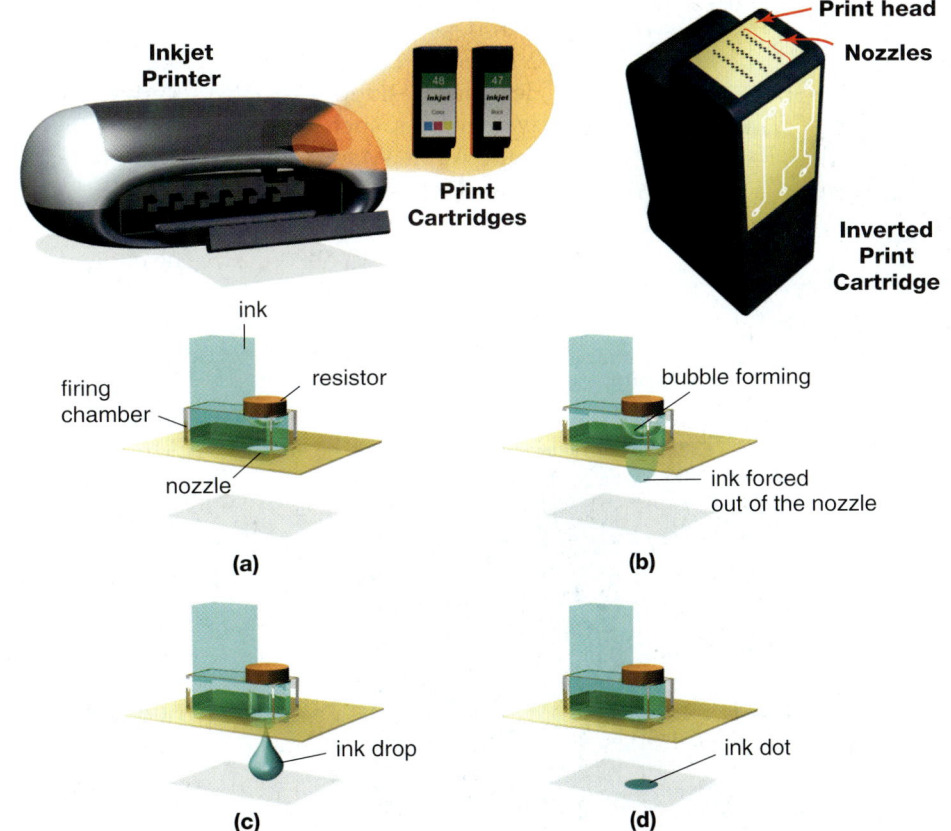

The print cartridge is positioned inside your inkjet printer so that the print head faces down towards the paper. The print head has 50 to several hundred nozzles, or small holes, through which ink droplets fall. These nozzles are narrower than a human hair. Inside the print head of color inkjet printers, there are three ink reservoirs that hold magenta (red), cyan (blue), and yellow ink. Depending on your printer, a fourth ink reservoir may be required to hold black ink, as well. (In non-color inkjet printers, there is only one ink reservoir for the black ink.

STEP (a): Once the printer receives the command to print, electrical pulses flow through thin resistors in the print head to heat the ink.

STEP (b): The heated ink forms a bubble. The bubble continues to expand until it is forced out of nozzle.

STEP (c): The ink drops onto the paper.

STEP (d): As the ink leaves the cartridge, the chamber begins to cool and contract, creating a vacuum to draw in the ink for the process to begin again.

The System Unit

We just looked at the components of your computer that you use to input and output data in some form. But where does the processing take place and where is the data stored? The **system unit** is the box that contains the central electronic components of the computer, including the computer's processor (its brains), its memory, and the many circuit boards that help the computer function. You'll also find the power source and all the storage devices (CD/DVD drive, floppy disk drive, and hard drive) in the system unit.

What styles of system units are there? The most common styles of system units on desktop computers are **desktop boxes**, which sit horizontally on top of your desk, and **tower configurations**, which typically sit vertically on the floor below your desk (see Figure 2.17). Some creatively designed desktop system units, such as the Apple iMac, house not just the computer's processor and memory, but its monitor as well. No matter the design, every computer—be it a desktop computer or a laptop—has a system unit.

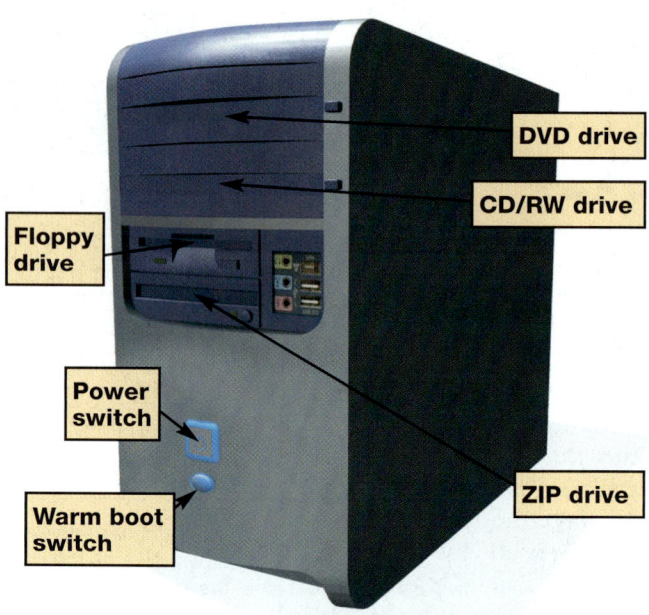

FIGURE 2.18
The front panel of your computer provides you with access to power controls as well as to the storage devices on your computer.

Which is the best system unit style? The choice between a desktop box and a tower configuration is ultimately up to you as both styles contain the parts needed to run the computer. The tower configuration may have a few advantages over the desktop style because tower configurations don't take up desktop space. In addition, tower configurations have more spaces reserved on the inside of the system unit that you can use to install additional devices that didn't come with your system, such as an additional DVD or CD drive.

ON THE FRONT PANEL

What's on the front panel of my computer? No matter whether you choose a desktop or tower design, the front panel of your computer provides you with access to power controls as well as to the storage devices on your computer. Figure 2.18 shows the front panel of a typical system. Although your system might be slightly different, chances are it includes many of the same features.

Power Controls
How do I turn my computer on and off? The button on the front panel of the system unit is the button you press to turn on (or power on)

FIGURE 2.17
System units come in several different designs. (a) The Apple iMac's sleek design takes up considerably less room than (b) a standard tower configuration.

the system. (You also find power-on buttons on some keyboards.) Although you use this button to turn on your system, you don't want to use it to turn off (or power off) your system. Modern operating systems want control over the shutdown procedure, so you turn off the power by clicking on a shutdown icon on the desktop, *not* by pushing the main power button.

If you do shut off the power using the main power button without shutting down your operating system first, nothing on your system will be permanently damaged. However, some files and applications may not close properly, so the operating system may need to do some extra work the next time you start your computer.

Why does my computer have two buttons on the front panel? Some desktop computers have two buttons on the front panel. The second button *restarts* the computer. Restarting the sys-tem while it's powered on is called a **warm boot**. You might need to perform a warm boot if the operating system or other software application stops responding. It takes less time to perform a warm boot than to completely power down and then restart all of your hardware. Doing a complete power down and restart is a **cold boot**. You do a cold boot each time you start the computer after it's been powered down completely. (Note that if your computer does not have a restart button, you can perform a warm boot by pressing the Ctrl, Alt, and Delete keys at the same time.)

Should I turn my computer off every time I'm done using it? Some people say you should leave your computer on at all times. They argue that turning your computer on and off throughout the day subjects its components to stress as the heating and cooling process forces the components to expand and contract repeatedly. Other people say you should shut down your computer when you're not using it. They claim that you'll end up wasting money on electricity to keep the computer running all the time. However, modern operating systems include power-management settings that allow the most power-hungry components of the system (the hard drive and monitor) to shut down after a short idle period.

So, if you use the computer sporadically throughout the day, it may be best to keep it on while you're apt to use it and power it down when you're done using it for the day. However, if you only use your computer for a little while each day, you'll be paying electricity charges during long periods of nonusage. If you're truly concerned about the "stresses" incurred from powering on and off your computer, you may want to buy a warranty with the computer, which will undoubtedly cost less than the extra power you'd use to keep your computer constantly running.

Can I "rest" my computer without turning it off completely? As mentioned above, your computer has power-management settings that help it conserve energy. These settings are called *standby mode* and *hibernation*. When your computer is in **standby mode**, its more power-hungry components, such as the monitor and hard drive, are put in idle. In essence, the computer is napping. To wake it up, you tap a key on the keyboard or move the mouse. When the computer is in **hibernation**, it is in a deeper sleep. When you wake the computer from hibernation (by pushing the power button), the computer reloads everything to your desktop so that it is exactly as it was before it went into hibernation. You can find the settings for standby and hibernation modes in the Performance and Maintenance option in the Control Panel menu (see Figure 2.19).

FIGURE 2.19

Using the hibernation and standby settings is not only good for the environment, but it is also good for your pocketbook.

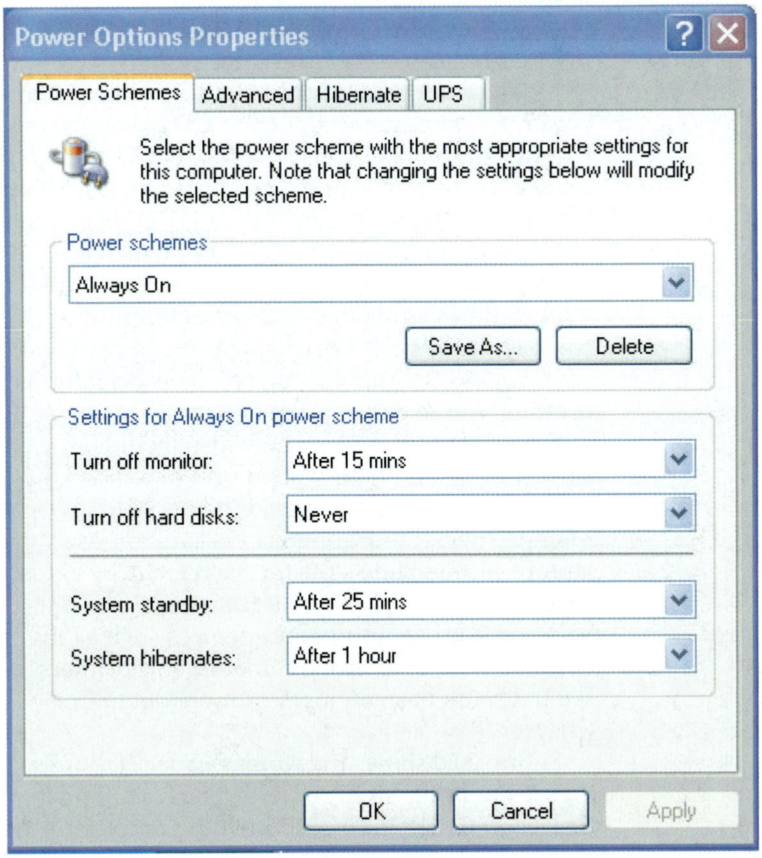

Drive Bays: Your Access to Storage Devices

What else is on the front panel? Besides the power button, the other features that can be seen from the front of your system unit are **drive bays**. These bays are special shelves reserved for storage devices, those devices that hold your data and applications when the power is shut off. There are two kinds of drive bays:

1. Internal drive bays cannot be seen or accessed from outside the system unit. Generally, internal drive bays are reserved for hard disk drives.
2. External drive bays can be seen and accessed from outside the system unit. External drive bays house floppy disk and CD drives, for example. Empty external drive bays are covered by a faceplate.

By looking at the front panel of your system unit, you can tell which devices have been installed, and often how many bays remain available for expansion.

What kinds of external drive bays do most PCs have? Most PCs still have a bay for a **floppy disk drive,** which reads and writes to easily transportable floppy disks. (Macs no longer include floppy disk drives.) Many computers also feature a **Zip disk drive**, which resembles a floppy disk drive but has a slightly wider opening. Zip disks work just like standard floppies but can carry much more data.

On the front panel, you'll also see one or two bays for storage devices such as CD drives. **CD-ROM drives** read CDs, whereas **CD-RW drives** can both read and write data to CDs. Most new computers also come with a **DVD drive** that allows the computer to read DVDs, which are the same size and shape as CDs but can hold more than 25 times as much data. Some computers also come with **DVD-RW** drives, which can read and write to DVDs. DVD-RW drives are especially useful if you're creating digital movies.

Several manufacturers now also include a slot on the front of the system unit in which you can insert portable *flash memory cards* such as Memory Sticks and Compact Flash cards. Many laptops also include slots for flash memory cards. **Flash memory cards** let you transfer digital data between your computer and devices such as digital cameras, PDAs, video cameras, and printers. We'll discuss flash memory in more detail in Chapter 8.

Figure 2.20 shows the storage capacities of the various portable storage media (such as floppy disks and CD-ROMs) used in your computer's drive bays. As you learned in Chapter 1, storage capacity is measured in bytes. A kilobyte (KB) equals about 1,000 bytes, a megabyte (MB) equals about a million bytes, and a gigabyte (GB) equals about a billion bytes.

ON THE BACK: PORTS

What's on the back of my system unit? Earlier in the chapter we looked at the devices you use to input and output data to a computer. These devices (such as monitors, printers, and keyboards) are among many other devices (called **peripheral devices**) that connect to the system unit through ports. **Ports** are the place on the system unit where peripheral devices attach to the computer so data can be exchanged between them. Because devices exchange data with the computer in various ways, a number of different ports have been created to accommodate these devices. Look at the back of your system unit. It's probably very similar to the one shown in Figure 2.21. Do you know what all the ports are for?

Which ports help me connect to input/output devices? Serial ports and parallel ports have long been used to connect input and output devices to the computer. **Serial ports** send data one bit (or piece of data) at a time and are often used to connect modems to the computer. Sending data one bit at a time is a slow way to communicate. Data sent over serial ports is transferred at a speed of 115 Kilobits per second (Kbps), or 115,000 bits per second. A **parallel port** sends data between devices in *groups* of bits at speeds of 500 Kbps, and is therefore much faster than a serial port. Parallel ports are often used to connect printers to computers.

Due to their abilty to transfer data at approximately 12 Mbps (that's 12,000 Kbps), **USB (Universal Serial Bus) ports** are fast replacing serial and parallel ports as the means to connect input and output devices to the computer. USB ports can connect a wide variety of peripherals to the computer, including keyboards, printers, Zip drives, and digital cameras.

Which ports help me connect with other computers? Another set of ports on your computer helps you communicate with other computers. Called **connectivity ports**, these ports give you access to networks and the Internet and enable your computer to function as a fax machine. To find connectivity ports, look for a port that resembles a standard phone jack. This jack is the **modem port**. It uses a traditional telephone signal to connect two computers. Most computers now come configured with a second connectivity port

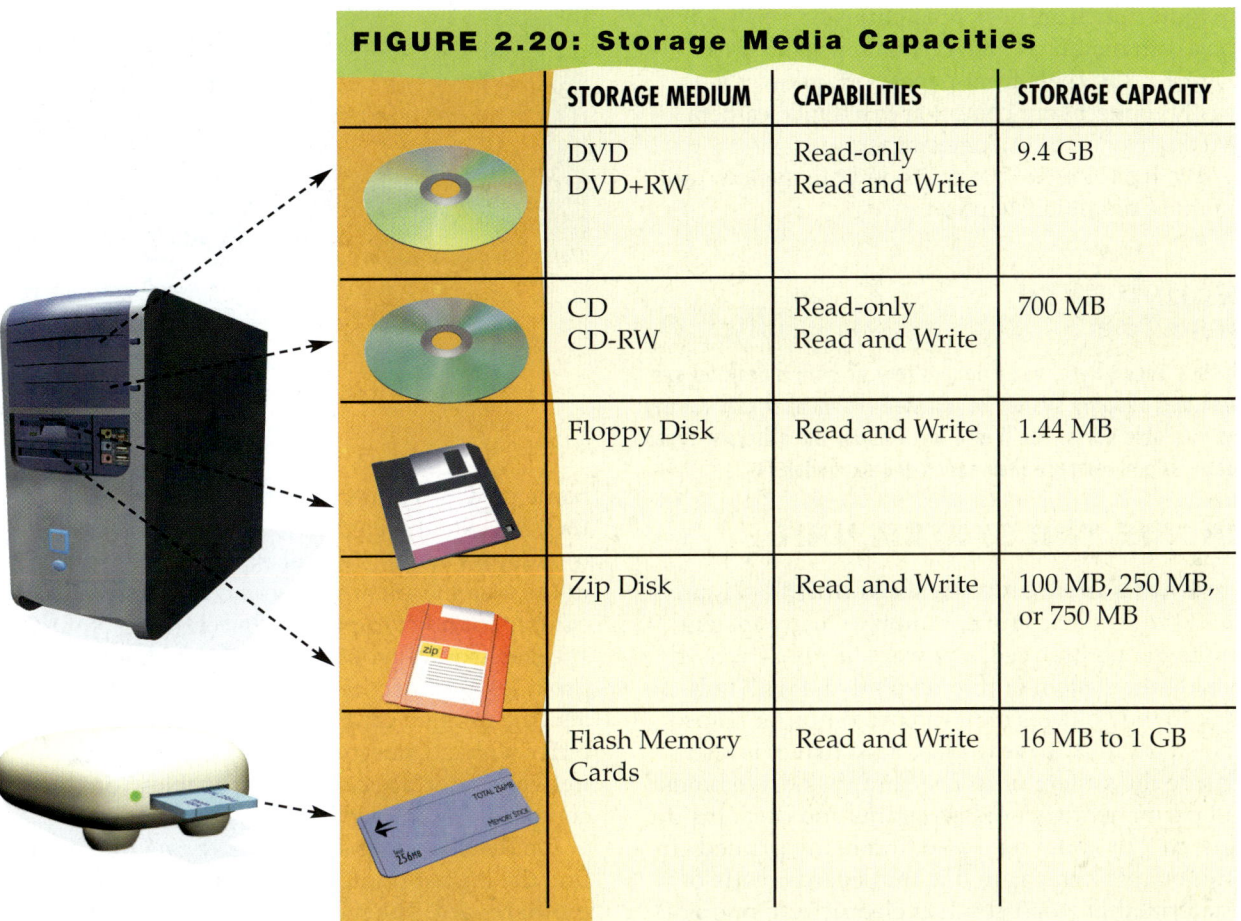

FIGURE 2.20: Storage Media Capacities

STORAGE MEDIUM	CAPABILITIES	STORAGE CAPACITY
DVD DVD+RW	Read-only Read and Write	9.4 GB
CD CD-RW	Read-only Read and Write	700 MB
Floppy Disk	Read and Write	1.44 MB
Zip Disk	Read and Write	100 MB, 250 MB, or 750 MB
Flash Memory Cards	Read and Write	16 MB to 1 GB

called an **Ethernet port**. This port is slightly larger than a standard phone jack and transfers data up to 100 Mbps. You use it to connect your computer to a cable modem or a network.

What are the fastest ports available? Newer interfaces such as **USB 2.0**, **FireWire** (or **IEEE 1394**), and the latest **FireWire 800** are the fastest ports available. Devices such as MP3 players, digital cameras, and digital camcorders all benefit from the speedy data transfer of USB 2.0 and FireWire. USB 2.0 transfers data at 480 Mbps and is approximately 40 times faster than the original USB port. The FireWire interface moves data at 400 Mbps, while FireWire 800 doubles the rate to 800 Mbps.

What are the ports on the front of my computer for? Traditionally, ports have been located on the back of the system unit. However, because so many portable devices (such as digital cameras) need to be plugged into the computer, and because ports on the front of the system unit are easier to access than are ports on the back, many

FIGURE 2.21

The back of your computer probably looks a lot like this one. There are several different ports because many devices exchange data with the computer in various ways.

manufacturers are designing ports (primarily USB ports) on the front of the system unit as well (see Figure 2.22). These front-panel ports allow you to easily connect your computer to a digital camera, MP3 player, or PDA, for example.

We'll explore all the ports on your system unit in more detail in Chapter 6.

FIGURE 2.22

Front-panel ports allow you to easily connect to your devices.

SOUND BYTE
PORT TOUR

In this Sound Byte, you'll take a tour of both a desktop system and a laptop system to compare the number and variety of available ports. You'll also learn about the different types of ports and compare their speed and expandability.

INSIDE THE SYSTEM UNIT

What's inside the system unit? Figure 2.23 shows the layout common to many system units. As you can see, a **power supply** is housed inside the system unit to regulate the wall voltage to the voltages required by computer chips. The **hard disk drive** (or just **hard drive**) is also inside the system unit. The hard disk drive holds all permanently stored programs and data. Inside the system unit, you'll also find many printed circuit boards, which are flat, thin boards made of material that won't conduct electricity. On top of this material, thin copper lines are traced, allowing designers to connect a set of computer chips.

The various circuit boards have specific functions that augment the computer's basic functions. Some provide connections to other devices, so they are usually referred to as **expansion cards** (or **adapter cards**). Typical expansion cards found in the system unit are the sound card and video card. A **sound card** provides a connection for the speakers and microphone, while a **video card** provides a connection for the monitor. There is also the **modem card**, which provides the computer with a connection to the Internet, and a **network interface card**, which enables your computer to connect with other computers.

On the bottom or side of the system unit, you'll find the largest printed circuit board, called the **motherboard**. The motherboard is named such because all of the other boards (video cards, sound cards, and so on) connect to it to receive power and to communicate—therefore, it's the "mother" of all boards.

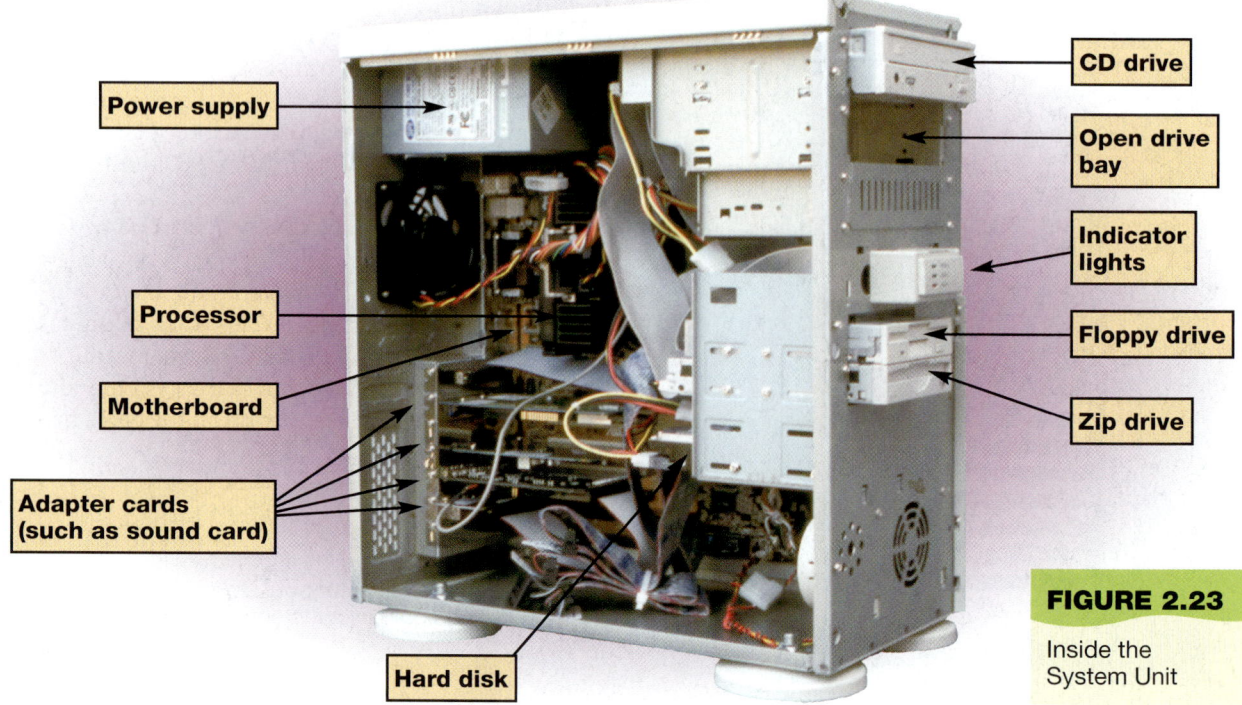

FIGURE 2.23

Inside the System Unit

THE SYSTEM UNIT 49

FIGURE 2.24

A motherboard contains the CPU, the memory (RAM) chips, and the slots available for expansion cards.

What's on the motherboard? The motherboard contains the set of chips that powers the system, including the central processing unit. The motherboard also houses the chips that provide the short-term memory for the computer as well as a set of slots available for expansion cards (see Figure 2.24).

What is the CPU? The **central processing unit** (**CPU** or **processor**) is the largest and most important chip in the computer. It is sometimes referred to as the "brains" of the computer as it controls all the functions performed by the computer's other components and processes all the commands issued to it by software instructions. Modern CPUs can perform three billion tasks a second without error, making them extremely powerful components.

What exactly is RAM? If you look at a motherboard, you'll see **random access memory** (**RAM**) as a series of small cards (called *memory cards* or *memory modules*) plugged into slots on the motherboard. Since the CPU processes data so rapidly, there needs to be a way to store data and commands nearby so they can be fed to the CPU very quickly. RAM is that storage space. The CPU can request the contents of RAM, which can be located, opened, and delivered to the CPU for processing in a few billionths of a second (or *nanoseconds*).

Sometimes RAM is referred to as "primary storage" for this reason, but it should not be confused with other types of *permanent* storage devices such as the hard disk drive. Because all the contents of RAM are erased when you turn off the computer, RAM is the *temporary* or **volatile storage** location for the computer. To save data more permanently, you need to save it to the hard drive or to another permanent storage device such as a floppy disk, CD, or Zip disk.

Does the system unit contain any other kinds of memory besides RAM? In addition to RAM, the motherboard also contains a form of memory called **ROM** (**read only memory**). ROM holds all the instructions the computer needs to start up. Unlike data stored in RAM, the instructions stored in ROM are permanent, making ROM a **nonvolatile storage** location. This means it does not get erased when the power is turned off.

What kind of data is saved on the hard disk drive? As noted earlier, the hard disk drive is the storage device you use to permanently hold the data and instructions your computer needs, even after the computer is turned off. This makes the hard disk a nonvolatile storage location. Today's hard drives, with capacities of up to 250 gigabytes (GB), can hold hundreds of billions of pieces of data.

As noted above, the hard disk drive is generally installed inside the system unit with all the other drive bays (see Figure 2.25). However, unlike the other drive bays, you can't access the hard disk drive from the outside of the system unit, making it a form of *nonportable* permanent storage.

FIGURE 2.25

You generally don't see the hard disk drive as it's enclosed in a sealed protective case. The hard disk drive holds all the data and instructions that the computer needs, even after the power is turned off.

BITS AND BYTES
Opening Up Your System Unit

Many people use a computer for years without ever needing to open their system unit. But there are two reasons you might want or need to do so: to replace a defective expansion card or device or to upgrade your computer. If your hard drive or CD-ROM drive fails, it's quite simple to open the system unit yourself and replace it. Adding additional memory or adding a DVD-RW drive are simple upgrade procedures that you can do safely at home. However, it's important that you follow the device's specific installation instructions. These instructions will detail any safety procedures you'll need to observe, such as unplugging the computer or grounding yourself to avoid static electricity, which can damage internal components.

SOUND BYTE
VIRTUAL COMPUTER TOUR

This Sound Byte will take you on a video tour inside your system unit. From opening the cover to locating the power supply, CPU, and RAM, to examining the expansion slots available, this video guide will teach you to be more familiar with what's inside your computer.

Setting It All Up: Ergonomics

It's important that you understand not only your computer's components and how they work together, but also how to set up these components safely. *Merriam-Webster's Dictionary* defines **ergonomics** as "an applied science concerned with designing and arranging things people use so that the people and things interact most efficiently and safely." In terms of computing, ergonomics refers to how you set up your computer and other equipment to minimize your risk of injury or discomfort.

Why is ergonomics important? Workplace injuries related to musculoskeletal disorders occur frequently in the United States. Approximately 1.8 million workers experience such disorders annually, with 600,000 needing to take time off from work. The Occupational Safety and Health Administration, or OSHA (www.osha.gov), reports that these types of injuries cost businesses and taxpayers up to $20 billion annually in workers' compensation and up to $40 billion in other expenses such as medical care.

How can I avoid injuries when I'm working at my computer? The following are some guidelines that can help you avoid discomfort, eyestrain, or injuries while you're working at your computer:

- **Position your monitor correctly.** Studies suggest it's best to place your monitor at least 25 inches from your eyes. You may need to decrease the screen resolution to make text and images more readable at that distance. Also, experts recommend the monitor be positioned so that it is at an angle 20 to 50 degrees below your line of sight.

- **Purchase an adjustable chair.** Adjust the height of your chair so that your feet touch the floor. (You may need to use a footrest to get the right position.) Back support needs to be adjustable so that you can position it to support your lumbar (lower back) region. You should also be able to move the seat or adjust the back so you can sit without exerting pressure on your knees. If your chair doesn't adjust, placing a pillow behind your back can provide the same support.

- **Assume a proper position while typing.** A **repetitive strain injury (RSI)** is a painful condition caused by repetitive or awkward movements of a part of the body. Improperly positioned keyboards are one of the leading causes of RSIs in computer users. Your wrists should be flat (unbent) with respect to the keyboard and your forearms parallel to the floor. You can either adjust the height

FIGURE 2.26

Ergonomic keyboards that curve and contain built-in wrist rests help you maintain proper hand position to minimize strain on your wrists.

of your chair or install a height-adjustable keyboard tray to ensure a proper position. Specially designed ergonomic keyboards like the one in Figure 2.26 can help you achieve the proper position of your wrists.

- **Take breaks from computer tasks.** Remaining in the same position for long periods of time increases stress on your body. Shift your position in your chair and stretch your hands and fingers periodically. Likewise, staring at the screen for long periods of time can lead to eyestrain, so rest your eyes by periodically taking them off the screen and focusing them on an object at least 20 feet away.
- **Ensure the lighting is adequate.** Assuring proper lighting in your work area is a good way to minimize eyestrain. To do so, eliminate any sources of direct glare (light shining directly into your eyes) or reflected glare (light shining off the computer screen) and ensure there is enough light to read comfortably.

Figure 2.27 illustrates how you should arrange your monitor, chair, body, and keyboard to avoid injury or discomfort while you're working on your computer.

SOUND BYTE
HEALTHY COMPUTING

In this Sound Byte, you'll see how to set up your workspace in an ergonomically correct way. You'll learn the proper location of the monitor, keyboard, and mouse, as well as ergonomic features to look for in choosing the most appropriate chair.

FIGURE 2.27

Achieving comfort and a proper typing position is the way to avoid repetitive strain injuries and other aches and pains while working at a computer. To achieve this, obtain equipment that boasts as many adjustments as possible. Every person is a different shape and size, requiring each workspace to be individually tailored.

TRENDS IN IT

EMERGING TECHNOLOGIES: Tomorrow's Displays

Most of us grew up with CRT monitors on our computers. Large and heavy but providing excellent resolution and clarity, these monitors were the main display for computers until the early 2000s. Soon, though, you may only find these clunky dinosaurs in the Smithsonian.

Today, LCD screens are where it's at. First introduced in 2000, these monitors have taken the desktop market by storm, and with prices steadily dropping, they now outsell CRTs. Lighter and less bulky than CRT monitors, they can be easily moved and take up less real estate on a desk. Still, current LCD technology does have limitations. LCD screens are relatively fragile (although less so than CRT monitors) and viewing angles are limited. In addition, LCDs can't display full-motion video as well as CRT monitors, making them unpopular with hard-core gamers.

Despite their limitations, LCDs will continue to be the predominant display device for computers, cell phones, and PDAs in the next few years. But according to sources like *PC Magazine,* new technologies are being developed that take LCD displays to the next level.

FLEXIBLE SCREENS

The most promising displays currently under development are *organic light-emitting displays,* (or *OLEDs*). These displays, currently used in some Kodak cameras, use organic compounds that produce light when exposed to an electric current. OLEDs tend to use less power than other flat-screen technologies, making them ideal for portable battery-operated devices. However, most research is being geared toward *FOLEDs* (*flexible OLEDs*). Unlike LCDs and CRTs, which use rigid surfaces such as glass, FOLED screens would be designed on lightweight, inexpensive, flexible material such as transparent plastics or metal foils. As shown in Figure 2.28, the computer screen of the future might roll up into an easily transported cylinder the size of a pen!

FOLEDs would allow advertising to progress to a new dimension. Screens could be hung where posters are hung now (such as on billboards). And wireless transmission of data to these screens would allow advertisers to display easily updatable full-

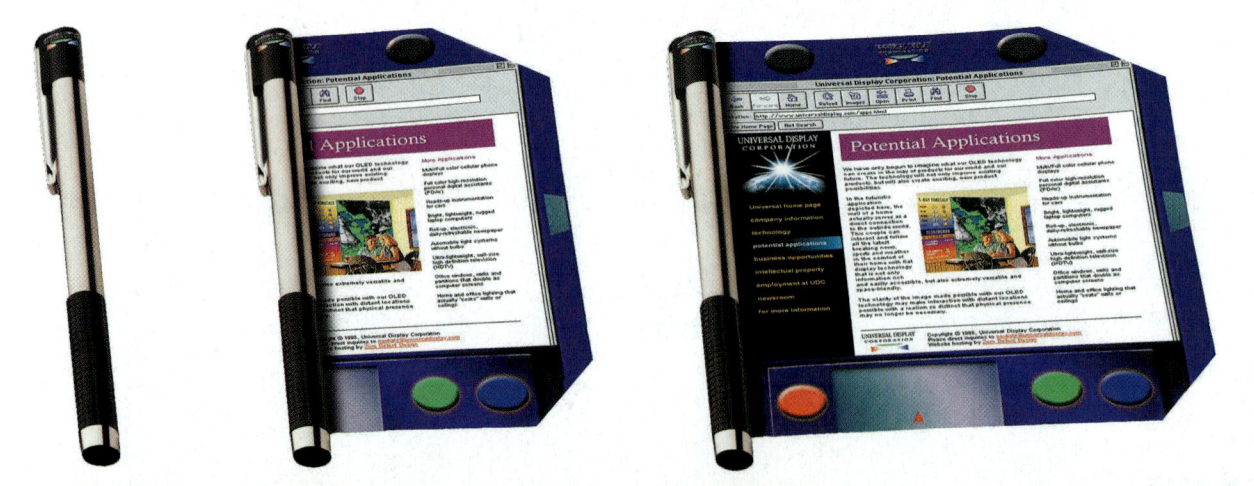

FIGURE 2.28

There's no need to lug around a heavy computer monitor. With FOLED technology, you'll be able to unroll a computer screen wherever you need it from a container the size of a pen. The prototype shown is currently being developed by Universal Display Corporation and may be available within two to three years.

motion images. Combining transparency and flexibility would also allow these displays to be mounted on windshields or eyeglasses. Who needs paper maps when you can examine directions right on your windshield without taking your eyes off the road?

There are some obstacles to be overcome before these screens will be widely available. Currently, the compounds that create blue hues age much faster than the ones that produce reds and greens. This makes it difficult to maintain balanced colors over long time periods. And the compounds used to form the OLEDs can be contaminated by exposure to water vapor or oxygen. However, once these problems are overcome, flexible displays should be popping up everywhere.

WEARABLE SCREENS

Who needs a screen when you can just wear one? Microdisplays are screens that, measured diagonally, are 1 inch or less in size. These displays are currently in use or are in development and can be used in head-mounted displays, such as the glasses shown in Figure 2.29. Eventually, these displays could replace heavier screens on laptops, desktops, and even PDAs.

"BISTABLE" SCREENS

Your computer screen constantly changes its images when you are surfing the Internet or playing a game. However, PDA and cell phone screens don't necessarily change that often. Therefore, something called a "bistable" display may one day be used in these devices. A *bistable display* has the ability to retain its image even when the power is turned off. In addition, bistable displays are lighter than LCD displays and reduce overall power consumption, resulting in longer battery life. As the market for portable devices continues to explode, you can expect to see bistable technologies in mobile computer screens.

So when will the current LCD technology be displaced once and for all? Since scientific research still must be conducted before these technologies will be commercially viable, it is difficult to say. But certainly within the next 10 years these and other as yet undreamed of technologies should replace current LCDs.

FIGURE 2.29
A prototype developed by MicroOptical Corporation features a microdisplay embedded in a pair of glasses. To users, it appears as though the display is projected right in front of them. Within a few years, you'll be able to buy a microdisplay when you buy glasses at retail stores. Microdisplays such as this can significantly cut the weight of portable computers.

Summary

1. **What input devices do you use to get data into the computer?** An input device enables you to enter data (text, images, and sounds) and instructions (user responses and commands) into the computer. You use keyboards to enter typed data and commands, whereas you use the mouse to enter user responses and commands. Keyboards are distinguished by the layout of the keys as well as the special keys found on the keyboard. The most common keyboard is the QWERTY keyboard. Most computers come with a standard two-button mouse, but you can also find optical mice, trackball mice, and wireless mice. Microphones are devices used to input sounds, while scanners and digital cameras input text and images.

2. **What output devices do you use to get data out of the computer?** Output devices enable you to send processed data out of your computer. This can take the form of text, pictures, sounds, and video. Monitors display soft copies of text, graphics, and video while printers create hard copies of text and graphics. Speakers are the output devices for sound. There are two popular types of computer monitors: cathode ray tube (CRT) monitors that look like TVs and liquid crystal display (LCD) monitors that have a flat-panel design. Popular printers include inkjet and laser printers.

3. **What's on the front of your system unit?** The system unit is the box that contains the central electronic components of the computer. On the front of the system unit, you'll find the power source as well as access to the storage devices (CD/DVD drive, floppy disk drive, and Zip disk drive) in your computer. Some computers also include ports on the front panel.

4. **What's on the back of your system unit?** On the back of the system unit you'll find a wide variety of ports that allow you to hook up peripheral devices (such as your monitor and keyboard) to your system. Serial, parallel, and USB ports are the most common. Serial ports send data 1 bit (or piece of data) at a time at speeds of 56 Kbps and are sometimes used to connect mice and keyboards to the computer. Parallel ports send data between devices in *groups* of bits at speeds of 92 Kbps. USB ports, fast replacing serial and parallel ports, transfer data at approximately 12 Mbps. Ports that provide even faster data transfer include Ethernet, USB 2.0, and FireWire.

5. **What's inside your system unit?** The system unit contains the main electronic components of the computer, including the motherboard, the main circuit board of the system. On the motherboard is the computer's central processing unit (CPU), which coordinates the functions of all other devices on the computer. RAM, the computer's volatile memory, is also located on the motherboard. RAM is where all the data and instructions are held while the computer is running. ROM, a permanent type of memory, is responsible for housing instructions to help start up the computer. The hard drive (the permanent storage location) and other storage devices (CD/DVD drives, floppy drives) are also located inside the system unit, as are circuit boards (such as sound, video, modem, and network interface cards) that help the computer perform special functions.

6. **How do you set up your computer to avoid strain and injury?** Ergonomics refers to how you arrange your computer and equipment to minimize your risk of injury or discomfort. This includes positioning your monitor correctly, buying an adjustable chair that ensures you have good posture while using the computer, assuming a proper position while typing, and making sure the lighting is adequate. Other good practices include taking frequent breaks as well as using other ergonomically designed equipment such as keyboards.

Materials on the Web

In addition to the review materials presented here, you'll find extra materials on the book's companion Web site (**www.prenhall.com/techinaction**) that will help reinforce your understanding of the chapter content. These materials include:

Sound Byte Lab Guides

For each Sound Byte mentioned in the chapter, there is a corresponding lab guide located on the book's companion Web site. These guides review the material presented in the Sound Byte and direct you to various Web resources that examine the material. The Sound Byte Lab Guides for this chapter include:

- Port Tour
- Virtual Computer Tour
- Healthy Computing

True/False and Multiple Choice Quizzes

The book's Web site includes a True/False and a Multiple Choice quiz for this chapter. You can take these quizzes, automatically check the results, and e-mail the results to your instructor.

Web Research Projects

The book's Web site also includes a number of Web research projects for this chapter. These projects ask you to search the Web for information on computer-related careers, milestones in computer history, important people and companies, emerging technologies, and the applications and implications of different technologies.

OBJECTIVES

After reading this chapter, you should be able to answer the following questions:

- What is the history of the Internet? (p. 66)

- How does data travel on the Internet? (pp. 66–67)

- What are my options for connecting to the Internet? (pp. 68–73)

- How do I choose an Internet service provider? (pp. 73–75)

- What is a Web browser? (pp. 75–76)

- What is a URL and what are its parts? (pp. 76–78)

- How can I use hyperlinks and other tools to get around the Web? (pp. 78–79)

- How do I search the Internet using search engines and subject directories? (pp. 79–83)

- What are Boolean operators and how do they help me search the Web more effectively? (pp. 82–83)

- How can I communicate through the Internet with e-mail, chat, IM, and newsgroups? (pp. 83–90)

- What is e-commerce and what e-commerce safeguards protect me when I'm online? (pp. 90–92)

- What are cookies and what risks do they pose? (p. 92)

- What are the various kinds of multimedia files found on the Web? (p. 94)

- What kind of software do I need to enjoy multimedia on the Web? (p. 95)

CHAPTER 3

Introducing the Internet:
Making the Most of the Web's Resources

TECHNOLOGY IN ACTION: INTERACTING WITH THE INTERNET

It's 10:00 p.m. as Max sits down to begin his online coursework. Although it's been a long day, he likes taking the online course because it lets him finish his degree and still keep his full-time job. While downloading the assignment file from the course Web site, he switches to his Instant Messenger (IM) account to see if any of his friends are online. Seeing his friend Tom is logged on, he chats with him for a while. Like a lot of his friends, Max has become a fanatic of IM. He uses it almost as much as he does e-mail.

Still waiting for his file to finish downloading, he visits ESPN.com to check out the score of the Red Sox game. He then goes to his favorite search engine, Google, and starts researching the topic he plans to write about for his class. As he conducts his searches, he experiments with some Web search techniques he learned about in his online class. He's amazed at how much information he can find when he searches the Web, yet how hard it is to find truly *useful* information.

Finally, Max sees that the assignment file has finished downloading. Tired of waiting what seems like forever for files to download over his modem, he vows to get a faster Internet connection, probably DSL or cable. He checks his e-mail one last time and finds the usual spam as well as a message from the student loan office reminding him his payment is due. He opens Microsoft Internet Explorer and clicks on the link for the bank from his Favorites list. With a few more clicks, he transfers enough money from his savings account to his checking account to cover his payment.

Does this level of Internet interaction sound at all like yours? If you're like many Americans, you use the Internet as much as you do your television, maybe even more. But do you really know how to get the most out of your Internet experience and which connection option is best for you? In this chapter, you'll learn what you should know about the Internet in order to use it to your best advantage. We'll look at the options you have for connecting to the Internet, as well as what you should know about the companies that provide you with Internet access. We'll then discuss how you can navigate and search the Web effectively so that your time spent on the Internet is useful. Finally, we'll investigate other Internet features you probably use, such as communication technologies (e-mail, IM, and the like), e-commerce, and multimedia experiences. But first, let's start by looking at the history of the Internet and how data travels across this big network.

Internet Basics

You've no doubt been on the Internet countless times. According to the U.S. Census Bureau, over 41 percent of American homes are connected to the Internet, a number that grows each day. But what exactly *is* the Internet? The **Internet** is the largest network in the world, actually a network of networks, connecting millions of computers from more than 65 countries. Today, most people use the Internet to communicate with others, although online shopping and entertainment activities follow close behind. Yet these uses are a far cry from the original intention of the Internet.

A BRIEF HISTORY OF THE INTERNET

Why was the Internet created? To understand why the Internet was created, you need to understand what was happening in the early 1960s in the United States. At this time, the Cold War with Russia was raging, and military leaders and civilians alike were concerned about a Russian nuclear or conventional attack on the United States. Meanwhile, the U.S. armed forces were becoming increasingly dependent on computers to coordinate and plan their activities. For the U.S. armed forces to operate efficiently, computer systems located in various parts of the country needed to have a reliable means of communication—one that could not be disrupted easily. Thus, the U.S. government funded much of the early research into the Internet to facilitate computer communications for the military.

At the same time, researchers also hoped the Internet would address the problems involved with getting different computers to communicate with each other. Although computers had been networked together since the early 1960s, there was no reliable way to connect computers from different manufacturers because these computers used different proprietary methods of communication. What was lacking was a common communications method that *all* computers could use, regardless of the differences in their individual designs. The Internet was therefore created to respond to these two concerns: to establish a safe form of military communications and to create a means by which all computers could communicate.

Who invented the Internet? The modern Internet evolved from an early "internetworking" project called the **Advanced Research Projects Agency Network (ARPANET)**. Funded by the U.S. government for the military in the late 1960s, ARPANET was the first attempt to allow computers to communicate over vast distances in a reliable manner. What started as a group of four computers networked together in the ARPANET has grown into millions of computers connected to the Internet today.

Although, many people participated in the creation of the ARPANET, two men who worked on the project, Vinton Cerf and Robert Kahn, are generally acknowledged as the "fathers" of the Internet. They earned this honor because they were primarily responsible for developing the communications protocols (or standards) still in use on the Internet today.

THE WEB VS. THE INTERNET

So are the Web and the Internet the same thing? Because the **World Wide Web** (**WWW** or the **Web**) is the part of the Internet we use the most, we sometimes think of the "Net" and the "Web" as being interchangeable. However, the Web is only a *part* of the Internet. Other parts of the Internet include communications systems such as e-mail and instant messaging, and information exchange technologies such as File Transfer Protocol (FTP) and newsgroups, all of which we'll discuss later in this chapter. What distinguishes the Web from the rest of the Internet is (1) its use of a special language (called *HTML*, or *Hypertext Markup Language*) that allows different computers to talk to and understand each other while connected to the Web and (2) its use of special links (called *hyperlinks*) that enable users to jump from one place to another in the Web.

THE INTERNET'S CLIENTS AND SERVERS

How does the Internet work? Computers connected to the Internet communicate (or "talk") to each other in a similar fashion as we do when we ask a question and get an answer. Thus, a computer connected to the Internet acts in one of two ways: it is either a **client**, a computer that asks for data, or a **server**, a computer that receives the request and returns the data to the client. Because the Internet uses clients and servers, it is referred to as a **client/server network**. (We'll discuss client/server networks in more detail in Chapter 7.)

How do computers talk to each other? Suppose you want to access the Web to check out snow conditions at your favorite ski area. As Figure 3.1 illustrates, when you type the Web site address of the ski area in your **Web browser**

INTERNET BASICS 67

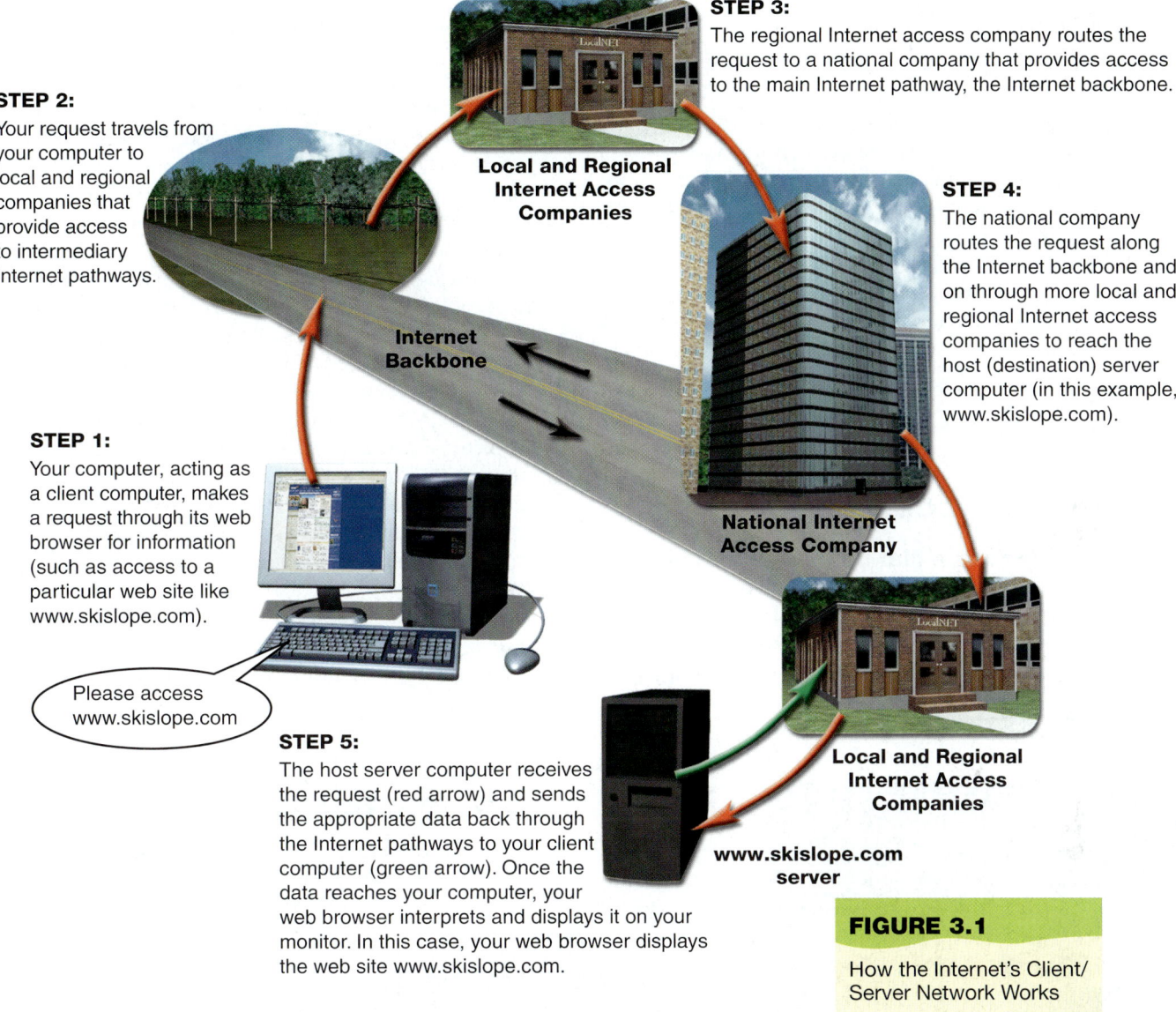

FIGURE 3.1

How the Internet's Client/Server Network Works

STEP 1: Your computer, acting as a client computer, makes a request through its web browser for information (such as access to a particular web site like www.skislope.com).

Please access www.skislope.com

STEP 2: Your request travels from your computer to local and regional companies that provide access to intermediary Internet pathways.

STEP 3: The regional Internet access company routes the request to a national company that provides access to the main Internet pathway, the Internet backbone.

STEP 4: The national company routes the request along the Internet backbone and on through more local and regional Internet access companies to reach the host (destination) server computer (in this example, www.skislope.com).

STEP 5: The host server computer receives the request (red arrow) and sends the appropriate data back through the Internet pathways to your client computer (green arrow). Once the data reaches your computer, your web browser interprets and displays it on your monitor. In this case, your web browser displays the web site www.skislope.com.

(software that allows you to access the Web) your computer acts as a *client computer* because you are asking for data from the ski area's Web site. Your browser's request for this data travels along several pathways, similar to interstate highways. The largest and fastest pathway is the main artery of the Internet, called the **Internet backbone**, to which all intermediary pathways connect. All data traffic flows along the backbone and then on to smaller pathways until it reaches its destination, which is the *server computer* for the ski area's Web site. The server computer returns the requested data to your computer via the same pathway system the data traveled initially. Your Web browser then interprets the data and displays it on your monitor.

How does the data get sent to the correct computer? Each time you connect to the Internet, your computer is assigned a unique identification number. This number, called an **IP (Internet Protocol) address**, is a set of four numbers separated by dots, such as 123.45.678.91. IP addresses are the means by which all computers connected to the Internet identify each other.

Similarly, each Web site is assigned an IP address that uniquely identifies it. However, because the long strings of numbers that make up IP addresses are difficult for humans to remember, Web sites are given "text" versions of their IP addresses. So, the ski area Web site mentioned above may have an IP address of 234.59.180.37 and a text name of **www.skislope.com**. When you type **www.skislope.com** into your browser window, your computer (with its own unique IP address) looks for the ski area's IP address (234.59.180.37). Data is exchanged between the ski area's server computer and your computer using these unique IP addresses.

Connecting to the Internet

To take advantage of the resources the Internet offers, you need a means to connect your computer to it. Home users have several connection options available. The most common method is a **dial-up connection**. With dial-up connections, you connect to the Internet using a standard telephone line. Other connection options, collectively called **broadband connections**, offer faster means to connect to the Internet. Broadband connections include cable, satellite, and various high-speed uses of traditional phone wires, such as DSL and ISDN.

DIAL-UP CONNECTIONS

How does a dial-up connection work? A dial-up connection is the least costly method of connecting to the Internet, and the hookup is quite simple (see Figure 3.2). With a dial-up connection, you use a standard phone line and a modem to connect your computer to the Internet. A **modem** is a device that converts (*mod*ulates) the digital signals the computer understands to the analog signals that can travel over phone lines. In turn, the computer on the other end must also have a modem to translate (*dem*odulate) the received analog signal back to a digital signal for the receiving computer to understand.

Although external modems do exist, modern desktop computers generally come with internal modems built into the system unit. Laptops usually use either internal modems or small credit card–sized devices called **PC cards** (sometimes called **PCMCIA cards**, or **Personal Computer Memory Card International Association**) that are inserted into a special slot on the laptop. Current modems have a maximum data transfer rate of 56 kilobits per second (Kbps, usually referred to as 56K). **Data transfer rate**, or **throughput**, is the measurement of how fast data travels between computers. It is also informally referred to as *connection speed*. When you use a 56K modem, you actually connect to the Internet

FIGURE 3.2

Dial-Up Connection Modem Options. (a) A dial-up Internet connection often uses a standard phone cord that connects a phone jack and your computer's internal modem, which is a card located inside the system unit. (b) External modems are peripheral devices that sit outside the system unit. Although you can still find external modems, most computers now come with internal modems. (c) Laptops use small credit-card–sized devices, called PC cards, to connect to the Internet.

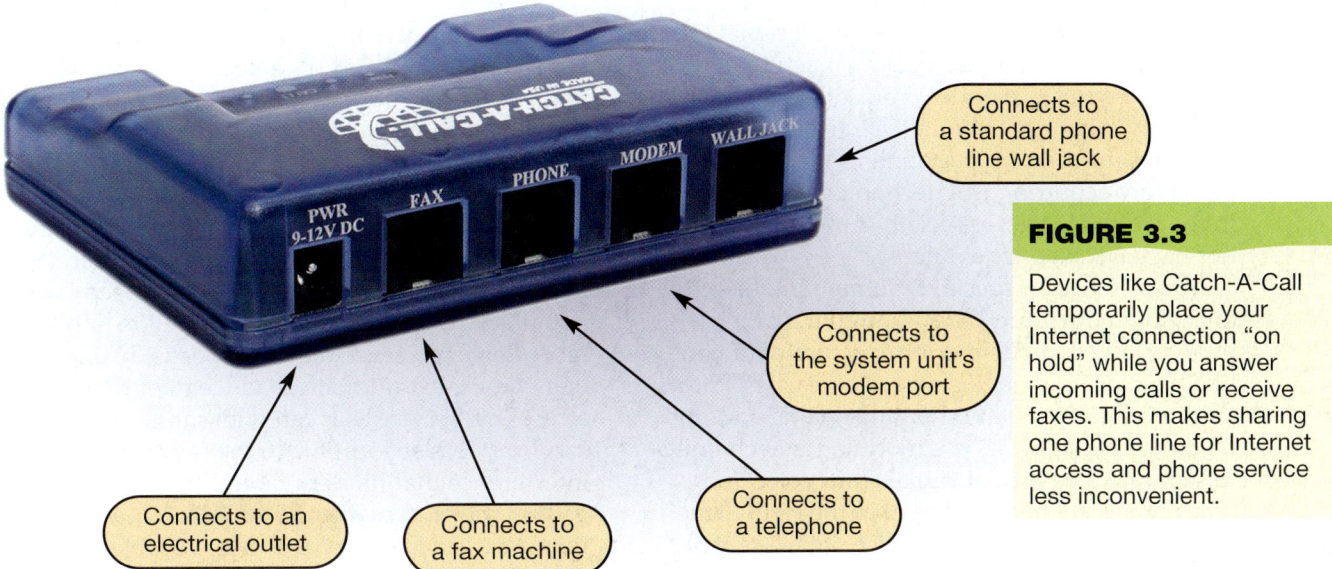

FIGURE 3.3

Devices like Catch-A-Call temporarily place your Internet connection "on hold" while you answer incoming calls or receive faxes. This makes sharing one phone line for Internet access and phone service less inconvenient.

at a speed lower than 56 Kbps due to interferences such as line noise.

What are the advantages of a dial-up Internet connection? A dial-up connection is the least costly way to connect to the Internet. Although slower than broadband connections, dial-up connections, with today's 56K modems, are often fine for casual Internet users who do not need a very fast connection.

What are the disadvantages of dial-up? In a word: speed. Even at 56 Kbps, moving through the Internet with a dial-up connection can be a slow and frustrating experience. Web pages can take a long time to load, especially if they contain multimedia files. Similarly, if you visit many Web sites at the same time or receive or send large files while you're on the Internet, you'll find that a dial-up connection is very slow. Another disadvantage to dial-up is that when you're on the Internet, you tie up your phone line if you don't have a separate line.

Can I avoid tying up my phone line when I use a dial-up connection? One solution to this problem is to install an additional phone line for your Internet connection. This can be costly, however. Alternatively, if you have a call-waiting feature with your phone service, you can use a device such as Catch-A-Call, shown in Figure 3.3. These devices enable you to receive incoming phone calls without disconnecting from the Internet. When you receive a call, the device displays a flashing red light. To accept the call, you pick up your phone and the device automatically puts your Internet connection "on hold" for several seconds. If neither of these options works for you, you may want to pursue other means of connecting to the Internet, such as a broadband connection.

BROADBAND CONNECTIONS

What broadband options do I have? The two leading broadband home Internet connection technologies are *Digital Subscriber Line (DSL)*, which uses a standard phone line to connect your computer to the Internet, and *cable*, which uses your television's cable service provider to connect to the Internet. Another viable solution to quicker Internet connection is *Integrated Services Digital Network (ISDN)*, which also uses standard telephone wire. Finally, some users connect to the Internet via *satellite*.

DSL

How does DSL work? Similar to a dial-up connection, **DSL** uses telephone lines to connect to the Internet. However, unlike dial-up, DSL allows phone and data transmission to share the same line, thus eliminating the need for an additional phone line. Phone lines are made of twisted copper wires known as **twisted pair wiring**. Think of this twisted copper wiring as a three-lane highway with only one lane being used to carry voice data. DSL uses the remaining two lanes to send and receive data separately, at much higher frequencies. Thus, although it uses a standard phone line, a DSL connection is much faster than a dial-up connection.

Can anyone with a phone line have DSL? Just because you have a traditional phone line in your house doesn't mean that you have access to DSL service. Your local phone company must have special DSL technology to offer you the service. Although more phone companies are acquiring DSL technology, many areas in the United States, especially rural ones, still do not have DSL service available.

You also need a special **DSL modem**, like the one shown in Figure 3.4. Although it's called a modem, a DSL modem doesn't actually function like the dial-up modems described earlier. Rather than modulating/demodulating analog and digital data, DSL modems use modulation techniques to separate the types of signals into voice and data signals so they can travel in the right "lane" on the twisted pair wiring. Voice data is sent at the lower speed, while digital data is sent at frequencies ranging from 128 Kbps to 1.5 megabits per second (Mbps).

Are there different types of DSL service? The more typical DSL transmissions download (or receive) data from the Internet faster than they can upload (or send) data. Such transmissions are referred to as **Asymmetrical Digital Subscriber Line (ADSL)**. Other DSL transmissions, called **Symmetrical Digital Subscriber Line (SDSL)**, upload and download data at the same speed. If you upload data to the Internet often (if you design and update your own Web site, for example), you may want to investigate the range of sending-speed capabilities of your DSL connection, or check to see if your DSL provider offers SDSL service.

What are the advantages to DSL? With data transfer rates that can exceed 8 Mbps, DSL beats the slow speeds of dial-up by a great deal. And with DSL service, you can connect to the Internet without tying up your phone line, so you can avoid the costs associated with installing a second phone line. In addition, unlike cable and satellite, DSL service does not share the line with other network users in your area. Therefore, in times of peak Internet usage, DSL speed is not affected, whereas cable and satellite hook-ups often experience reduced speeds during busy times. Additionally, bad weather does not affect DSL service as it can with satellite, and DSL service is less susceptible to the radio frequency interference that hinders cable.

Are there drawbacks to DSL? As mentioned earlier, DSL service is not available in all areas. If you do have access to DSL service, the quality and effectiveness of your service depend on your proximity to a phone company central office (CO). A CO is the place where a receiving DSL modem is located. Data is sent through your DSL modem to the DSL modem at the CO, and then it connects to the Internet. For the DSL service to work correctly, you need to be within approximately three miles of a CO because the signal quality and speed weaken drastically at distances beyond 18,000 feet. A simple call to your local phone company can determine your proximity to a CO and whether DSL service is available. You can also find out whether DSL is available in your area by check-ing www.DSL.com or www.Getspeed.com.

Cable

If I have cable TV in my home, do I have access to cable Internet? While cable TV and a **cable Internet connection** both use coaxial cable, they are separate services. In fact, even though you may have cable TV in your home, cable Internet service may not be available in your area. Cable TV is a one-way service in which the cable company feeds your television programming signals. In order to bring two-way Internet connections to homes, cable companies must upgrade their networks for two-way data transmission capabilities and with **fiber-optic lines**, which transmit data at close to the speed of light along glass fibers or wires. Because data sent through fiber-optic lines is transmitted at the speed of light, transmission speeds are much faster than other conventional copper wire technologies.

What do I need to hook up to cable Internet? Cable Internet connection requires a **cable modem**, as shown in Figure 3.5. Generally, the modem is located somewhere near your computer. The cable modem is then connected to an

FIGURE 3.4

You need a special DSL modem like this one to connect to the Internet using DSL.

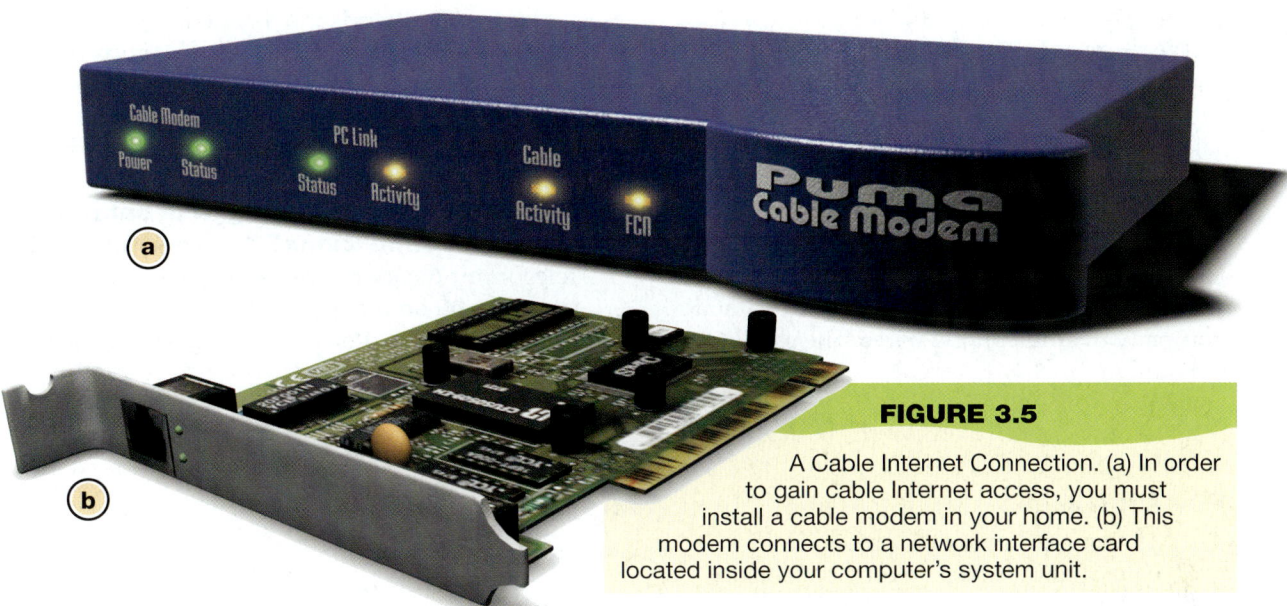

FIGURE 3.5

A Cable Internet Connection. (a) In order to gain cable Internet access, you must install a cable modem in your home. (b) This modem connects to a network interface card located inside your computer's system unit.

expansion (or adapter) card called a **network interface card (NIC)**, located inside your system unit. The cable modem works similarly to a traditional dial-up modem in that it modulates and demodulates the cable signal into digital data and back again. Since the cable TV signal and Internet data can share the same line, you can watch cable TV and surf the Internet simultaneously.

Why would I choose a cable connection? The speed of a cable Internet connection is about the same as the speed of a DSL connection. With cable Internet, you can receive data at speeds up to 4 Mbps and send data at approximately 300 Kbps. As technology improves, these transfer rates also will improve. However, the availability of cable Internet over DSL may be the main reason you choose cable. Although cable service is not available in all areas, it is quickly rolling out. Check your local cable TV provider to determine whether cable Internet is available where you live and the transfer rates available in your area.

Cost may also be a consideration in choosing cable. Installation fees tend to be lower for cable, especially with the do-it-yourself kits found in most computer accessory stores. Monthly charges for cable are also slightly lower than for DSL, and packaged deals for cable TV subscribers may also be available, reducing costs even more.

Are there any disadvantages to cable Internet? Unlike DSL, where you have a dedicated line to the CO, you share your cable connection with your neighbors. Therefore, you may experience periodic decreases in connection speeds during peak usage times with cable Internet connections. While your connection speeds are still faster than a dial-up alternative, your ultimate speed depends on how many other users are trying to transmit data at the same time as you are.

ISDN

How does an ISDN Internet connection work? Integrated Services Digital Network (ISDN) is a high-speed digital technology that allows you to transmit and receive digital data at speeds up to 128 Kbps. To use ISDN, you need an **ISDN modem**, which sends and receives the digital data to and from your computer over a traditional phone line. Because this modem is already working only with digital data, it doesn't need to translate analog data to digital data like a traditional modem. Instead, it is really a "terminal adapter" as it translates data between different devices that connect to the line.

What are the advantages of ISDN service? ISDN is at least twice as fast as dial-up. ISDN service also allows you to connect two devices (a computer and a phone, for example) to the ISDN "modem" and use the devices simultaneously. Therefore, unlike a traditional dial-up connection, you can talk on the phone and surf the Internet at the same time over the same phone line.

What are the disadvantages to ISDN? Although the cost of ISDN is generally less than the other broadband options, the speed of service is limited to 128 Kbps, making ISDN a less preferred broadband option for most users. In fact, because of these limitations and the increased availability of other broadband services, ISDN will soon become obsolete.

Satellite

What is satellite all about? Satellite Internet is a way to connect to the Internet when other high-speed options are not available. To take advantage of satellite Internet, all you need is a small satellite dish, which is placed outside your home and connects to your computer with **coaxial cable**, the same type of cable used for cable TV. As shown in Figure 3.6, you use a traditional modem to send data to the satellite company over telephone wires. The satellite company then sends the data to a satellite orbiting the earth. The satellite, in turn, sends the data back to your satellite dish and to your computer. The major provider for satellite Internet is DirecPC, the same company that offers DirecTV, satellite television.

What are the advantages to satellite Internet connections? Since several major telecommunications companies maintain satellites in orbit above the equator, almost anyone in the United States can receive satellite service. It is

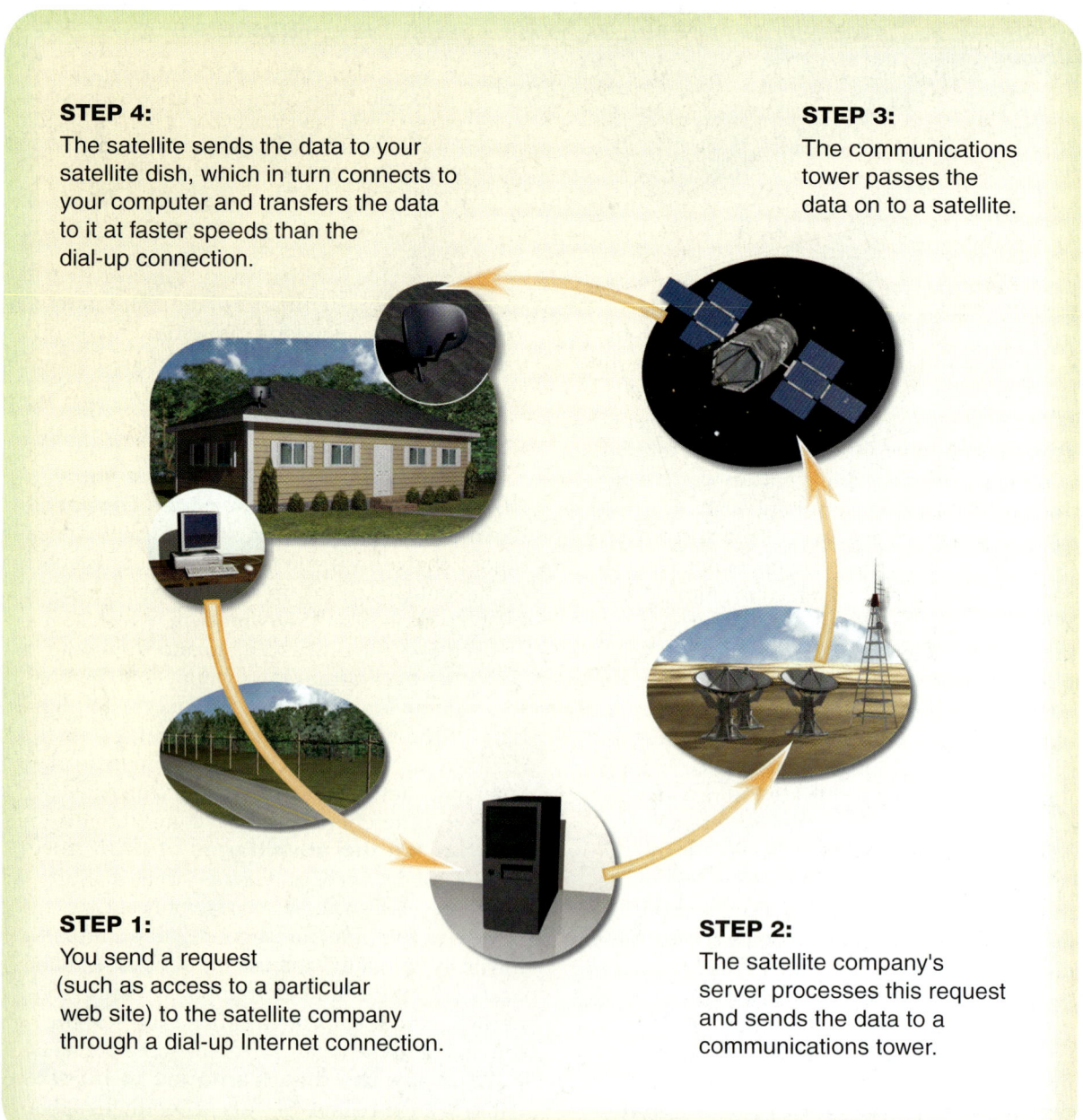

STEP 4:
The satellite sends the data to your satellite dish, which in turn connects to your computer and transfers the data to it at faster speeds than the dial-up connection.

STEP 3:
The communications tower passes the data on to a satellite.

STEP 1:
You send a request (such as access to a particular web site) to the satellite company through a dial-up Internet connection.

STEP 2:
The satellite company's server processes this request and sends the data to a communications tower.

FIGURE 3.6

A satellite Internet connection relies on standard telephone wires to connect to the Internet. However, data returned to your computer is sent via satellite at much faster speeds.

FIGURE 3.7 Comparing Internet Connection Options

CONNECTION OPTION	DATA TRANSFER RATE (APPROXIMATE)	MONTHLY COSTS (APPROXIMATE)	ADDITIONAL EQUIPMENT	ADDITIONAL FEES (APPROXIMATE)
DIAL-UP	56 Kbps (max)	Local phone rates	Internal or external modem	None
DSL (ADSL)	128 Kbps upload 8.45 Mbps download	$40–$320	DSL modem	Installation $100–$200
DSL (SDSL)	1.5 Mbps upload/download	$40–$370	DSL modem	Installation $100–$200
CABLE	128–384 Kbps upload 1–2.5 Mbps download	$30–$70	Cable modem	Installation $75–$200
ISDN	128 Kbps (max)	$10–$40	ISDN modem	Installation
SATELLITE	56 Kbps upload 400 Kbps download	$40–$50	Dish	Installation

Note: The data transfer rates listed in this table are approximations. As technologies improve, so too do data transfer rates. Costs and fees are also subject to change.

---

therefore a particularly popular choice for those who live in rural areas of the country where neither cable nor DSL service is available.

Are there any drawbacks to satellite? Because other broadband services may not be available, satellite broadband may be your only alternative to dial-up. However, restrictions to this service do apply. First, while satellite download rates of 400 Kbps are faster than the other broadband options, and significantly faster than the 56-Kbps dial-up speed, upload to the Internet is still at the slower 56-Kbps dial-up speed since the data travels over traditional phone wires. Second, since download transmissions are not "wired," but rather are sent as radio waves, the strength and reliability of the signal are more vulnerable to interference. In addition, your satellite dish must face south for the best line of sight to the satellites circling the earth's equator. If high buildings, mountains, or other tall objects obstruct your southern exposure, your signal may be blocked. Unfavorable weather conditions can also block or interfere with the satellite transmission signal.

CHOOSING THE RIGHT INTERNET CONNECTION OPTION

How do I choose which Internet connection option is best for me? Dial-up is the most common means of Internet connection, but also the slowest. If you need a faster connection, you should choose one of the broadband alternatives. Your location plays the biggest role in choosing among broadband options. As mentioned earlier, some areas aren't able to offer DSL and/or cable Internet service. Additionally, you may need to consider what other services you want bundled into your payment, such as cable or satellite TV, and what service inconveniences you are willing to live with, such as the slower peak time rates associated with cable and the weather interferences associated with satellite. Figure 3.7 shows the data transfer rates, monthly costs, and additional equipment and fees associated with each of the Internet connection options.

Finding an Internet Service Provider

Once you have chosen the method by which you'll connect to the Internet, whether it's dial-up or broadband, you need a way to *access* the Internet. **Internet service providers (ISPs)** are national, regional, or local companies that connect individuals, groups, and other companies to the Internet. EarthLink, for example, is a well-known national ISP.

As mentioned earlier, the Internet is a network of networks, and the central component of the

TRENDS IN IT

EMERGING TECHNOLOGIES: What Is the Next Generation of Broadband?

In the 1990s, most homeowners connected to the Internet with a dial-up modem. In the early days, dial-up modem data transfer rates crawled along at a speed of 14.4 Kbps. These speeds have improved so that you can now connect to the Internet with a dial-up modem at speeds of 56 Kbps. Meanwhile, DSL technology has increased the data transfer rate of phone lines, cable TV provides similar broadband speeds, and satellite is available where other technologies are not. In fact, today, nearly 25 percent of all American homes have some form of broadband access to the Internet. Estimates are that the demand for broadband access will increase to nearly 50 percent in only a few years.

Yet despite these advances, consumers are demanding even faster Internet connections, and new technologies are being developed to satisfy our need for speed. Fiber-optic technology, currently available only to corporate and urban America, is slowly making its way to the suburbs and rural areas. Fiber-optic technology not only enables the transmission of data at close to the speed of light, but also provides an uninterrupted data pathway that makes the delivery of communications more reliable. The backbone to this fiber-optic revolution is already being formed, as networks of fiber-optic cable are being laid throughout the country.

But there is more on the horizon than just increased fiber-optic reach. Numerous companies are exploring a technology that sends data, voice, and video signals over normal electric powerlines at speeds that may one day exceed even the fastest fiber-optic systems. Called "powerline connectivity," this technology is being tested in Europe and Australia, but still has a lot of research ahead of it before it will be considered viable. With a powerline Internet connection, you could one day transfer exobits of data—that's 1 with eighteen zeros after it—per second through ordinary powerlines. Now that's *fast*.

Proponents of powerline Internet connectivity say that the technology has benefits beyond speed. Unlike cable or DSL, powerline Internet access requires no new wires, and connections would be available in every room in which there is a power outlet. As long as you're connected to a powerline, you would have high-speed Internet access. Thus, powerline technology resolves the "last mile" delivery dilemma currently encountered by fiber-optic technology as fiber-optic networks struggle to lay the "last mile" of cable into homes.

In addition, hopes are that powerline Internet access will be a cheaper means of connecting to the Internet. And with access to the Internet through powerlines, we may see more household appliances (such as refrigerators and ovens) being connected to the Internet. A totally networked home may be closer than you think.

Internet network is the *Internet backbone*. The Internet backbone is the main pathway of high-speed communications lines through which all Internet traffic flows. Large communications companies, such as AT&T, MCI, and Sprint, are "backbone providers" that control access to the main lines of the Internet backbone. These backbone providers supply Internet access to ISPs, which, in turn, supply access to other users.

Large businesses and educational facilities connect to the Internet through one of the *regional* ISPs that then connect to the backbone. Local cable and telephone companies fit into this category of regional ISPs. Home or small business users connect to the Internet through *local* ISPs or through **online service providers (OSPs)**, which are Internet access providers such as America Online (AOL) that have their own proprietary online content.

Where do I find an ISP? If you have a broadband connection, your broadband provider *is* your ISP. If you're accessing the Internet from a dial-up connection, you need to determine what ISPs are available in your area. Look in the phonebook, check ads in the newspaper, or ask friends which ISP they use. You're no doubt familiar with many of them already. Additionally, you can go to sites such as The List (**www.thelist.com**) or All Free ISP (**www.all-free-isp.com**) for listings of national and regional ISPs.

How exactly are OSPs different from ISPs? OSPs and ISPs both provide you with access to the Internet so you can visit Web sites and send e-mail to your friends. However, OSPs (such as CompuServe, AOL, and Microsoft's

MSN) go a step further than standard ISPs by offering unique content and special services and areas that only their subscribers can access. Initially, when getting around the Internet was more cumbersome, the streamlined content and directory-like searching mechanisms OSPs offered were worth the higher monthly fees. Now, with efficient sites such as Yahoo!, many people find it is more cost-effective and equally convenient to connect to the Internet with a local or national ISP.

CHOOSING AN ISP

What factors should I consider in choosing an ISP? If you're in the market for an ISP, you'll need to consider the following:

- How much does the ISP cost for monthly Internet access and what other services does it offer?
- How are services paid for and how are renewals handled? (For example, does the ISP automatically charge your credit card each month?)
- Does the ISP have a local access number so that you can avoid long-distance phone charges while you use the Internet?
- Will you need more than one e-mail account? If so, how many accounts does the ISP provide?
- If you travel a lot, does the ISP have local access in the areas where you'll be traveling or an available 800 number to connect to?
- Does the ISP allow you to access your e-mail via the Web?
- Are you planning on having a Web site? If so, does the ISP have available space for your Web site on its server?
- Is there a trial period? Trial periods enable you to check out availability during peak and off-peak hours before committing to the ISP long-term.
- How is the ISP's customer service? The best ISPs provide customer service that is accessible by phone and the Web.

It's important that you carefully select the right ISP for your needs. While one ISP may be perfect for your friends, their needs may be different from yours. Remember, too, that changing your ISP can be both time-consuming (you may need to change connections) and inconvenient (if your e-mail address changes you'll need to notify everyone).

SOUND BYTE
CONNECTING TO THE INTERNET

In this Sound Byte, you'll learn the basics of connecting to the Internet from home, including the various types of Internet connections as well as useful information on selecting the right ISP.

Navigating the Web: Web Browsers

Once you're connected to the Internet and have an ISP to access its services, you're free to explore the many features the Internet has to offer, including the Web. However, in order to explore the Web, you need more than an ISP—you also need a *Web browser*. As we defined earlier, a Web browser is software installed on your computer system that allows you to locate, view, and navigate the Web. The most common browsers in use today are **Netscape Navigator** and Microsoft's **Internet Explorer (IE)**. These browsers are "graphical" browsers, meaning they can display pictures (graphics) in addition to text, as well as other forms of multimedia, such as sound and video.

BITS AND BYTES

Connecting Personal Digital Assistants (PDAs) and Cell Phones to the Internet

Ever wonder how wireless devices such as PDAs and cell phones connect to the Internet? Connecting your PDA or cell phone requires that you have a **wireless Internet service provider**, such as Verizon or T-Mobile. The connection is by no means fast—wireless devices connect at a maximum speed of 14.4 Kbps. However, to make such Internet connections worthwhile, Web sites are beginning to create content specifically designed for wireless devices. This specially designed content is text-based, contains no graphics, and is designed so that it fits the display screens of cell phones and PDAs. You'll learn more about how mobile devices connect to the Internet in Chapter 8.

BITS AND BYTES

Alternative Web Browsers

Although Netscape and IE are the market-leading Web browsers, you do have other options.

- Opera (**www.opera.com**) is a browser that has a small but dedicated following. Opera's advantage over Netscape and IE is that it can preserve your surf sessions: when you launch Opera, it loads and opens all the Web sites you had open when you last used it.
- Lynx (**http://lynx.isc.org**) is an alternative text-only browser that you navigate by highlighting emphasized words on the screen with the up and down arrow keys, and then pressing Enter.
- KIWE, the Kid's Internet World Explorer browser (**www.kiwe.net**), provides a safe online environment for children. Although KIWE charges a fee, many parents feel that the cost is justified in exchange for knowing their children are interacting in a safe environment.

BROWSER FEATURES

What features do browsers offer? As you can see in Figure 3.8, Navigator and IE offer nearly identical features with which to navigate the Web. The biggest difference between the browsers is that IE integrates almost seamlessly with all Microsoft products, including the Windows operating system and Office applications. Because of this integration, PC manufacturers preinstall IE on their computers. However, many fans of Netscape still take the time to download this free browser. Because of Netscape's affiliation with AOL, Netscape offers services that IE doesn't, such as free access to AOL's popular Instant Messenger. Because of each browser's unique features, many users use *both* IE and Netscape.

Getting Around the Web: URLs, Hyperlinks, and Other Tools

Unlike text in a Word document, which is linear (meaning you read it from top to bottom, left to right, one page after another), the Web is anything but linear. As its name implies, the Web is a series of connected paths or links that connect you to different **Web sites**, or locations on the Web. You gain initial access to a particular Web site by typing in its unique address, or **Uniform Resource Locator** (**URL**, pronounced "you-are-ell"). For example, the URL of the Web site for *Popular Science* magazine is **http://www.popsci.com**. By typing in a URL, you connect to the **home page**, or main page, of the Web site. Once in the home page, you can move all around the site by clicking on specially formatted pieces of text called *hyperlinks*. Let's look at these and other navigation tools in more detail.

BITS AND BYTES

What's the Difference between FTP and HTTP?

The HTTP protocol allows files to be transferred from a Web server and seen on your computer using a browser. FTP, on the other hand, is used to upload and download files from one computer to another. FTP files use an FTP file server, whereas HTTP files use a Web server. Most FTP servers require that you have a user ID and a password to connect to the server. An FTP site can include text, graphics, audio, video, and application files. **ftp://ftp.uwp.edu** is a typical FTP address. (In this case, the site is used to transfer files to the University of Wisconsin at their Parkside campus.) To upload and download files from FTP sites, you need file transfer software, such as Fetch or CuteFTP.

URLS

What do all the parts of the URL mean? As noted above, a URL is a Web site's address. And, like a regular street address, a URL is composed of several parts that help identify the Web document for which it stands, as shown in Figure 3.9. The first part of the URL indicates the set of rules (or the **protocol**) used to retrieve the specified document. The protocol is generally followed by a colon, two forward slashes, www

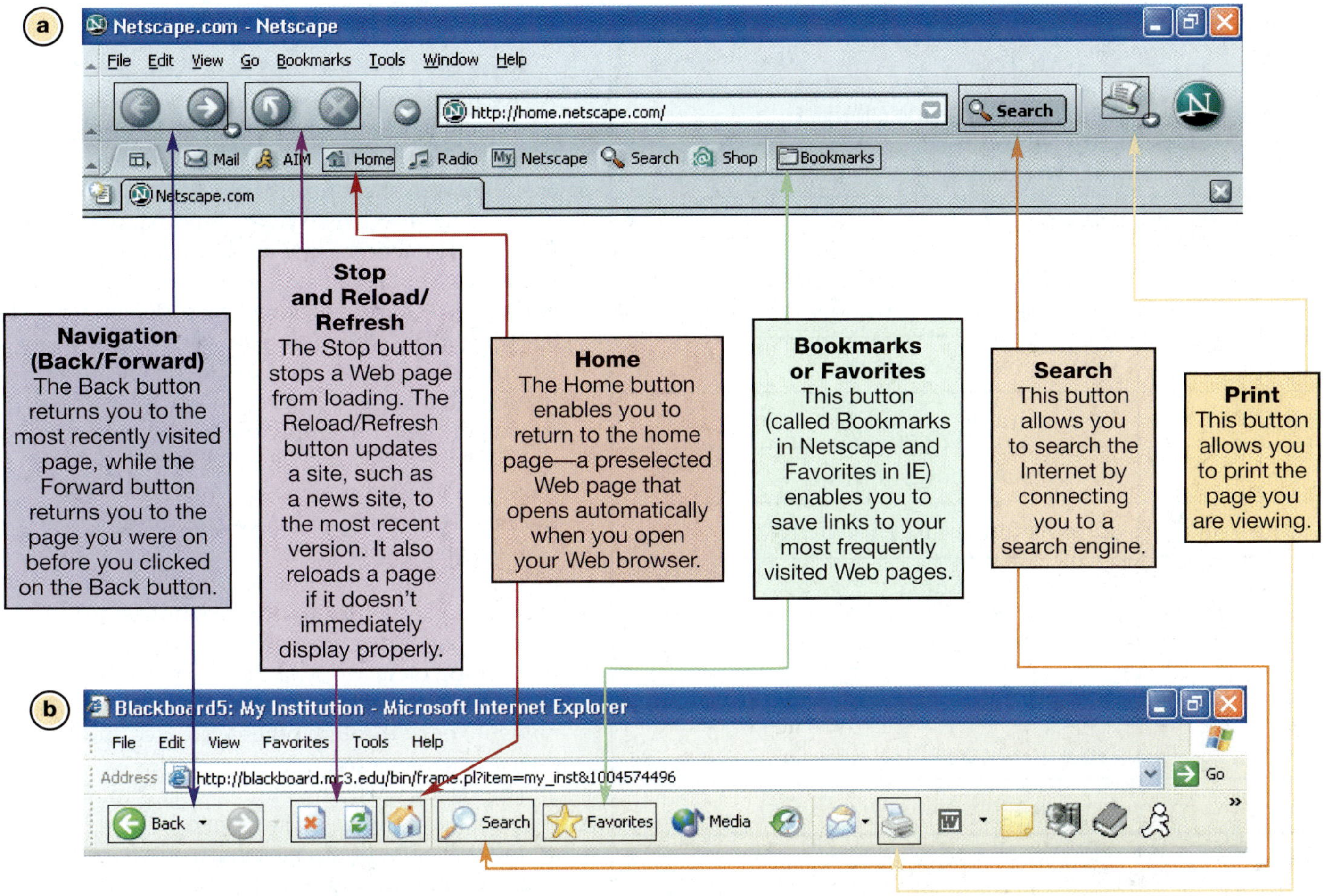

GETTING AROUND THE WEB: URLS, HYPERLINKS, AND OTHER TOOLS

FIGURE 3.8

Although the names are different, both Netscape (a) and IE (b) offer similar navigational tools.

Navigation (Back/Forward) The Back button returns you to the most recently visited page, while the Forward button returns you to the page you were on before you clicked on the Back button.

Stop and Reload/Refresh The Stop button stops a Web page from loading. The Reload/Refresh button updates a site, such as a news site, to the most recent version. It also reloads a page if it doesn't immediately display properly.

Home The Home button enables you to return to the home page—a preselected Web page that opens automatically when you open your Web browser.

Bookmarks or Favorites This button (called Bookmarks in Netscape and Favorites in IE) enables you to save links to your most frequently visited Web pages.

Search This button allows you to search the Internet by connecting you to a search engine.

Print This button allows you to print the page you are viewing.

(indicating World Wide Web), and then the **domain name**.

What's the protocol? For the most part, URLs begin with the protocol **http**, which is short for the **Hypertext Transfer Protocol**. Another common protocol used to transfer files over the Internet is **FTP (File Transfer Protocol)**.

What's in a domain name? Domain names consist of two parts: The first part indicates who the site's **host** is. For example, in the URL **www.berkeley.edu**, berkeley.edu is the domain name and berkeley is the host. The three-letter suffix in the domain name (such as .com or .edu) is called the **top-level domain (TLD)**. This suffix indicates the kind of organization the host is. Figure 3.10 lists the top-level domains that are currently approved and in use.

In addition to the domains listed in Figure 3.10, there are also TLDs for each country in the world. These are two-letter designations such as .UK for the United Kingdom and .US for the United States. Within a country-specific domain, further subdivisions can be made for regions or states. For instance, the .US domain contains subdomains for each state, using the two-letter abbreviation of the state. For example, the URL for the state of Pennsylvania's Web site is **www.state.pa.us**.

FIGURE 3.9

The Parts of a URL

FIGURE 3.10 Current Top-Level Domains and Their Authorized Users

DOMAIN NAME	WHO CAN USE THE DOMAIN NAME
.aero	Members of the air transport industry
.biz	Businesses
.com	Originally for commercial sites, can be used by anyone now
.coop	Cooperative associations
.edu	Degree granting institutions
.gov	United States government
.info	Information service providers
.mil	United States military
.museum	Museums
.name	Individuals
.net	Networking organizations
.org	Organizations (often nonprofits)
.pro	Credentialed professionals

What's the information after the domain name that I sometimes see? When the URL is only the domain name (such as **www.nytimes.com**), you are requesting a site's home page. However, at times, a forward slash and additional text follow the domain name, such as **www.nytimes.com/pages/cartoons**. The information after the slash indicates a particular file or **path** (or **subdirectory**) within the Web site. In this example, you would connect to the cartoon pages in the *New York Times* site.

SOUND BYTE
WELCOME TO THE WEB
In this Sound Byte, you'll visit the Web via a series of guided tours of useful Web sites. This tour serves as an introductory guide to Web newcomers as well as a great resource for more experienced users.

HYPERLINKS AND BEYOND

How do I get around in a Web site? As we mentioned earlier, once you've reached a Web site, you can jump from one location, or Web page, to another within the Web site or to an-other Web site altogether by clicking on specially coded text called **hyperlinks**, shown in Figure 3.11. Generally, text that operates as a hyperlink appears in a different color (often blue) and/or is underlined. Sometimes images also act as hyperlinks. When you pass your cursor over a hyperlinked image, the cursor changes to a hand with a finger pointing upward. To access the hyperlink, you simply click on the image.

To get back to your original location or a Web page you viewed previously, you can also use the browser's **Back** and **Forward buttons** (shown previously in Figure 3.8). If you want to back up more than one page, you can use the down arrow next to the Back button to access a list of most recently visited Web sites. By clicking on any one of these sites in the list, you can return to that page without having to navigate back through other Web sites and Web pages you've visited.

The **History list** on your browser's toolbar is also a handy feature. The History list shows all the Web sites and pages that you've visited over a certain period of time. These Web sites are organized according to date and can go back as far as three weeks.

As another way to retrace your steps, some sites also provide a **breadcrumb list**—a list of sites you've visited that usually appears at the top of a page. Figure 3.11 shows an example of a breadcrumb list. Breadcrumbs get their name from the Hansel and Gretel fairy tale in which the children dropped breadcrumbs on the trail to find their way back out of the forest.

FAVORITES AND BOOKMARKS

How do I mark a site I want to return to in my Web browser? If you want an easy way to return to a specific Web page, you can do so by using your browser's **Favorites** or **Bookmark** feature (shown previously in Figure 3.8). (IE calls this feature Favorites; Netscape calls the same feature a Bookmark.) These features place a marker of the site's URL in an easily retrievable list in your browser's toolbar. To add a Web page to your list of bookmarked Favorites in IE, from within the site you wish to mark,

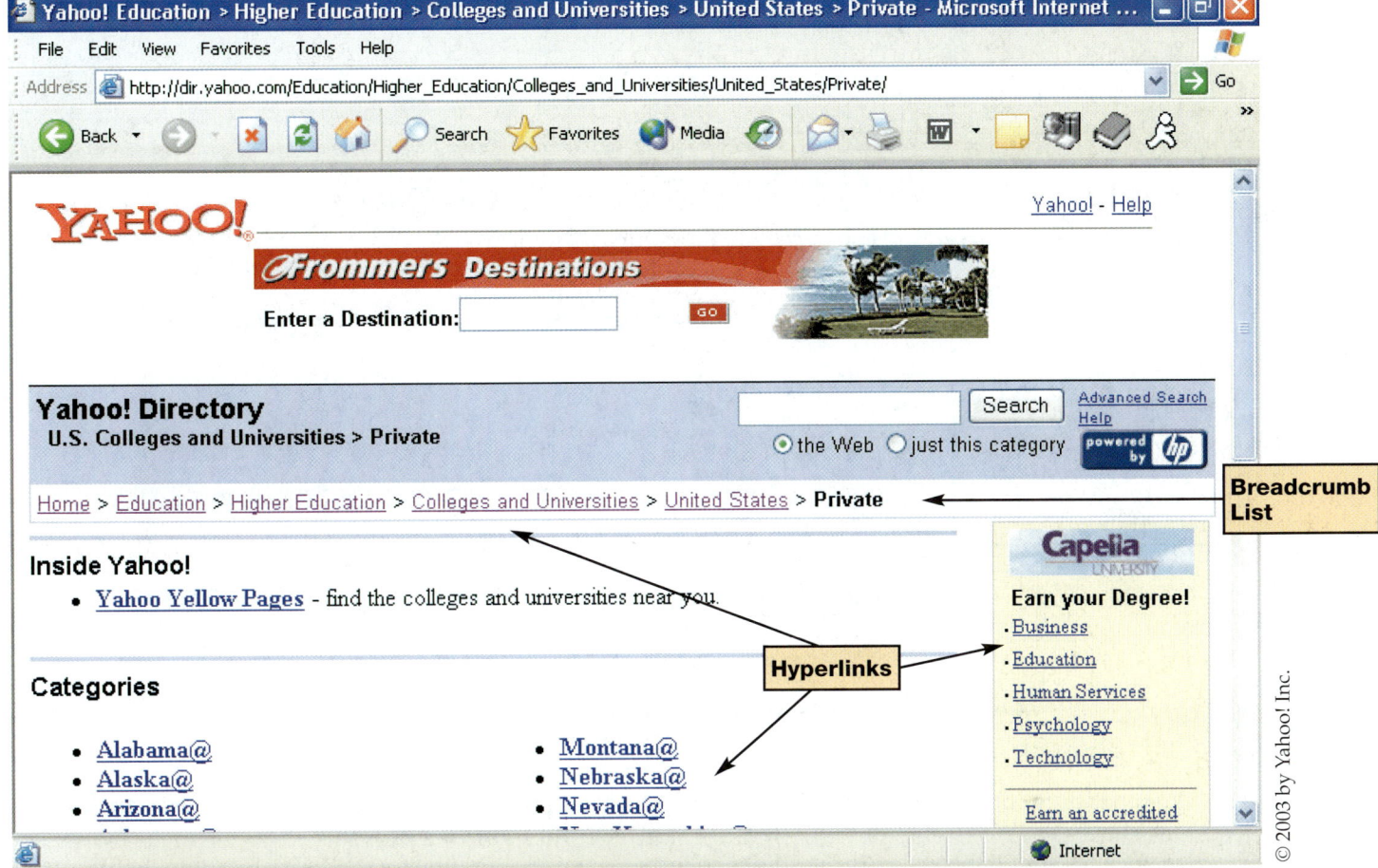

FIGURE 3.11

When you click on a hyperlink, you jump from one location in a Web site to another. When you click on the links in a breadcrumb list, you can navigate your way back through a Web site.

click on the Favorites menu and select Add to Favorites. As shown in Figure 3.12, you can modify the name of the Web page on your Favorites list to make it more meaningful. (The process for adding and modifying a Bookmark in Netscape is very similar.) If your list of Favorites or Bookmarks becomes long, you can create folders to better organize these sites into categories.

Searching the Web: Search Engines and Subject Directories

The Internet, with its billions of Web pages, offers its visitors access to masses of information on virtually any topic. There are two main tools you can use to find information on the Web: A **search engine** is a set of programs that searches the Web for specific words (or **keywords**) you wish to query (or look for) and then returns a list of the Web sites on which those keywords are found. Popular search engines include Google and AlltheWeb. You can also search the Web using a **subject directory**, which is a structured outline of Web sites organized by topics and subtopics. Yahoo! is a popular subject directory. Figure 3.13 lists popular search engines and subject directories and their URLs.

SEARCH ENGINES

How do search engines work? Search engines have three parts. The first part is a program called a **spider** (also known as a **crawler** or **bot**). The spider constantly collects data on the Web, following links in Web sites and reading

FIGURE 3.12

Using the IE Favorites feature (or the Netscape Bookmarks feature) makes returning to an often used or hard-to-find Web page much easier.

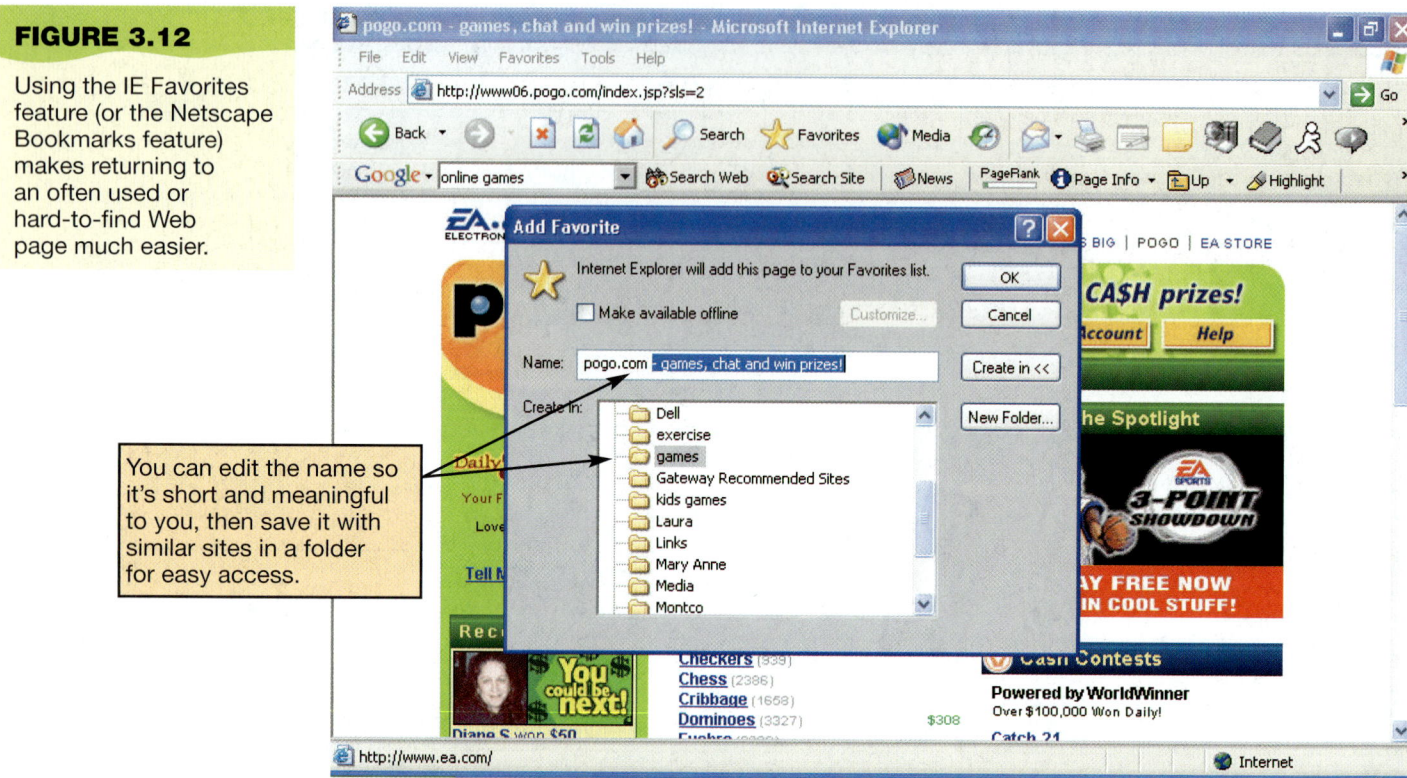

You can edit the name so it's short and meaningful to you, then save it with similar sites in a folder for easy access.

Web pages. Spiders get their name because they crawl over the Web using multiple "legs" to visit many sites simultaneously. As the spider collects data, the second part of the search engine, an **indexer** program, organizes the data into a large database. When you use a search engine, you interact with the third part: the search engine software. This software searches the indexed data, pulling out relevant information according to your search. The resulting list appears in your Web browser as a list of **hits**, or sites that match your search.

Why don't I get the same results from all search engines? Each search engine uses a unique formula, or algorithm, to formulate the search and create the resulting index, as shown in Figure 3.14. In addition, search engines differ in how they rank the search results. Most search engines rank their results based on the *frequency* of the appearance of your queried keywords in Web sites as well as the *location* of those words in the sites. Thus, sites that include the key-words in their URL or site name most likely appear

FIGURE 3.13: Popular Search Engines and Subject Directories			
SEARCH ENGINES		**SUBJECT DIRECTORIES**	
AlltheWeb	www.alltheWeb.com	CompletePlanet	www.completeplanet.com
AltaVista	www.altavista.com	LookSmart	www.looksmart.com
Dogpile	www.dogpile.com	Lycos	www.lycos.com
Excite	www.excite.com	MSN	http://search.msn.com
Google	www.google.com	Open Directory Project	www.dmoz.org
Teoma (Ask Jeeves)	www.teoma.com	Yahoo!	www.yahoo.com

at the top of the hit list. After that, results vary due to differences in each engine's proprietary formula.

In addition, search engines differ as to which sites they search. For instance, Google and AlltheWeb search nearly the entire Web, whereas specialty search engines search only sites that have been specifically identified as being relevant to the particular specialty. Specialty search engines exist for almost every industry or interest. For example, **www.DailyStocks.com** is a search engine used primarily by investors that searches for corporate information to help them make educated decisions.

What are the advantages of using the different kinds of search engines?
Using a search engine with a large index such as Google can be advantageous in conducting a search on hard-to-find information because it searches that many more indexed Web sites. However, if you're looking for only a few sites that have a high relevancy to your search, a search engine that has a smaller database, such as the Ask Jeeves search engine Teoma, may be more helpful. If you can't decide which search engine is best, you may want to try a **meta search engine**, such as Dogpile (**www.dogpile.com**). Meta search engines search other search engines rather than individual Web sites.

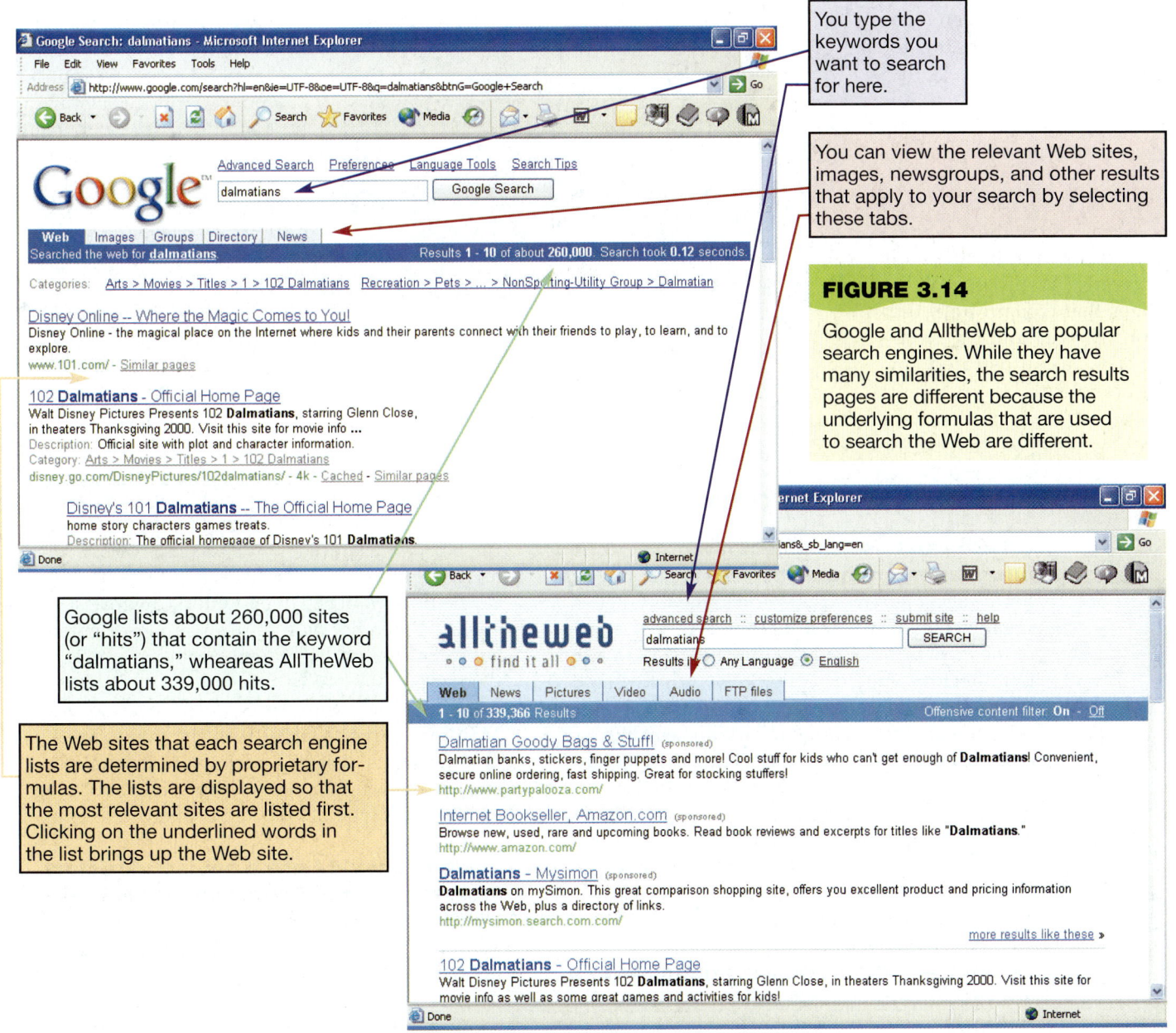

FIGURE 3.14

Google and AlltheWeb are popular search engines. While they have many similarities, the search results pages are different because the underlying formulas that are used to search the Web are different.

DIG DEEPER

Refining Your Web Searches: Boolean Operators

When you conduct Web searches, you often receive a list of "hits" that includes thousands—even millions—of Web pages that have no relevance to the topic you're trying to search. **Boolean operators** are words you can use to refine your searches, making them more effective. These words—AND, NOT, and OR—describe the relationships between keywords in a search.

Narrowing Searches

Using the Boolean AND operator helps you narrow (or limit) the results of your search. When you use the AND operator to join two keywords, the search engine returns only those documents that include *both* keywords (not just one). For example, if you type *Norway AND Sweden* into the search engine's search box, it will list only Web sites with pages that contain **both** the word *Norway* **and** the word *Sweden*, as illustrated in Figure 3.15.

You can also narrow your search by using the NOT operator. When you use the NOT operator to join two keywords, the search engine doesn't show the results of any pages containing the word following NOT. For example, as illustrated in Figure 3.16, if you want information on buying cars but you don't want any information on Fords, you could type *cars NOT Ford* into the search box.

Be aware, however, that when you use the NOT operator, you may eliminate documents that contain the unwanted keyword but that also contain important information that may have been useful to you.

Expanding Searches

The OR operator expands a keyword search so that the search results include both keywords. For example, if you type *laptop OR notebook* into the search box, it will list Web sites with pages that contain **either** the word *laptop* **or** the word *notebook* **or** both, as shown in Figure 3.17. Boolean OR searches are particularly helpful if there are a variety of synonymous keywords you could use in your search.

Other Helpful Search Strategies

Combining terms produces more specific results. To do so, though, you also need to use parentheses to add order to your search. For example, if you are looking for tutorials or lessons to better use the program Excel, you can search for *(Tutorials OR Lessons) AND Excel*. Similarly, if you want to know how to better use the entire Office suite with

Norway 822,000 hits

Norway and Sweden 14,200 hits

Sweden 1,260,000 hits

FIGURE 3.15

Using the AND operator will narrow your search as the search engine will return only those pages that include both the words Norway and Sweden, indicated by the shaded area in the diagram

SUBJECT DIRECTORIES

How can I use a subject directory to find information on the Web? As mentioned earlier, a *subject directory* is a guide to the Internet organized by topics and subtopics. Yahoo! is one of the most popular subject directories. With a subject directory, you do not use keywords to search the Web. Instead, after selecting the main subject from the directory, you narrow your search by successively clicking on subfolders that match your search until you have reached the appropriate information. For example, to find previews on newly released movies in Yahoo's subject directory, you would click on the main category of Entertainment, select the subcategory Movies and Films, select the further subcategory Preview, and then open one of the listed Web sites.

Can I find the same information with a subject directory as I can with a search engine? Most subject directories are more commercial and consumer-oriented than academic-

or research-based. If you look at the main categories in the subject directory of Yahoo! in Figure 3.18, for example, you'll see categories such as Entertainment, Computers & Internet, and Recreation & Sports. Even within categories such as Reference, you find consumer-oriented subcategories such as Phone Numbers and Quotations.

Many subject directories, such as Yahoo! and MSN, are part of a larger Web site that focuses on offering its visitors a variety of information, such as the weather, news, sports, and shopping guides. This type of Web site is referred to as a **portal**.

When should I use a subject directory instead of a traditional search engine? Directory searches are great for finding information on general topics (such as sports and hobbies) rather than narrowing in on a specific or unusual piece of information. For example, conducting a search on the keyword *hobbies* on a search engine does not provide you with

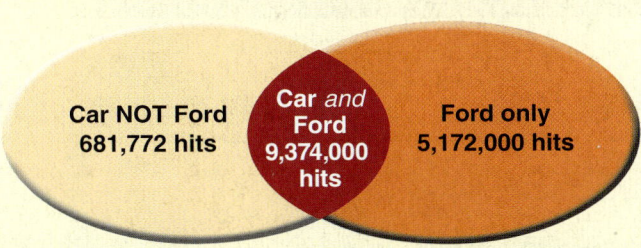

FIGURE 3.16

Using the NOT operator will narrow your search as the search engine will not return those pages that include the word following NOT. In this case, the search engine will only list those hits containing the word car but not the word Ford, as indicated in the shaded area.

FIGURE 3.17

Using the OR operator will broaden your search as the search engine will return pages that include either of the keywords. In this case, the search engine will list all hits containing the words laptop and notebook, as indicated by the shaded area.

the exception of Access, you can search for *(Tutorials OR Lessons) AND (Office NOT Access)*.

To search for an exact phrase, you simply place quotation marks around your keywords. The search engine will look for only those Web sites that contain the words in that *exact order*. For example, if you want information on the book series *Lord of the Rings* and you type in these words without quotation marks, your search results will contain Web pages that include any of the words "Lord," "of," "the," and "Rings," although not necessarily in that order. Typing in "Lord of the Rings" in quotes guarantees all search results will include this exact phrase.

Some search engines also let you use the plus sign (+) and minus sign (–) instead of the words AND and NOT, respectively. Additionally, you can use the asterisk (*) to replace a series of letters and the percent sign (%) to replace a single letter in a word. These symbols, called **wildcards**, are helpful when you're searching for a keyword but are unsure of its spelling, or if a word can be spelled in different ways or may contain different endings. For example, if you're doing a genealogy project and are searching for the name *Goldsmith*, you might want to use *Goldsm&th* to take into consideration alternate spellings of the name (such as *Goldsmyth*). Similarly, if you're searching for sites related to psychiatry and psychology and you type *psych**, the search results will include all pages containing the words *psychology, psychiatry, psychedelic*, and so on.

Using Boolean search techniques can make your Internet research a lot more efficient. With the simple addition of a few words, you can narrow your search results to a more manageable and more meaningful list.

a convenient list of hobbies, as does a subject directory. And although most directories tend to be commercially oriented, there are academic and professional directories whose sites subject experts select and annotate. These directories are created specifically to facilitate the research process. The Librarians' Index to the Internet (**www.lii.org**), for example, is an academic directory whose index lists librarian-selected Web sites that have little if any commercial content.

Communicating Through the Internet: E-Mail and Other Technologies

For better or worse, **e-mail** (short for **electronic mail**) is fast becoming the primary source of communication in the 21st century. However, it is not the only form of Internet-based communication: *chat rooms, instant messaging,* and *newsgroups* are

SOUND BYTE
FINDING INFORMATION ON THE WEB

In this Sound Byte, you'll learn how and when to use search engines and subject directories. Through guided tours, you'll learn effective search techniques, including how to use Boolean operators and meta search engines.

also popular forms. Like any other means of communication, you need to know how to use these tools efficiently to get the best out of them.

E-MAIL

Why did e-mail catch on so quickly?
The quick adoption of e-mail is due, in part, to the fact that it's fast and convenient and reduces the costs of postage and long-distance telephone charges. In addition, with e-mail, the sender and receiver don't have to be available at the same time in order to communicate. Because of these

FIGURE 3.18

Subject directories such as Yahoo! are best for searches that are more commercial and consumer in nature.

and other reasons, e-mail has become a primary means of communication. In fact, more than 90 percent of Americans who access the Internet claim that their main activity is sending and receiving e-mail. The question facing e-mail users today is not *how* to use e-mail, but how best to *manage* e-mail.

How can I organize my e-mail? If you connect to your e-mail through your ISP and use a software program such as Outlook or Eudora to view your e-mail, you can use special features of these programs to manage your e-mail inbox. As you can see in Figure 3.19, you can choose to organize your e-mail by task, sender, or priority using color codes or distributing your messages to designated folders within your inbox. Additionally, you can automatically filter out unwanted e-mail and sort the remaining e-mail into topic-specific folders.

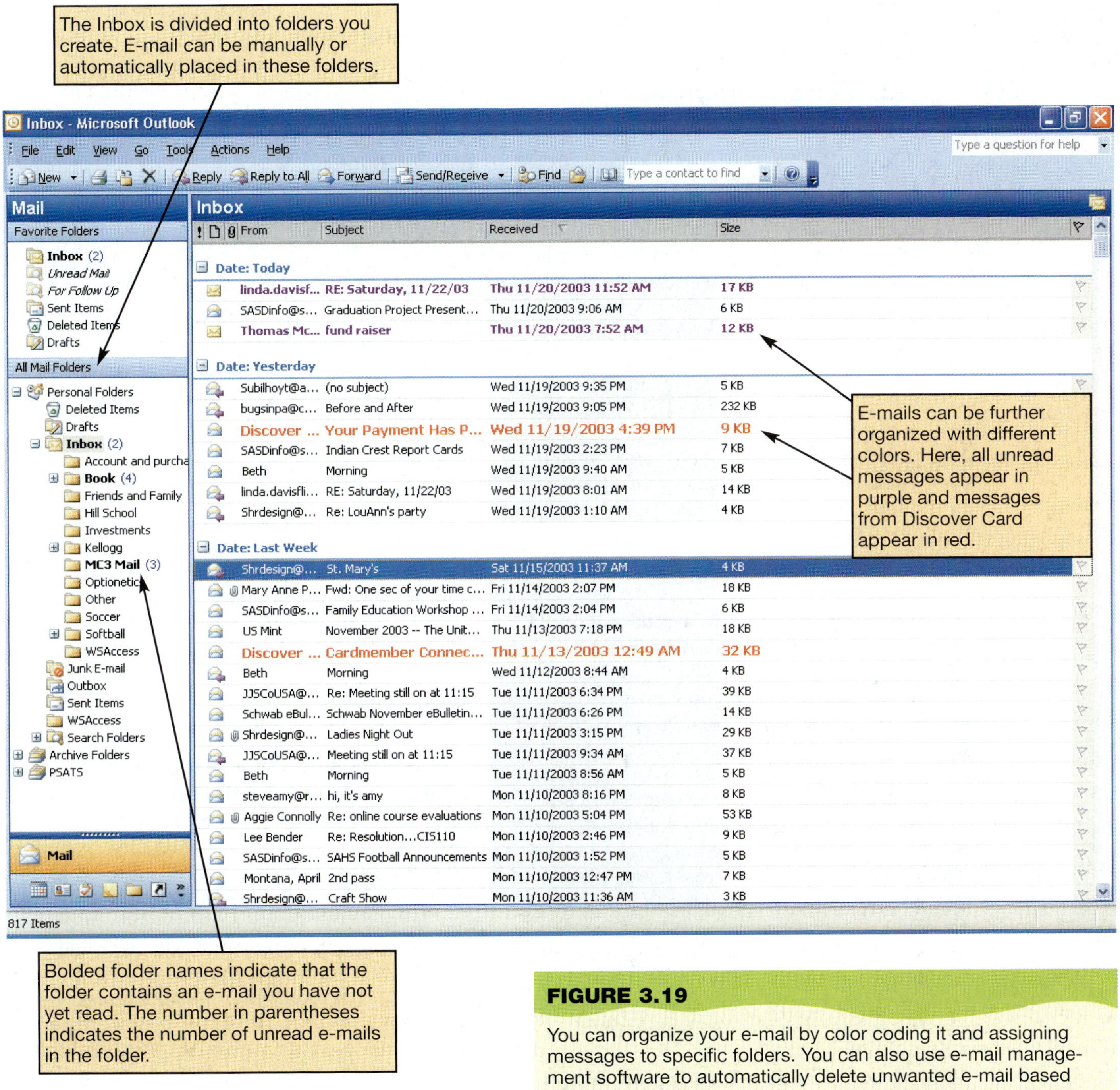

COMMUNICATING THROUGH THE INTERNET: E-MAIL AND OTHER TECHNOLOGIES

The Inbox is divided into folders you create. E-mail can be manually or automatically placed in these folders.

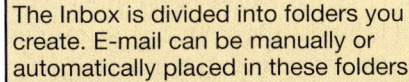

E-mails can be further organized with different colors. Here, all unread messages appear in purple and messages from Discover Card appear in red.

Bolded folder names indicate that the folder contains an e-mail you have not yet read. The number in parentheses indicates the number of unread e-mails in the folder.

FIGURE 3.19

You can organize your e-mail by color coding it and assigning messages to specific folders. You can also use e-mail management software to automatically delete unwanted e-mail based on criteria you select.

BITS AND BYTES

Why Is It Called "Spam"?

Why and how unwanted e-mail has been dubbed "spam" is not really known, but one theory is that the name came from the Monty Python song that praises the canned processed meat product SPAM. The song is an endless repetition of the term *spam*, much like the endless repetition of useless text that constitutes the unwanted e-mail. Also, electronic spam, like its canned ham namesake, is rarely asked for but often served. When served, it's rarely eaten and generally pushed out of the way, similar to electronic spam that is often received but rarely read.

BITS AND BYTES

Removing Temporary Internet Files

When you're on the Internet, your browser keeps track of the Web sites you've visited so it can load them faster the next time you visit them. This "cache" (or hiding place) of the HTML text pages, images, and video files from recently visited Web sites can make your Internet surfing more efficient, but it can also congest your hard drive. To keep your system running efficiently, delete your Temporary Internet Cache periodically. For Internet Explorer 6, select Tools, Internet Options, then in the Temporary Internet Files area, select Delete Files. For Netscape Navigator 6, select Edit, Preferences, Advanced, Cache, and then select Clear Disk Cache.

Do I need more than one e-mail account? Your primary e-mail account is most likely a **client-based e-mail** account. Client-based e-mail is dependent on an e-mail account provided by your ISP and a client software program, such as Microsoft's Outlook or Eudora. To access client-based e-mail accounts, you need to connect with your ISP, such as EarthLink, and configure your computer with Outlook or Outlook Express.

Web-based e-mail uses the Internet as the "client"; therefore, you can access a Web-based e-mail account from any computer that has access to the Web—no special client software is needed. Yahoo! and Hotmail offer free Web-based e-mail accounts. AOL is another popular Web-based e-mail provider, though not free. Many ISPs such as EarthLink now provide Web-based e-mail options as well.

Why should I have a Web-based e-mail account if I already have a client-based account? Web-based e-mail accounts are good to have for several reasons. As we just mentioned, if you have a Web-based e-mail account, your e-mail is accessible from any computer as long as you have access to the Internet. This is helpful if you travel and aren't able to connect to your client-based e-mail accounts on your home computer. In addition, a secondary Web-based e-mail account provides you with a "permanent" e-mail address. Your other e-mail accounts and addresses may change when you switch ISPs or change employers, so having a permanent e-mail address is important. Finally, if you need to send personal e-mails from work, using a Web-based e-mail account keeps your private e-mail off of the company's system.

With all these benefits, why wouldn't I just want a free Web-based e-mail account? One drawback to free Web-based e-mail accounts is that they have storage limits. For example, Yahoo! gives users 4 MB (megabytes) of free storage, while Hotmail gives users just 2 MB of free storage. If you want more storage, you have to pay an annual fee—ranging from around $20 for 10 MB to $60 for 100 MB of storage. Also, with Web-based e-mail you don't have the organizing and management capabilities you get with Outlook or Eudora. Thus, if you want those types of features, it's best to stick with client-based e-mail.

Spam

Would having a Web-based e-mail account help reduce the spam I receive? Companies that send out **spam**, unwanted or "junk" e-mail, find your e-mail address from either a list they purchase or with software that looks for e-mail addresses on the Internet. If you've used your e-mail address when purchasing anything online or opening up an online account, or if you've participated in a newsgroup or a chat room, your e-mail address will eventually appear on one of the lists spammers get. You can use your Web-based e-mail address when you fill out forms on the Web and avoid having your primary account filled with spam. Additionally, both Hotmail and Yahoo! prescreen for spam. But if your Web-based e-mail account is saturated with spam, you can abandon that account with little inconvenience. It's much harder to abandon your primary e-mail address.

How else can I prevent spam? There are several ways you can prevent spam:

1. Install antispam software. At **www.download.com**, for instance, you'll find several free programs that block spam.
2. Before registering on a Web site, read its privacy policy to see how it uses your

SOUND BYTE
CREATING A WEB-BASED E-MAIL ACCOUNT

In this Sound Byte, you'll see a step-by-step demonstration explaining how to create a free Yahoo! Web-based e-mail account. You'll also learn the options available with such accounts.

COMMUNICATING THROUGH THE INTERNET: E-MAIL AND OTHER TECHNOLOGIES

e-mail address. Don't give the site permission to pass on your e-mail address to third parties.

3. Don't reply to spam to remove yourself from the spam list. By replying, you are confirming your e-mail address is active. Instead of stopping spam, you may receive more.

4. Subscribe to an e-mail forwarding service such as **www.Emailias.com** or **www.Sneakemail.com**. These services "screen" your e-mail messages, forwarding only those messages you designate as being OK to accept.

For more information on preventing spam, see the Technology in Focus feature "Protecting Your Computer and Backing Up Your Data."

Internet Hoaxes

Besides spam, what other potential headaches could I face with e-mail? Internet hoaxes are e-mails that contain information that is untrue. Hoax messages may request that you send money to cover medical costs for an impoverished and sick child, or ask you to pass on bogus information, such as how to avoid a virus. Chain e-mail letters are also considered a form of Internet hoax.

Why are hoaxes so bad? The sheer number of e-mails generated by hoaxes can cost millions in lost opportunity costs due to time spent reading, discarding, or resending the message, and they can clog up the Internet system. If you receive an e-mail you think might be a hoax, don't pass it on. First determine whether it is a hoax by visiting the U.S. Department of Energy's Hoaxbusters site at **http://hoaxbusters.ciac.org**.

CHAT ROOMS

What's a chat room? A chat room is an area on the Web where many people come together to communicate online. The conversations are in real time and are visible to everyone in the chat room. Usually, chat rooms are created to address a specific topic or area of interest, and chances are you can find an active chat room on any subject of interest to you. Yahoo.com is a good source to locate chat rooms.

Do people know who I am in a chat room? When you enter a chat room, you sign in with a username and password. It's best to not disclose your true identity but rather "hide" behind a username, thus protecting your privacy. On the other hand, the people you are chatting with are also hiding their identities. Some chatters use this veil of privacy to cover dishonest intentions. Undoubtedly, you have heard stories of individuals, especially young teenagers, being deceived (and sometimes harmed) by someone they've "met" in a chat room. A number of Web sites, such as **www.chatdanger.com**, try to protect vulnerable people such as children from malicious chat room users (see Figure 3.20).

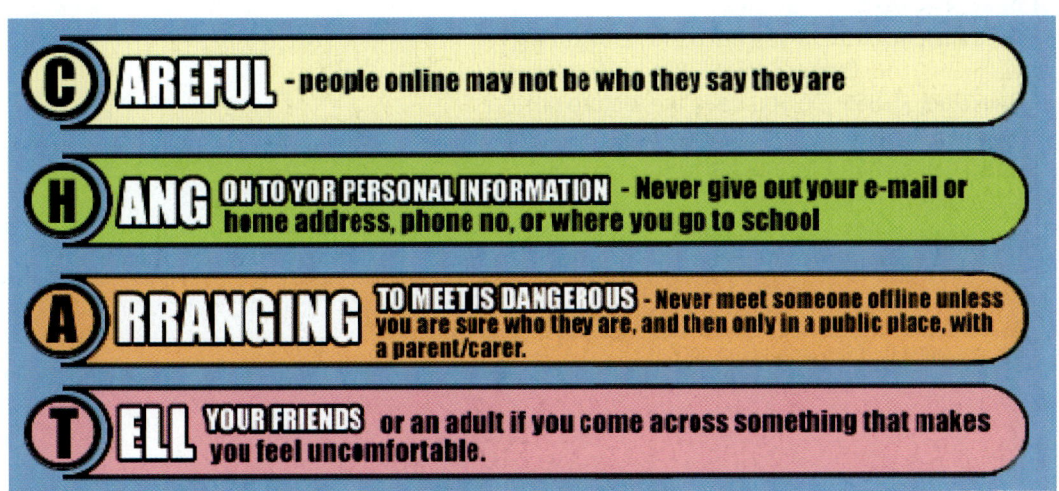

FIGURE 3.20

Chatdanger.com is produced by Childnet International, a non-profit organization working to help make the Internet safe for children.

BITS AND BYTES

Talking (Not Typing) on IM

Would you rather talk to your Buddy than have a typed conversation? You each need only AOL IM or MSN Messenger and a microphone. After you have selected the Buddy you want to talk to, right-click on the Buddy's screenname, select Connect to Talk from the drop-down menu, and then select Connect. The Talk box will contact your Buddy, see if he or she wants to talk, then try to make a connection. When a connection has been made, you can begin to talk to each other. Talking over IM is like talking on a walkie-talkie. While you are talking, you need to hold down the Push to Talk button and release it to hear your Buddy.

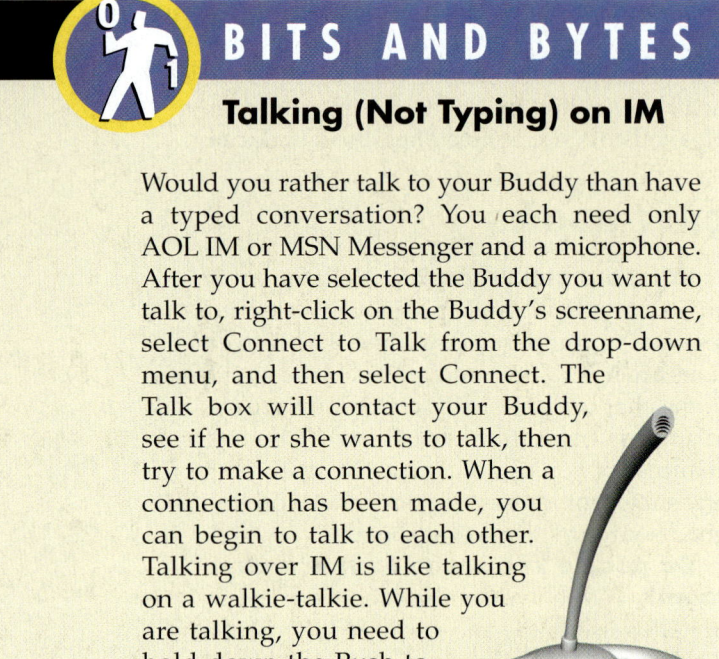

Netiquette

Are there special ways to behave in a chat room? General rules of etiquette (often referred to as "**netiquette**") exist across chat rooms and other online forums, including obvious standards of behavior such as introducing yourself when you enter the room and specifically addressing the person you are talking to. Chat room users are also expected to refrain from swearing, name-calling, and using explicit or prejudiced language, and are not allowed to harass other participants. In addition, chat room users cannot repeatedly post the same text with the intent to disrupt the chat. (This behavior is called *scrolling*.) Similarly, users shouldn't type in all capital letters, as this is interpreted as shouting.

INSTANT MESSAGING

How does instant messaging work? **Instant messaging (IM) services** are programs that enable you to communicate in real time with friends who are also online. AOL's Instant Messenger (AIM or IM), shown in Figure 3.21,

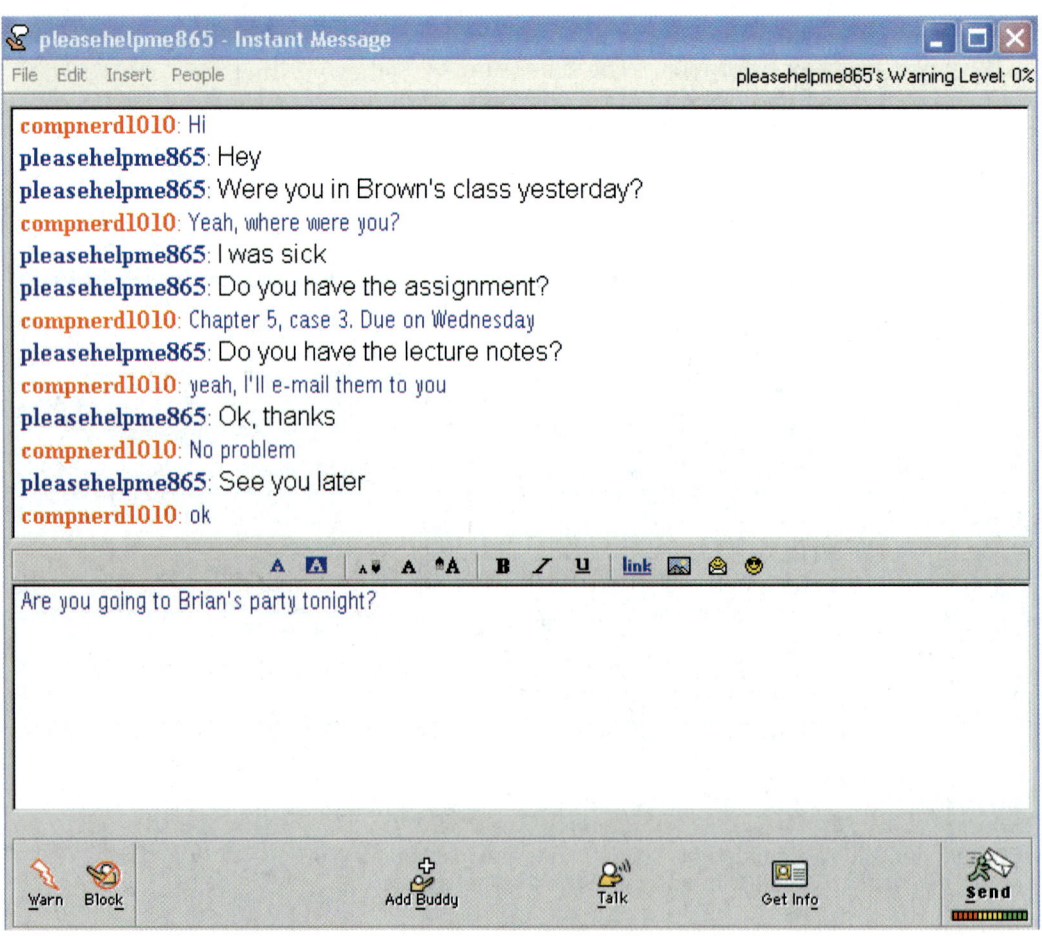

FIGURE 3.21

Instant messaging services such as AOL Instant Messenger enable you to have real-time online conversations with friends and family.

Key Terms

academic fair use	p. 89	Integrated Services Digital Network (ISDN)	p. 71
Advanced Research Projects Agency Network (ARPANET)	p. 66	Internet	p. 66
Asymmetrical Digital Subscriber Line (ADSL)	p. 70	Internet backbone	p. 67
		Internet Explorer (IE)	p. 75
Back button	p. 78	Internet hoaxes	p. 87
Bookmark	p. 78	Internet service providers (ISPs)	p. 73
Boolean operators	p. 82	IP (Internet Protocol) address	p. 67
breadcrumb list	p. 78	ISDN modem	p. 71
broadband connections	p. 68	keywords	p. 79
Buddy List	p. 90	meta search engine	p. 81
business-to-business (B2B)	p. 90	modem	p. 68
business-to-consumer (B2C)	p. 90	multimedia	p. 94
cable Internet connection	p. 70	netiquette	p. 88
cable modem	p. 70	Netscape Navigator	p. 75
chat room	p. 87	network interface card (NIC)	p. 71
click-and-brick businesses	p. 90	newsgroup (discussion group)	p. 90
client	p. 66	online service providers (OSPs)	p. 74
client-based e-mail	p. 86	path (subdirectory)	p. 78
client/server network	p. 66	PC cards (PCMCIA cards)	p. 68
coaxial cable	p. 72	plagiarism	p. 89
consumer-to-consumer (C2C)	p. 90	plug-in (or player)	p. 95
cookies	p. 92	portal	p. 82
copyright violation	p. 89	protocol	p. 76
data transfer rate (throughput)	p. 68	satellite Internet	p. 72
dial-up connection	p. 68	search engine	p. 79
Digital Subscriber Line (DSL)	p. 69	server	p. 66
domain name	p. 77	spam	p. 86
DSL modem	p. 70	spider (crawler or bot)	p. 79
e-commerce (electronic commerce)	p. 90	streaming audio	p. 94
		streaming video	p. 94
e-mail (electronic mail)	p. 83	subject directory	p. 79
Favorites	p. 78	Symmetrical Digital Subscriber Line (SDSL)	p. 70
fiber-optic lines	p. 70	top-level domain (TLD)	p. 77
Forward button	p. 78	twisted pair wiring	p. 69
FTP (File Transfer Protocol)	p. 77	Uniform Resource Locator (URL)	p. 76
History list	p. 78	Web-based e-mail	p. 86
hits	p. 80	Web browser	p. 66
home page	p. 76	Web sites	p. 76
host	p. 77	wildcards	p. 83
http (Hypertext Transfer Protocol)	p. 77	wireless Internet service provider	p. 75
hyperlinks	p. 78		
indexer	p. 80	World Wide Web (WWW or the Web)	p. 66
instant messaging (IM) services	p. 88		

Buzz Words

Word Bank

- Internet service provider
- dial-up
- AOL
- hyperlink
- plug-in
- wildcard
- newsgroup
- Bookmark
- portal
- DSL
- satellite
- browser
- breadcrumb list
- search engine
- spam
- keyword
- chat room
- ISDN
- cable modem
- PC card
- host
- subject directory
- Buddy List
- cookie(s)
- URL

Instructions: Fill in the blanks using the words from the Word Bank above.

The day finally arrived when Juan no longer was a victim of slow Internet access through traditional (1)_____ connection. He could now hook up to the Internet through his new high-speed (2)_____. He had been investigating broadband access for a while, and thought that connecting through his existing phone lines with (3)_____ would be convenient. Unfortunately, it was not available in his area. Where Juan lived, a clear southern exposure did not exist, so he did not even entertain the idea of a (4)_____ connection. While Juan relished the speedy access, he was faced with changing from (5)_____, his online service provider, to a different (6)_____ through his cable company. While he needed to change his e-mail address, he was glad he didn't have to give up instant messaging, as his (7)_____ of online contacts had grown to be quite extensive. Juan knew he could access instant messaging through the (8)_____ Netscape Navigator.

Juan clicked on his list of favorite Web sites and found the movie review site he had saved as a (9)_____ yesterday. He prefers to use this feature rather than entering in the (10)_____ of the sites he often visits. Juan navigated through the site, clicking on (11)_____ that take him immediately to the pages he is most interested in. Finding the movie he wants to see, Juan orders tickets online. The credit card information he input during an earlier visit to the site automatically appears. This time, Juan is glad that Web sites use (12)_____ to capture personal information.

Then, using the (13)_____ at the top of the Web site, he traces his steps back to his starting point. Juan next types in the address for Google, his preferred (14)_____ and types in the (15)_____ to begin his search for a good restaurant in the area.

Organizing Key Terms

Instructions: This chapter introduced many new terms and concepts. In the illustration below, fill in each of the blank boxes with keywords from the chapter in order to show how categories of ideas fit together.

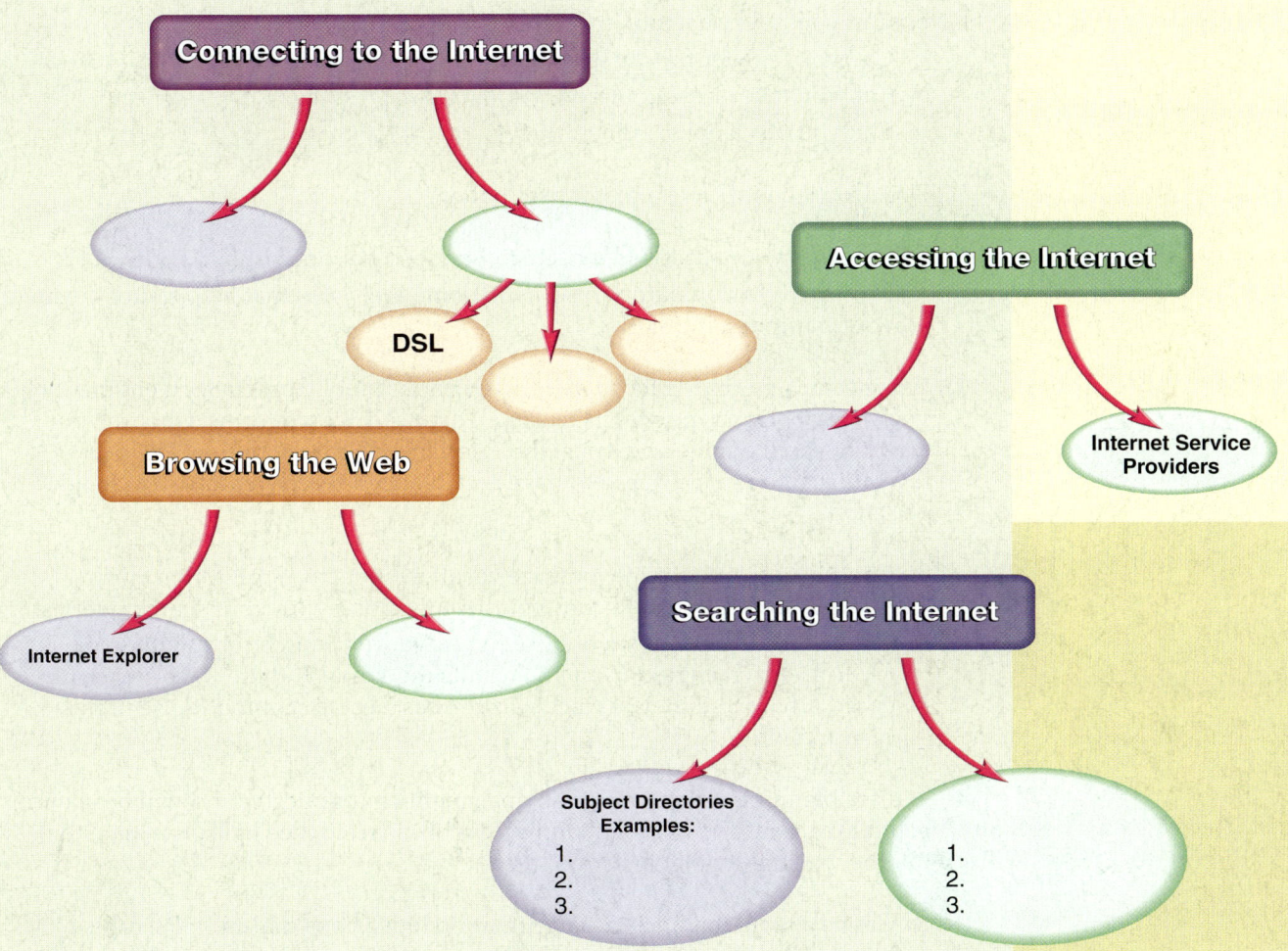

Making the Transition to . . . Next Semester

1. Your school most likely has many online support facilities. Do you know what they are? Go to your school's Web site and search for online support.

   a. Is there online tutoring?
   b. Can you reserve a book from the library online?
   c. Can you register for classes online?
   d. Can you take classes online?
   e. Can you buy books online?

2. Does your school have a plagiarism policy?

   a. Research your school's Web site to find it. What does it say?
   How well do you paraphrase? Find some Web sites that help test or evaluate your paraphrasing skills.

3. Using search engines effectively is an important tool. Some search engines help you with Boolean type searches using Advanced Search forms. Choose your favorite search engine and select the "Advanced Search" option. (If your favorite search engine does not have an advanced search feature, try Google or Yahoo!.)

   a. Conduct a search for inexpensive vacation spots for Spring Break using Boolean search terms. Record your results along with your search queries.
   b. Conduct the same search but use the Advanced Search form with your favorite search engine. Were the results the same? If there were any differences, what were they? Which was the best search method to use in this case and why?

4. You are planning on moving to an apartment next semester and will be leaving behind the comforts of broadband access of the residence halls. Evaluate the Internet options available in your area.

   a. Create a table that includes information on various dial-up ISPs, cable Internet, DSL, and satellite broadband service providers. The table should include the name of the provider, the cost of the service, the upload and download transfer rates, and the installation costs (service and parts). Also include whether a Web-based e-mail account will be available. Include the URL of each ISP or broadband service provider's Web site.
   b. Based on the table you create, write a brief paragraph describing which service you would choose and why.

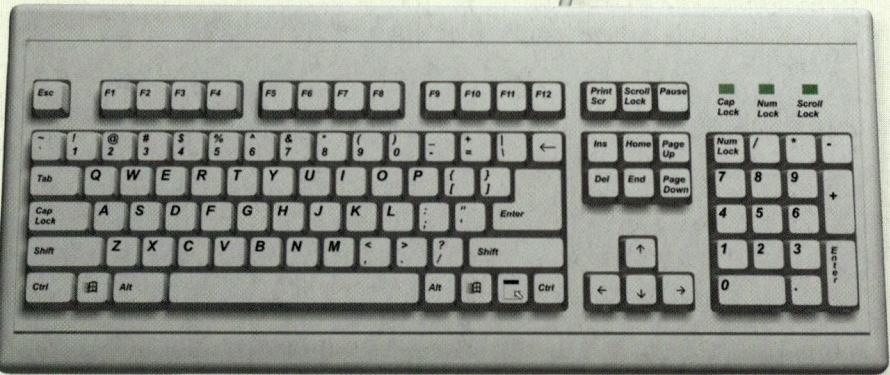

Making the Transition to . . . The Workplace

1. Using a search engine, locate several Web resources that offer assistance in writing a resume. For example, the University of Minnesota (**www.umn.edu/ohr/ecep/resume**) has a ResumeTutor that guides you as you write your resume.

   a. What other Web sites can you find that help you write a resume?
   b. Do they all offer the same services and have the same features?
   c. Which Web site features do you think work best?

2. Your resume will need to be accompanied by a cover letter. Research Web sites that offer advice for and samples of cover letters.

   a. What Web sites do you feel offer the best advice on how to write a cover letter?
   b. What style cover letter works best for you?
   c. What do the Web sites say you should include in your cover letter and why?

3. You have noticed that your coworkers are using the Internet to conduct research. However, they are not careful to check the validity of the Web sites they find before using the information.

   a. Research the Internet for Web site evaluation guidelines. Print out your sources and findings.
   b. Using the material from Step (a) above, create a scorecard or set of guidelines that will help others determine whether a Web site is reliable.

4. You would like to know how fast your Internet connection speed is. Your coworker in the IT (Information Technology) department recommended the following sites for you to check out.

   www.testmyspeed.com **www.bandwidthplace.com**
   www.computingcentral. **www.pcpitstop.com**
   msn.com/internet/speedtest.asp

   a. List reasons why you would be interested in measuring your Internet connection speed.
   b. List four factors that can affect your connection speed.
   c. Discuss why "defragging" your hard drive may help improve your connection speed.

5. "An e-mail is no more private than a postcard."

   a. Search the Internet for resources that can help you support or oppose the above statement. Print out sources for both sides of the argument.
   b. Write a paragraph that summarizes your position.

Critical Thinking Questions

Instructions: *Albert Einstein used "Gedanken experiments," or critical thinking questions, to develop his theory of relativity. Some ideas are best understood by experimenting with them in our own minds. The following critical thinking questions are designed to demand your full attention but require only a comfortable chair—no technology.*

1. The Internet was initially created in part to enable scientists and educators to share information quickly and efficiently. It is evident the advantages the Internet brings to our lives, but does Internet access also cause problems?

    a. What advantages and disadvantages does the Internet bring to your life?
    b. What positive and negative effects has the Internet had on our society as a whole?

2. File-swapping site Napster's unprecedented rise to fame came to a quick halt due to accusations of copyright infringements. However, even with Napster's close, downloading music from the Internet still occurs.

    a. What's your opinion on having the ability to download music files of your choice? Do you think the musicians who oppose online music sharing make valid points?
    b. Discuss the differences you see in sharing music files online and sharing CDs with your friends.

3. Google is the largest and most popular search engine on the Internet today. Because of its size and popularity, some people claim that Google has enormous power to influence a Web user's search experience solely by its Web site ranking processes. What do you think about this potential power? How could it be used in negative or harmful ways?

    a. Some Web sites pay search engines to list them near the top of their results pages. These "sponsors" therefore get priority placement. What do you think of this policy?
    b. What affect (if any) do you think that Google has on Web site development? For example, do you think Web site developers intentionally include frequently searched words in their pages so that they will appear in more "hits" lists?
    c. When you "google" someone, you type their name in the Google search box to see what comes up. What privacy concerns do you think such "googling" could present? Have you ever googled yourself or your friends?

4. Billions of dollars exchange hands every year over the Internet. Should an Internet sales tax be charged on goods purchased over the Internet? If so, to whom should the tax proceeds go? Who would be responsible for collecting the tax? How much should be charged?

5. Should there be a charge placed on sending e-mail or on having IM conversations? What would be an appropriate charge? If a charge is placed on e-mail and IM conversations, what would happen to their use?

6. In the United States, Internet content is not censored or restricted due to the First Amendment right of free speech. However, in some countries, censorship is permitted. Search engines such as Google must restrict access to particular Web sites from searches originating in certain countries.

   a. Do you think such censorship of Web sites should be allowed? Should search engines have to follow these laws?
   b. What kinds of problems exist for search engines that must abide by such censorship laws?
   c. What other kinds of companies do you think face the same type of legal issues as search engines?

7. Currently, a number of states in the United States are attempting to pass bills outlawing spam or fining spam senders. What do you think of such efforts to stop spam? Will they work? Do you think spam should fall under free speech protection? If so, can you think of any creative ways to stop spam?

8. Search engines provide an easy and efficient means to get information from the Web. Some argue that conducting searches on the Internet provides answers but does not inspire thoughtful research.

   a. In your opinion, can the Internet inspire thoughtful research?
   b. Should use of the Internet be banned, or at least limited, for research projects in schools? Why or why not?

9. Companies that use cookies claim they help make your Web site visits "more efficient" by tailoring the content you see and the advertisements that appear on them. However, cookies also raise privacy concerns for many Internet users. Do you think companies should be allowed to use cookies? If so, what information should companies be able to collect using cookies?

10. The safety of children on the Internet is a topic of constant debate. To address concerns of many parents, Congress is considering mandating the creation of a "safe haven" on the Internet for children under the age of 13. This kids-only domain would include only age-appropriate content that would be screened and filtered. This domain would filter out all pornography sites, unmonitored chat rooms, and other Web sites deemed inappropriate for minors, or any site that "lacks serious, literary, artistic, political or scientific value for minors."

    a. Is this an appropriate action for Congress to take? Is such an action necessary?
    b. If such a domain were created, would it help "protect" children? Why or why not?

Team Time

Problem:

With millions of sites on the Internet, finding useful information can be a daunting—at times, impossible—task. However, there are methods to make searching easier, some of which have been discussed in this chapter. In this Team Time, each team will search for specific items or pieces of information on the Internet and compare search methodologies.

Task:

Split your group into two or more teams depending on class size. Each group will search for the same items. No restrictions on search processes are to be made. The only requirement is to document the entire search process, as described below.

Search Items:

- What was America's first penny candy to be individually wrapped?
- What fraternity in which college was the movie *Animal House* based on?
- What are the previous names for the American League baseball team the Anaheim Angels?
- What is the cheapest price to purchase a copy of the latest version of Microsoft Office Professional?
- Where can you go to buy the least expensive pink fuzzy bathrobe?

Process:

STEP 1: Teams are positioned at computers connected to the Internet.

STEP 2: Each team is given the list of search items. Teams can use whatever search strategies they feel will best reach the desired goal with the most accuracy in the least amount of time.

STEP 3: Teams compare notes as to what search methods they used to find each item.

Conclusion:

Were subject directories better than search engines for certain searches? What methods were used to narrow choices down? How were final answers determined?

Becoming Computer Fluent

Instructions: Using keywords from the chapter, write a letter to your local cable company imploring them to bring cable modem service to your neighborhood. In the letter, include your dissatisfaction with dial-up as well as your opinion on why cable is better than DSL (which is currently being offered in your neighborhood) and satellite. Also include the activities on the Internet you think people in the community could benefit from using high-speed cable access.

Materials on the Web

In addition to the review materials presented here, you'll find extra materials on the book's companion Web site (**www.prenhall.com/techinaction**) that will help reinforce your understanding of the chapter content. These materials include:

Sound Byte Lab Guides

For each Sound Byte mentioned in the chapter, there is a corresponding lab guide located on the book's companion Web site. These guides review the material presented in the Sound Byte and direct you to various Web resources that examine the material. The Sound Byte Lab Guides for this chapter include:

- Connecting to the Internet
- Welcome to the Web
- Finding Information on the Web
- Creating a Web-Based E-mail Account
- The Best Utilities for Your Computer

True/False and Multiple Choice Quizzes

The book's Web site includes a True/False and a Multiple Choice quiz for this chapter. You can take these quizzes, automatically check the results, and e-mail the results to your instructor.

Web Research Projects

The book's Web site also includes a number of Web research projects for this chapter. These projects ask you to search the Web for information on computer-related careers, milestones in computer history, important people and companies, emerging technologies, and the applications and implications of different technologies.

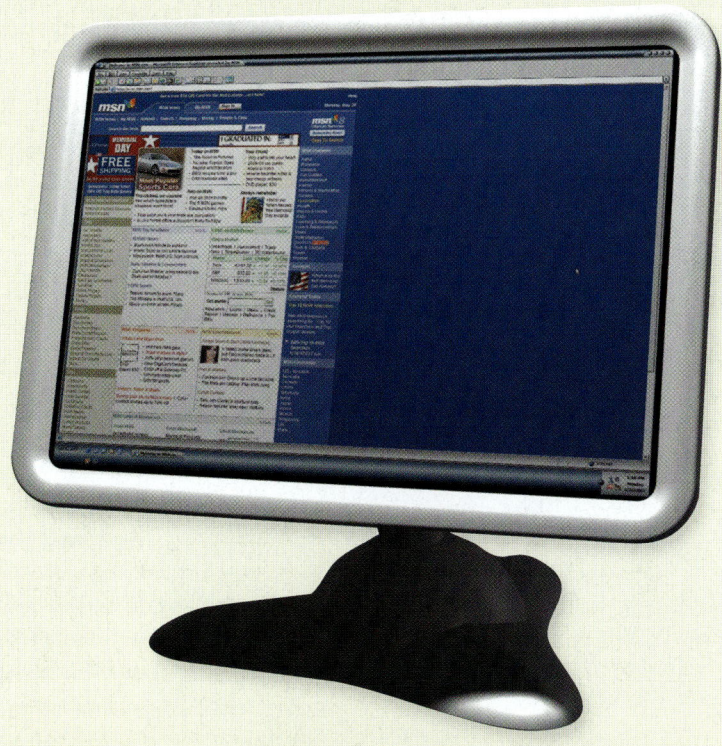

OBJECTIVES

After reading this chapter, you should be able to answer the following questions:

- What's the difference between application software and system software? (p. 108)

- What kinds of applications are included in productivity software? (pp. 108–120)

- What kinds of software do businesses use? (pp. 120–123)

- What are the different kinds of graphics and multimedia software? (pp. 123–128)

- What is educational and reference software? (pp. 129–130)

- What are the different types of entertainment software? (p. 131)

- What kinds of software are available for communications? (pp. 132–133)

- Where can I go for help when I have a problem with my software? (p. 133)

- How can I purchase software or get it for free? (pp. 134–136)

- How do I install and uninstall software? (pp. 136–138)

example, "b" and "m" are both phonemes that distinguish the words "bad" and "mad" from each other in the English language. A typical language such as English comprises thousands of different phonemes. And due to differences in pronunciation, some phonemes may actually have several different corresponding matching sounds.

Once all the sounds are assigned to phonemes, word and phrase construction can begin. The phonemes are matched against a word list that contains transcriptions of all known words in a particular language. Since pronunciation can vary (for example, "the" can be pronounced so that it rhymes with "duh" or "see"), the word list must contain alternate pronunciations for many words. Each phoneme is worked on separately; the phonemes are then chained together to form words that are contained in the word list. Because a variety of sounds can be put together to form many different words, the software analyzes all the possible values and picks the one value that it determines has the best probability of correctly matching your spoken word. The word is then displayed on the screen or is acted upon by the computer as a command.

Why are there problems with speech-recognition software? We don't always speak every word the same way, and accents and regional dialects result in great variations in pronunciations. Therefore, speech recognition is not perfect and requires significant "training." Training entails getting the computer to recognize your particular way of speaking, a process that involves reading prepared text into the computer so the phoneme database can be adjusted to your specific speech patterns.

Another approach to speech inconsistencies is to restrict the word list to a few key words or phrases and then have the computer guess the probability that a certain phrase is being said. This is how cell phones that respond to voice commands work. The phone doesn't really figure out you said "call home" by breaking down the phonemes. It just determines how likely it is that you said "call home" as opposed to "call office." This cuts down on the processing power needed and reduces the chance of mistakes. However, it also restricts the words you can use to achieve the desired results.

While not perfect, speech-recognition software programs can be of invaluable service for individuals who can't type very well or who have physical limitations that prevent them from using a keyboard or mouse. For those whose careers depend on a lot of typing, using speech-recognition software reduces the chances of their incurring debilitating repetitive strain injuries. Additionally, since most people can speak faster than they can write or type, speech-recognition software can help you work more efficiently. It can also help you to be productive during generally nonproductive times. For example, you can dictate into a digital recording device while doing other things such as driving, then later download the digital file to your computer and let the program type up your words for you.

Speech recognition should continue to be a hot topic for research over the next decade. Aside from the obvious benefits to the disabled, many people are enamored with the idea of talking to their computers!

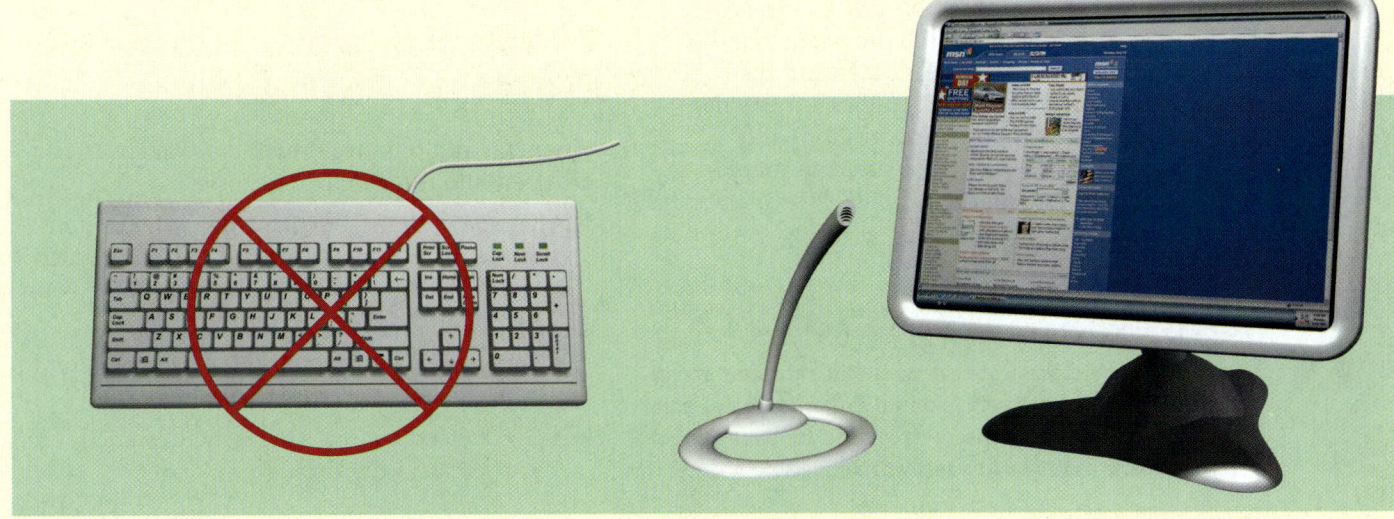

FIGURE 4.11 Software Suites

	WORD PROCESSING	SPREADSHEET	PRESENTATION	PIM	DATABASE	SPEECH RECOGNITION
MICROSOFT OFFICE	WORD	EXCEL	POWERPOINT	OUTLOOK	ACCESS	INTEGRATED IN OFFICE 2003
PROFESSIONAL	x	x	x	x	x	x
STANDARD	x	x	x	x		x
ACADEMIC	x	x	x	x		x
DEVELOPERS	x	x	x	x	x	x
COREL SUITE	WORDPERFECT	QUATTRO PRO	COREL PRESENTATIONS	COREL CENTRAL	PARADOX	SCANSOFT DRAGON NATURALLYSPEAKING
STANDARD	x	x	x	x	x	x
PROFESSIONAL	x	x	x	x	x	x
ACADEMIC	x	x	x	x	x	x
LOTUS SMARTSUITE	WORDPRO	LOTUS 1-2-3	FREELANCE	ORGANIZER	APPROACH	
SMART SUITE	x	x	x	x	x	

What are the most popular productivity software suites? As illustrated in Figure 4.11, there are three primary developers of productivity software suites: Microsoft, Corel, and Lotus. Microsoft and Corel offer different bundled packages with different combinations of software applications, whereas Lotus only offers SmartSuite.

Why would I buy a software suite instead of individual programs? Most people buy software suites because they're cheaper than buying each program individually. In addition, because the programs bundled together in a software suite come from the same developer, they work well together (that is, they provide for better integration) and share common features, toolbars, and menus. For example, say you own Corel's WordPerfect, a stand-alone word processing software application, and Microsoft's stand-alone spreadsheet program Excel. If you want to incorporate an Excel chart into a WordPerfect document, you may have trouble since the two programs come from different developers. However, with a software suite such as Microsoft Office, you can easily incorporate an Excel chart into a Word document simply by clicking an icon.

Financial and Business-Related Software

Financial and business-related software can be grouped into three main categories:
- Personal financial software that helps you perform "businesslike" tasks at home, such as preparing your taxes and managing your personal finances
- General business software used in different capacities across industries
- Specialized business software designed for particular industries

SOUND BYTE
SPEECH-RECOGNITION SOFTWARE

In this Sound Byte, you'll see a demonstration of speech-recognition software included with Microsoft Office. You'll also learn how to access and train speech-recognition software so that you can create and edit documents without typing.

Materials on the Web

In addition to the review materials presented here, you'll find extra materials on the book's companion Web site (**www.prenhall.com/techinaction**) that will help reinforce your understanding of the chapter content. These materials include:

Sound Byte Lab Guides

For each Sound Byte mentioned in the chapter, there is a corresponding lab guide located on the book's companion Web site. These guides review the material presented in the Sound Byte and direct you to various Web resources that examine the material. The Sound Byte Lab Guides for this chapter include:

- Sound Byte: Using Speech Recognition
- Sound Byte: Enhancing Photos with Image-Editing Software

True/False and Multiple Choice Quizzes

The book's Web site includes a True/False and a Multiple Choice quiz for this chapter. You can take these quizzes, automatically check the results, and e-mail the results to your instructor.

Web Research Projects

The book's Web site also includes a number of Web research projects for this chapter. These projects ask you to search the Web for information on computer-related careers, milestones in computer history, important people and companies, emerging technologies, and the applications and implications of different technologies.

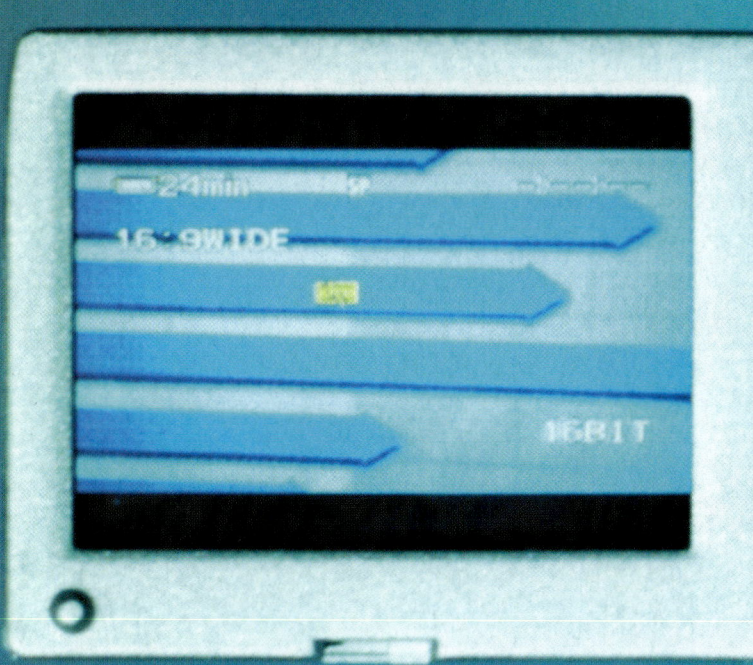

When did everything go "digital"? It used to be that you'd only find "analog" forms of entertainment. Today, no matter what you're interested in—music, movies, television, radio—a "digital" version exists (see Figure 1). MP3 files encode digital forms of music, while digital cameras and video camcorders are now ommonplace. In Hollywood, feature films are being shot with digital equipment. (For example, *Star Wars: Episode II Attack of the Clones* was filmed entirely in digital format and played in special digital release at digitally ready theaters.)

Digital
TECHNOLOGY IN FOCUS
ENTERTAINMENT

Even radio and television are going digital. Satellite radio systems like XM radio and HD radio are digital formats, and High-Definition Television (HDTV), a digital encoding of television signals, will become the national standard in 2006. In this Technology in Focus, we'll look at the most popular form of digital entertainment: digital photography. But first, let's consider what makes digital special.

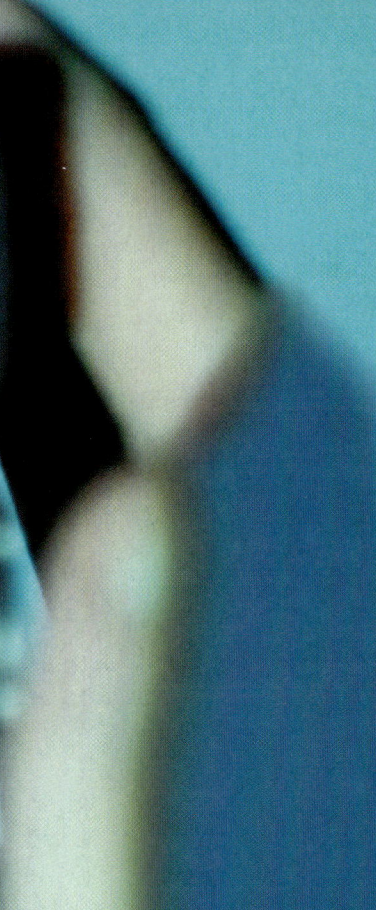

WHAT'S SO SPECIAL ABOUT "DIGITAL"?

So what *is* so special about "digital"? Think about the information captured in music and film: sounds and images. Sound is carried to your ears by sound waves, which are actually patterns of pressure changes in the air. Images are our interpretation of the changing intensity of light waves around us. These sound and light waves are called *analog* or continuous waves. They illustrate the loudness of the sound or the brightness of the colors in the image at a given moment in time. They are continuous signals because you would never have to lift your pencil off the page to draw them: they are just one long continuous line.

The first generation of recording devices (such as vinyl records and analog television) was designed to reproduce these sound and light waves. The needle in a groove of a vinyl record vibrates in the same pattern as the original sound wave. Television signals are actually waves that tell your TV how to display the same color and brightness as seen in the original studio. But it's difficult to describe a wave, even mathematically. Very simple sounds, like the C note of a piano, have a very simple shape, like that shown in Figure 2a. However, something like the word "hello" generates a very complex pattern, like that shown in Figure 2b.

Digital formats are descriptions of these signals as a long string of numbers. This is the main reason why digital recording has such an advantage over analog. Digital gives us a simple way to describe sound and light waves *exactly*, so sounds and images can be reproduced perfectly each time. We already have easy ways to distribute digital information (on CDs, DVDs, or using e-mail, for example). But how could a digital format, a sequence of numbers, act as a convenient way to express these complicated wave shapes?

The answer is provided by something called **analog-to-digital conversion**. In analog- to-digital

FIGURE 1   ANALOG VS. DIGITAL ENTERTAINMENT

	ANALOG	DIGITAL
MUSIC	Vinyl albums Cassette tapes	CDs MP3 files
PHOTOGRAPHY	35 mm SLR (single lens reflex) cameras Photos stored on film	Digital cameras Photos stored as digital files
VIDEO	8 mm, Hi8, or VHS camcorders Film stored on VHS tapes	Digital video (DV) camcorders Film stored as digital files; often distributed on DVDs
RADIO	AM/FM radio	HD radio XM radio
TELEVISION	Current broadcast TV	HDTV

TECHNOLOGY IN FOCUS

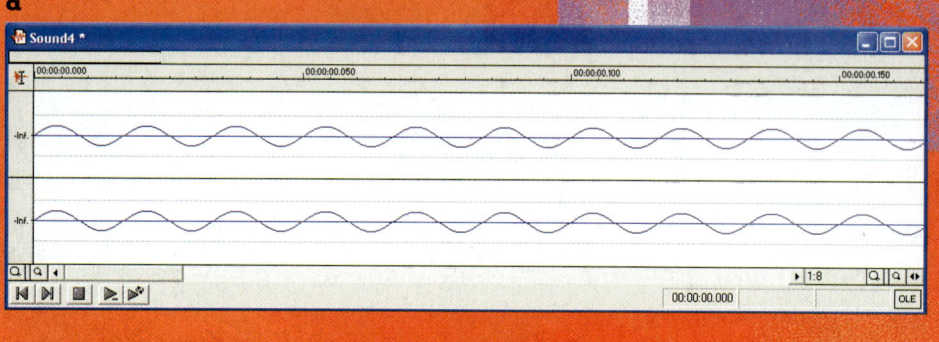

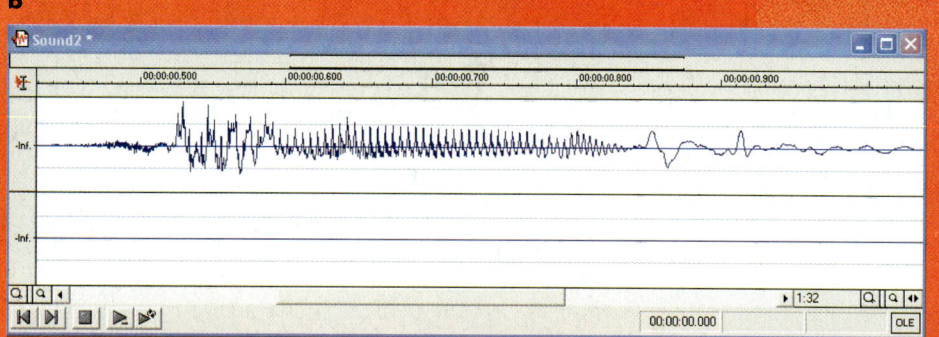

FIGURE 2 (a) This is an analog wave showing the simple, pure sound of a piano playing middle C. (b) This is the complex wave produced when a person says "hello."

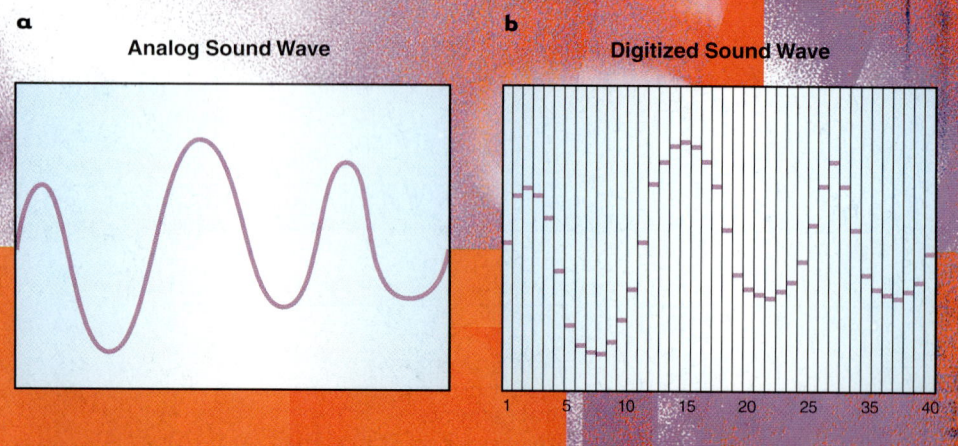

FIGURE 3 (a) Here you see a simple analog wave. (b) Here you see a digitized version of the same wave.

conversion, the incoming analog signal is measured many times each second. The strength of the signal at each measurement is recorded as a simple number. The series of numbers produced by the analog-to-digital conversion process gives us the digital form of the wave. Figure 3 shows an analog and digital version of the same wave. In Figure 3a, you see the original continuous analog wave. You could draw the wave in Figure 3a without lifting your pencil from the page. In Figure 3b, the wave has been digitized and now is not a single line but rather is represented as a series of points or numbers.

So how does this all work? Let's take music as an example. Figure 4 shows how the process of creating digital entertainment begins with the physical act of playing music, which creates analog waves. Next, a chip inside the recording device called an *analog-to-digital converter* (ADC) "digitizes" these waves into a series of numbers. This series of numbers can be recorded onto CDs, DVDs, or sent electronically. On the receiving end, a playback device, like a CD or DVD player, is fed that series of numbers. There, a *digital-to-analog converter* (DAC), a chip that converts the digital numbers to a continuous wave, reproduces the original wave exactly.

Rather, the digital wave will be *close* to exact. How accurate it is, how close the digitized wave is in shape to the original analog wave, depends on the **sampling rate** of the ADC. The sampling rate specifies the number of times the analog wave is measured each second. The higher the sampling rate, the more accurately the original wave can be recreated. However, higher sam-

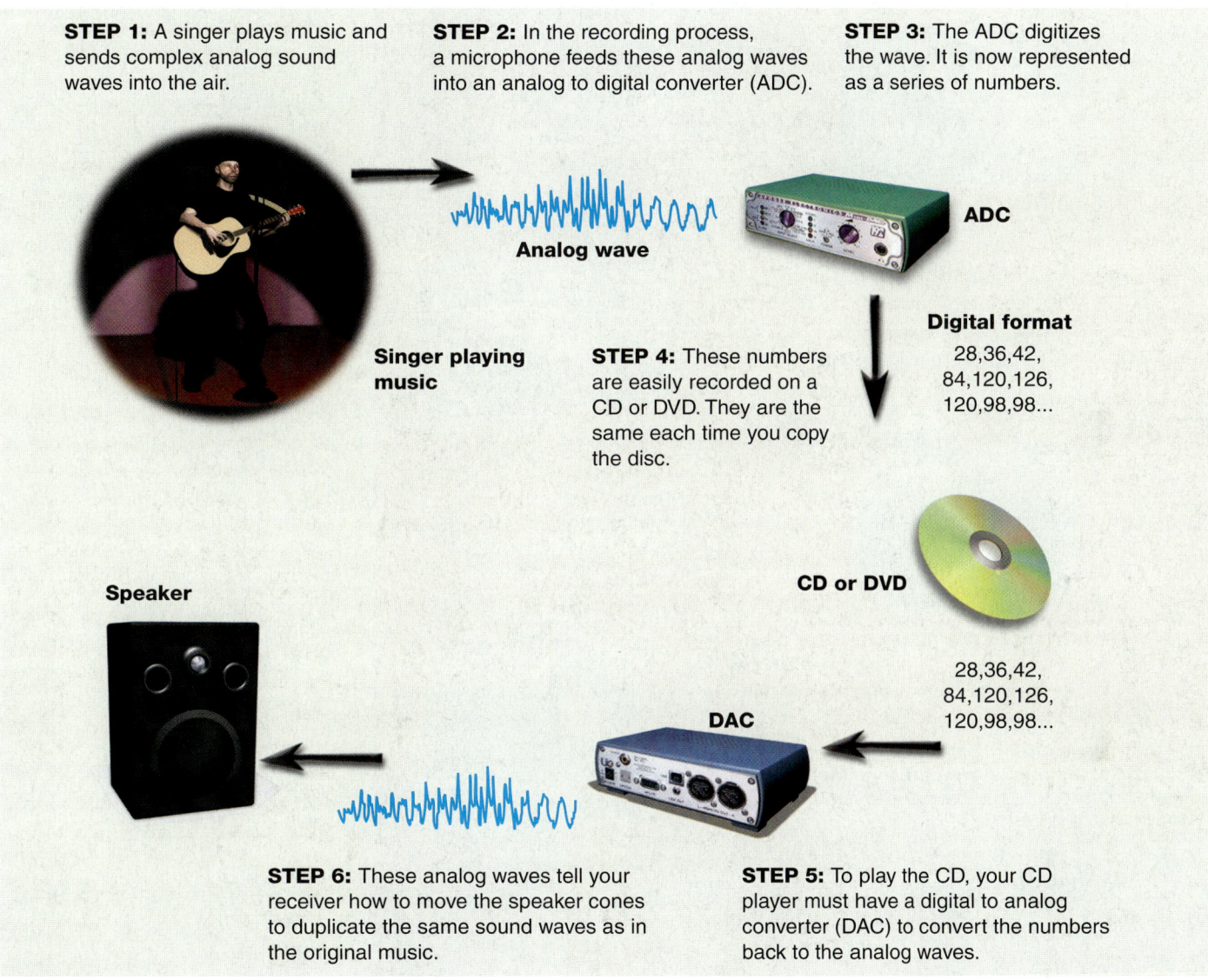

FIGURE 4

pling rates also produce much more data, and therefore result in bigger files. For example, sound waves on CDs are sampled at a rate of 44,000 times a second. This produces a huge list of numbers—44,000 of them each second!

So, when sounds or image waves are "digitized," it means that analog data is changed into digital data—from a wave into a series of numbers. The digital data is perfectly reproducible and can be distributed easily on CDs, DVDs, or through the airwaves. It can also be easily processed by a computer.

These digital advantages have revolutionized photography, music, movies, television, and radio. For example, digital television has a sharper picture and superior sound quality. However, there is a cost in the shift from analog to digital technologies. When all television signals go digital in 2006, consumers will be forced to choose between upgrading to digital HDTV sets or purchasing a converter for older sets. The digital revolution in television will bring better quality and additional conveniences, but at a cost, as the older analog equipment is phased out.

The same tension exists in the migration from analog to digital technology in photography. Let's take a look at this form of entertainment and explore the advantages and investment required in migrating from an analog to a digital format.

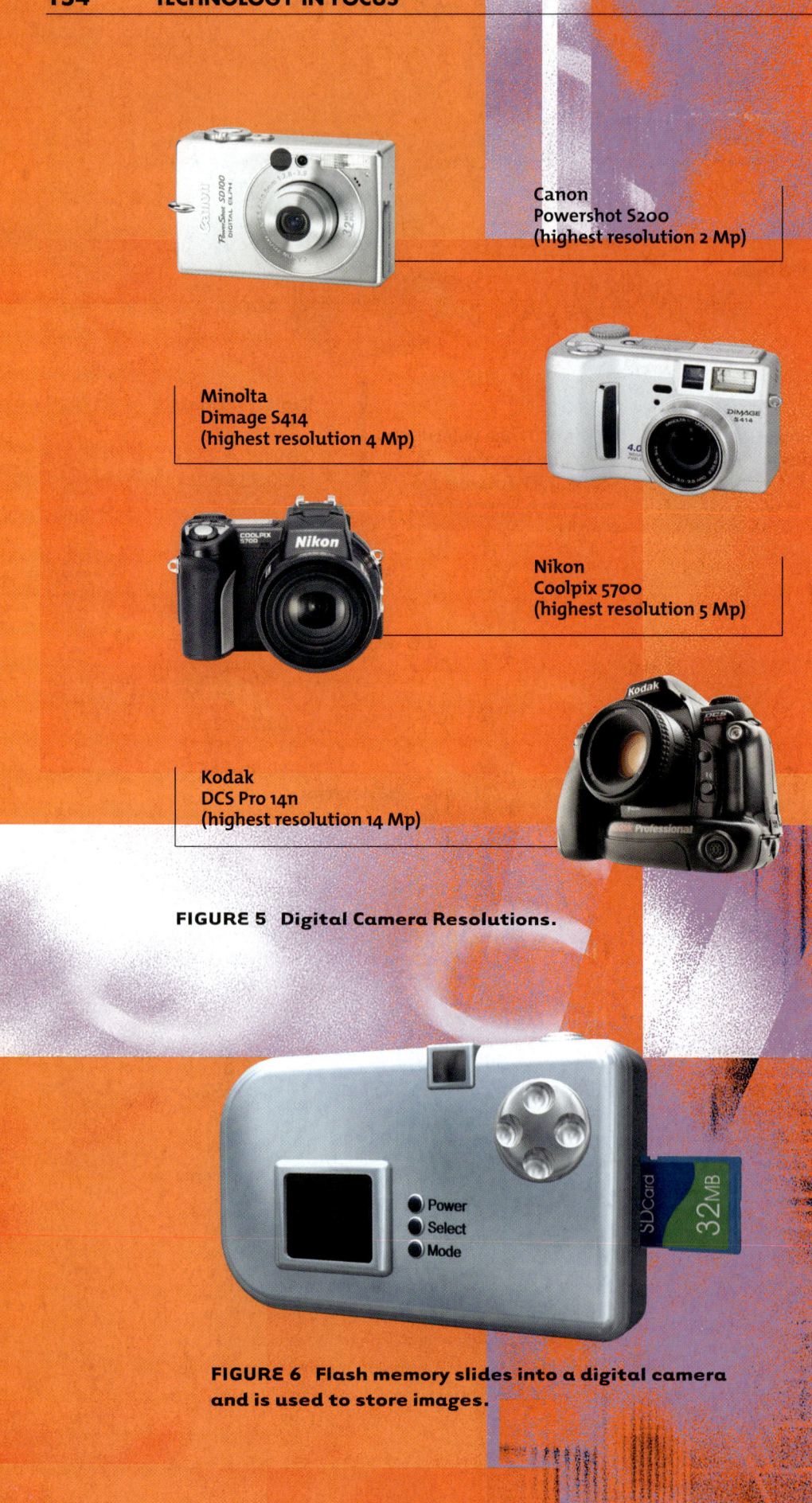

FIGURE 5 Digital Camera Resolutions.

FIGURE 6 Flash memory slides into a digital camera and is used to store images.

DIGITAL PHOTOGRAPHY

Before digital cameras hit the market, most people used some form of 35-mm single lens reflex (SLR) camera. When you take a picture using a traditional SLR camera, an aperture (a small "window" in the camera) opens, allowing light to hit the 35-mm film inside. Chemicals coating the film react when exposed to light. Later, additional chemicals develop the image on the film and it is printed on special light-sensitive paper. A variety of lenses and processing techniques, special equipment, and filters are needed to create printed photos from traditional SLR cameras.

Digital cameras, on the other hand, do not use film. Instead, they capture images and immediately convert those images to digital data, a long series of numbers that represents the color and brightness of millions of points in the image. Unlike traditional cameras, digital cameras also allow you to see your images the instant you shoot them.

Digital Camera Quality

Part of what determines the quality of a digital camera is its **resolution**, or the sharpness of the images it records. A digital camera's resolution is measured in megapixels (MP). The prefix *mega* is short for millions. The word *pixel* is short for picture element, or a single dot in a digital image. The higher the number of megapixels, the higher the quality of the camera and the images it takes.

Popular camera models come in a range of resolutions. An inexpensive pocket-sized camera like the Canon

DIGITAL PHOTOGRAPHY 155

FIGURE 7 FILE TYPES COMMONLY USED IN DIGITAL CAMERAS

FILE TYPE	COMPRESSED	QUALITY	SAMPLE FILE SIZE	# IMAGES THAT WILL FIT ON A 128-MB FLASH CARD
TIFF	No	HIGH: Contains all the original data	6.0 MB	21
JPEG (at highest camera resolution)	Yes	MEDIUM: Moderate compression; some lost quality	2.4 MB	53
JPEG (at lowest camera resolution)	Yes	LOW: More compression; more lost quality	0.4 MB	320

Note: The file sizes in this table refer to image storage on a Canon EOS 10D camera.

PowerShot S200 is a 2-MP camera, which means that every photo it takes contains 2 megapixels, or 2 million picture elements. More expensive consumer cameras like the Minolta Dimage S414 measure 4 MP. Professional photographers are moving to digital cameras as well. Professional digital cameras like the Kodak DCS Pro 14n take photos at resolutions up to 14 MP, but sell for over $4,500. Figure 5 shows some popular digital camera models and the number of pixels they record at their maximum resolution.

If you're interested in an inexpensive digital camera and only plan to make 5 x 7 or 8 x 10 prints, a 2-MP camera is fine. However, these cameras do not record enough pixels to print larger-size prints. If you did print an 11 x 14 enlargement from a 2-MP shot, the image would look grainy—you would see individual dots of color instead of a clear, sharp image.

Camera prices continue to drop as new models with higher resolutions are introduced, so 4-MP and 5-MP cameras are becoming affordable. With such resolutions, you can print larger photos (11 x 14 and up) and still have sharp, detailed images.

Digital Camera Storage

When a digital camera takes a photo, it stores the images on a flash memory card inside the camera, as shown in Figure 6. Flash memory cards are small, powerful, and allow you to transfer digital information between your camera and your computer or printer. Flash memory therefore takes the place of film used in traditional cameras.

To fit more photos on the same size flash memory card, digital cameras allow you to choose from several different file types in order to *compress*, or squeeze, the image data into less space. When you choose to compress your images, you will lose some of the detail, but in return you'll be able to fit more images on your flash card. Figure 7 shows the most common file types supported by digital cameras: TIFF and JPEG. TIFF files record all of the original image information and so are larger than the compressed JPEG files. JPEG files can be compressed just a bit, keeping most of the details, or compressed a great deal, losing some detail. Most cameras allow you to select from a few different JPEG compression levels.

Preparing Your Camera and Taking Your Photos

Preparing your camera includes ensuring that your camera's batteries are charged and the settings are correct. Digital cameras consume a great deal of power, so you might want to carry a spare charged battery pack. Also make sure the flash

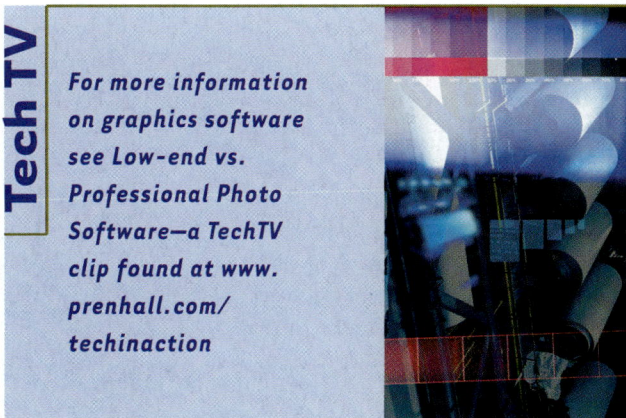

TechTV *For more information on graphics software see Low-end vs. Professional Photo Software—a TechTV clip found at www.prenhall.com/techinaction*

card is installed and that it has enough space for the number of photos you plan to take.

Next, set the resolution on your camera. Most cameras offer two or three different resolution settings. For example, a 6-MP camera might be able to shoot images at 6 MP, 2.7 MP, or 1.5 MP. If you're taking a photo that will be enlarged and that needs to be at a very high quality, use the full power of your camera. Shoot the image at 6 MP and save the image as uncompressed data (a TIFF file) at the highest resolution. If you're planning to use the image for a Web page, where having a smaller file would be helpful, use a lower resolution and the space-saving compressed JPEG format. If you're unsure how you're going to use your images, use the maximum resolution and save them as TIFF files as long as you have enough space on your flash card.

Most cameras include an autofocus feature and automatically set the aperture and correct shutter speed. This makes taking a digital photo as simple as pressing a button. The great thing about digital cameras is that they let you instantly examine your photos in an LCD window on the camera. If you don't like a certain photo, you can delete it immediately, freeing up space on your flash card.

Transferring Your Photos to Your Computer

If you just want to print your photos, you may not need to transfer them to your computer. Many photo printers can make prints directly from your camera or from the flash memory card itself. However, transferring the photos to your computer allows you to store them and frees up your flash card for reuse. Once they're on your computer, you can transfer them to CDs, e-mail them, use them in Web pages, or edit them.

Transferring your photos to your computer is simple. All digital cameras have a built-in USB port. Using a USB cable, you can connect the camera to your computer and perform the transfer. Another option is to transfer the flash card from your camera to the computer. Some desktops have flash card slots on the

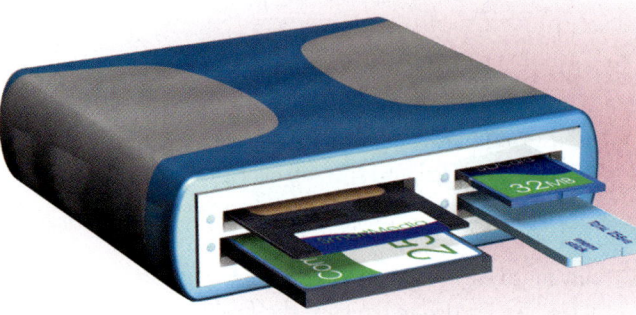

FIGURE 8 If your computer does not have a built-in flash card reader, you can buy an external reader that attaches to your computer via a USB port

HOW DO MY OLD PHOTOS BECOME DIGITAL?

Obviously, not every document or image you have is in an electronic form. What about all the photographs you have already taken? Or an article from a magazine or a hand-drawn sketch? How can these be converted into digital format?

Digital SCANNERS like the one shown in Figure 9 are devices that convert paper text and images into digital formats. You can place any flat material on the glass surface of the scanner and then convert it into a digital file. Most scanner software allows you to store the converted images as TIFF files or in compressed form as JPEG files. And some scanners include optional hardware that allows you to scan film negatives or slides as well.

Scanner quality is measured by its resolution, which is given in dots per inch (dpi). Most modern scanners can digitize a document at 2,400 x 4,800 dpi, in either color

FIGURE 9 Scanners can convert paper documents, photo prints, or strips of film negatives into digital data.

or gray-scale modes. You can easily connect a scanner to your computer using USB 2.0 or FireWire ports. Scanners also typically come with software supporting optical character recognition (OCR). OCR software converts pages of handwritten or typed text into electronic files. You can then open and edit these converted documents with traditional word processing programs.

DIGITAL PHOTOGRAPHY 157

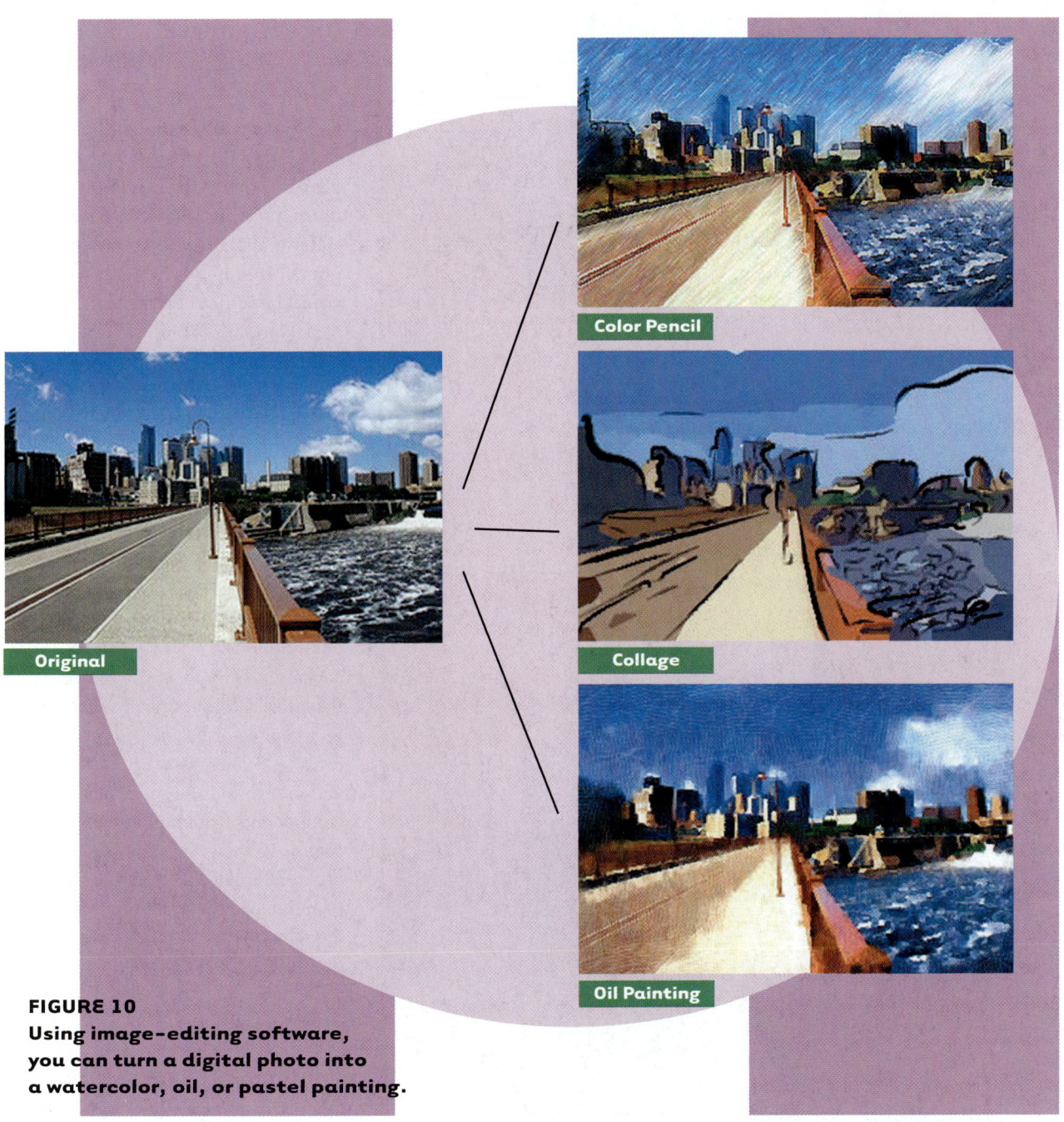

FIGURE 10
Using image-editing software, you can turn a digital photo into a watercolor, oil, or pastel painting.

front of the system unit. However, if yours does not, you can buy an external memory card reader like the one shown in Figure 8 and attach it to your computer using an available USB port.

When you connect your camera to your computer, with the Windows XP operating system, the rest of the transfer is automatic. You'll hear a "ding-dong" sound, telling you the computer and camera are connected and can communicate. Next, a series of prompts appear, asking you which images you'd like to transfer and where you'd like to store them. You now have TIFF or JPEG image files on your computer. If you're satisfied with the photos, you can send them to your friends as e-mail attachments, for example. If you're not satisfied with them, you can process them further.

TECHNOLOGY IN FOCUS

FIGURE 11 Many photo printers incorporate displays to let you see the images before you print. (a) The Canon S530D inkjet printer allows you to insert flash cards directly into the printer. (b) The Olympus P-400 dye-sublimation printer produces professional quality images that will last without fading.

Processing Your Photos and Adding Special Effects (Optional)

Once you're a seasoned digital photographer, you may want to process your photos, cropping them, for example, or adding special effects. Traditional photographers often invest in special equipment and chemicals needed to develop 35-mm film. The photographer can then resize or crop the photo, add different filtering effects, or combine photos. With digital photography, you can do all of this using inexpensive image-editing software.

There are hundreds of image-editing programs available, from freeware to very sophisticated suites. Adobe Photoshop and Jasc Paint Shop Pro are two popular packages. They allow you to remove flaws like red-eye; crop images; correct poor color balance; apply filtering effects like mosaics, charcoal, and impressionistic style; and merge components from multiple images. Figure 10 shows just a few examples of the filtering effects you can apply to an image. The exact set of filtering effects you will have depends on the software you're using.

Printing Your Photos (Optional)

Once you've processed your photos, you can print them using a professional service or your own printer. A variety of online film labs offer digital printing services. These labs accept original or edited image files and print them on professional photo paper with high-quality inks. The paper and ink used at processing labs are higher quality than what is available for home use and produce heavier, glossier prints that won't fade. In addition, Kodak and Sony have kiosks in department stores and photography stores that accept image files directly from flash cards and print them on the spot. You can do a small amount of editing at these kiosks as well, cropping the image or correcting red-eye.

Photo printers for home use are available in two technologies: inkjet and dye sublimation (see Figure 11 on the following page). Most popular and inexpensive are inkjet printers. As noted in Chapter 2, some inkjet printers are capable of printing high-quality color photos, although they vary in speed and quality. Some include an LCD so you can review the image as you stand at the printer, while others are portable, allowing you to print your photos wherever you are. Some printers even allow you to crop the image right at the printer, without having to use special image-editing software.

A competing technology is called *dye-sublimation* printing. Dye-sublimation printers produce images using a heating element instead of an inkjet nozzle. The heating element passes over a ribbon of translucent film that has been dyed with bands of colors. By controlling the temperature of the element, dyes are vaporized from a solid into a gas. The gas vapors penetrate the photo paper before they cool again to solid form, producing glossy, high-quality images. If you're interested in a printer to use only for printing photographs, a dye-sublimation printer is a good choice. However, some models only print specific photo sizes, such as 4 x 6 prints, so be sure the printer you buy will fit your long-term needs.

Transferring your images to the printer is similar to transferring them to your computer. If you have a direct-connection camera, you can plug the camera directly into the printer with a cable. Some printers have slots that accept different

FIGURE 12 DIGITAL PHOTOGRAPHY ADVANTAGES

	TRADITIONAL SLR CAMERA	DIGITAL ADVANTAGE
DEVELOPING	Send out or pay more for one-hour developing	Immediate processing
	Forced to develop the entire roll	Print only the shots you like
	Special equipment required for applying special effects before printing	Easy to apply filters and special effects using inexpensive software
STORAGE	Need to purchase and carry large number of rolls of film; temperature and x-ray sensitive	No need to purchase film; large flash memory cards provide room for many shots
	Can only be stored on hard drive, CD, or DVD after scanning the negative or print	Can easily be stored on hard drive, CD, or DVD with no scanning necessary
	Negatives must be protected in special sleeves	Images stored as data files; only need to back up these files to ensure they are protected
DISTRIBUTION	Paper prints can be mailed; however, must be scanned into a digital format for electronic distribution	Images can be printed on home printers and professionally; easy to distribute image files as e-mail attachments or on CD; can also post images to Web easily
QUALITY	Sets the standard for quality	Excellent quality comparable to traditional photos

types of flash memory cards. You can also transfer your images to the printer from your computer if you have stored them there.

Some printers support a system known as DPOF (Digital Print Order Format). Using a combination of a DPOF camera and a DPOF printer, you can review all the shots on your camera and build an "order" of how many copies and what sizes you would like to print. You then just insert the flash card into the printer and your entire order is printed automatically.

The Digital Advantage

Is a digital camera right for you? Figure 12 lists just a few of the advantages digital cameras have over traditional cameras. Digital cameras give you the power to create images that only professional photographers with expensive processing studios could produce a few years ago.

DIGITAL VIDEO

Personal video cameras have been popular for a long time. The first camcorders were analog video cameras, like the one shown in Figure 13a. These were large, heavy units that held a full-size VHS tape. The push to produce smaller, lighter models led to the introduction of compact VHS tapes and then to 8-mm and Hi8 formats. Still, all of these are analog formats, and each records to its own specific type of tape.

The newest generation of video equipment for home use is the *digital video,* or DV, format shown in Figure 13b on the following page. Introduced in 1995, the digital video standard led to a new generation of recording equipment. Today, digital video cameras offer many advantages over their VHS counterparts. They are incredibly small and light and use tapes of a new format called MiniDV. These tapes can hold one to three hours of video but are just about 2 square inches in size, and allow manufacturers to design stylish and sleek cameras.

Not only that, by using digital video cameras you can easily transfer video files to your computer. Then, using simple video-editing software, you can edit the video at home, cutting out sections, resequencing segments, and adding titles. To do the same on analog videotape would require thousands of dollars of complex audio/video equipment available only in video production studios. And with digital video, you can save (or *write*) your final product on a CD or DVD and play it in your home DVD system or on your computer.

Although you have terrific convenience and control with digital video, there are costs in making the move from analog to digital. Digital video cameras come in a wide range of prices, but they are more expensive than the analog models still

BUT WHAT IF I ALREADY HAVE AN ANALOG CAMCORDER?

If you have an older analog video camera, you can still begin to play with digital video if you purchase a special unit called a *video capture device.* A VIDEO CAPTURE DEVICE digitizes and compresses analog video and then passes it on to your computer. To use one, you connect your analog video camera to the video capture unit, which in turn connects to your computer. Your camera then feeds its video signals to the capture device, which then sends the video to the computer in a digitized form. You can use this device to convert and send existing analog tapes into your computer or to process new videos you shoot with your analog camcorder into digital computer files. Figure 14 shows an example of an external video capture device. Note that video capture devices also are available as expansion (adapter) cards that you install in an unused expansion slot inside your computer.

FIGURE 14 You can still create digital video with an older analog camera by connecting your camera to a video capture device, which in turn connects to your computer.

available. This is reasonable since digitizing video and audio requires a lot of processing power. Video for motion pictures is recorded as high-resolution images at a rate of 30 frames per second (fps). This means 1,800 images need to be digitized for each minute of video. In addition to the video, the audio also must be digitized. Digital video cameras do the analog-to-digital conversion right in the camera itself. They are equipped with FireWire ports so you can later send the digital data quickly from the camera directly to your computer.

For many people, the advantages of working with digital video are worth the extra cost. Let's look at how you would use a digital video camera and see if the investment would be worthwhile for you.

Preparing Your Camera and Shooting Your Video Footage

Preparing your digital video camera involves making sure you have enough battery power and tape capacity. Batteries for digital video cameras are rechargeable and can provide between one and nine hours of shooting time. Longer-lasting batteries cost and weigh more, so you'll want to think about how you use your camera before deciding what batteries to purchase. Most videographers recommend carrying two spare batteries, though having one spare is fine if you can recharge it while you're using the second.

Digital video cameras record on tape, so make sure you have enough tape to cover the event. Some models also let you record short segments onto a flash memory card. If you want to record a short clip that

FIGURE 13 (a) First-generation home video camcorders recorded on full-size VHS cassettes. (b) Modern digital video camcorders can be smaller and lighter.

TECHNOLOGY IN FOCUS

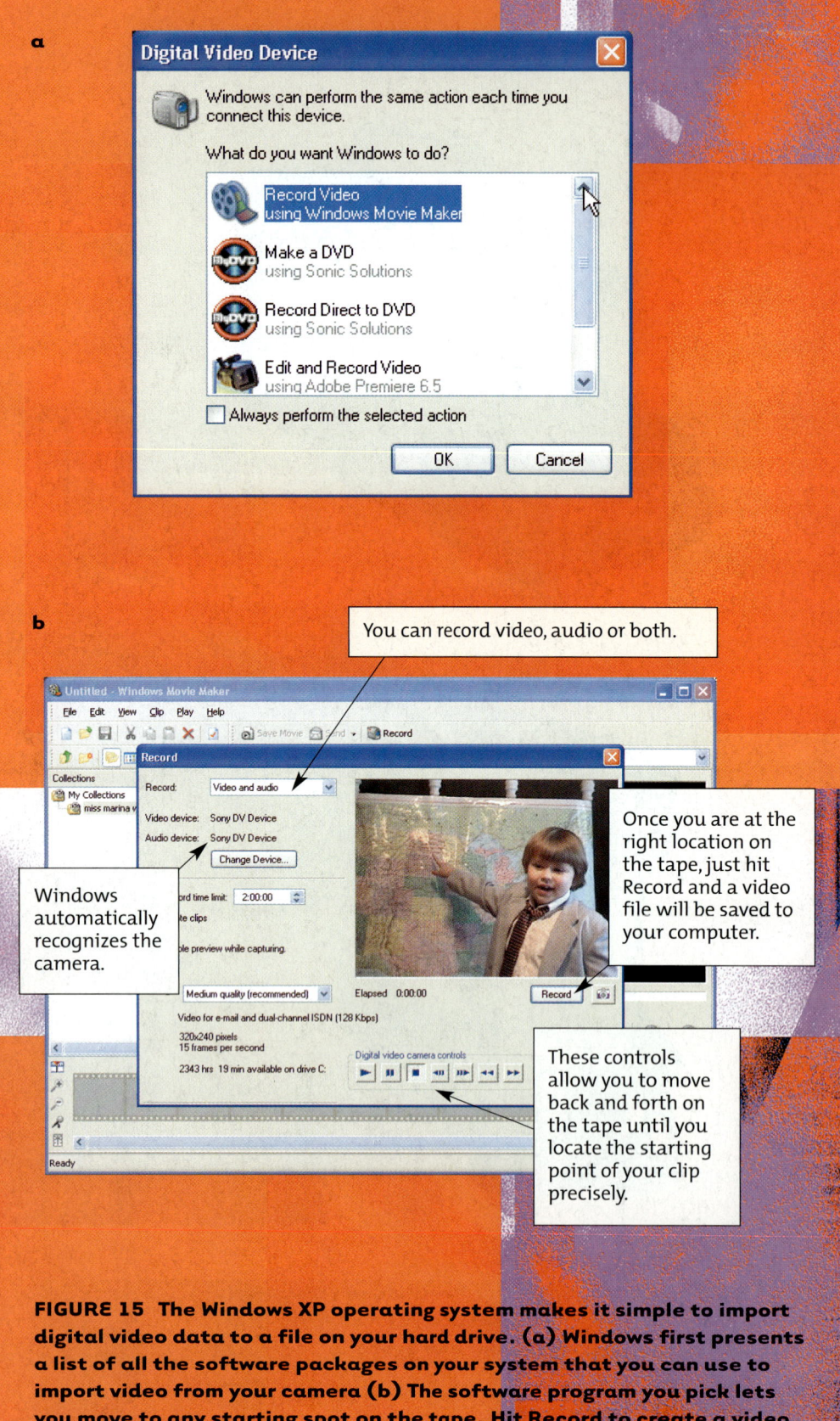

FIGURE 15 The Windows XP operating system makes it simple to import digital video data to a file on your hard drive. (a) Windows first presents a list of all the software packages on your system that you can use to import video from your camera (b) The software program you pick lets you move to any starting spot on the tape. Hit Record to create a video file on the hard drive.

you can quickly transfer to your computer without having to hook up any cables, make sure you have a large enough flash memory card with you. Check the user guide for your particular camera to see how much flash memory you need to store video. For example, on the Sony DCR-TRV80, you can store about 10 minutes of video on a 64-MB Memory Stick.

Shooting video with a digital video camera is similar to shooting video with an analog camera. Automated programs control the exposure settings for different environments (nighttime shots, action events, and so on), while automatic focusing and telephoto zoom lens features are common as well. Many cameras include an "antishake" feature that stabilizes the image when you're using the camera without a tripod. Using these features you can capture great footage by just pointing and hitting Record.

Transferring Your Video to Your Computer

Digital video cameras already hold your video as digital data, so transferring the data to your computer is simple. All you need is a FireWire port on your computer and a FireWire cable. While every FireWire port is the same, there are two different types of connectors used on FireWire cables: 4-pin and 6-pin. Digital video cameras usually have a port that matches the 4-pin connector, while desktop computers may have a port matching either the 4- or 6-pin connectors.
Be sure to check both your camera and computer
and buy a cable with the matching connector on each end.

Once you connect your camera to your computer, Windows XP automatically identifies it, recognizing its manufacturer and model. Windows then scans the software on your system and presents a list of all the programs you can use to import your video. Windows Movie Maker is one such program and is included with the Windows XP Home Edition. Other digital video–editing programs like Adobe Premiere and Pinnacle Studio DV can import video as well. These are more powerful, full-featured programs that you purchase separately. Figure 15a shows the lists that pops up for a computer with Windows Movie Maker, Sonic Solutions, and Adobe Premiere installed.

The software allows you to fast forward, pause, and rewind, moving to the segment you wish to transfer (or record) to your hard drive. In Figure 15b, we've chosen to record (transfer) both video and audio. You can use the digital video camera control arrows on the bottom right-hand side of the Record dialog box to locate the exact piece of footage you want to transfer. Click the Record button and the video file transfers to the hard drive.

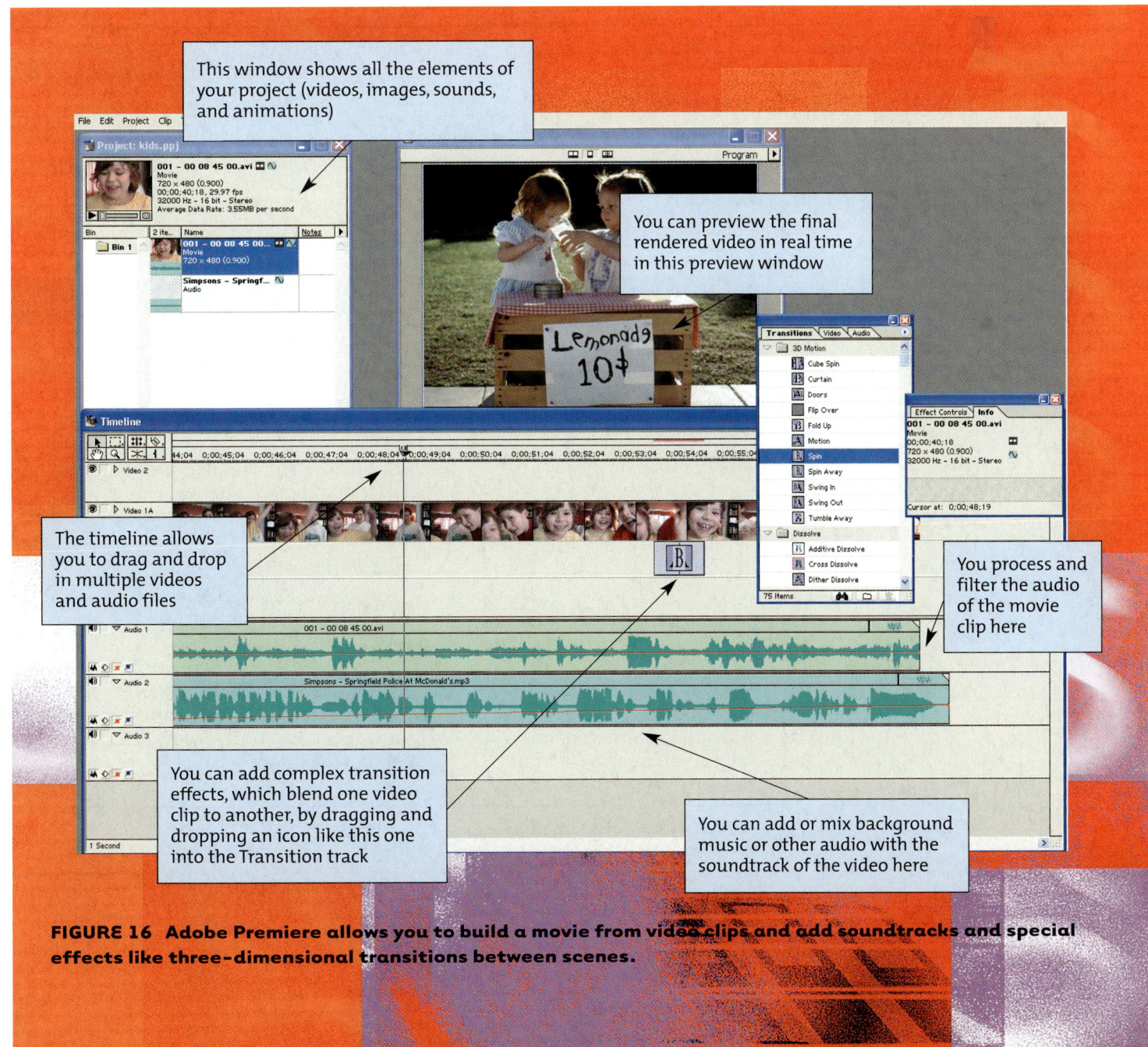

FIGURE 16 Adobe Premiere allows you to build a movie from video clips and add soundtracks and special effects like three-dimensional transitions between scenes.

FIGURE 17 TYPICAL FILE FORMATS FOR DIGITAL VIDEO

FORMAT	FILE EXTENSION	NOTES
QuickTime	.mov .qt	You can download QuickTime player for free from www.apple.com/quicktime. The pro version allows you to build your own QuickTime files.
MPEG (Moving Picture Experts Group)	.mpg .mpeg	MPEG-4 video standard adopted internationally in 2000; recognized by most video player software.
Windows Media Video	.wmv	Microsoft file format recognized by Windows Media Player (included with Windows operating system)
Microsoft Video for Windows	.avi	Microsoft file format recognized by Windows Media Player (included with Windows operating system)
RealMedia	.rm	Format from RealNetworks is popular for streaming video. You can download the player for free at www.real.com.

FIGURE 18 DIGITAL VIDEO ADVANTAGES

	ANALOG VIDEO CAMERA	DIGITAL ADVANTAGE
Editing	Home users cannot edit their videos without expensive equipment	Editing video is easy using inexpensive software
Storage	Film is stored on analog tapes (VHS, 8 mm, or Hi8 tapes)	Video is stored on smaller MiniDV tapes. Short video clips can be saved directly to flash memory cards. You can easily transfer and store video files onto a computer hard drive, CD, or DVD.
	Tapes have a limited lifetime and are sensitive to heat, magnetic fields, water damage, and mechanical breakdown	There is no loss of information over time
Distribution	You can make copies of tapes but only if you own multiple VHS VCRs or Hi8 players. To make digital versions of analog tapes, you need a separate video capture device.	You can make DVD copies using a simple DVD-RW drive. You can also post videos to Web sites or attach them to e-mail messages.
Quality	Excellent video and audio quality	Excellent video quality and CD-quality audio

DIGITAL VIDEO

Editing Your Video and Adding Special Effects

Once the digital video data is in a file on your hard drive, the fun really begins. Video-editing software presents a storyboard or time line with which you can manipulate your video file, as shown in Figure 16. Using this software, you can review your clips frame by frame or trim them at any point. You can order each segment on the time line in whatever sequence you like and correct segments for color balance, brightness, or contrast.

In addition, you can add transitions to your video like those you're used to seeing on TV—fades to black, dissolves, and so on. Figure 16 shows how easy it is to add transitions in Adobe Premiere: just select the type of transition you want from the drop-down list and drag that icon into the time line where you want the transition to occur.

Video-editing software also lets you add titles, animations, and audio tracks to your video, including background music, sound effects, and additional narration. In Figure 16 there are two audio tracks, the original voices on the video as well as an additional audio clip. You can adjust the volume of each audio track to switch from one to the other or have both playing together. Finally, you can preview all of these effects in real time.

Outputting (Exporting) Your Video

Once you're done editing your video file, you can save (or export) it in a variety of formats. Figure 17 (on page 164) shows some of the popular video file formats in use today, along with the file extensions they use. (*File extensions* are the letters that follow the period in a file name, such as Movie1.mpg. These extensions indicate the type of data inside the file.)

Your choice of file format for your finished video will depend on what you want to do with your video. For example, the RealMedia streaming file format is a great choice if your file is very large and you'll be posting it on the Web. The Microsoft AVI format is a good choice if you're sending your file to a wide range of users since it's very popular and commonly accepted as the standard video format on Windows machines.

When you export your video, you have control over every aspect of the file you create, including its format, window size, frame rate, audio quality, and compression level. You can customize any of these if you have specific production goals, but most often just using the default values works well.

When would you want to customize some of the audio and video settings? If you're trying to make the file as small as possible so it will download quickly or so it can fit on a single CD, you would select values that trade off audio and video quality for file size. For example, you could drop the frame rate to 15 fps, shrink the window size to 320 x 240 pixels, and switch to mono audio instead of stereo.

You can also try different compression choices to see which one does a better job of compressing your particular file. **Codecs** (**com**pression/ **dec**ompression) are rules implemented in either software or hardware that squeeze the same audio/video information into less space. Some information will be lost using compression, and there is a variety of different codecs to choose from, each claiming better performance than its competitors. Commonly used codecs include MPEG, Indeo, and Cinepak. There is no one codec that is always superior—a codec that works well for a simple interview may not do a good job compressing a live-action scene.

If you'd like to save your video onto a DVD, you can use special DVD authoring software such as Ulead's DVD Workshop or Adobe's Encore DVD. These programs create final DVDs that have animated menu systems and easy navigation controls, allowing the viewer to quickly move from one movie or scene to another. Home DVD players as well as gaming systems like Playstation 2 and Xbox can read these DVDs, so your potential audience is even greater!

The Digital Advantage

Is a digital video camera right for you? Figure 18 lists just a few of the advantages of digital video cameras over traditional analog camcorders. Analog video changed how we communicate. It became possible to make a video of a baby's first steps and send it to relatives all across the country. Digital video allows you even more creative control over the videos you produce, enabling you to edit them at home and easily share them with others.

OBJECTIVES

After reading this chapter, you should be able to answer the following questions:

- What software is included in system software? (p. 168)

- What are the different kinds of operating systems? (pp. 168–170)

- What are the most common desktop operating systems? (pp. 170–173)

- How does the operating system provide a means for users to interact with the computer? (pp. 175–179)

- How does the operating system help manage the processor? (pp. 176–177)

- How does the operating system manage memory and storage? (pp. 178–179)

- How does the operating system manage hardware and peripheral devices? (p. 179)

- How does the operating system interact with application software? (p. 179)

- How does the operating system help the computer start up? (pp. 180–184)

- What are the main desktop and windows features? (pp. 185–186)

- How does the operating system help me keep my computer organized? (pp. 186–192)

- What utility programs are included in system software and what do they do? (pp. 192–200)

CHAPTER 5

Using System Software:
The Operating System, Utility Programs, and File Management

**TECHNOLOGY IN ACTION:
WORKING WITH SYSTEM SOFTWARE**

Franklin begins his workday as he does every morning, powering on his computer and watching it boot up. Once he sees the welcoming image of his desktop, he opens Microsoft Outlook to check his e-mail, Internet Explorer to access his company's Web site, and Microsoft Word to bring up the proposal he needs to finish. As he reads his e-mail, a warning pops up alerting him that one message may contain a file with a virus. He deletes the file without opening it, glad that his antivirus software had been automatically updated the night before.

Clicking back to Word, Franklin searches for a proposal he worked on last year. Fortunately, he knows where to look as he has been creating folders for his projects and diligently saving his files in their proper folder. He learned the hard way that keeping his files organized in folders is worth the effort it takes to create them. Last year, his desktop was a complete mess. He was constantly losing time trying to find files because he couldn't remember where he saved them and he gave them names he easily forgot. His organized folders now make finding his files a snap.

Before calling it a day, Franklin has one more thing to do. Recently, his computer has been running sluggishly, so he is hoping to improve its performance. Last night, he ran Disk Cleanup, a utility program that removes unneeded files from the hard drive, as well as ScanDisk, a utility program that checks for disk errors. Although he had seen an improvement in his computer's performance, he decides to use the defrag utility to defrag his hard drive, hoping it will give him more space. As he's leaving work, Franklin hears the clicking of the hard drive as the defrag utility goes to work.

Are you as familiar with your system as Franklin is? In this chapter, you'll learn all about system software and how vital it is to your computer. We'll start by examining the operating system (OS), looking at the different operating systems on the market as well as the tasks the OS manages. We'll then look at how you can use the OS to keep your files and folders on your desktop organized so that you can use your computer more efficiently. Finally, we'll look at the many utility programs included as system software on your computer. Using these utility programs, you'll be better able to take care of your system and extend its life.

System Software Basics

As you learned in the last chapter, there are two basic types of software in your computer: application software and system software. **Application software** is the software you use to do everyday tasks at home and at work. It includes programs such as Microsoft Word and Excel. **System software** is the set of software programs that helps run the computer and coordinates instructions between application software and the computer's hardware devices. From the moment you turn on your computer to the time you shut it down, you are interacting with system software.

System software consists of two primary types of programs: the operating system and utility programs. The **operating system (OS)** is the main program that controls how your computer system functions. It manages the computer's hardware, including the processor (CPU), memory, and storage devices, as well as peripheral devices such as the monitor and printer. The operating system also provides a consistent means for software applications to work with the CPU. Additionally, it is responsible for the management, scheduling, and interaction of tasks as well as system maintenance. Finally, the OS provides a *user interface* through which users can interact directly with the computer.

In addition to the operating system, system software includes **utility programs.** These are small programs that perform many of the general housekeeping tasks for the computer, such as system maintenance and file compression.

Do all computers have operating systems? Every computer, from the smallest notebook to the largest supercomputer, has an operating system. Even tiny personal digital assistants (PDAs) as well as some appliances have operating systems. The role of the OS is critical; the computer cannot operate without it.

Operating System Categories

Although most computer users can only name a few operating systems, hundreds exist. As Figure 5.1 illustrates, these operating systems can be classified into four categories, depending on the number of users they service and the tasks they perform. Some operating systems coordinate resources for many users on a network (multi-user operating system), while other operating systems, such as those found in some household appliances and car engines, don't require the intervention of any users at all (real-time operating system). Some operating systems are available

FIGURE 5.1 Operating System Categories

CATEGORY OF OPERATING SYSTEM	EXAMPLES OF OPERATING SYSTEM SOFTWARE	EXAMPLES OF DEVICES USING THE OPERATING SYSTEM
Real-Time Operating System	There are no commercially available RTOS programs. Noncommercially available programs include QNX 4 and Lynx	Scientific instruments Automation and control machinery Video games
Single-User, Single-Task Operating System	Palm OS Pocket PC (Windows CE) Windows Mobile 2003 MS-DOS Symbian OS	PDAs Embedded computers in cell phones, cameras, appliances, and toys
Single-User, Multitask Operating System	Windows family (XP, 2000, ME, 98, 95, NT) Mac OS Linux	Personal desktop computers/laptops
Multiuser Operating System	UNIX Novell NetWare Windows 2003 Server OS/2	Networks Mainframes Supercomputers

commercially, for personal and business use (single-user, multitask operating system), while others are proprietary systems developed specifically for the devices they manage (single-user, single-task operating system).

REAL-TIME OPERATING SYSTEMS

Do machines with built-in computers need an operating system? Machinery that is required to perform a repetitive series of specific tasks in an exact amount of time requires a **real-time operating system (RTOS)**. This type of operating system is a program with a specific purpose and must guarantee certain response times for particular computing tasks, or the machine's application is useless. For example, instruments such as those found in the scientific, defense, and aerospace industries that need to perform regimented tasks or record precise results require real-time operating systems.

Real-time operating systems are also found in many types of robotic equipment. Television stations use robotic cameras with real-time operating systems that glide within a suspended cable system to record sports events from many angles. You also encounter real-time operating systems in devices you use in your everyday life, such as fuel-injection systems in car engines, video game consoles, and in many home appliances (see Figure 5.2).

Real-time operating systems require minimal user interaction. The programs are written specifically to the needs of the devices and their functions. Therefore, there are no commercially available "standard" RTOS software programs.

SINGLE-USER OPERATING SYSTEMS

What type of operating system controls my desktop computer? Because your computer can only handle one person working on it at a time, but it can perform a variety of tasks simultaneously, it uses a **single-user, multitask operating system**. The Windows family of operating systems and the Macintosh operating system (Mac OS) are examples of this type of operating system. Usually, when you buy a desktop or laptop computer, its operating system software will already be installed on the computer's hard disk. Sometimes you may need to install the OS yourself if you change or upgrade to a different version, or reinstall it in the case of a system problem.

Does the same kind of operating system also control my PDA? All computers on which one user is performing just one task at a time require a **single-user, single-task operating system**. PDAs currently can perform only one task at a time by a single user, so they require single-user, single-task operating system software such as Pocket PC or Palm OS.

Microsoft's Pocket PC is an application that includes both operating system software (Windows CE) and application components bundled specifically for PDAs (see Figure 5.3). Besides the address book, date book, memo pad, and to-do list that are standard with Windows CE, the bundled Pocket PC software also includes versions of Word, Excel, Outlook, and Internet Explorer that are designed specifically for PDAs. The latest version of Pocket PC, Windows Mobile 2003, comes with advanced wireless Internet connection capabilities.

FIGURE 5.2

Devices from the space shuttle to cars to stethoscopes to cell phones use real-time operating systems.

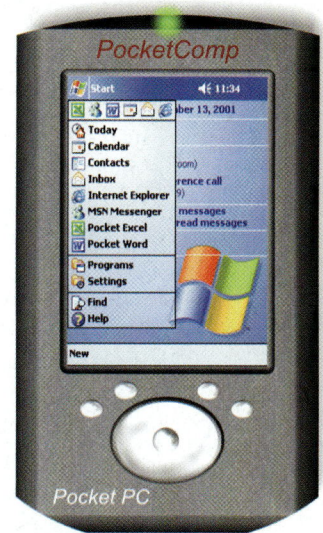

FIGURE 5.3

Although PDAs use a single-user, single-task operating system in which only one user can perform one task at a time, the operating system has a similar look to that of a traditional desktop operating system.

Palm OS, on the other hand, is found in a number of devices but is strictly an operating system. If your PDA uses Palm OS as its operating system, you must purchase and install any additional software you want to run on your PDA.

Cell phones also use a single-user, single-task operating system that not only manages the functions of the phone itself, but also provides other functionality, such as built-in phone directories, games, and calculators. Symbian is the leading OS software for mobile phones.

Are there any other single-user, single-task operating systems? Microsoft's **Disk Operating System (MS-DOS)** is another example of a single-user, single-task operating system. DOS was the first widely installed operating system in personal computers. Compared to the operating systems we are familiar with today, DOS was a highly user "unfriendly" OS. To use it, you needed to type in specific commands. For example, to copy a file named "letter" from the hard drive to a floppy disk you would type in the following command after the C prompt:

```
C:\>copy letter.txt A:
```

Although DOS is infrequently used today as a primary operating system, IT professionals still use it to edit and repair system files and programs.

MULTIUSER OPERATING SYSTEMS

What kind of operating system do networks use? A **multiuser operating system** (also known as a **network operating system**) enables more than one user to access the computer system at one time by efficiently juggling all the requests from multiple users. Networks require a multiuser operating system because many users access the server computer at the same time and share resources such as printers. A network operating system is installed on the server computer and manages all user requests, ensuring they do not interfere with each other. For example, in a network where users share a printer, the printer can only produce one document at a time. The OS is therefore responsible for managing all the printer requests and making sure they are processed one at a time. Examples of network operating systems include UNIX, Novell NetWare, and Windows 2003 Server.

What other kinds of computers require a multiuser operating system? Large corporations with hundreds or thousands of employees often use powerful computers known as mainframes. These computers are responsible for storing, managing, and simultaneously processing data from all users. Mainframe operating systems fall into the multiuser category. Examples include IBM's OS/2 and z/OS.

Supercomputers also use multiuser operating systems. Scientists and engineers use supercomputers to solve complex problems or to perform massive computations. Some supercomputers are single computers with multiple processors, while others consist of multiple computers that work together.

Desktop Operating Systems

As mentioned above, desktop computers (and laptops) use single-user, multitask operating systems, of which there are several on the market, including Windows and Mac OS. The type of processor in the computer determines which operating system a particular desktop computer uses. The combination of operating system and processor is referred to as a computer's **platform**.

For example, Microsoft's Windows operating systems are designed to coordinate with a series of processors from Intel Corporation that share the same or similar sets of instructions, whereas Apple's Macintosh operating systems work primarily with processors from Motorola Corporation designed specifically for Apple computers. The two operating systems, as well as application programs designed for those operating systems, are not interchangeable. If you attempt to load a Windows OS on a Mac, for example, the Mac processor will *not* understand the operating system and will not function properly.

MICROSOFT WINDOWS

What is the most popular operating system for desktop computers? **Microsoft Windows** is the market leader in operating system sales, maintaining an approximate 90 percent market share. Although Windows XP is the most recent version on the market, many computers still run earlier versions, such as Windows 95, Windows 98, Windows Millennium Edition (ME), and Windows 2000. Windows XP comes in a number of versions to suit different users, including Windows XP Home Edition, Windows XP Professional, and Windows XP Tablet PC.

DESKTOP OPERATING SYSTEMS

What is the difference between the various Windows operating systems? Figure 5.4 presents a time line of the evolution of Microsoft Windows. As you can see, with each new version, Microsoft made improvements. What was once only a single-user, single-task operating system is now also a powerful multiuser operating system. Over time, Windows

FIGURE 5.4 Windows Timeline

1985	WINDOWS 1.0	Introduces point-and-click commands with a mouse and includes modest multitasking capabilities and desktop applications.
1987	WINDOWS 2.0	Includes better graphics capabilities and introduces keyboard shortcuts and the ability to overlap windows.
1990	WINDOWS 3.0	Added programs to manage applications, files and print jobs as well as improved icons.
1992	WINDOWS 3.1	First widely used PC graphical user interface (GUI) operating system. Improved point-and-click mouse operations and multitasking capabilities.
1993	WINDOWS NT 3.1	Fundamentally different operating system with increased security, power, performance, and multitasking scheduler.
1995	WINDOWS 95	Provides major enhancements over Windows 3.1. This operating system runs faster and more efficiently, introduces Plug and Play capabilities, long file names, short-cut right-click menus, and a cleaner desktop. Sells more than 1 million copies within 4 days.
1996	WINDOWS NT 4.0	Includes a similar feel to that of Windows 95 but with enhanced network support and security features.
1997	WINDOWS CE	Released to compete with the Palm OS for Personal Digital Assistants (PDAs). It has the same look and features of Windows 95.
1998	WINDOWS 98	This upgrade to Windows 95 includes additional file protection features and incorporates Internet Explorer 4.0, a customizable taskbar, and desktop features that let you customize backgrounds as well as live web content such as a stock ticker or weather map.
2000	WINDOWS 2000 PROFESSIONAL	This upgrade to Windows NT offers improvements to file security and Internet support.
2000	WINDOWS MILLENNIUM EDITION (ME)	This upgrade to Windows 95 and Windows 98 includes system backup and multimedia capabilities (such as Media Player).
2001	WINDOWS XP HOME AND PROFESSIONAL	Offers a new multi-user desktop as well as improved digital media features and Internet capabilities.
2001	WINDOWS XP TABLET PC	Designed specifically for new Tablet PC notebooks, this operating system incorporates a digital pen that enables users to write directly on the Tablet screen and perform mouse functions. It also includes built-in wireless technologies.

improvements have concentrated on increasing user functionality and "friendliness," improving Internet capabilities, and enhancing file privacy and security.

MAC OS

How is Mac OS different from Windows? Although Apple's **Mac OS** and the Windows operating systems are not compatible, they are very similar in terms of functionality. In 1984, Mac OS became the first operating system to incorporate the user-friendly "point-and-click" technology in a commercially affordable computer. Both operating systems now have similar "windows" work areas on the desktop that house individual applications and support users working in more than one application at a time (see Figure 5.5).

Despite their similarities, there are many subtle and not-so-subtle differences that have created loyal fans of both products. Macs have long been recognized for their superior graphics display and processing capabilities. Users also attest to Mac's greater system reliability and better document recovery. Despite these advantages, there are fewer software applications that are available for the Mac platform (although this is beginning to change) and Mac systems tend to be a bit more expensive than Windows-based PCs.

The most recent version of the Mac operating system, Mac OS X, is based on the UNIX operating system. Previous Mac operating systems had been based on their own proprietary program. Mac OS X includes a new user interface and larger icons among other features.

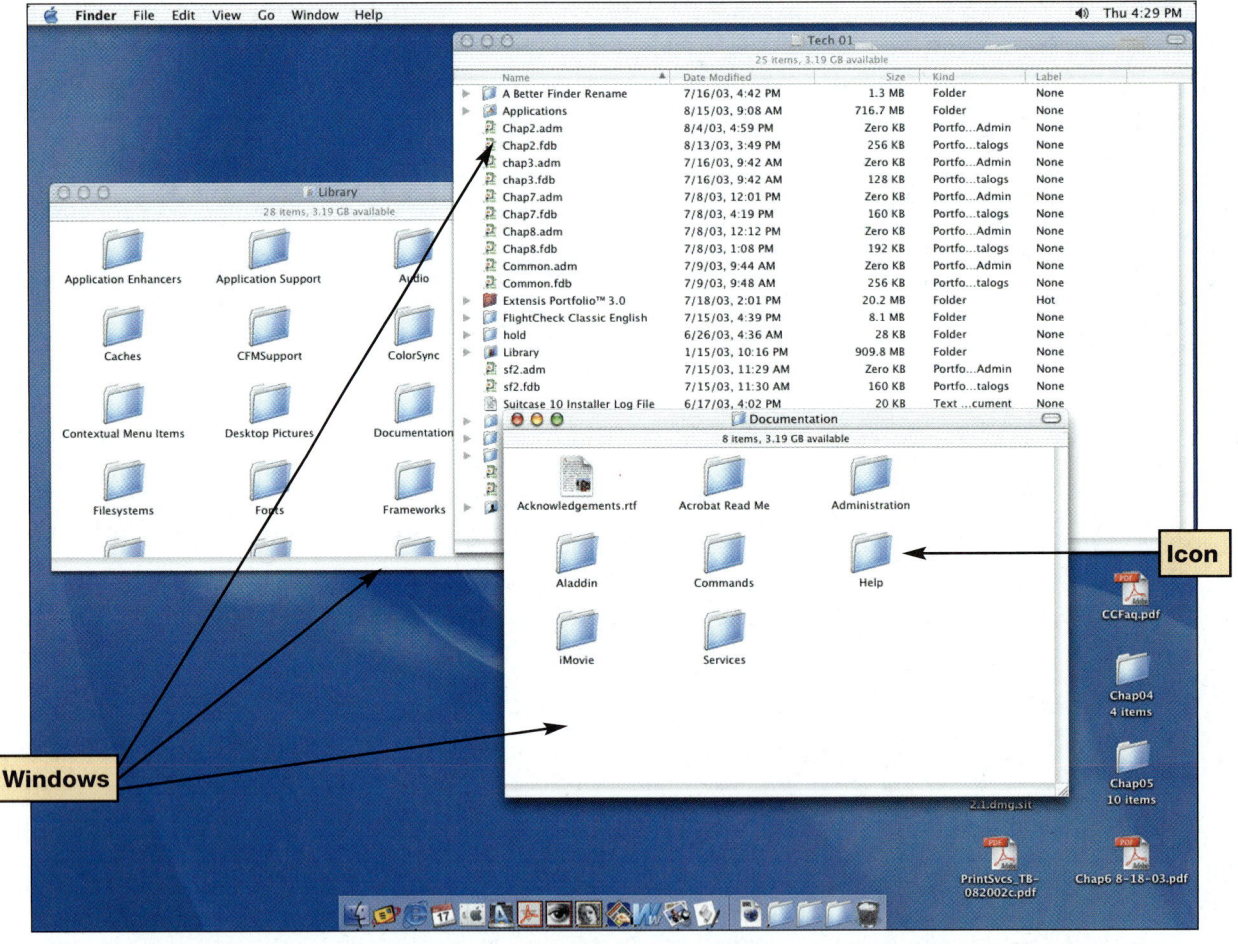

FIGURE 5.5

The most recent version of the Mac operating system, Mac OS X, is based on the UNIX operating system. Mac OS X also incorporates a new user interface called Aqua, which boasts larger icons and brighter colors.

BITS AND BYTES

Why a Penguin?

If you've seen Linux products, you know its logo is a penguin. Why would Linus Torvalds choose a penguin? It's not exactly the image most people think of when considering operating systems. However, Torvalds liked the image of a penguin because he felt that it represented the idea that "the world is a good place to be." With regard to the penguin portraying the true appeal of Linux, Torvalds writes on the Linux Web site: "Some people have told me they don't think a fat penguin really embodies the grace of Linux, which just tells me they have never seen an angry penguin charging at them in excess of 100 miles per hour. They'd be a lot more careful about what they say if they had."

Chances are you'll be seeing more of this penguin around as Linux continues to make strides as a legitimate competitor to market leader Windows XP

UNIX

What is UNIX? UNIX is an operating system originally conceived in 1969 by Ken Thompson and Dennis Ritchie of AT&T's Bell Labs. In 1974, the UNIX code was rewritten in the standard programming language C. Initially, the code was not proprietary—in other words, no company like Microsoft or Apple owned it. Rather, any programmer was allowed to use the code and modify it to meet his or her needs. By the late 1970s, a multitude of vendors were marketing the UNIX code in a variety of forms, none of which were compatible with any other. In the mid-1980s, a group of UNIX vendors formed the X/Open company, based on an "open systems" concept, to standardize UNIX specifications among the various vendors.

This triggered AT&T, which still "owned" the original version of UNIX but had ignored it commercially, to join with Sun Microsystems to further develop its strain of UNIX. Eventually, AT&T sold its UNIX system to Novell, which later sold it to SCO. SCO continues to develop UNIX today. Other slightly different versions of UNIX also exist. Although you'll most often find UNIX in use as a network operating system, some versions can be used on desktop systems.

LINUX

What is Linux? Linux is an open-source operating system based on UNIX and designed primarily for use on personal computers (although some versions can also be used on networks). An **open-source program** is one that is available for developers to use or modify as they wish. Linux began in 1991 as a part-time project by a Finnish university student named Linus Torvalds, who wanted to create a free operating system to run on his home computer. He posted his operating system program code to the Web for others to use and modify. It has since been tweaked by scores of programmers as part of the Free Software Foundation GNU (or "GNU's not UNIX") project.

Today, Linux is gaining a reputation as a stable operating system that is not subject to crashes and failures. Because the code is open and available to anyone, Linux is quickly tweaked to meet virtually any new operating system need. For example, when Palm PDAs emerged, the Linux OS was promptly modified to run on this new device. Similarly, only a few weeks were necessary to get the Linux OS ready for the new Intel Xeon processor, a feat unheard of in proprietary operating system development. Linux is also gaining popularity among computer manufacturers, which have begun to ship it with some of their latest PCs.

Linux's user interface, called GNOME (pronounced "gah-NOHM"), actually allows you to select which desktop appearance (Windows or Mac) you'd like your system to display. This means that if you're using Linux for the first time, you don't have to learn a new interface: you just use the one you're most comfortable with already. GNOME also includes a word processing program, spreadsheet program, database manager, presentation developer, a Web browser, and an e-mail program.

TechTV For information on running Linux see Review: Lindows 4.0—a TechTV clip found at www.prenhall.com/techinaction

TRENDS IN IT

EMERGING TECHNOLOGIES:
Open-Source Software: Why Isn't Everyone Using It?

Proprietary software, such as Microsoft Windows, is developed by corporations and sold for profit. This means that the *source code*, the actual lines of instructional code that make the program work, is not accessible. Without being able to access the source code, it's difficult to modify the software or see exactly how the program author constructed various parts of the system.

Restricting access to the source code protects companies from having their programming ideas stolen, and prevents customers from using modified versions of the software. This benefits the companies that create the software since their software code can't be pirated (or stolen). However, in the late 1980s, computer specialists became concerned over the fact that large software companies (such as Microsoft) were controlling a large portion of market share and driving out competitors. They also felt that proprietary software was too expensive and contained too many bugs (errors).

These people felt that software should be developed without a profit motive and distributed with its source code free for all to see. The theory was that if many computer specialists examine, improve, and change the source code, a more full-featured, bug-free product would result. Hence the open-source movement was born.

Open-source software is freely distributed (no royalties accrue to the creators), contains the source code, and can in turn be redistributed freely to others. Most open-source products are created by teams of programmers and modified (updated) by hundreds of other programmers around the world. You can download the products for free off the Internet. Linux is probably the most widely recognized name in open-source software, but other products such as MySQL (a database program) and OpenOffice (a suite of productivity applications) are also gaining in popularity.

So if an operating system such as Linux is free, why does Windows (which you must pay for) have such a huge market share? Corporations and individuals have grown accustomed to one thing that proprietary software makers can provide: technical support. It is almost impossible to provide technical support for open-source software since it can be freely modified, and there is no one specific developer to take responsibility for technical support. Therefore, corporations have been reluctant to install open-source software extensively due to the cost of the internal staff of programmers that must support it.

Companies like Red Hat have been combating this problem. Red Hat has been packaging and selling versions of Linux since 1994 (see Figure 5.7). The company provides a warranty and technical support for its version of Linux (which Red Hat programmers modified from the original source code). Packaging open source software in this manner has made using it much more attractive to businesses. Today, many Web servers are hosted on computers running Linux.

So when will free versions of Linux (or another open-source operating system) be the dominant OS on home computers? The answer is maybe never. Most casual computer users won't feel comfortable without technical support, therefore any open-source products for home use would probably need to be marketed the way Red Hat markets Linux. Also, many open-source products are not easy to install and maintain.

However, companies such as Lindows (**www.lindows.com**) are making operating systems with easy-to-use visual interfaces that work with the Linux operating system. If one of these companies can develop an easy-to-use product and has the marketing clout to challenge Microsoft, you may see more open-source software deployed in the home computer market in the future.

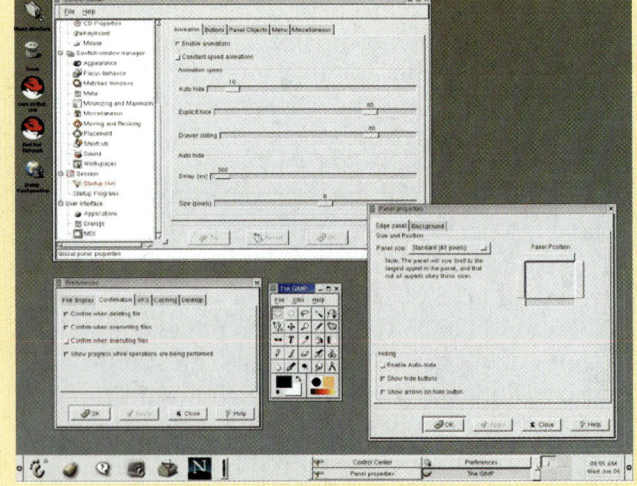

FIGURE 5.7

Red Hat packages and sells a version of Linux that includes technical support.

What the Operating System Does

As shown in Figure 5.8, the operating system is like a traffic cop that coordinates the flow of data and information through the computer system. In doing so, the OS performs several specific functions:

- It provides a way for the user to interact with the computer.
- It manages the processor, or central processing unit (CPU).
- It manages the memory and storage.
- It manages the computer system's hardware and peripheral devices.
- It provides a consistent means for software applications to work with the CPU.

In this section, we'll look at each of these functions in detail.

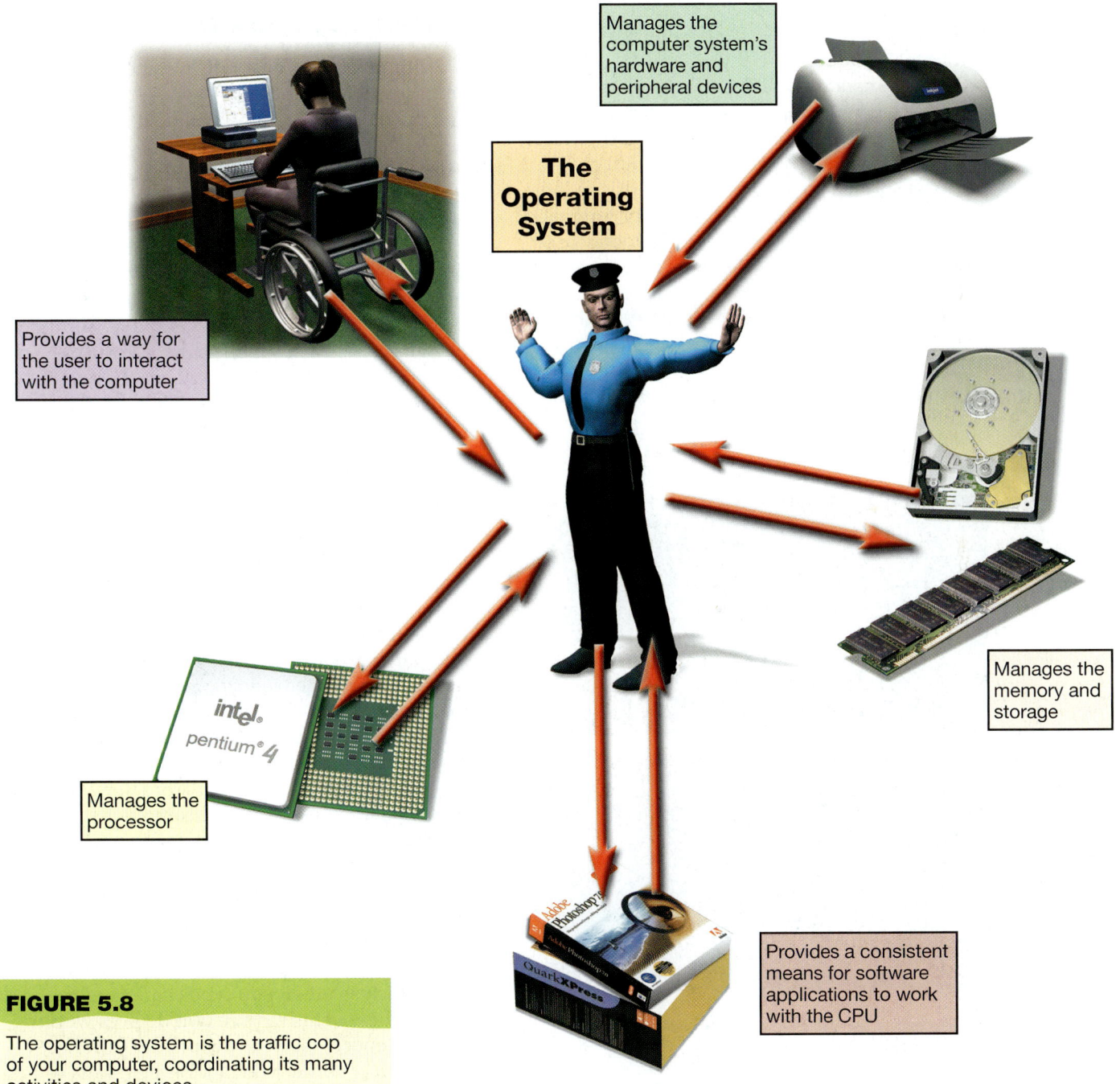

FIGURE 5.8

The operating system is the traffic cop of your computer, coordinating its many activities and devices.

THE USER INTERFACE

How does the operating system control how I interact with software? The operating system provides a **user interface** that enables you to interact with the computer. As noted earlier, the first personal computers had a DOS operating system with a command-driven interface, as shown in Figure 5.9a. A **command-driven interface** is one in which you enter commands in order to communicate with the computer system. The commands themselves were not always easy to understand and therefore the interface proved to be too complicated for the average user. Therefore, PCs were used primarily in business and by professional computer operators.

The command-driven interface was later improved by incorporating a menu-driven interface, as shown in Figure 5.9b. A **menu-driven interface** is one in which you choose a command from menus displayed on the screen. Menu-driven interfaces eliminated the need to know every command since you could select most of the commonly used commands from a menu. However, they were still not easy enough for most people to use.

What kind of interface do programs use today? Most operating systems today, such as the Mac OS and Microsoft Windows, use a **graphical user interface,** or **GUI** (pronounced "gooey"). Unlike the command- and menu-driven interfaces used earlier, GUIs display graphics and use the point-and-click technology of the mouse and cursor, making them much more user friendly. As illustrated in Figure 5.10, a GUI uses **windows** (rectangular boxes that contain programs displayed on the screen), **menus** (lists of commands that appear on the screen), and **icons** (pictures that represent an object such as a software application or a file or folder). Since users no longer have to enter commands to interact with the computer, GUIs are a big reason why desktop computers are now such popular tools.

PROCESSOR MANAGEMENT

Why does the operating system need to manage the processor? When you use your computer, you are usually asking it to perform several tasks at once. For example, you might be printing a Word document, waiting for a file to download from the Internet, listening to a CD from your CD drive, and working on a PowerPoint presentation, all at the same time—or at least what *appears* to be at the same time. Although the processor is the powerful brains of the computer, processing all of its instructions and performing all of its calculations, it can only perform one action at a time. Therefore, it needs the operating system to arrange for the execution of all these activities in a systematic way to give the appearance that everything is happening simultaneously.

To do so, the operating system assigns a slice of its time to each activity requiring the processor's attention. The OS must then switch between different processes thousands of times a second to make it appear that everything is happening in a seamlessly fluid manner. Otherwise, you wouldn't be able to listen to a CD and print at the same time without experiencing delays in the process. When the operating system allows you to perform more than one task at a time, it is said to be **multitasking**.

How exactly does the operating system coordinate all the activities? When you type and print a document in Word, for example, many different devices in the computer system are involved, including your keyboard, mouse, and printer. Every keystroke, every mouse click, and each signal to the printer creates an action, or **event,** in the respective device (keyboard, mouse, or printer) to which the operating system responds.

While sometimes these events occur sequentially (such as when you type characters one at a time), other events require two devices working simultaneously (such as the printer printing while you continue to type). While it *looks* as though the keyboard and printer are working at the same time, as mentioned above, the processor

(a)
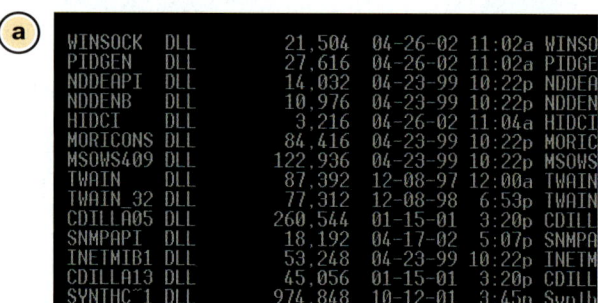

(b)
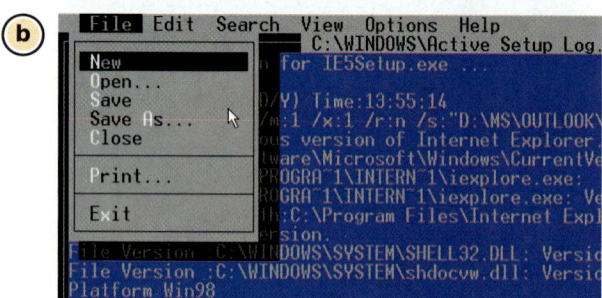

FIGURE 5.9

(a) Command-driven and (b) menu-driven interfaces were not user friendly.

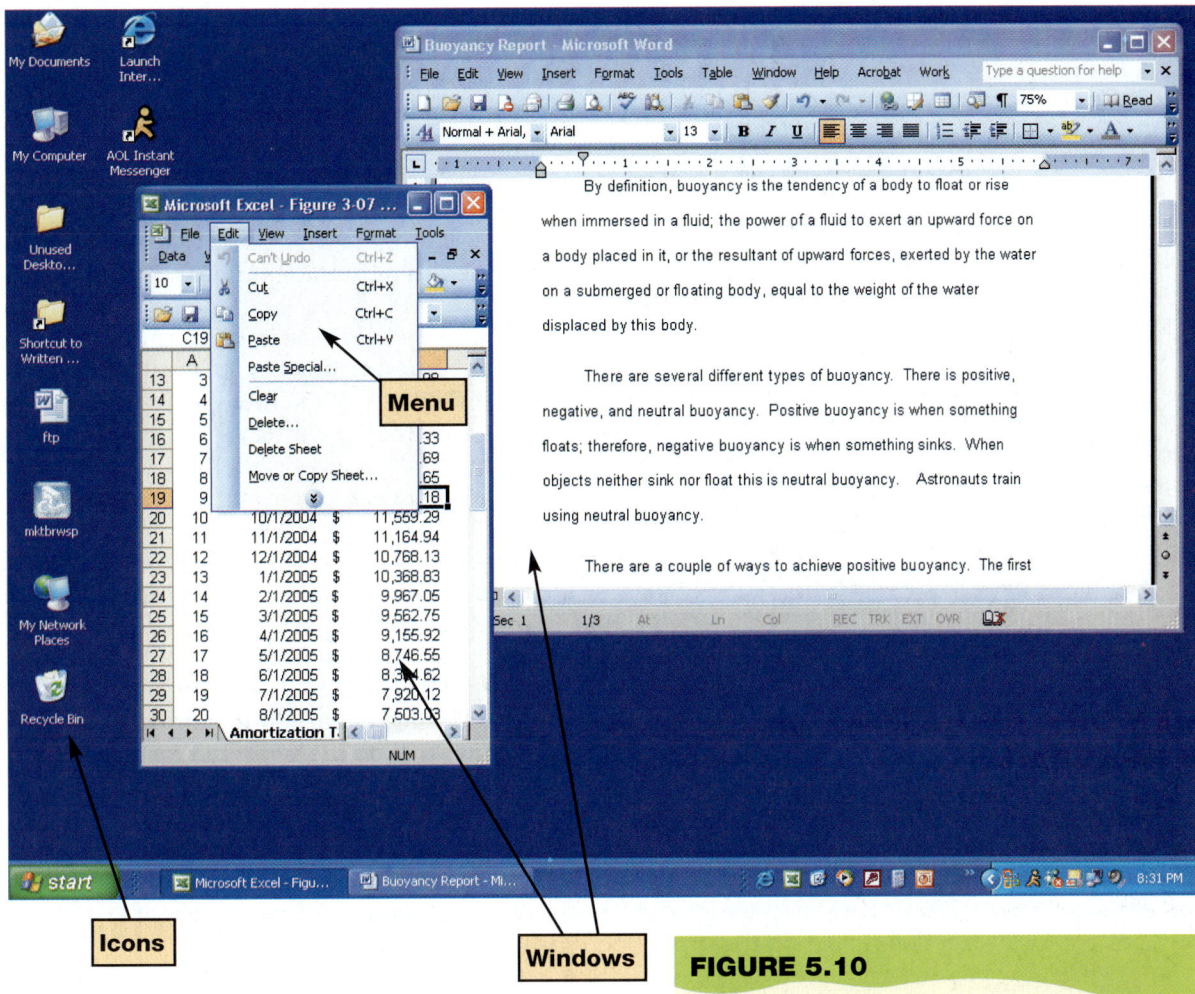

FIGURE 5.10

Today's operating systems include a graphical user interface (GUI). These user interfaces are much more user friendly than the DOS command-driven interfaces used previously.

actually can only handle one event at a time and switches back and forth between processes. To do this, the operating system controls the timing of events the processor works on.

For example, assume you are typing and you want to print another document. When you tell your computer to print your document, the printer generates a unique signal called an **interrupt** that tells the operating system that it is in need of immediate attention. Every device has its own type of interrupt, which is associated with an **interrupt handler**, a special numerical code that prioritizes the requests. These requests are placed in the **interrupt table** in the computer's primary memory (or random access memory, RAM).

In our example, the operating system pauses the CPU from its typing activity when it receives the interrupt from the printer, and puts a "memo" in a special location in RAM called a *stack*. The "memo" is a reminder of where the CPU was before it left off so that it can work on the printer request. The CPU then retrieves the printer request from the interrupt table and begins to process it. Upon completion of the printer request, the CPU goes back to the stack, retrieves the "memo" it placed about the keystroke activity, and returns to that task until it is interrupted again.

What happens if there is more than one document waiting to be printed? The operating system also coordinates multiple activities for peripheral devices such as printers. When the processor receives a request to send information to the printer, it first checks with the operating system to ensure that the printer is not already in use. If it is in use, the OS puts the request in another temporary storage area in RAM called the *buffer*. It will wait in the buffer until the **spooler**, a program that helps coordinate all print jobs currently being sent to the computer printer, indicates the printer is available. If more than one print job is waiting, a line, or **queue**, is formed so that the printer can process the requests in order.

BITS AND BYTES

One OS for More Than One CPU?

Mainframes and supercomputers have long used multiple processors to divide the huge quantities of calculations and tasks they perform into manageable pieces. Although personal computers have traditionally had only one CPU, the trend toward machines with multiprocessors is quickly growing. In computers with multiple processors, rather than having two operating systems, each servicing a CPU, a single operating system coordinates the activities of both CPUs at the same time. Windows XP, Mac OS, and Linux can all support multiprocessor systems.

MEMORY AND STORAGE MANAGEMENT

Why does the operating system have to manage the computer's memory? As the operating system coordinates the activities of the processor, it uses RAM as a temporary storage area for instructions and data the processor needs. The processor then accesses these instructions and data from RAM when it is ready to process them. The OS is therefore responsible for coordinating the space allocations in RAM to ensure that there is enough space for all the waiting instructions and data. It then clears the items from RAM when the processor no longer needs them.

Can my system ever run out of RAM space? RAM has limited capacity. The average computer system has anywhere from 128 megabytes (MB) to 2 gigabytes (GB) of memory in RAM. Although 2 GB of RAM seems like a lot of space, if you're running numerous multimedia-intensive applications at the same time, you can easily use up the RAM in your computer.

What happens if my computer runs out of RAM? When there isn't enough room in RAM for the operating system to store the required data and instructions, the operating system borrows room from the more spacious hard drive. This process of optimizing RAM storage by borrowing hard drive space is called **virtual memory**. As shown in Figure 5.11, when more RAM space is needed, the operating system swaps out from RAM the data or instructions that have not been recently used and moves them to a temporary storage area on the hard drive called the **swap file** (or **page file**). If the data and/or instructions in the swap file are needed later, the operating system swaps them back into active RAM and replaces them in the hard drive's swap file with less active data or instructions. This process of swapping is known as **paging**.

Can I ever run out of virtual memory? Only a portion of the hard drive is allocated to virtual memory. You can manually change this setting to increase the amount of hard drive space allocated, but eventually your computer system

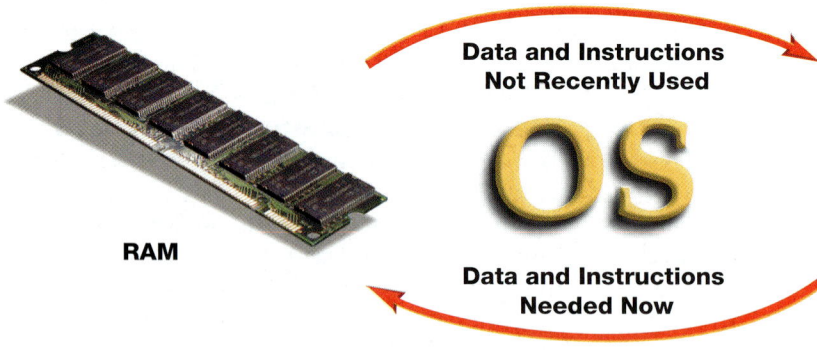

FIGURE 5.11

Virtual memory borrows excess storage capacity from the hard drive when there is not enough capacity in RAM.

will become sluggish as it is forced to page more and more often. This condition of excessive paging is called **thrashing**. The solution to this problem is to increase the amount of RAM in your system so that you can avoid it having to send data and instructions to virtual memory. (You'll learn how to monitor your RAM and virtual memory requirements in Chapter 6.)

How does the operating system manage storage? If it weren't for the operating system, the files and applications you save to the hard drive and other storage locations would be a complete mess. Fortunately, the OS has a file management system that keeps track of the names and locations of each file you save and programs you install. We'll talk more about file management later in the chapter.

HARDWARE AND PERIPHERAL DEVICE MANAGEMENT

How does the operating system manage the hardware and peripheral devices? Each device attached to your computer comes with a special program called a **device driver** that facilitates the communication between the device and the operating system. Since the OS must be able to communicate with every device in the computer system, the device driver translates the specialized commands of the device to commands that the operating system can understand, and vice versa. Thus, devices will not function without the proper device driver, as the OS would not know how to communicate with them.

How can I get device drivers? Manufacturers preinstall device drivers for the original hardware that comes with your computer, such as your mouse, monitor, and CD-ROM drive. However, if you replace any of these devices or purchase additional devices, such as a scanner, printer, or DVD player, you must install separate device drivers. Hardware manufacturers generally include device drivers with their devices, but sometimes you have to download the driver from the manufacturer's Web site. Should you purchase a device secondhand and not receive the device driver, you can often contact the manufacturer for a copy. You can also check out Web sites such as **www.driverzone.com** to locate drivers.

Is Plug and Play a device driver? Plug and Play (PnP) is not a driver. Instead, it is a software and hardware standard that Microsoft created with the Windows 95 operating system. This standard is designed to facilitate the installation of a new piece of hardware in personal computers. Plug and Play enables users to "plug" in their new device to a port on the system, turn on the system, and immediately "play," or use, the device. The OS automatically recognizes the device and its driver without any further user manipulations to the system.

SOFTWARE APPLICATION COORDINATION

How does the operating system help software applications run on the computer? Software applications feed the CPU the instructions it needs to process data. These instructions take the form of computer code. Every software application, no matter what its type or manufacturer, needs to interact with the CPU. For programs to work with the CPU, they must contain code that the CPU recognizes. Rather than having the same blocks of code for similar procedures in each software application, the operating system itself includes the blocks of code that software applications need in order to interact with it. These blocks of code are called **application program interfaces (APIs).** Microsoft DirectX, for example, is a group of multimedia APIs built into Windows that improve graphics and sounds when you're playing games or watching video on your PC.

To create programs that can communicate with the operating system, software programmers need only *refer* to the API code blocks in their individual application programs, rather than including the entire code in the application itself. Not only do APIs avoid redundancies in software code, they also make it easier for software developers to respond to changes in the operating system.

Large software developers such as Microsoft have many software applications under their corporate umbrella, and use the same APIs in all or most of their software applications. Since APIs coordinate with the operating system, all applications that have incorporated these APIs have common interfaces such as similar toolbars and menus. Therefore, the software applications have the same look to many of their features. An added benefit to this system is that applications sharing these same formats can also easily exchange data between different programs. As such, it's easy to create a chart in Microsoft Excel from data in Microsoft Access and incorporate the finished chart into a Microsoft Word document.

The Boot Process: Starting Your Computer

Although it only takes a minute or two, a lot of things happen very quickly between the time you turn on the computer and when it is ready for you to enter your first command. As you learned earlier, all data and instructions (including the operating system) are stored in RAM while your computer is on. When you turn your computer off, RAM is wiped clean of all its data (including the OS). So, how does the computer know what to do when you turn your computer on if there is nothing in RAM? It runs through a special process, called the **boot process** (or start-up process), to load the operating system into RAM.

What are the steps involved in the boot process? The boot process, illustrated in Figure 5.12, consists of four basic steps:

1. The basic input/output system (BIOS) is activated by powering on the CPU
2. The BIOS checks that all attached devices are in place (called a Power-On Self-Test, or POST)
3. The operating system is loaded into RAM
4. Configuration and customization settings are checked

As the computer goes through the boot process in Windows operating systems, indicator lights on the keyboard and disk drives will illuminate

STEP 1: The CPU activates the basic input/output system (BIOS). The BIOS is located on a ROM chip. Data stored in ROM does not get erased when the computer is turned off.

STEP 2: BIOS checks that all attached devices are in place and working and that the video card and memory are responding correctly. This test is called a power-on self test (POST). If the results of the POST are okay, the boot process continues.

STEP 4: The registry is checked for further configurations and customizations. If the entire system is checked out and loaded properly, the desktop appears on your screen. The computer system is now ready to accept your first command.

STEP 3: BIOS first looks in the floppy disk drive for the system files. When it doesn't find the operating system in the floppy disk drive, it looks in the hard disk drive. It then loads the operating system from its permanent storage location on the hard disk to RAM.

FIGURE 5.12

The Boot Process

THE BOOT PROCESS: STARTING YOUR COMPUTER

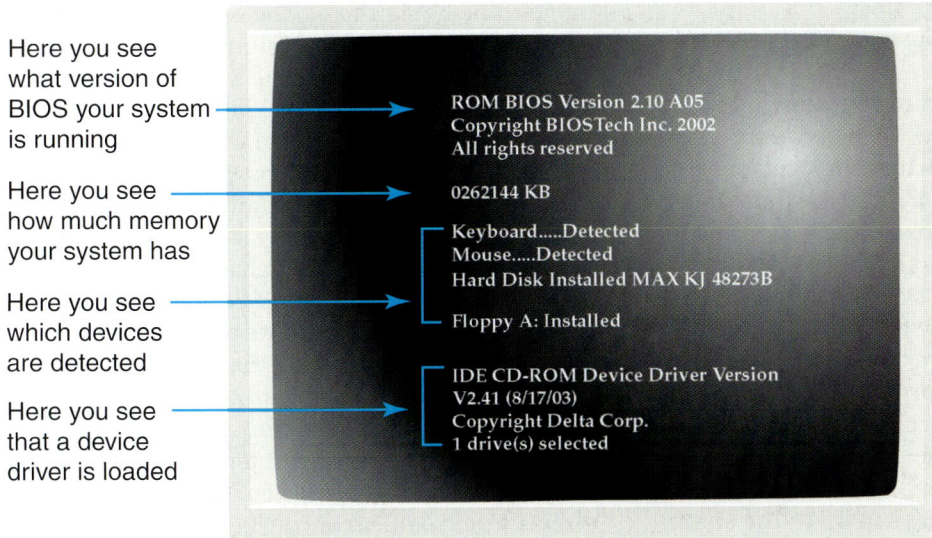

FIGURE 5.13

If you have a version of Windows before XP, you'll see a screen similar to this one during the boot process.

- Here you see what version of BIOS your system is running
- Here you see how much memory your system has
- Here you see which devices are detected
- Here you see that a device driver is loaded

and the system will emit various beeps. If you have a version of Windows before XP, text like that shown in Figure 5.13 will scroll down the screen as well. When you boot up on a Mac, you won't hear any beeps or see any keyboard lights illuminate. Instead, a Welcome screen will appear, indicating the progress of the start-up process. Once the boot process has completed these steps, it is ready to accept commands and data. Let's look at each of these steps in more detail.

STEP 1: ACTIVATING BIOS

What's the first thing that happens after I turn on my computer? In the first step of the boot process, the CPU activates the BIOS. The **basic input/output system,** or **BIOS** (pronounced "bye-OSE"), is a program that manages the data between the operating system and all the input and output devices attached to the system, hence its name. BIOS is also responsible for loading the OS from its permanent location on the hard drive to RAM.

BIOS itself is stored on a special ROM (read-only memory) chip on the motherboard. Unlike data stored in RAM, data stored in ROM is permanent and does not get erased when the power is turned off.

STEP 2: PERFORMING THE POWER-ON SELF-TEST

How does the computer determine whether the hardware is working properly? The first job BIOS performs is to ensure that essential peripheral devices are attached and operational. This process is called the **power-on self test,** or **POST**. The POST consists of a test on the video card and video memory, a BIOS identification process (during which the BIOS version, manufacturer, and data are displayed on the monitor), and a memory test to ensure memory chips are working properly.

The BIOS compares the results of the POST with the various hardware configurations that are permanently stored in CMOS (pronounced "see-moss"). CMOS is a special kind of memory that uses almost no power. A little battery provides enough power so its contents will not be lost after the computer is turned off. CMOS contains information about the system's memory, types of disk drives, and other essential input and output hardware components. If the results of the POST compare favorably to the hardware configurations stored on CMOS, the boot process continues.

BITS AND BYTES

How Did "Boot" Get Its Name?

The term *boot*, used to describe the process of starting a computer, gets its name from the term *bootstrap*. In the olden days, men used straps of leather, called bootstraps, to help them pull on their boots. The use of bootstraps in this way created the expression to "pull oneself up by the bootstraps." In computing terms, the *bootstrap loader* is a very small program that begins the process of loading a much larger and more powerful program that then controls the rest of the system.

STEP 3: LOADING THE OPERATING SYSTEM

How does the operating system get loaded into RAM? Next, BIOS looks through the storage disks for the **system files**, the main files of the operating system. The first place it looks is the floppy disk drive. When it doesn't find the OS there, it looks in the hard disk drive. It then loads the operating system from its permanent storage location on the hard drive to RAM.

Once the system files are loaded into RAM, the **kernel**, or **supervisor program**, is loaded. The kernel is the essential component of the operating system. It is responsible for managing the processor and all other components of the computer system. Because it stays in RAM the entire time your computer is powered on, the kernel is called *memory resident*. So to not take up all the RAM, other parts of the OS that are less critical stay on the hard drive and are copied over to RAM on an as-needed basis. These programs are called *nonresident*. Once the kernel is loaded, the operating system takes over the control of the computer's functions.

STEP 4: CHECKING FURTHER CONFIGURATIONS AND CUSTOMIZATIONS

When are the other components and configurations of the system checked? Although CMOS checks the configuration of memory and essential peripherals in the beginning of the boot process, the operating system continues to check the configuration of other system components in this last phase of the boot process. The **registry** contains all the different configurations (settings) used by the OS as well as by other applications. It contains the customized settings you put into place, such as mouse speed, the display settings for your monitor and desktop, as well as instructions as to what programs should be loaded first.

Why do I sometimes need to enter a password at the end of the boot process? In a networked environment, such as that found at most colleges, the operating system services many users. In order to determine whether a user is authorized to use the system (that is, whether a user is a paid student or college employee), authorized users are given a login name and password. Typing in your login name and password at the end of the boot process is called **authentication**. The authentication process blocks unauthorized users from entering the system.

You may also need to insert a password following the boot process to log into your account on your home computer. The newest version of the Windows operating system, Windows XP, is a multiuser system. Even in a home environment, all users with access to a Windows XP computer (such as family members or roommates) have their own user accounts. Users can set up a password to protect their account from being accessed by another user without permission.

How do I know if the boot process is successful? The entire boot process takes only a minute or two to complete. If the entire system is checked out and loaded properly, the successful process completes by displaying the desktop. The computer system is now ready to accept your first command.

HANDLING ERRORS IN THE BOOT PROCESS

What can go wrong during the boot process? During the boot process, if you come across a message like the following:

```
Non-system disk or disk error

Replace and strike any key when ready
```

check to see whether you've left a diskette in the floppy disk drive. As explained earlier, when the BIOS is performing its system check, it first looks

BITS AND BYTES

Are Booting and Installing the Same Thing?

Like most software programs, the operating system is saved to the hard disk, or *installed*, only once. Typically, computer manufacturers preinstall the operating system. However, booting is done *every* time you turn the computer on, either from an off position (called a *cold boot*) or when you restart the system after it's already on (called a *warm boot*). You might need to perform a warm boot if the operating system or other software application stops responding. You do a cold boot each time you start the computer after you've turned it off completely.

in the floppy drive for the operating system. With early computers, operating systems were not permanently stored on the hard drive but were instead loaded into the system from a floppy disk. (That's how DOS, or Disk Operating System, got its name.)

If BIOS doesn't find a floppy in the floppy drive, it proceeds to the hard drive. However, if it *does* find a floppy in the floppy drive, it will attempt to find the OS on that floppy. When it does not find the operating system software on the floppy, the boot process stops and displays the error message "Non-system disk or disk error." When this happens, simply remove the floppy disk and press any key to resume the boot process.

How can I tell if there are other errors during the boot process? During the boot process, a string of text scrolls down your monitor. This text contains a list of devices attached to your system as well as their settings. If there is a problem with loading a device during the POST, a separate error message (generally on a blue background) appears on your screen.

Sometimes, however, the problem occurs before the video (display) card that controls your monitor has been activated, making a display message impossible. In those instances, you will hear a series of beeps. (These beeps are in place of a single beep you would hear if everything was loading properly.) Each device in your computer system is assigned a specific beep code. Since different BIOS manufacturers have different beep codes, you can identify the error by listening to the number of beeps and then comparing them to the beep codes listed in your computer's user manual or on the BIOS manufacturer's Web site.

Figure 5.14 shows sample beep codes for the Phoenix BIOS. Being able to identify an "error beep" during the boot process will facilitate conversations you may later have with a technical assistant to diagnose the problem.

What should I do if my keyboard or other device doesn't work after I boot my computer? Sometimes during the boot process BIOS skips a device (such as a keyboard) or improperly identifies it. You won't hear any beeps or see any error messages when this happens. Your only indication that this sort of problem has occurred is that the device won't respond after the system has been booted. When that happens, you can generally resolve the problem by rebooting. If the problem persists, you may want to check the operating system's Web site for any "patches" (or software fixes) that may resolve the issue. If there are no patches

FIGURE 5.14 Sample Phoenix BIOS Beep Codes	
SEQUENCE OF BEEPS YOU HEAR	**WHAT THE BEEPS INDICATE**
1-1-3	Your computer isn't able to read the configuration information stored in the CMOS chip.
1-1-4	There is something wrong with BIOS.
1-4-2	Some of the memory in your computer is not functioning correctly.
4-2-2	Your computer is not able to communicate with the keyboard.

or the problem persists, you may want to get technical assistance.

What is "Safe Mode"? Sometimes Windows does not boot properly and you end up with a grayish-looking screen with the words "Safe Mode" in the corners, as shown in Figure 5.15. **Safe Mode** is a special diagnostic mode designed for troubleshooting errors that occur during the boot process. While in Safe Mode, only the "essential" devices of the system (such as the mouse, keyboard, and monitor) function.

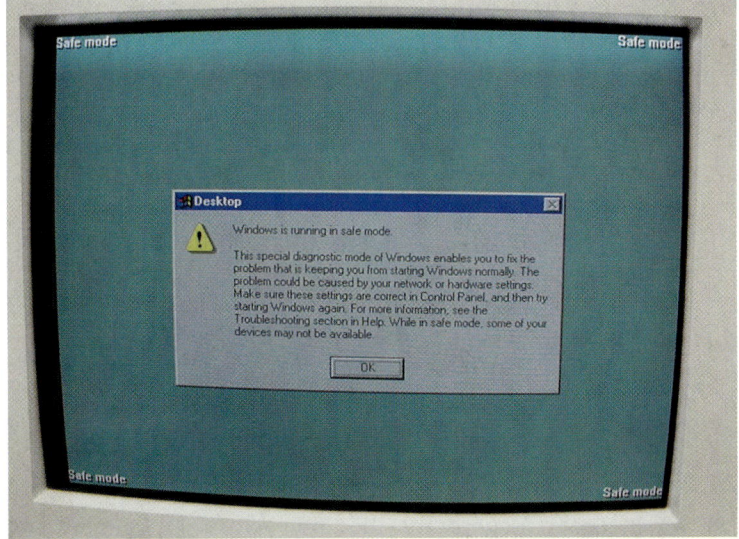

FIGURE 5.15

If your screen looks like this, your computer has booted into Safe Mode. This means that something did not function properly during the boot process. Safe Mode provides you with enough functionality so that you can accomplish diagnostic testing.

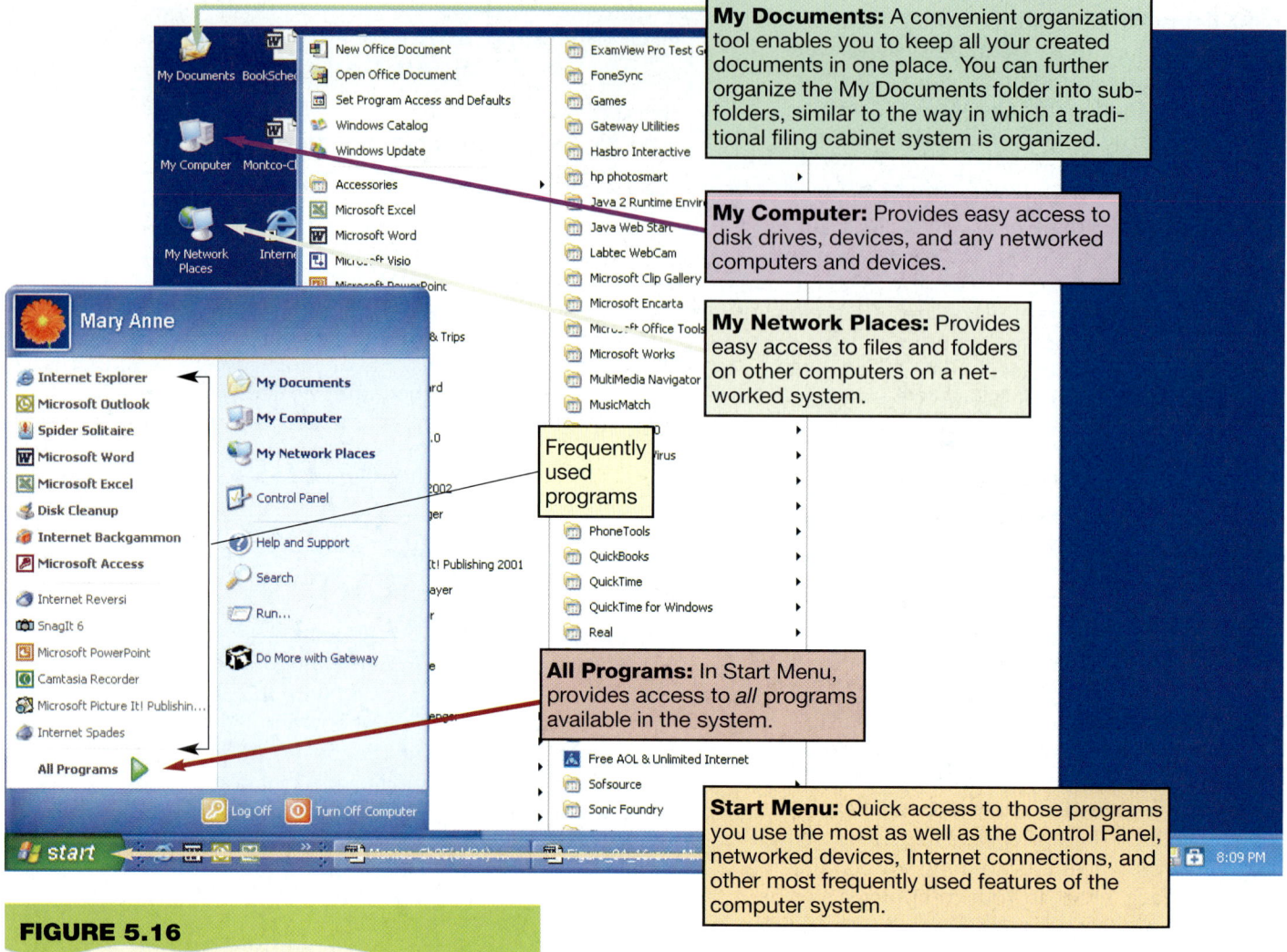

FIGURE 5.16

The Windows desktop puts the most commonly used features of the operating system at your fingertips.

Even the regular graphics device driver will not be activated in Safe Mode. Instead, the system runs in the most basic graphics mode, resulting in a grayish-toned screen.

What should I do if my operating system boots into Safe Mode? Sometimes, Safe Mode indicates that there is a problem with the loading of a device or software application. Try rebooting the machine before doing anything else. If you still end up in Safe Mode and if you have recently installed new software or a new hardware device, try uninstalling it. (Make sure you use the Add/Remove feature in the Control Panel.) If the problem then goes away after rebooting, you have determined the cause of the problem. You can then reinstall the device or software. If the problem does not go away, you should consult a technical support person for further diagnosis.

The Desktop and Windows Features

The **desktop** is the first interaction you have with the operating system and the first image you see on your monitor. As its name implies, your computer's desktop puts at your fingertips all of the elements necessary for a productive work session that are typically found on or near the top of a traditional desk, such as files and folders.

What are the main features of the desktop? The very nature of a desktop is that it enables you to customize it to meet your individual needs. As such, the desktop on your computer may be different from the desktop on your friend's computer. However, most desktops share common features, some of which are illustrated in Figure 5.16.

THE DESKTOP AND WINDOWS FEATURES

What are common features of a window? As noted earlier, one feature introduced by the graphical user interface is *windows* (with a lowercase "w"), the rectangular panes on your computer screen that display applications running on your system. Windows provide for a flexible, user-friendly, multitasking environment. Figure 5.17 illustrates some of the features of windows, including **toolbars** (groups of icons collected together in a small box) and **scrollbars** (bars that appear at the side or bottom of the screen that control which part of the information is displayed on the screen). Using the Minimize, Maximize and Restore, and Close buttons, you can open, close, resize, and move windows anywhere around the desktop.

How can I see more than one window on my desktop at a time? You can easily arrange the windows on a desktop by "tiling" them, which means arranging separate windows so that they sit next to each other either horizontally or vertically. You can also arrange windows by "cascading" them so that they overlap one another, or simply resize two open windows so they appear on the screen at the same time.

Tiling windows makes accessing two or more active windows more convenient. For example, as shown in Figure 5.18, you can input stock prices from a Web site into an Excel spreadsheet without clicking back and forth between the browser and Excel windows by tiling them horizontally. To untile the windows, or to bring a window back to its full size, click on the Restore button in the top right-hand corner of the window.

Can I move or resize the windows once they are tiled? Regardless of whether the windows are tiled, you can resize and move them around the desktop. You can reposition windows on the desktop by pointing to the title bar at the top of the window with your cursor, and while holding down the left mouse button, drag them to a different location. To resize a window, place your cursor on any side or corner of a window until it changes to a double-headed arrow [↕]. You can then left-click and drag the window to the new desired size.

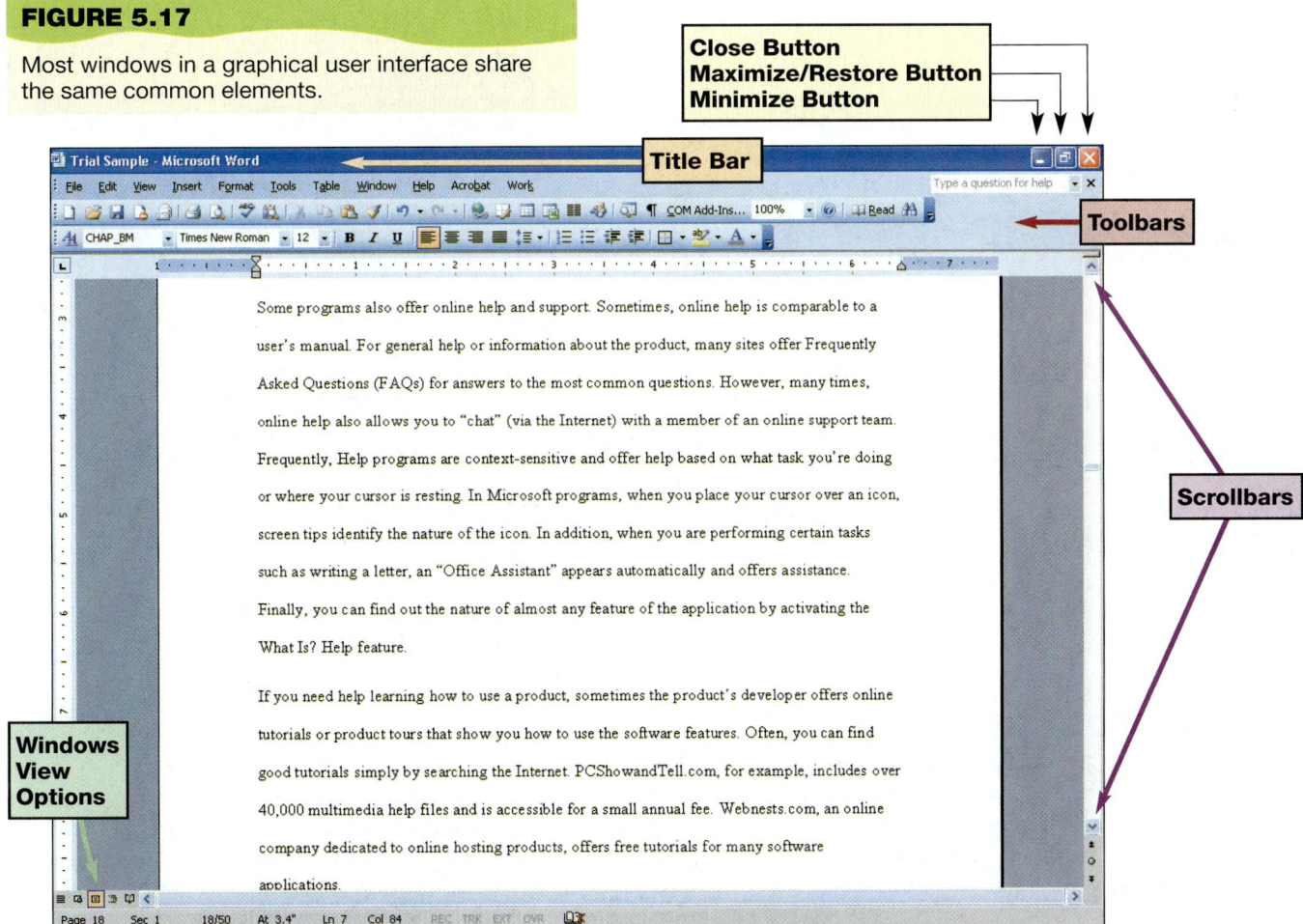

FIGURE 5.17

Most windows in a graphical user interface share the same common elements.

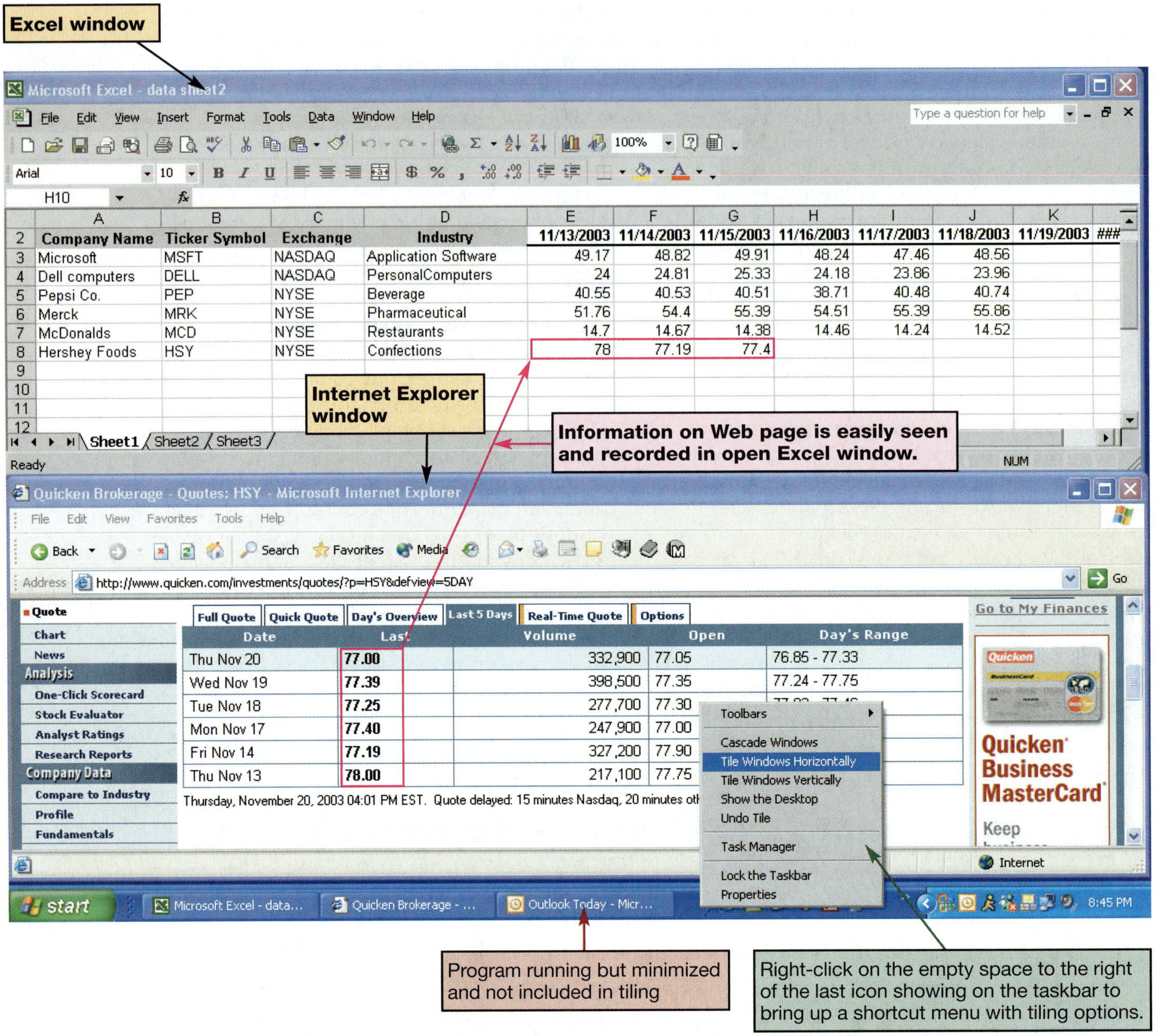

CHAPTER 5 USING SYSTEM SOFTWARE: THE OPERATING SYSTEM, UTILITY PROGRAMS, AND FILE MANAGEMENT

FIGURE 5.18
Tiling windows is a great way to see two windows at the same time. Having both the Internet Explorer and Excel windows visible at the same time makes recording data from the Internet to the spreadsheet much simpler.

SOUND BYTE
CUSTOMIZING WINDOWS XP
In this Sound Byte, you'll find out how to customize your desktop. You'll learn how to configure the desktop, set up a screen saver, change pointer options, customize the Start menu, and manage user accounts.

Organizing Your Computer: File Management

You have learned so far that the operating system is responsible for managing the processor, memory, storage, and devices, and that it provides a mechanism for applications and users to interact with the

ORGANIZING YOUR COMPUTER: FILE MANAGEMENT

computer system. An additional function of an operating system is to enable **file management**, which entails providing organizational structure to the computer's contents. The OS allows you to organize the contents of your computer in a hierarchical structure of **directories** that includes *files*, *folders*, and *drives*. In this section, we'll discuss how you can use this hierarchical structure to create a more organized and efficient computer.

ORGANIZING YOUR FILES

What exactly is a file? Technically, a **file** is a collection of related pieces of information stored together for easy reference. A file in an operating system is a collection of program instructions or data stored and treated as a single unit. Files can be generated from an application, such as a Word document or Excel spreadsheet. Additionally, files can represent an entire application, a Web page, an audio file, or an image file. Files are stored to the hard drive, a floppy disk, or other storage medium for "permanent" storage. As the number of files you save increases, it is important to keep them organized in **folders**, or collections of files.

How does the operating system organize files? Windows organizes the contents of the computer in a hierarchical structure with drives, folders, subfolders, and files. The hard drive, represented as the C: drive, is where you permanently store most of your files, while the floppy drive is the A: drive. You may have additional drives (D:, E:, and/or F:) depending on whether you have any additional storage devices (Zip, CD, or DVD drives) installed in your computer.

The C: drive is like a large filing cabinet in which all files are stored. As such, the C: drive is the top of the filing structure of the computer system and is referred to as the **root directory**. All other folders and files are organized within the root directory. There are areas in the root directory that the operating system has filled with folders holding special OS files. The programs within these files help run the computer and generally shouldn't be touched. The Windows operating system also creates other folders, such as My Documents and My Pictures, which are available for you to begin to store and organize your text and image files, respectively.

How can I easily locate and see the contents of my computer? If you use a Windows PC, **Windows Explorer** is the program that helps you manage your files and folders by showing the location and contents of every drive, folder, and file on your computer. You can access Windows Explorer by right-clicking the My Computer desktop icon or the Start button and selecting Explore from the shortcut menu. (If you use a Mac, the Finder is the program that enables you to manage your files and folders.) As illustrated in Figure 5.19, Windows Explorer is divided into two sections.

The left-hand side shows the contents of your computer in a hierarchical "tree" structure. It displays all the drives of the system as well as other commonly accessed areas such as the Desktop and the My Documents folder. You can open the folders on the left-hand side to reveal their contents by clicking on the plus (+) sign next to the folder name. Once you open a folder, the plus sign changes to a minus (–) sign and the contents of that folder are displayed in the right-hand side of the Explorer window. You can choose to expand or collapse from view the contents of a folder by clicking on the plus and minus sign. If a folder does not have a plus or minus sign, there are no other folders contained within the folder, although it may contain files.

How should I organize my files? Creating folders is the key to organizing your files, as folders keep related documents together. Again, think of your computer as a big filing cabinet to which you can add many separate filing drawers, or subfolders. Those drawers, or subfolders, have the capacity to hold even more folders, which can hold other folders or individual files. For example, you can create one folder for your class work called Classes. Inside the Classes folder, you can create folders for each of your classes (such as Intro to Computers, Bio 101 and British Literature). Inside each of those folders you can create subfolders for each class's assignments, completed homework, research, notes, and so on.

Grouping related files together in folders allows you to more easily identify and find files. Which would be easier, going to the Bio101 folder to find a file or searching through the 143 individual files in My Documents hoping to find the right one? Grouping files in a folder also allows you to move them more efficiently, so you can quickly transfer critical files needing frequent backup to a floppy disk, for instance.

SOUND BYTE
FILE MANAGEMENT

In this Sound Byte, you'll examine the features of file management and maintenance. You'll learn the various methods of creating folders, how to turn a jumble of unorganized files into an organized system of folders, and how to maintain your file system.

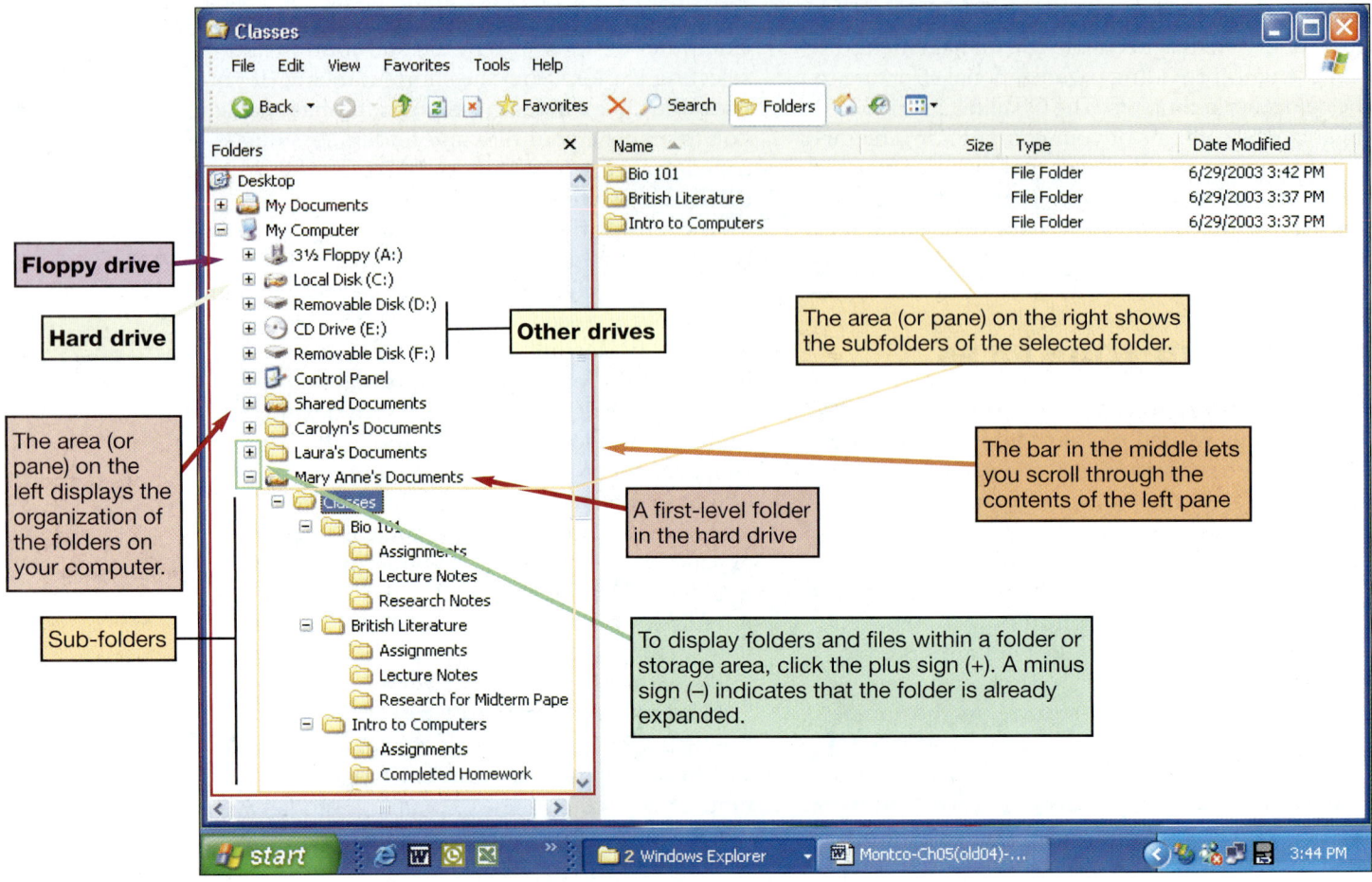

FIGURE 5.19

Windows Explorer lets you see the contents of your computer.

VIEWING AND SORTING FILES AND FOLDERS

Are there different ways I can view and sort my files and folders? In Windows XP, when you are in a folder, such as My Documents, you can use any of the viewing options located on the View menu to arrange and view your files and folders:

- **Tiles view** displays files and folders as icons in list form. Each icon includes the filename, the application associated with the file, and the file size. The display information is customizable. The Tiles view also displays picture dimensions, a handy feature for Web-page developers.

- **Icon view** also displays files and folders as icons in list form, but the icons are smaller and no other file information beside the filename is listed. However, file information is displayed when you place your cursor over the file icon.

- **List view** is another display of even smaller icons and file names. This is a good view if you have a lot of content in the folder and need to see most or all of it at once.

- **Thumbnails view,** illustrated in Figure 5.20, shows the contents of folders as small images. Thumbnails view is therefore the best view to use if your folder contains picture files. For those folders that contain collections of MP3 files, you can download the cover of the CD or an image of the artist to display on any folder to further identify that collection.

- **Details view** is the most interactive view. As shown in Figure 5.21, the files and folders are displayed in list form, but the additional file information is displayed in columns alongside the file name. You can sort and display the contents of the folder by any of the column headings, so you may sort the contents alphabetically by filename or type, or hierarchically by date last modified or by file size.

ORGANIZING YOUR COMPUTER: FILE MANAGEMENT

FIGURE 5.20

Thumbnails view is an especially good way to display folders containing picture files since some of the pictures in the folder are shown on the cover of the folder.

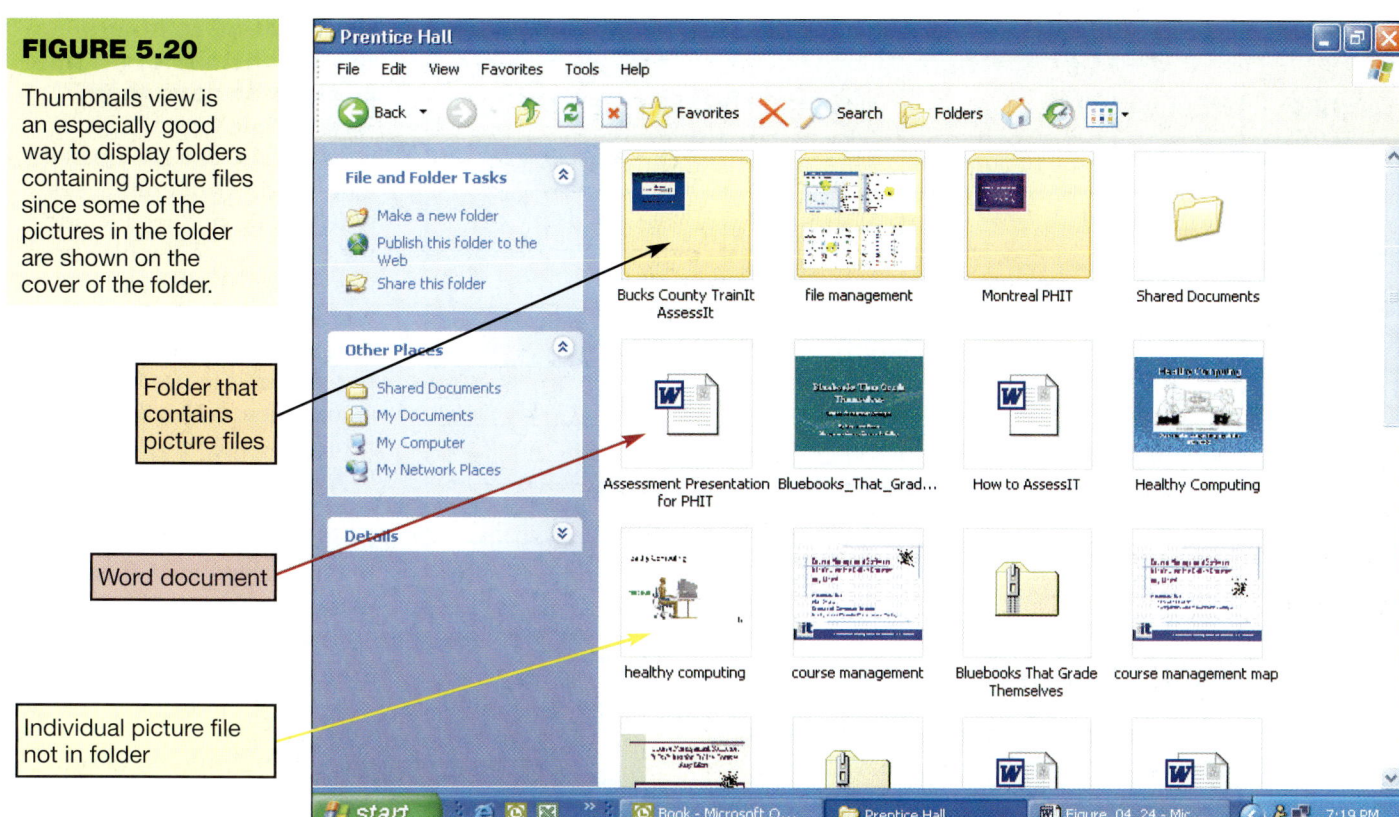

- Folder that contains picture files
- Word document
- Individual picture file not in folder

FIGURE 5.21

Details view allows you to sort and list your files in a variety of ways to further assist you in quickly finding the correct file.

Sort files and folders by clicking on these headers.

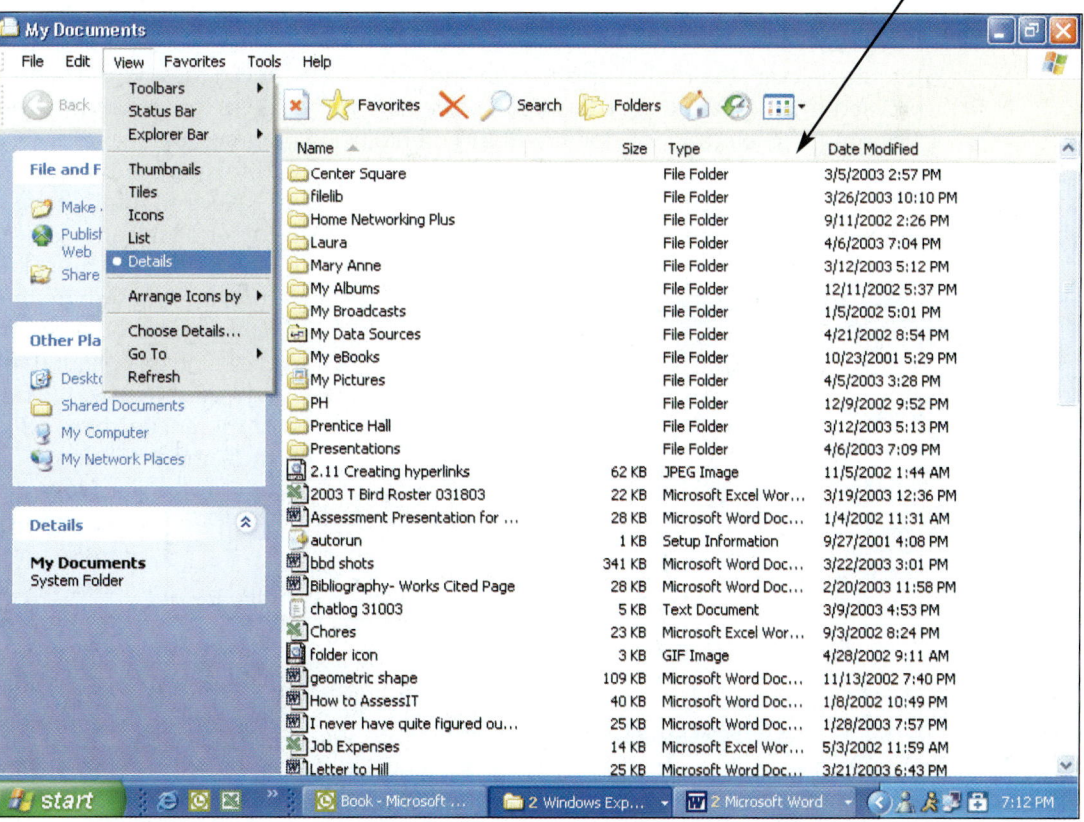

FIGURE 5.22 Filename Extensions

EXTENSION	TYPE OF DOCUMENT	APPLICATION THAT USES THE EXTENSION
.doc	Word processing document	Microsoft Word; Corel Word Perfect
.xls	Workbook	Microsoft Excel
.ppt	PowerPoint presentation	Microsoft PowerPoint
.mdb	Database	Microsoft Access
.bmp	Bitmap image	Windows
.zip	Compressed file	WinZip
.pdf	Portable Document Format	Adobe Acrobat
.htm	Web page	Hypertext Markup Language

NAMING FILES

Are there special rules I have to follow when I name files? Files have names just like people. The first part of a file, or the **filename**, is similar to our first names and is generally the name you assign to the file when you save it. For example, "bioreport" may be the name you assign a report you have completed for biology. In a Windows application, following the filename and after the dot (.) comes a three-letter **extension**, or **file type**. Like our last name, this extension identifies what kind of family of files the file belongs to or what application should be used to read the file. For example, if the "bioreport" file is a Word document, it will have a .doc extension and be named "bioreport.doc." Figure 5.22 lists the common file extensions and the types of documents they indicate.

Do I need to know the extensions of all files in order to save them? As shown in Figure 5.23, when you save a file created in a Windows operating system, you do not need to add the extension to the filename; it is automatically added for you. With Mac and Linux operating systems, you are not required to use file extensions when saving files. This is because the information as to the type of application the computer should use to open the file is stored inside the file itself. However, if you're using these operating systems and will be sending files to Windows users, you should add an extension to your filename so that they can more easily open your files.

Are there things I shouldn't do when naming my file? Each operating system has its own naming conventions, or rules, which are listed in Figure 5.24. Beyond those conventions, it's important that you name your files so that you can easily identify them. A filename like "research.doc" may be descriptive to you if you're only working on one research paper. However, if you create other research reports later and need to quickly identify the contents of these files, you'll soon wish you had been more descriptive. Giving your files more descriptive names, such as "bioresearch.doc" or, better yet, "bio101research.doc," is a good idea.

Keep in mind, however, every file in the same folder or storage device (hard disk, floppy disk, CD, and so on) must be *uniquely* identified. Therefore, files may share the same filename (such as *bioreport*.doc or *bioreport*.xls), or they may share the same extension (bioreport.xls or budget.xls); however, no two files stored on the same device or folder can share *both* the same file name *and* the same extension.

How can I tell where my files are saved? When you save a file for the first time, you give the file a name and designate where you want to save it. The location of the file is

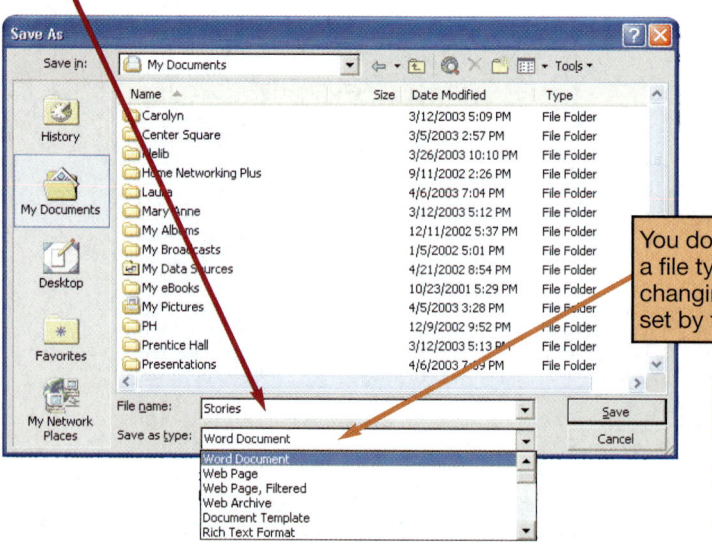

Type in the file name in this box.

You don't need to select a file type unless you are changing the default type set by the application.

FIGURE 5.23

When you save a file in Windows, you type in the filename in the Save As dialog box. The extension is added automatically, though you can change it if necessary.

FIGURE 5.24 File Naming Conventions

	MACINTOSH	WINDOWS
FILE AND FOLDER NAME LENGTH	Up to 31 characters	Up to 255 characters
CASE SENSITIVE?	Yes	No
FORBIDDEN CHARACTERS	The colon (:)	" / \ * ? < > | :
SPACES ALLOWED?	Yes	Yes
THREE-LETTER FILE EXTENSIONS NEEDED?	No	Yes
PATH SEPARATOR	:	\

defined by a **file path** that identifies the exact location of the file, starting with the drive in which the file is located, and including all folders, subfolders (if any), the filename, and extension. For example, if you were saving a picture of Emily Bronte for a term paper for an English Comp course, the file path might be:

`C:\My Documents\Spring 2003\English Comp\Term Paper\Illustrations\EBronte.jpg`

C: indicates the drive the file is stored in (in this case, the hard drive), and My Documents is the file's primary folder. Spring 2003, English Comp, Term Paper, and Illustrations are successive subfolders within the My Documents main folder. Last is the file name, "EBronte," separated from the file extension (in this case, jpg) by a period. Notice that in between the drive, primary folder, subfolders, and filename are backslash marks (\). These backslash marks, used by Windows and DOS, are referred to as **path separators**. Mac files use a colon (:) and UNIX and Linux use the forward slash (/) as the path separator.

WORKING WITH FILES

Once I create files, how can I do things like renaming and deleting them? Once you've located your file with Windows Explorer, you can perform many other file management actions, such as opening, copying, moving, renaming, and deleting files. Figure 5.25 explains how you can perform all of these actions.

FIGURE 5.25 File Management Commands and Actions

COMMAND	ACTION
OPEN	You can open a file by clicking on the file from its storage location. The operating system then determines which application needs to be loaded to open the requested file and opens the file within the correct application automatically.
COPY	You can copy a file to another location using the Copy command. When you copy a file, a duplicate file is created and the original file remains in its original location.
MOVE	You can move, or transfer, a file from one location to another using the Move command. When you move a file, the original file is erased from its original location.
RENAME	You can rename your files using the Rename command.
DELETE	You can delete a file using the Delete command. When you delete a file located on the hard drive, it is moved to the Recycle Bin. Files from other storage locations such as the floppy or CD drives are deleted permanently.

BITS AND BYTES
Recovering Deleted Recycle Bin Files

Once you empty the Recycle Bin, because you don't see the filename anymore, it looks as if the file has been erased from the hard drive. However, only the *reference* to the deleted file is deleted permanently, so the operating system has no easy way to find the file. The file data actually remains on the hard drive, until otherwise written over. Should you delete a file from the Recycle Bin in error, you can immediately restore the deleted file by clicking the undo arrow on the toolbar. Programs such as RestoreIT! or Roxio's GoBack allow you to recover longer-term deleted files—but the longer you wait to recover a deleted file, the chances of a fully recovery decrease. That's because the probability that your file has been overwritten increases.

Where do deleted files go? One of the improvements made to Windows 95 is the **Recycle Bin,** which is a folder on the desktop where files deleted *from the hard drive* reside until you permanently purge them from your system. Unfortunately, files deleted from other drives, such as the floppy drive, CD, or network drive, do not go to the Recycle Bin, but are deleted from the system immediately. (Mac systems have something similar to the Recycle Bin, called the Trash Can. To delete files on a Mac system, you simply drag the file to the Trash Can on the desktop. Then select Empty Trash from the Finder menu in OSX, or from the Special menu in earlier versions.

Is it possible to retrieve a file that I've accidentally deleted? The benefit of the Recycle Bin is that you can restore the files you place there. To do so, open the Recycle Bin, locate the file, and select Restore. To permanently delete your files from the Recycle Bin, select Empty the Recycle Bin after clicking on the desktop icon.

Utility Programs

You have learned that the operating system is the single most essential piece of software in your computer system because it coordinates all the system's activities and provides a means by which other software applications and users can interact with the system. However, there is another set of programs included in system software. **Utility programs** are small applications that perform special functions. Some utility programs help manage system resources (such as disk defragmenter utilities), others help make your time and work on the computer more pleasant (such as screensavers), and still others improve efficiency (such as file compression utilities).

Some of these utility programs are incorporated into the operating system. Other utility programs, such as antivirus programs, have become so large and require such frequent updating that they are sold as stand-alone "off-the-shelf" programs in stores or as Web-based services available for an annual fee. Sometimes utility programs are offered as software suites, bundled together with other useful maintenance and performance-boosting utilities. Still other utilities are offered as shareware programs and are available as free downloads from the Web. Figure 5.26 illustrates the various types of utility programs available within the Windows operating system as well as those available as off-the-shelf programs in stores.

In this section, we'll explore many of the utility programs you'll find installed on a Windows operating system. Unless otherwise noted, you can find these utilities in the Control Panel or in the Start menu by selecting Programs, Accessories, and then System Tools. (We'll also take a brief look at some Mac utilities.) We'll discuss antivirus and personal firewall utility programs in Chapter 7.

DISPLAY UTILITIES

How can I change the appearance of my desktop? The Display folder in the Control Panel has all the features required to change the appearance of your desktop, providing different options for the desktop background, screensavers, windows colors, font sizes, and screen resolution. Although Windows

FIGURE 5.26 Utility Programs Available within Windows and as Standalone Programs

WINDOWS UTILITY PROGRAM	OFF-THE-SHELF (STAND-ALONE) UTILITY PROGRAM	FUNCTION
FILE MANAGEMENT		
Add/Remove Programs	Aladdin Systems Easy Uninstall	Properly installs/uninstalls software
Windows Explorer File Compression	WinZip	Reduces file size
WINDOWS SYSTEM MAINTENANCE AND DIAGNOSTICS		
Backup	Norton Ghost	Backs up important information
Disk Cleanup	Ontrack System Suite	Removes unnecessary files from hard drive
Disk Defragmenter	Norton SystemWorks	Arranges files on hard drive in sequential order
ScanDisk	Norton CleanSweep	Checks hard drive for unnecessary or damaged files
System Restore	FarStone RestoreIT!	Restores system back to a previously established set point

has many different background themes and screensaver options available, there are hundreds of downloadable options available on the Web. Just search for backgrounds or screensavers on your favorite search engine to customize your desktop. To access the background and screensaver options and all display utilities, choose Control Panel from the Start menu, then select the Display folder. (Note: if you have Windows XP and are showing the Categories view, the Display folder is in the Appearance and Themes category.)

Can I make the display on my LCD monitor clearer? If you use a laptop computer or have a flat-panel LCD monitor, you may be interested in the Clear Type feature Windows XP offers. Turning this feature on smoothes the edges of screen fonts to make words easier to read. Note that Clear Type is not very effective with cathode ray tube (CRT) monitors.

Do I really need to use a screensaver? **Screensavers** are animated images that appear on a computer monitor when no user activity has been sensed for a certain time. Originally, screensavers were used to prevent *burn-in*, the result of the same image being rescanned into the phosphor inside the monitor's cathode ray tube. In the early days of personal computers when the cursor was left blinking in the same spot for hours, burn-in was a concern. However, today's CRT display technology has changed, thus burn-in is not probable. Screensavers are now used almost exclusively for decoration.

You can also control how long your computer sits idle before the screensaver starts. For example, if you don't want people looking at what's on your screen when you leave your computer unexpectedly for a time, you may want to program your screensaver to run after only a minute or two of inactivity. However, if you find that you let your computer sit inactive for a while but need to look at the screen image (while you study or read a screen image, for example), you may want to extend the period of inactivity a bit.

BITS AND BYTES
Putting Pictures on Your Desktop

You can have most any picture displayed as your desktop background with only a few clicks of your mouse. If you have a digital photograph or an image that you want displayed as your desktop background, simply right-click on the image, select "Set as Background," and you're done. Your image will appear immediately as the desktop background. On a Mac, you would select the Apple Menu, Control Panels, Appearance and then select the Desktop tab. Click the Place Picture button, select the image you want to display on your desktop, click Open, then click Set Desktop.

THE ADD OR REMOVE PROGRAMS UTILITY

What is the correct way to add new programs to the system? These days, when you install a new program to your system, the program automatically runs a wizard that walks you through the installation process. If a wizard does not initialize automatically, however, you should go to the Add or Remove Programs folder in the Control Panel. This prompts the operating system to look in the CD or floppy disk drive for the setup program of the new software and starts the installation wizard.

What is the correct way to remove unwanted programs from my system? Some people think that deleting a program from the Program Files folder on the C: drive is the best way to remove a program from your system. However, most programs include support files such as a help file, dictionaries, and graphics files that are not located in the main program folder found in Program Files. Depending on the supporting file's function, it can be scattered throughout various folders within the system. You would normally miss these files by just deleting the main program file from the system. By selecting the Add and Remove Programs folder from the Control Panel, you not only delete the main program file, you also delete all supporting files as well.

FILE COMPRESSION UTILITIES

What is file compression? A **file compression utility** is a program that takes out redundancies in a file to reduce the file size. File compression is helpful because it makes a large

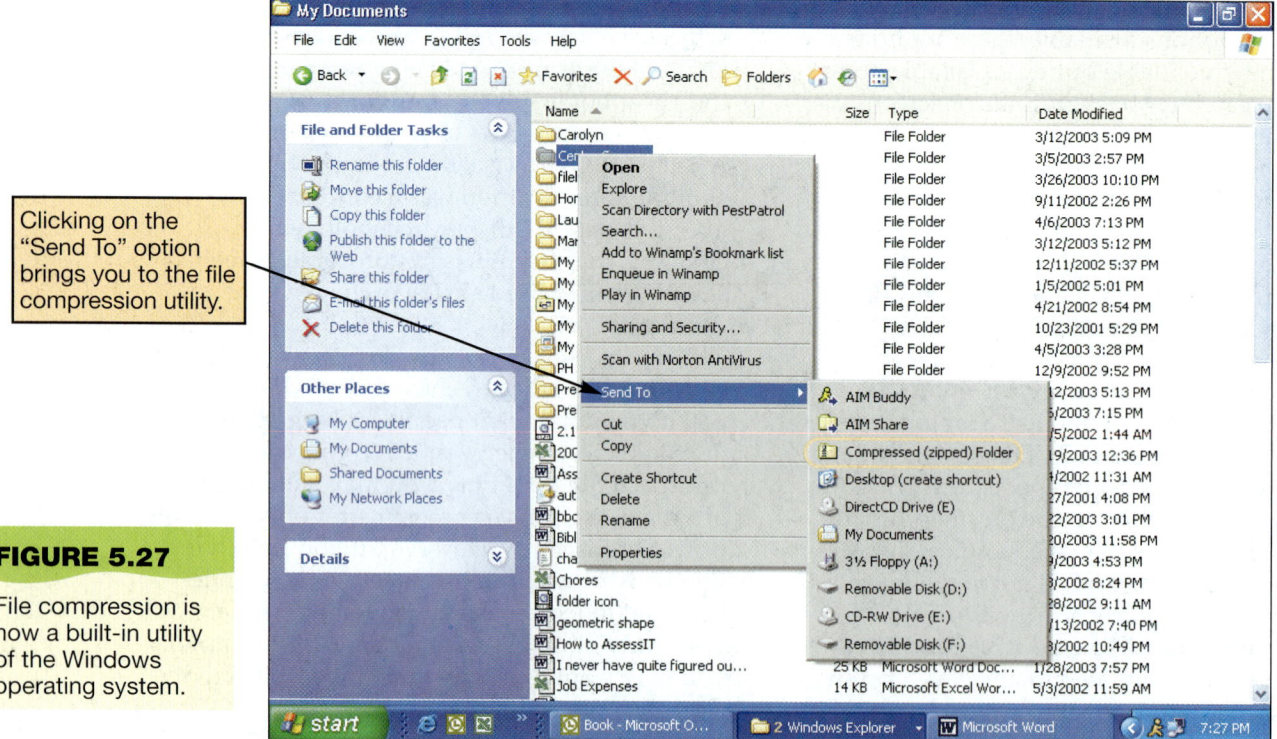

Clicking on the "Send To" option brings you to the file compression utility.

FIGURE 5.27

File compression is now a built-in utility of the Windows operating system.

file more compact, making it easier and faster to send over the Internet, upload to a Web page, or save onto a disk. As shown in Figure 5.27, Windows XP has built-in compression (or zip) file support. There are also several stand-alone freeware and shareware programs, such as WinZip (for Windows) and StuffIt (for Windows or Mac) that you can obtain to compress your files.

How does file compression work? Compression programs look for repeated patterns of letters and replace these patterns with a shorter placeholder. The repeated patterns and the associated placeholder are cataloged and stored temporarily in a separate file, called the dictionary. For example, in the following sentence, you can easily see the repeated patterns of letters:

The rain in Spain falls mainly on the plain.

Although in this example there are obvious repeated patterns (**ain** and **the**), in a large document, the repeated patterns may be more complex. The compression program's algorithm therefore runs through the file several times to determine the optimal repeated patterns to obtain the greatest compression.

How effective are file compression programs? The effectiveness of file compression—that is, how much a file's size is reduced—depends on several factors, including the type and size of the individual file and the compression method used. Current compression programs can reduce text files by as much as 50 percent. However, some files, such as database files, already contain a form of compression and therefore do not compress further. Other file types, especially graphics and audio formats, have gone through a compression process that reduces file size by permanently discarding "unnecessary" data. For example, image files such as JPEG (Joint Photographic Experts Group), GIF (graphics interchange format), and PNG (Portable Network Graphics) files discard small variations in colors that the human eye may not pick up. Likewise, MP3 files permanently discard sounds that the human ear cannot hear.

How do I decompress a file I've compressed? When you want to restore the file to its original state, you simply "decompress"

SOUND BYTE
FILE COMPRESSION
In this Sound Byte, you'll learn about the advantages of file compression and how to use Windows XP to compress and decompress files. If you own an earlier operating system, this Sound Byte will teach you how to find and install file compression shareware software programs.

the file and the pieces of file that the compression process temporarily removed are restored to the document. Generally, the same program you used to compress the file has the capability to decompress the file as well.

SYSTEM MAINTENANCE UTILITIES

Are there any utilities that make my system work faster? Over time, as you add or delete information to a file, the file pieces are saved in scattered locations on the hard disk. Locating all the pieces of the file takes extra time, making the operating system less efficient. **Disk defragmenter utilities** regroup related pieces of files together on the hard disk, allowing the OS to work more efficiently. You can find the Windows Disk Defragmenter utility under System Tools in the Accessories folder of the Start menu. On Macs, you can defrag your hard drive with Norton Utilities or Mac Tools. Depending on your usage, you should defrag your hard drive at least once a month.

What else can I do if my system runs slowly? **Disk Cleanup** is a Windows utility that cleans unnecessary files off your hard drive. These files include files that have accumulated in the Recycle Bin as well as temporary files, which are files created by Windows to store data temporarily while a program is running. Windows usually deletes these temporary files when you exit the program, but sometimes it "forgets" or doesn't have time if your system freezes up or incurs a problem prevent-ing you from properly exiting out of a program. Disk Cleanup also removes temporary Internet files (Web pages stored on your hard drive for quick viewing) as well as offline Web pages (pages that are stored on your computer so you can view them without being connected to the Internet). If not deleted periodically, these unnecessary files can deter efficient operating performance.

DIG DEEPER

How Disk Defragmenter Utilities Work

To understand how disk defragmenter utilities work, you must first understand the basics of how a hard disk drive stores files. A hard disk drive is composed of several platters, or round thin plates of metal, that are covered with a special magnetic coating that records the data. The platters are about 3.5 inches in diameter (approximately the width of a floppy disk) and are stacked onto a spindle. There are usually two or three platters in any hard disk drive, with data being stored on one or both sides. Data is recorded on hard disks in concentric circles, called *tracks*, which are further broken down into pie-shaped wedges called *sectors* (see Figure 5.28). The data is further identified by *clusters*, which are two or more sectors.

When you want to save (or *write*) a file, the bits that make up your file are recorded onto one or more sectors of the drive. In order to keep track of which sectors hold which files, the drive also stores an index of all sector numbers in a table called the **File Allocation Table (FAT)**. To save a file, the computer will look in the FAT for sectors that are not already being used and will then record the file information on those sectors. When you open (or *read*) a file, the computer searches through the FAT for the sectors that hold the desired file and reads that file. Similarly, when you delete a computer file, you are actually not deleting the file itself, but rather the reference in the FAT to the file.

Windows XP offers a different file system than the FAT, called the **New Technology File System (NTFS)**. NTFS was developed with the Windows NT version and has been used in Windows 2000 and Windows XP. When you install Windows XP, you may choose whether to use FAT32 (FAT32 is the latest version of FAT and supports larger cluster sizes than the original FAT) or NTFS as the file system, unless the manufacturer has already preinstalled NTFS.

Most system manufacturers are choosing to use the NTFS file system as a default system, but some have continued to preinstall FAT32. The benefits of NTFS over FAT32 are that NTFS supports hard drive capacities larger than 32 GB and files sizes larger than 4 GB. Most of today's hard drives have capacities that exceed 32 GB and with

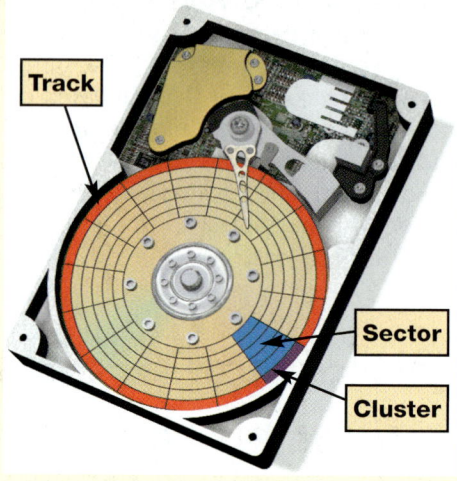

FIGURE 5.28

On a hard disk platter, data is recorded onto tracks, which are further broken down into sectors and clusters.

multimedia capabilities, it's not uncommon to see a file size greater than 4 GB. Additionally, NTFS was designed to be more secure and more efficient and will ultimately nudge out FAT structure with subsequent operating system versions.

So how does a disk become fragmented? When only part of an older

How can I control which files Disk Cleanup deletes? When you run Disk Cleanup, the program scans your hard drive to determine which folders have files that can be deleted, and calculates the amount of hard drive space that would be freed up by doing so. You check off which type of files you would like to delete, as shown in Figure 5.30.

How do I diagnose potential errors or damage on my storage devices? ScanDisk is a Windows utility that checks for lost files and fragments as well as physical errors on your hard drive. Lost files and fragments of files occur as you save, resave, move, delete, and copy files on your hard drive. Sometimes the system becomes confused, leaving references on the file allocation table to files that no longer exist or have been moved. Physical errors on the hard drive occur when the mechanism that reads the hard drive's data (which is stored as 1s or 0s) can no longer determine whether the area holds a 1 or a 0. These areas are called *bad sectors*. Sometimes ScanDisk can recover the lost data, but more often, it deletes the files that are taking up space unnecessarily. ScanDisk also makes a note of any bad sectors so the system will not use them again to store data.

file is deleted, the deleted section of the file creates a gap in the sector of the disk where the data was originally stored. In the same way, when new information is added to an older file, there may not be space sequentially near where the file was originally saved to save the new information. In that case, the system writes the added part of the file to the next available location on the disk, and a reference is made in the FAT or NTFS table as to the location of this file fragment. Over time, as files are saved, deleted, and modified, the bits of information for various files fall out of sequential order and the disk becomes fragmented.

Disk defragmentation is a problem because when a disk is fragmented, the operating system is not as efficient. It takes longer to locate a whole file since more of the disk must be searched for the various pieces. A fragmented hard drive can greatly slow down the performance of your computer.

How can you make the files line up more efficiently on the disk? At this stage, the disk defragmenter utility enters the picture. The defragmenter tool takes the hard drive through a defragmentation process in which pieces of files that are scattered over the disk are placed together and arranged sequentially on the hard disk. Also, any unused portions of clusters that were too small in which to save data before are grouped together, increasing the available storage space on the disk. Figure 5.29 shows before and after shots of a fragmented disk having gone through the defragmentation process.

You can watch a disk become fragmented, and then defragged, in the Hard disk Drive Sound Byte.

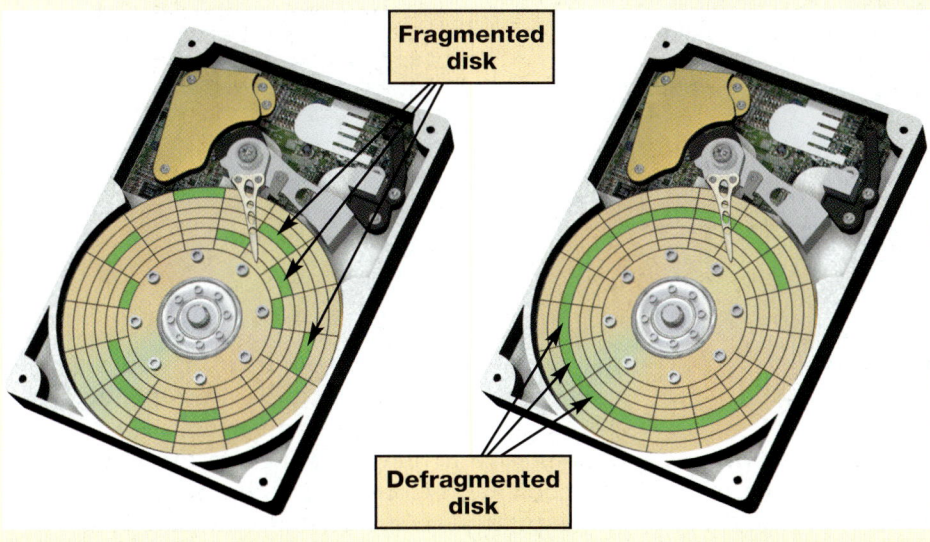

FIGURE 5.29

Over time, as files are saved, deleted, and modified, the fragments of information for various files fall out of sequential order on the hard disk and the disk becomes fragmented. Defragmenting the hard drive arranges file fragments so that they are located next to each other. This makes the hard drive run more efficiently.

Where can I find ScanDisk? Windows XP has moved the ScanDisk utility from System Tools (where it is located in previous versions of Windows) to Disk Properties. Right-click on the disk you want to diagnose, select Properties, then select Tools. It is now listed as Error Checking. Norton Disk Doctor is a stand-alone product included in Norton Utilities that performs the same check-and-repair routine. On Macs, you use the Disk First Aid utility to test and repair disks. You find this utility in the Utilities folder on your hard drive.

How can I check on a program that has stopped running? If a program on your system has stopped working, you can use the Windows **Task Manager utility** to check on the program or to exit out of the nonresponding program. Although you can access Task Manager from the Control Panel, it is more easily accessible by pressing Ctrl+Alt+Del at the same time, or by right-clicking on an empty space on the Task Bar at the bottom of your screen. The Applications tab of Task Manager lists all programs that you are using and indicates whether they are working properly ("running") or have stopped improperly ("not responding"). You can terminate programs that are not responding by pressing the End Task button on the dialog box.

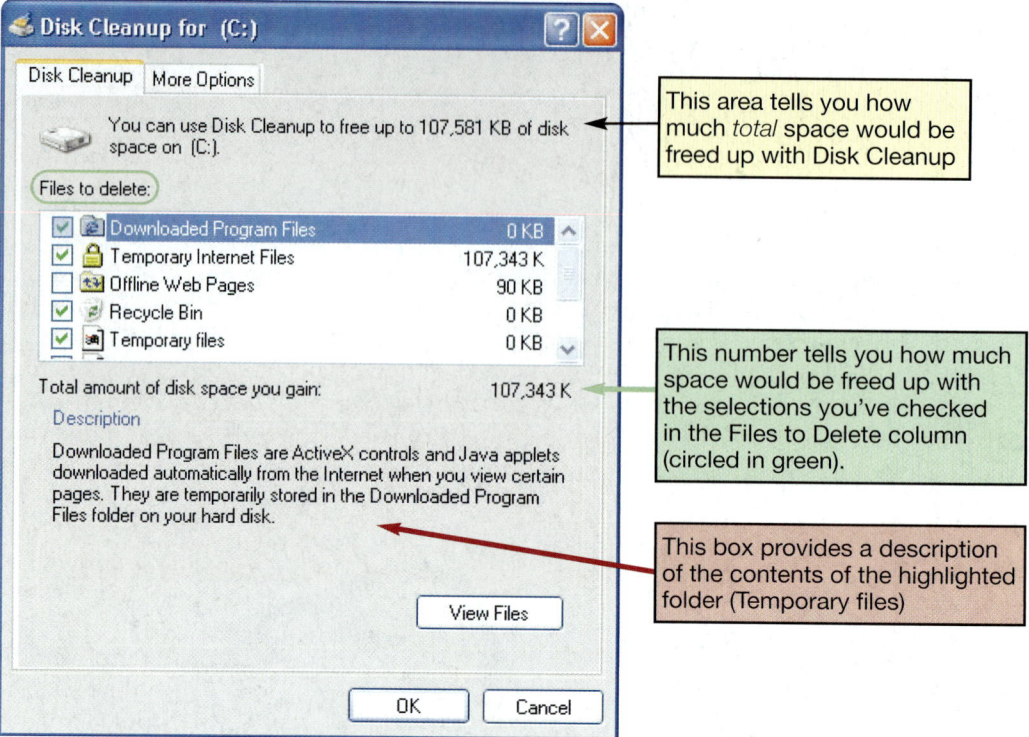

FIGURE 5.30

Using Disk Cleanup will help free up space on your hard drive.

SYSTEM RESTORE AND BACKUP UTILITIES

Is there an "undo" command for the system? Say you have just installed a new software program and your computer freezes. After rebooting the computer, when you try to start the application, the system freezes once again. You uninstall the new program, but your computer continues to freeze after rebooting. What can you do now?

Windows XP has a new utility called **System Restore** that lets you restore your system settings back to a specific date when everything was working properly. You can find the System Restore Wizard under System Tools in the Accessories Programs menu. In this case, since the computer was running just fine before you installed the software, you would restore your computer back to a date prior to the software installation, such as a day or two earlier. System Restore does not affect your personal data files (such as Microsoft Word documents, browsing history, drawings, favorites, or e-mail) so you won't lose changes made to these files when you use System Restore.

How does the computer remember its previous settings? Every time you start your computer, or when a new application or driver is installed, Windows XP automatically creates a snapshot of your entire system's settings. This snapshot is called a **restore point**. You can also create and name your own restore points at any time. Creating a restore point is a good idea before making changes to your computer such as installing hardware or software. If something goes wrong with the installation process, Windows XP can reset your system back to the restore point. As shown in Figure 5.31, Windows includes a Restore Point Wizard that walks you through the process of setting restore points.

How can I protect my data in the event something goes awry with my system? When you use the Windows **Backup utility**, you create a duplicate copy of all the data on your hard disk and copy it to another storage device, such as a Zip disk. A backup copy protects your data in the event your hard disk fails or files are accidentally erased. While you may not need to back up *every* file on your computer, you should back up the files that are most important to you and keep the backup copy in a safe location. Note that in Windows XP Home Edition, you must manually install the Backup utility from the ValueAdd folder on the CD-ROM. (For more information on backing up your files, see the Technology in Focus "Protecting Your Computer and Backing Up Your Data.")

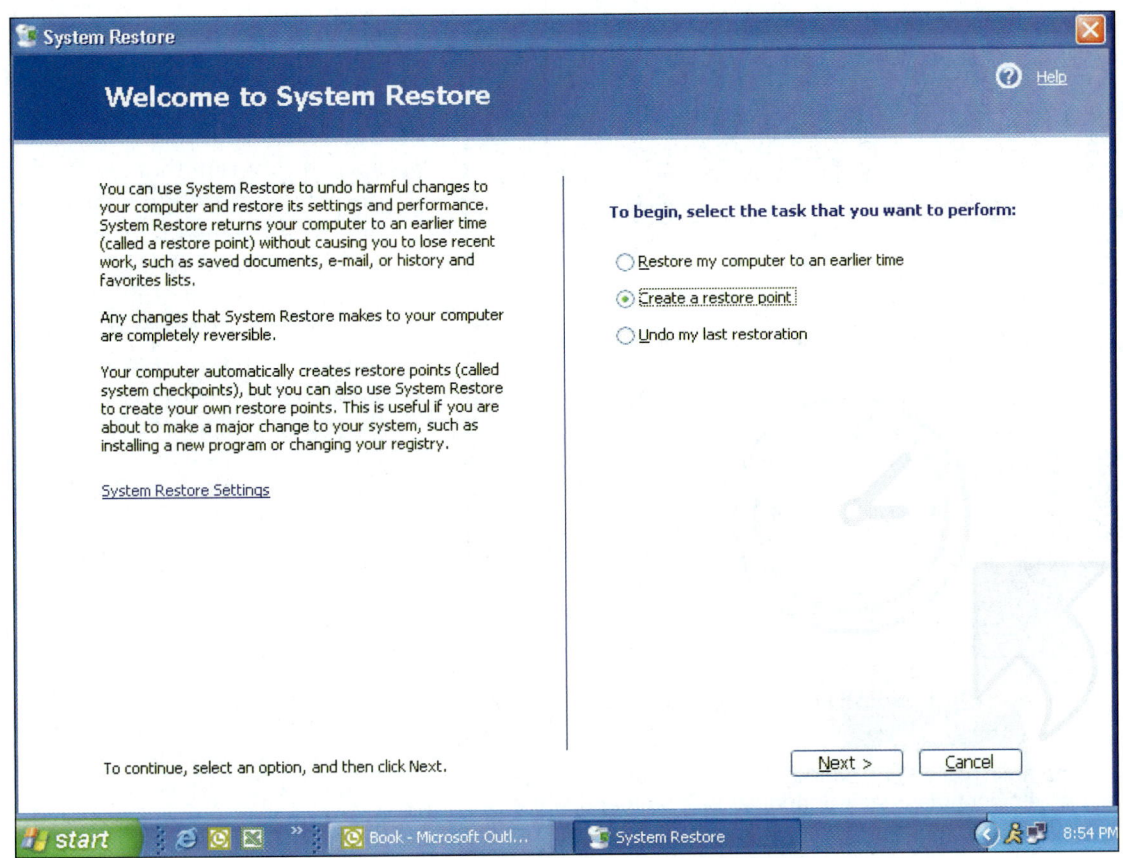

FIGURE 5.31
Setting a restore point is good practice before installing any hardware or software.

THE TASK SCHEDULER UTILITY

How can I remember to perform all these maintenance procedures? To keep your computer system in top shape, it is important to run some of the utilities described above routinely. For example, depending on your usage, you may want to defrag your hard drive every month or so and clean out temporary Internet files once a week. However, many computer users forget to initiate these tasks. Luckily, the Windows **Task Scheduler utility**, shown in Figure 5.32, allows you to schedule tasks to automatically run at predetermined times, with no interaction necessary on your part.

ACCESSIBILITY UTILITIES

Are there utilities designed for handicap access? **Utility Manager** is a utility found in the Accessories folder of Windows XP. Through the Utility Manager, you can magnify the screen image, have screen contents read to you, and display an onscreen keyboard. The accessibility features include the following:

- The Magnifier is a display utility that creates a separate window that displays a magnified portion of your screen. This feature makes the screen more readable for users who have impaired vision. Additionally, you can change the color scheme of the window with the Magnifier so that the screen colors

SOUND BYTE
LETTING YOUR COMPUTER CLEAN UP AFTER ITSELF
In this Sound Byte, you'll learn how to use the various maintenance utilities within the operating system. In addition, you'll learn how to use Task Scheduler to automatically clean up your hard disk. You'll also learn the best times of the day to schedule these maintenance tasks and why they should be done on a routine basis to make your system more stable.

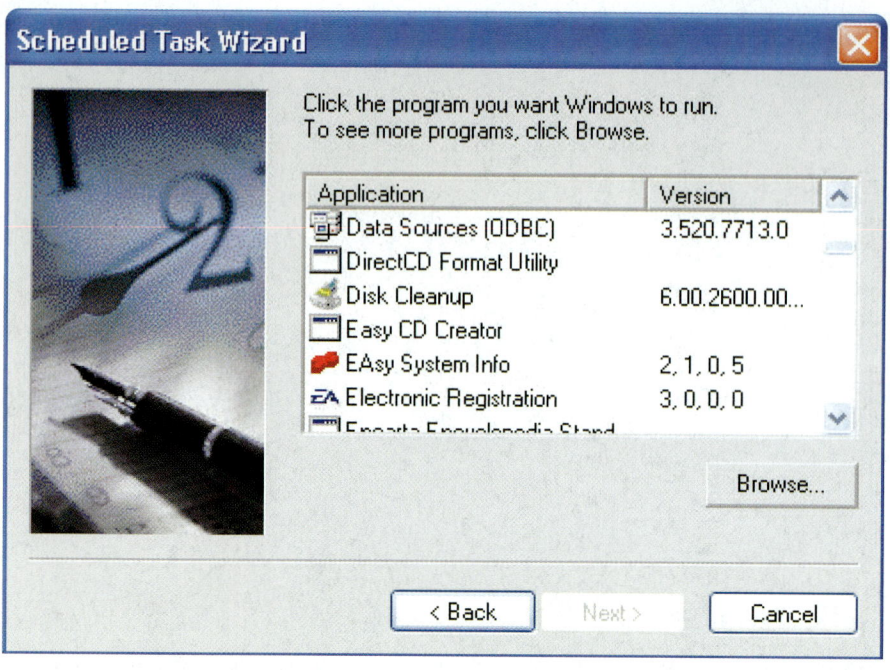

FIGURE 5.32

To keep your machine running in top shape, schedule a maintenance routine to automatically run at convenient times. Task Scheduler can be found in the Accessories folder in the Programs menu.

are inverted. Some visually impaired individuals find it easier to see white text on a dark background.

- The Narrator utility is a very basic speech program that reads what is onscreen, whether it's the contents of a window, menu options, or text you have typed. The Narrator coordinates with text utilities, such as Notepad and WordPad, as well as Internet Explorer, but may not work correctly with other programs. For this reason, Narrator is not meant for individuals who must rely solely on a text-to-speech utility to operate the computer.

- The Onscreen Keyboard displays a keyboard on the screen. You type by clicking on or hovering over the keys with a pointing device (mouse or trackball) or joystick. Similar to the Narrator, this utility is not meant for everyday use for severely handicapped individuals. A separate program with more functionality is better in those circumstances.

Summary

1. **What software is included in system software?** System software consists of the operating system and utility programs. The operating system manages the computer's hardware and provides a means for software applications to work with the CPU. It also provides the user interface. Utility programs perform tasks such as system maintenance and file compression.

2. **What are the different kinds of operating systems?** Real-time operating systems are designed for systems with a specific purpose and response time. Single-user, single-task operating systems are designed for computers on which one user is performing one task at a time. Single-user, multi-task operating systems are designed for computers on which one user is performing more than one task at a time. Multiuser operating systems are designed for systems in which multiple users are working on more than one task at a time.

3. **What are the most common desktop operating systems?** Microsoft Windows and Mac OS are two popular operating systems. UNIX is an operating system often used on networks. Linux is an open-source operating system designed primarily for personal computers.

4. **How does the operating system enable users to interact with the computer?** The operating system provides a user interface that enables you to interact with the computer. Most operating systems use a graphical user interface (GUI). GUIs display graphics, use point-and-click technology, and include windows, menus, and icons.

5. **How does the operating system help manage the processor?** When the operating system allows you to perform more than one task at a time, it is multitasking. To provide for seamless multitasking, the operating system controls the timing of events the processor works on.

6. **How does the operating system manage memory and storage?** The operating system uses RAM as a temporary storage area for instructions and data the processor needs. The operating system is therefore responsible for ensuring that there is enough space for the instructions and data. If there isn't sufficient space, the operating system allocates files to virtual memory.

7. **How does the operating system manage hardware and peripheral devices?** Programs called device drivers translate the specialized commands of devices to commands that the operating system can understand, and vice versa.

8. **How does the operating system interact with application software?** For programs to work with the CPU, they must contain code the CPU recognizes. Rather than having the same blocks of code appear in each software application, the operating system includes blocks of code (called APIs) to which software applications refer. These blocks of code are called application program interfaces (APIs).

9. **How does the operating system help the computer start up?** When you start your computer, it runs through a process consisting of four steps: (1) The BIOS is activated. (2) In the power-on self-test, the BIOS checks that all attached devices are in place. (3) The operating system is loaded into RAM. (4) Configuration and customization settings are checked.

10. **What are the main desktop and windows features?** The desktop provides you with access to your computer's files, folders, and commonly used tools and applications. Common features of windows include toolbars and scrollbars.

11. **How does the operating system help me keep my computer organized?** The operating system allows you to organize the contents of your computer in a hierarchical structure that includes files, folders, and drives.

12. **What utility programs are included in system software and what do they do?** Windows utilities include those that allow you to add or remove programs, compress files, defrag your hard drive, clean files off your system, restore your system, and back up files.

Key Terms

application program interfaces (APIs)	p. 179	open-source program	p. 173
application software	p. 168	operating system (OS)	p. 168
authentication	p. 182	paging	p. 178
Backup utility	p. 198	path separators	p. 191
basic input/output system (BIOS)	p. 181	platform	p. 170
boot process	p. 180	Plug and Play (PnP)	p. 179
command-driven interface	p. 176	power-on self test (POST)	p. 181
desktop	p. 184	queue	p. 177
device driver	p. 179	real-time operating system	p. 169
directories	p. 187	Recycle Bin	p. 192
Disk Cleanup	p. 195	registry	p. 182
disk defragmenter utilities	p. 195	restore point	p. 198
Disk Operating System (DOS)	p. 170	root directory	p. 187
extension (file type)	p. 190	Safe Mode	p. 183
file	p. 187	ScanDisk	p. 196
File Allocation Table (FAT)	p. 196	screensavers	p. 193
file compression utility	p. 194	scrollbars	p. 185
file management	p. 187	single-user, multitask operating system	p. 169
file path	p. 191	single-user, single-task operating system	p. 169
filename	p. 190	spooler	p. 177
folders	p. 187	swap file (page file)	p. 178
graphical user interface (GUI)	p. 176	system files	p. 182
icons	p. 176	System Restore	p. 198
interrupt	p. 177	system software	p. 168
interrupt handler	p. 177	Task Manager utility	p. 197
interrupt table	p. 177	Task Scheduler utility	p. 199
kernel (supervisor program)	p. 182	thrashing	p. 179
Linux	p. 173	toolbars	p. 185
Mac OS	p. 172	UNIX	p. 173
menu-driven interface	p. 176	user interface	p. 176
menus	p. 176	Utility Manager	p. 199
Microsoft Windows	p. 170	utility programs	p. 192
multitasking	p. 176	virtual memory	p. 178
multiuser operating system (network operating system)	p. 170	windows	p. 176
New Technology File System (NTFS)	p. 196	Windows Explorer	p. 187

Buzz Words

Word Bank

- system software
- Windows Explorer
- utility programs
- real-time operating system
- single-user, multitask operating system
- single-user, single-task operating system
- platform
- defrag
- files
- Windows XP
- Linux
- Task Scheduler
- Mac OS Windows
- boot
- cold boot
- Task Manager
- Safe Mode
- window
- folders
- file management
- file compression
- tracks
- sectors
- FAT

Instructions: Fill in the blanks using the words from the Word Bank above.

Veena was looking into buying a new computer, and was trying to decide what (1)_____ to buy, a PC or a Mac. She had used PCs all her life, so she was more familiar with the (2)_____ operating system. Still, she liked the way the (3)_____ looked, and was considering switching. Her brother didn't like either operating system so used (4)_____ , a free operating system, instead.

After a little research, Veena decided to buy a PC. With it, she got the most recent version of Windows, (5)_____. She vowed that with this computer, she'd practice better (6)_____, since she often had a hard time finding files on her old computer. To view all of the folders on her computer, she opened (7)_____. She made sure that she gave descriptive names to her (8)_____ and placed them in organized (9)_____.

Veena also decided that with her new computer, she'd pay more attention to the the (10)_____, those little special function programs that help with maintainance and repairs. These special function programs, in addition to the OS, make up the (11)_____. Veena looked into some of the more frequently used utilities. She thought it would be a good idea to (12)_____ her hard drive regularly so that all the files lined up in contiguous (13)_____ and so that it was more efficient. She also looked into (14)_____ utilities, which would help her reduce the size of her files when she sent them to others over the Internet. Finally, she decided to use the Windows (15)_____ utility to schedule tasks automatically so that she wouldn't forget.

Organizing Key Terms

Instructions: This chapter introduced many new terms and concepts. In the illustration below, fill in each of the blanks with words from the Word Bank and the chapter in order to show how categories of ideas fit together.

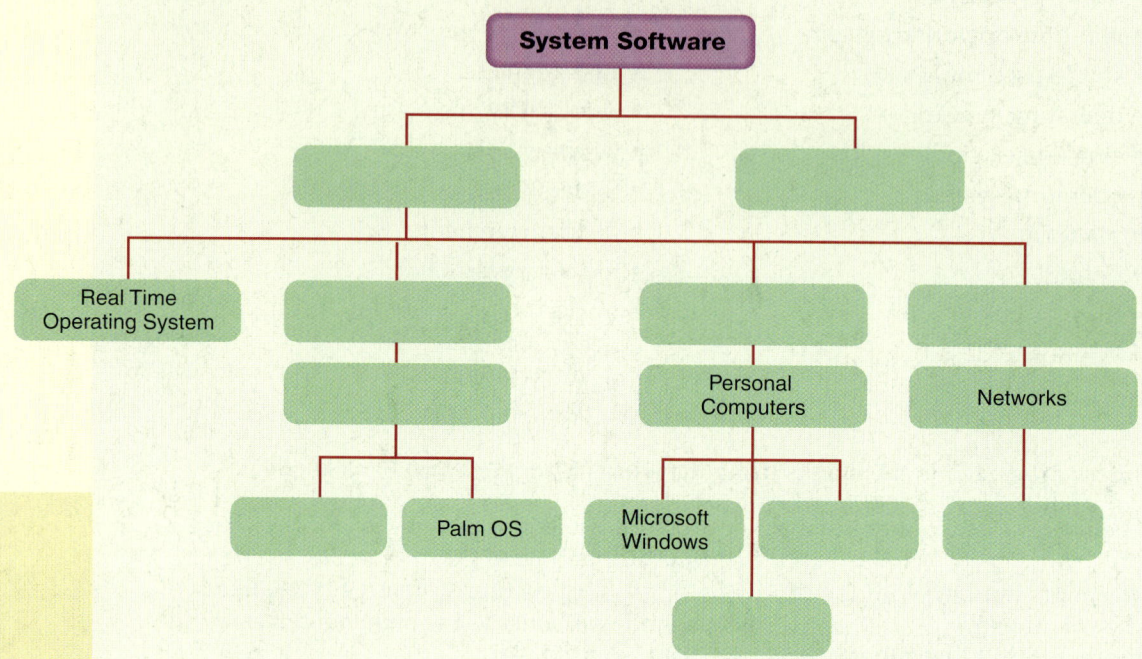

Making the Transition to . . . Next Semester

1. It's the beginning of a new semester, and you promise yourself that you are going to keep all files related to your schoolwork more organized this semester. Develop a plan that outlines how you'll set up folders and subfolders for each subject. Identify at least three different folders for each class. If time and schedule permit, discuss your organization scheme with your instructor.

2. Your school requires that you purchase a laptop to run on their system. The required machine runs on the Windows operating system. You have a reasonably new Apple computer at home.

    a. Research the compatibility issues between the two computers.
    b. How does a PDA running with a Palm OS fit into the equation?
    c. Can you "synch" the PDA with either or both machines?
    d. Explore the application "Virtual PC." What does it do? Would it be helpful in this situation?

3. It is the night before the major term paper for your philosophy class is due. Your best friend comes screaming down the hall begging for help. His only copy of his draft paper is on his desktop computer and it is suddenly booting up with the words "Safe Mode" in the corners of the screen. What would be the most useful questions to ask him? What steps would you take to debug the problem? If you cannot get the computer to come out of Safe Mode, is there a way to retrieve the draft? How many times will you say "Make Backups!" that evening?

4. This semester you have added six new applications to your laptop. You know what courses you will be taking next semester and realize they will require an additional eight major software applications. A friend who is in a similar position tells you she's not worried about putting that much software on her computer because she has a really big hard drive.

    a. Is hard disk storage your only concern? Should it be your main concern or should you worry more about having sufficient RAM? How does your use of the programs impact your answer?
    b. Does virtual memory management by your operating system allow you to ignore RAM requirements?

5. You decide to buy a new keyboard for next semester, a very fancy one that is wireless, features integrated volume and CD player controls, and an integrated trackball. You are also planning to upgrade your printer. Do you have to worry about having the correct device drivers for these peripherals if:

    a. You are using a "Plug and Play" operating system?
    b. You have an older PC but are using the latest Windows operating system?
    c. You have an older PC and its original operating system, Windows 95?

Making the Transition to . . . the Workplace

1. Your company has been having trouble with some of its Macintosh computers running inefficiently. Your boss asks you to research the utility programs your company could use on its Macs to make them run better. In particular, your boss would like you to find a disk defrag utility, a file compression utility, and a diagnostic utility you could run to check the hard drive for errors. Using the Internet, what utilities can you find? Will they run on all versions of the Mac OS?

2. The company that you work for has just announced a new internal accounting structure. From now on, each department will be charged individually for the costs associated with computer usage, such as backup storage space, Internet usage, and so on.

   a. Research how the operating system may be set up to monitor such activity by department.
   b. What other activities do you think the operating system can be set up to monitor?

3. Your new boss is considering moving some of the department operations to UNIX-based computer systems. He asks you to research the advantages and disadvantages of moving to UNIX, Linux, or Mac OS X. How would these choices impact his department in the area of

   a. Budget for technical support for the systems?
   b. Choice and budget for hardware for the systems?
   c. Costs of implementation?
   d. Possibility for future upgrades?

Critical Thinking Questions

Instructions: Albert Einstein used "Gedanken experiments," or critical thinking questions, to develop his theory of relativity. Some ideas are best understood by experimenting with them in our own minds. The following critical thinking questions are designed to demand your full attention but require only a comfortable chair—no technology.

1. Open-source programming is a philosophy that states programmers should make their code available to everyone rather than keeping it proprietary. The operating system Linux has had much success as an open-source code. The chapter mentioned some of the advantages of open-source code, such as quicker code updates in response to technological advances and changes.

    a. What are other advantages of open-source code?
    b. Can you think of disadvantages to open-source code?
    c. Why do you think that companies such as Microsoft maintain proprietary restrictions on their code?
    d. Are there disadvantages to maintaining proprietary code?

2. Operating system interfaces have evolved from a text-based console format to the current graphical user interface. What direction do you think they will move toward next? How could operating systems be organized and used in a manner that is more responsive to humans and better suited to how we think? Are there alternatives to hierarchical file structures for storage? Can you think of ways in which operating systems could be able to adapt and customize themselves based on your usage?

3. Suppose you are building a computer system from scratch and have complete discretion as to your choice of operating system. Which one would you install and why?

4. Which environment do you think is better for consumers: to have companies develop smaller, more inexpensive operating systems and then allow competing companies to develop and market utility programs, or to have very large full-featured operating systems that include most utilities as part of the operating system itself? Do you think that including utility programs with the operating system makes the cost of the operating system higher?

Team Time

Problem:

You have been hired to help set up the technology requirements for a small advertising company. The company is holding off buying anything until the decision has been made as to which platform the computers should run on. Obviously, one of the critical decisions is the operating system.

Task:

Recommend the appropriate operating system for the company.

Process:

STEP 1: Break up into three teams: each team will represent one of the three primary operating systems today: Windows, Mac, and Linux.

STEP 2: As a team, research the pros and cons of your operating system. What features does it have that would benefit your company? What features does it not have that your company would need? Why (or why not) would your operating system be the appropriate choice? Why is your OS better (or worse) than either of the other two options?

STEP 3: Develop a presentation that states your position with regard to your operating system. Your presentation should have a recommendation, with facts to back it up.

STEP 4: As a class, decide which operating system would be the best choice for the company.

Conclusion:

Since the operating system is the critical piece of software in the computer system, the selection should not be taken lightly. The OS that is best for an advertising agency may not be best for an accounting firm. It is important to make sure you consider all aspects of the work environment and the type of work that is being done to ensure a good fit.

Becoming Computer Fluent

Using key terms from the chapter, write a letter to your computer-illiterate aunt explaining the benefits of simple computer maintenance. First explain any symptoms her computer may be experiencing (such as a sluggish Internet connection), then include a set of steps she can follow in setting up a regimen to remedy the problems. Make sure you explain some of the system utilities described in this chapter, including, but not limited to, defrag, Disk Cleanup, and Task Scheduler. Include any other utilities she might need and explain why she should have them.

Materials on the Web

In addition to the review materials presented here, you'll find extra materials on the book's companion Web site (**www.prenhall.com/techinaction**) that will help reinforce your understanding of the chapter content. These materials include:

Sound Byte Lab Guides

For each Sound Byte mentioned in the chapter, there is a corresponding lab guide located on the book's companion Web site. These guides review the material presented in the Sound Byte and direct you to various Web resources that examine the material. The Sound Byte Lab Guides for this chapter include:

- Customizing Windows XP
- File Management
- File Compression
- Letting Your Computer Clean Up After Itself

True/False and Multiple Choice Quizzes

The book's Web site includes a True/False and a Multiple Choice quiz for this chapter. You can take these quizzes, automatically check the results, and e-mail the results to your instructor.

Web Research Projects

The book's Web site also includes a number of Web research projects for this chapter. These projects ask you to search the Web for information on computer-related careers, milestones in computer history, important people and companies, emerging technologies, and the applications and implications of different technologies.

OBJECTIVES

After reading this chapter, you should be able to answer the following questions:

- How can I determine whether I should upgrade my existing computer or buy a new one? (pp. 212–214)

- What does the CPU do and how can I evaluate its performance? (pp. 214–217)

- How does memory work in my computer and how can I evaluate how much memory I need? (pp. 217–220)

- What are the computer's main storage devices and how can I evaluate whether they match my needs? (pp. 220–227)

- What components affect the output of video on my computer and how can I evaluate whether they meet my needs? (pp. 228–231)

- What components affect my computer's sound quality and how can I evaluate whether they meet my needs? (pp. 231–232)

- What are the ports available on desktop computers and how can I determine what ports I need? (pp. 233–237)

- How can I ensure the reliability of my system? (pp. 237–239)

CHAPTER 6

Evaluating Your System:
Understanding and Assessing Hardware

**TECHNOLOGY IN ACTION:
A CASE OF PC ENVY**

After saving up for a computer, Natalie took the leap a few years ago and bought a new desktop PC. Now she is wondering what to do. Her friends with newer computers are burning CDs and DVDs, and they're able to hook their digital cameras right up to their computers and create multimedia. They seem to be able to do a hundred things at once without their computers slowing down at all.

Natalie's computer can't do any of these things—or at least she doesn't think it can. And lately it seems to take longer to open files and scroll through Web pages. Making matters worse, her computer freezes three or four times a day and takes a long time to reboot. Now she's wondering whether she should buy a new computer, but the thought of spending all that money again makes her think twice. As she looks at ads for new computers, she realizes she doesn't know what such things as "CPU" and "RAM" really are, or how they affect her system. Meanwhile, she's heard it's possible to upgrade her computer, but the task seems daunting. How will she know what she needs to do to upgrade, or whether it's even worth it?

How well is your computer meeting your needs? Are you unsure whether it's best to buy a new computer or upgrade your existing system? If you don't have a computer, do you fear purchasing one because computers are changing all the time? Do you know what all the terms in computer ads mean and how the different parts affect your computer's performance?

In this chapter, you'll learn how to evaluate your computer system to determine whether it is meeting your needs. You'll start by figuring out what you want your ideal computer to be able to do. You'll then learn about important components of your computer system (its CPU, memory, storage devices, audio and video devices, and ports) and how these components affect your system. Along the way, worksheets will help you conduct a system evaluation, and multimedia Sound Bytes will show you how to install various components in your system and how to increase its reliability. You'll also learn about the various utilities available to help speed up and clean up your system. If you don't have a computer, this chapter will provide you with important information you need about computer hardware to make an informed purchasing decision.

To Buy or To Upgrade: That Is the Question

There never seems to be a good time to buy a new computer. It seems that if you can just wait a year, computers will inevitably be faster and cost less. But is this actually true?

As it turns out, it is true. In fact, a rule of thumb often cited in the computer industry, called **Moore's Law,** describes the pace at which CPUs (the central processing unit)—the small chip that can be thought of as the "brains" of the computer—improve. This mathematical rule, named after Gordon Moore, the cofounder of the CPU chip manufacturer Intel, predicts that the number of transistors inside a CPU will increase so fast that CPU capacity will double every 18 months. (The number of transistors on a CPU chip helps determine how fast it can process data.)

As you can see in Figure 6.1, this rule of thumb has held true since 1965, when Moore first published his theory. Imagine if you could find a bank that would agree to treat your money this way. If you had put 10 cents in a savings account in 1965, you would have a balance of more than $3.3 million today!

In addition to the CPU becoming faster, other system components also improve dramatically. For example, the capacity of memory chips such as dynamic random access memory (DRAM)—the most common form of memory found on personal computers—increases about 60 percent every year, as shown in Figure 6.2. Meanwhile, hard disk drives have been growing in storage capacity by about 50 percent each year.

FIGURE 6.2 The Growing Capacity of Memory Chips

YEAR	CAPACITY OF MEMORY (DRAM) CHIP
1977	16 KB
1980	64 KB
1983	256 KB
1985	1,000 KB
1989	4,000 KB
1992	16,000 KB
1996	64,000 KB
2001	256,000 KB
2002	1,000,000 KB

Source: David Patterson and John Hennesey, *Computer Architecture: A Quantitative Approach,* 2nd edition, Morgan/Kaufman, p. 22.

So with technology advancing so quickly, which is better: upgrading a current computer or buying a new one? Certainly, no one wants to buy a new computer every year just to keep up with technology. Even if money weren't a consideration, the time it would take to transfer all of your files and to reinstall and reconfigure your software would make buying a new computer every year terribly inefficient.

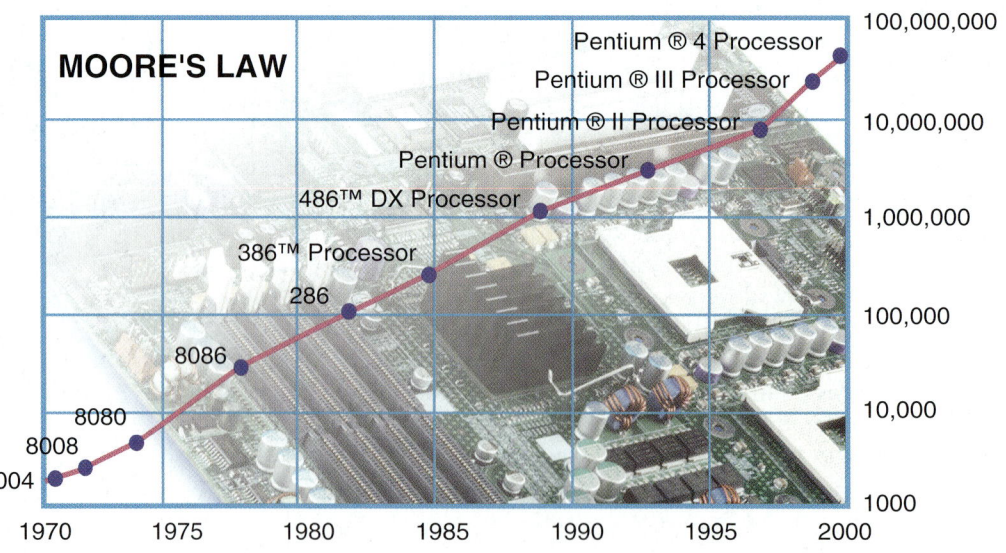

FIGURE 6.1

Moore's Law predicts that CPUs will continue to get faster. The number of transistors on a CPU chip helps determine how fast it can process data.

Of course, no one wants to keep doing costly upgrades that won't significantly extend the life of a system, either. So how can you determine if your system is suitable or needs upgrading? And how can you know which is the better option: upgrading or buying a new computer? In this chapter, you'll determine how to answer these questions by learning useful information about computer systems. The first step is figuring out what you want your computer to do for you.

What Is Your Ideal Computer?

As you decide whether you should upgrade or buy a new computer, it's important to know exactly what you want your ideal computer system to be able to do. Later, as you perform a system evaluation, you can compare your existing system to your ideal system. This will help you determine whether you should purchase hardware components to add to your system or buy a new system.

But what if I don't have a computer? Even if you're a new computer user, being able to understand and evaluate computer systems will make you a more informed buyer. What is a CPU and how does it affect your system? How much RAM do you need and what role does it play in your system? It's important that you're able to answer questions such as these before you buy a computer.

How do I know what my ideal system is? To determine your ideal system, consider what you want to be able to do with your computer. For example, do you want to be able to edit digital photos? Do you want to watch and record DVDs? Or are you just using your computer for word processing? The worksheet in Figure 6.3 lists a number of ways in which you may want to use your computer. In the second column, place a check next to those computer uses that apply to you.

Next, look at the list of desired uses for your computer and determine whether your current system can perform these activities. If there are things you can't do, you may need to purchase additional hardware or a better computer. For example, if you want to play CDs, all you need is a CD-ROM drive. However, you need a CD-R drive if you want to burn (record) CDs. Likewise, if you're going to edit digital video files or play games that include a lot of sounds and graphics, you may want to add more memory, buy a better set of speakers, and possibly invest in a new monitor. In this chapter, you'll learn

FIGURE 6.3 What Should Your Ideal Computer System Be Able to Do?

COMPUTER USES	DO YOU WANT YOUR SYSTEM TO DO THIS?	CAN YOUR SYSTEM DO THIS NOW?
ENTERTAINMENT USES		
Access the Internet/Send E-Mail		
Play CDs and DVDs		
Record (Burn) CDs and DVDs		
Produce Digital Videos		
Record and Edit Digital Music		
Edit Digital Photos		
Play Graphics-Intensive Games		
Transfer Digital Photos (or Other Files) to Your Computer Using Flash Memory Cards		
Connect All Your Peripheral Devices to Your Computer at the Same Time		
Download Music from the Internet		
EDUCATIONAL USES		
Perform Word Processing Tasks		
Use Other Educational Software		
Create CD or Zip Disk Backups of All Your Files		
Access Library and Newspaper Archives		
Create Multimedia Presentations		
BUSINESS USES		
Create Spreadsheets		
Create Databases		
Work on Multiple Software Applications Quickly and Simultaneously		
Conduct Online Banking/Pay Bills Online		
Conduct Online Job Searches/Post Resume		
"Synchronize" Your PDA and Desktop Computer		

about the hardware you may need to achieve your ideal system.

Note that you may also need new software and training to use new system components. However, many computer users forget to consider the training they'll need when they upgrade their computer. Missing any one of these pieces might be the difference between your computer enriching your life or it becoming another source of stress.

How do I know if I need training? Although computers are becoming increasingly user friendly, you still need to learn how to use them to your best advantage. Say you want to edit digital photos. You know image-editing software exists, but how do you know if your computer's hardware can support the software? What will happen if you can't get it installed or don't know how to use it? If you have questions like these, you know you need training. Training shouldn't be an afterthought. Consider the time and effort involved in learning about what you want your computer to do before you buy hardware or software. If you don't, you may have a wonderful computer system but lack the skills necessary to take full advantage of it.

BITS AND BYTES

Where Can You Go for More Training?

In addition to taking this class and asking friends and relatives for help, you have many options for acquiring computer training:

- Online newsgroups—discussion forums where people post and respond to questions about a certain topic—can help answer your hardware or software questions. To find a newsgroup, visit **www.tile.net**.
- Computer manufacturers offer training courses. For example, Dell Computer's EducateU Web site (**www.learndell.com**) offers training on various hardware and software.
- Booksellers offer tutorials for many software applications as well as resources that provide training on hardware components. Often, these tutorials include a separate CD with examples, video demonstrations, and other resources.
- Most software companies provide online tutorials and documentation as well as built-in tutorials with their software, usually found under the Help menu.

Assessing Your Hardware: Evaluating Your System

With a better picture of your ideal computer system in mind, you can make a more informed assessment of your current computer. To determine whether your computer system has the right hardware components to do what you ultimately want it to do, you need to conduct a **system evaluation**. To do so, you look at your computer's subsystems, what they do, and how they perform. These subsystems include:

- The CPU subsystem
- The memory subsystem (your computer's RAM)
- The storage subsystem (your hard drive and other drives)
- The video subsystem (your video card and monitor)
- The audio subsystem (your sound card and speakers)
- Your computer's ports

In the rest of this chapter, we'll examine each of these subsystems. At the end of each section, you'll find a small worksheet you can use to evaluate each subsystem on your computer. When you've finished your evaluation, you'll know a lot more about your hardware so you can better decide whether you need to upgrade your system or buy a new one.

Evaluating the CPU Subsystem

As mentioned earlier, your computer's **CPU** (processor) is very important, as it is the "brains" of the computer. The CPU processes instructions, performs calculations, manages the flow of information through a computer system, and is responsible for processing the data you input into information. The CPU, as shown in Figure 6.4, is located on the **motherboard**, the primary circuit board of the computer system. There are two main types of processors on the market: Intel processors (such as the Pentium family), which are used on PCs, and Motorola processors (such as the G4), which are used on Macintosh computers.

How does the CPU work? The CPU is composed of two units: the control unit and the **arithmetic logic unit (ALU)**. The **control unit** coordinates the activities of all the other computer

EVALUATING THE CPU SUBSYSTEM 215

FIGURE 6.4
The CPU is a small chip that sits on the motherboard inside your system unit.

components. The ALU is responsible for performing all the arithmetic calculations (addition, subtraction, multiplication, and division). Additionally, the ALU makes logic and comparison decisions, such as comparing items to determine if one is greater than, less than, equal to, or not equal to another.

Every time the CPU performs a program instruction, it goes through the same series of steps: First, it fetches the required piece of data or instruction from RAM, the temporary storage location for all the data and instructions the computer needs while it is running. Next, it decodes the instruction into something the computer can understand. Once the CPU has decoded the instruction, it executes the instruction and stores the result to RAM before fetching the next instruction. This process is called a **machine cycle**. (We'll discuss the machine cycle in more detail in Chapter 9.)

How is CPU speed measured? The computer goes through these machine cycles at a steady and constant pace. This pace, known as **clock speed**, is controlled by the system clock, which works like a metronome in music. The system clock keeps a steady beat, regulating the speed at which the processor goes through machine cycles. Processors work incredibly fast, going through millions or billions of machine cycles each second. Processor speed is measured in units of **megahertz (MHz)**, or 1 million hertz, and **gigahertz (GHz)**, or 1 billion hertz. Hertz (Hz) means "machine cycles per second," so a 3.0-GHz processor performs work at a rate of 3 billion machine cycles per second.

How can I tell how fast my CPU is? You can easily identify the speed of your current system's CPU. To do so, locate the My Computer icon on your desktop, right-click, and select Properties. As shown in Figure 6.5, the General tab of the System Properties dialog box shows you which CPU is installed in your system as well as its speed.

How fast should my CPU be? At a minimum, your CPU should meet the requirements of your system's software and hardware. If your system is older and you are buying new software and peripheral devices, your CPU may not be able to handle the load.

For example, say your computer is three years old. For the past three years, you've been using it primarily for word processing and to surf the Internet. You recently purchased a digital camera. Now you want to edit your digital photos, but your system doesn't seem to be able to handle this. You check the system requirements on the photo-editing software you just installed and realize that the software runs best with a more powerful processor. In this case, if everything else in your system is running properly, a faster CPU would help improve the software's performance. Pentium

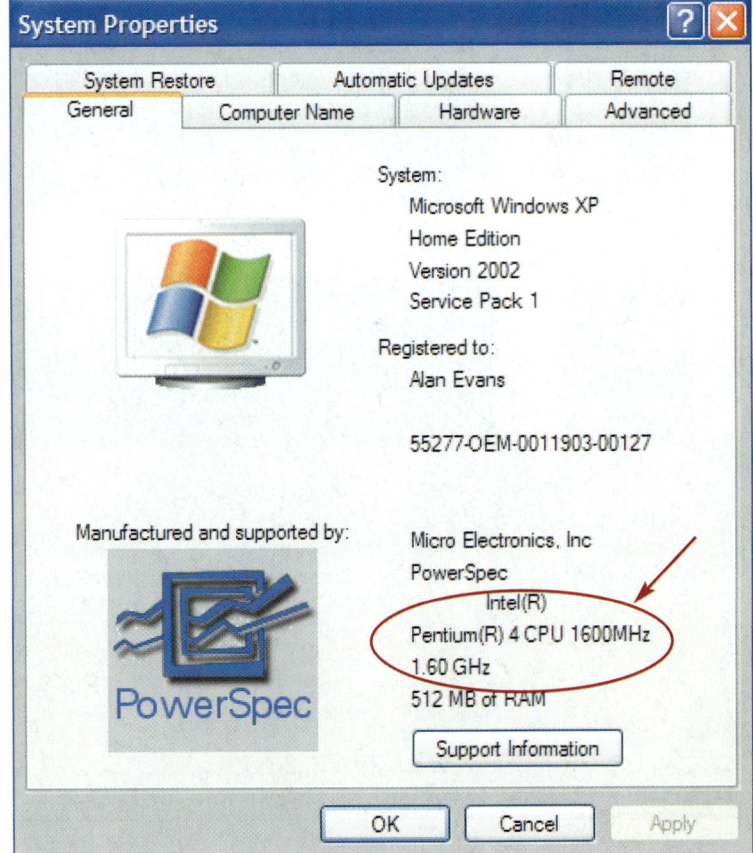

FIGURE 6.5
The General tab of the System Properties dialog box identifies which CPU you have as well as its speed. This computer contains an Intel Pentium 4 CPU running at 1.60 GHz.

SOUND BYTE
USING WINDOWS XP TO EVALUATE CPU PERFORMANCE

In this Sound Byte, you'll learn how to use the utilities provided by Windows XP to evaluate your CPU's performance. You'll also learn about shareware utilities (software that you can install and try before you purchase it) that expand on the information the Task Manager provides.

III processors running at 700 MHz or higher and Pentium 4 processors running at any speed are good processors for the average user, enabling you to run most programs at a decent speed.

How can I tell whether my CPU is fast enough? The workload your CPU experiences varies considerably depending on what you're doing. Even though your CPU meets the minimum requirements specified for a particular software application, if you are running other software (in addition to the operating system, which is always running), you'll need to check to see how well the CPU is handling the entire load. You can tell whether your CPU speed is limiting your system performance if you periodically watch how busy it is as you work on your computer. The percentage of time that your CPU is working is referred to as **CPU usage**.

To view information on your CPU usage, right-click on an open area of the System toolbar, select Task Manager (a utility program that comes with Windows XP), and click on the Performance tab, shown in Figure 6.6. The CPU Usage graph records your CPU usage for the past several seconds. Of course, there will be periodic peaks of high CPU usage, but if you see that your CPU usage levels are greater than 90 percent during most of your work session, a new CPU would contribute a great deal to your system performance.

UPGRADING YOUR CPU

Is it expensive or difficult to upgrade a CPU? Replacement CPUs are expensive. In addition, although it is reasonably easy to install a CPU, it can be difficult to determine which CPU to install. Not all CPUs are interchangeable, and the replacement CPU must be compatible with the motherboard. Some people opt to upgrade the entire motherboard, but motherboards are a lot more difficult to install. As we'll discuss at the end of this chapter, if you plan on upgrading your computer in other ways in addition to upgrading the CPU, you may just want to consider buying a new computer.

Are the fastest CPUs the best to use? If you decide to upgrade your CPU, consider buying one that is *not* the most recently released with the fastest speed, but rather a slightly slower one of the same type. For example, if the Intel Pentium 4 family has just released a 3-GHz CPU, you will pay a premium to buy this newest processor. However, its release will drive down the prices on the earlier Pentium 4 chips running at speeds of 2.8 GHz, 2.6 GHz, and 2.4 GHz. Buying a slightly slower CPU and investing the savings in other system components (such as additional RAM) will often result in a better performing system. (The same is true if you're buying a new computer: Often, you'll save money

FIGURE 6.6

The Performance tab of the Windows Task Manager utility shows you how busy your CPU actually is when you're using your computer. In this case, current CPU usage level is at 10 percent. If CPU usage levels are above 90 percent for long periods of time, you may want to consider getting a faster, more powerful processor.

without losing a great deal of performance by buying a computer with a CPU slightly slower than the fastest one on the market.)

What else do I need to consider if I want to upgrade my CPU? You first need to determine whether a new CPU is compatible with your system's motherboard. (**PowerLeap.com** has a great tool to walk you through CPU and motherboard compatibility.) Check your computer's manual for the make and model of the components on your motherboard.

In addition, because the CPU generates a lot of heat, a small cooling device called a *heat sink* sits directly on top of the CPU to absorb excess heat. If you upgrade your CPU, you need to make sure you purchase the correct heat sink for your processor (although often, heat sinks come in a kit with the replacement CPU).

Be mindful that while replacing the CPU itself is not terribly difficult (as shown in Figure 6.7), it is important that you pick the right replacement part and process. It may be best to get the help of a professional should you decide to upgrade your CPU.

FIGURE 6.7

It can be fairly simple to replace a CPU: you line up the slots and drop it in. However, because it's important that you pick the right replacement part and process, many computer owners seek the help of professionals when upgrading their CPU.

Will replacing the CPU be enough to improve my computer's performance? You may think that if you have the fastest processor, you will have a system with the best performance. However, upgrading your CPU will only affect the *processing* portion of the system performance, not how quickly data can move to or from the CPU. Your system's overall performance depends on many other factors, including the amount of RAM installed as well as hard disk speed. Therefore, replacing or upgrading the CPU may not offer significant improvements to your system's performance if there is insufficient RAM or hard drive capacity.

Does your CPU subsystem need to be upgraded?		
	CURRENT SYSTEM	UPGRADE REQUIRED?
CPU Speed (in MHz or GHz)		
CPU Usage at Appropriate Level?		

Evaluating RAM: The Memory Subsystem

RAM (random access memory) is your computer's temporary storage space. Although we refer to RAM as a form of storage, RAM is really the computer's short-term memory. As such, it remembers everything that the computer needs to process the data into information, such as inputted data and software instructions, but only while the computer is on. This means that RAM is an example of **volatile storage**. When the power is off, the data stored in RAM is cleared out. This is why, in addition to RAM, systems always include **nonvolatile storage** devices for permanent storage of instructions and data when the computer is powered off. Hard disks provide the greatest nonvolatile storage capacity in the computer system.

Why not use a hard drive to store the data and instructions? It's about one million times faster for the CPU to retrieve a piece of data from RAM than from a hard disk drive. The time it takes the CPU to retrieve data from RAM is measured in *nanoseconds* (billionths of seconds) whereas retrieving data from a fast hard drive takes an average of 10 *milliseconds* (or ms, thousandths of seconds). This difference is influential in designing a balanced computer system and can have a tremendous impact on system performance. Therefore, it's critical that your computer has more than enough RAM.

Where is RAM located? You can find RAM inside the system unit of your computer on the motherboard. **Memory modules** (or memory cards), the small circuit boards that hold a series of RAM chips (see Figure 6.8), fit into special slots on the motherboard. There are three different types of memory modules: SIMMs (Single Inline Memory Modules), DIMMs, (Dual Inline Memory Modules), and RIMMs (Rambus Inline Memory Modules). Most computers today will either have DIMM or RIMM modules. As a consumer, you need to know this information only if you are upgrading your RAM (you need to stay with the same kind of module when you upgrade).

FIGURE 6.8

Memory modules like this one hold a series of RAM chips and fit into special slots on the motherboard.

Are there different types of RAM? To add to the confusion new computer users face, several different types of RAM are available. DRAM, SRAM (static RAM), and SDRAM (synchronous DRAM) are all slightly different in how they function, and in the speed at which memory can be accessed. If you add RAM to your system, you need to determine what type your system needs. To do so, consult your user's manual or the manufacturer's Web site. In addition, many online RAM resellers (such as **www.crucial.com**) can help you determine the type of RAM your system needs based on the model number and brand of your computer. We'll discuss the different kinds of RAM in more detail in Chapter 9.

How can I tell how much RAM I have installed in my computer? The amount of RAM that is actually sitting on memory modules in your computer is your computer's **physical memory**. The easiest way to see how much RAM you have is to look at the General tab of the System Properties dialog box. This is the same tab you looked at to determine your system's CPU type and speed, and is shown in Figure 6.5. RAM capacity is measured in megabytes (MB) or gigabytes (GB). The computer in Figure 6.5 has 512 MB of RAM installed.

More detailed information on physical memory is displayed in the Physical Memory table in the Performance tab of Windows Task Manager, shown in Figure 6.9. The Physical Memory table shows both the total amount of physical memory you have installed as well as the *available* physical memory you have at this moment. You can see this computer has approximately 1,047 MB (1,047,564 KB) of total memory installed, and approximately 680 MB is available (which means 367 MB is being used). The amount of available memory will always be less than the amount of total memory because a certain portion of the physical memory is always tied up running the operating system.

How much memory does the operating system need to run? The memory that your operating system uses is referred to as **kernel memory**. This memory is listed in a separate Kernel Memory table in the Performance tab. In Figure 6.9, the Kernel Memory table tells you that approximately 74 MB (74,208 KB) of the total 1,047 MB of RAM is being used to run Windows XP.

As you know from Chapter 5, the operating system is the main software application that runs the computer. Without it the computer would not work. At a minimum, the system needs enough RAM to run the operating system. Therefore, the amount of kernel memory that the system

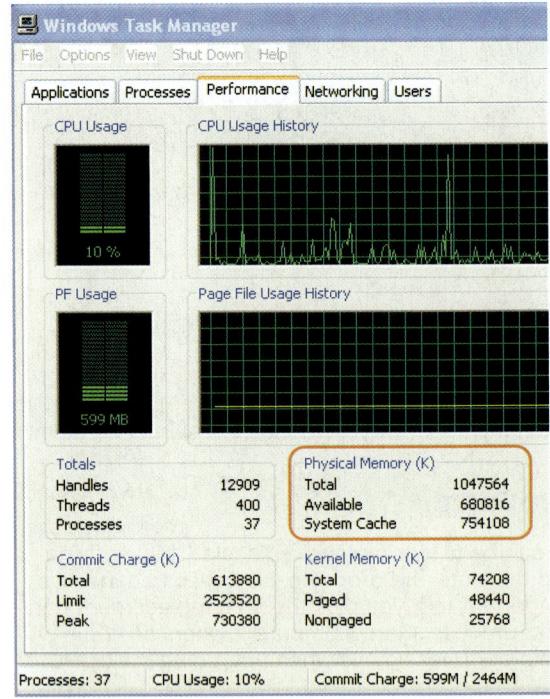

FIGURE 6.9

The Performance tab of the Windows Task Manager shows you how much physical memory is installed in your system, as well as how much is currently being used and how much is available. The computer shown here has approximately 680 MB of memory still available from a total of 1,047 MB.

is using is the *absolute minimum* amount of RAM that your computer can run on. However, because you are running additional applications, you need to have more RAM than the minimum.

How much RAM do I need? Since RAM is the temporary holding space for all the data and instructions that the computer uses while it is on, most computer users need quite a bit of RAM. In fact, it's not unusual to have 1 GB of RAM on a newer home system. The amount of RAM your system needs depends on how you use it. At a minimum, you need enough RAM to run the operating system (as explained above), plus whatever other software applications you're using, then a bit of additional RAM to hold the data you're inputting.

To determine how much RAM you need, list all the software applications you might be running at one time. Figure 6.10 shows an example of RAM requirements. In this example, if you are running your operating system, word processing and spreadsheet programs, a Web browser, a music player, and photo-editing software simultaneously, you would need a *minimum* of 348 MB RAM.

However, it's a good idea to have more than the minimum amount of RAM, so you can use more programs in the future. When upgrading RAM, the rule of thumb is to buy as much as you can afford but no more than your system will handle.

VIRTUAL MEMORY

Would adding more RAM also improve my system performance? If your system is **memory bound**—that is, limited in how fast it can send data to the CPU because there's not enough RAM installed—it will become sluggish, freeze more often, or just shut down during the day. If this is the case, adding more RAM to your system will have an immediate impact on performance.

How do I know whether my system is memory bound? You learned in Chapter 5 that if you don't have enough RAM to hold all of the programs you're currently trying to run, the operating system will begin to store the data that doesn't fit in RAM into a space on the hard disk called **virtual memory**. When it is using virtual memory, your operating system builds a file called the **page file** on the hard drive to allow processing to continue. This enables the system to run more applications than can actually fit in your computer's RAM.

So far, this system of memory management sounds like a good idea, especially since hard disk drives are much cheaper than RAM per megabyte of storage. The drawback is speed. Remember

FIGURE 6.10 Sample RAM Requirements

APPLICATION	MINIMUM RAM REQUIRED
Windows XP	128 MB
Microsoft Word	8 MB
Microsoft Excel	8 MB
Internet Explorer	12 MB
Windows Media Player	64 MB
Microsoft Picture It!	128 MB
Total RAM Required If Running All Programs Simultaneously	348 MB RAM

that accessing data from the hard drive to send it to the CPU is more than one million times slower than accessing data from RAM. Another drawback is that some applications do not run well on virtual memory. So, using virtual memory is a method of last resort. If your system is running with a large page file (that is, if it is using a large amount of virtual memory), adding more RAM will dramatically increase performance.

How do I determine how much virtual memory I'm using? As shown in Figure 6.11, the Performance tab of the Task Manager provides information about your computer's virtual memory (referred to as the *PF Usage*). By watching the

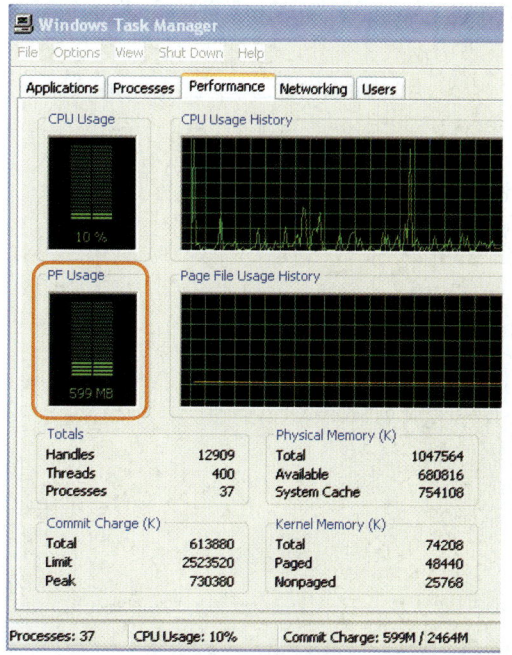

FIGURE 6.11

You can see how much page file (virtual memory) is being used at any given point in time by looking at the PF Usage chart in the Performance tab of Task Manager (in this case, 599 MB is being used).

information displayed in the PF Usage section, you can determine how much virtual memory is being used. If you consistently have a large page file in use (that is, more than 1.5 percent of the amount of RAM installed in your system) and very little physical memory available, adding more RAM will significantly improve your system performance.

SOUND BYTE
INSTALLING RAM

In this Sound Byte, you'll learn how to select the appropriate type of memory to purchase, how to order memory online, and how to install it yourself. As you'll discover, the procedure is a simple one and can add great performance benefits to your system.

Do you need more memory?	CURRENT SYSTEM	UPGRADE REQUIRED?
Amount of RAM		
Type of RAM		
Maximum Amount of RAM You Need		
Total RAM You Can Add to Your Computer		
Total Number of Memory Slots		
Number of Open Memory Slots		

ADDING RAM

Is there a limit to how much RAM I can add to my computer? Every computer is designed with a maximum limit on the amount of RAM it can support. In addition, each computer is designed with a specific amount of slots on the memory board in which the memory cards fit, and each slot may have a limitation on the amount of RAM it can support. To determine these limitations, check your owner's manual or the manufacturer's Web site.

Once you know how much RAM your computer can support, you can determine the best configuration of memory cards to achieve the greatest amount of RAM. For example, say you have a total of four memory card slots: two are already filled with 128-MB RAM cards and the other two are empty. Maximum RAM allowed for your system is 512 MB. This means you can buy two more 128-MB RAM modules for the two empty slots, for a total of 512 MB (4 x 128 MB) of RAM. If all the memory card slots are already filled, you may be able to replace the old modules with greater-capacity RAM modules, depending on your maximum allowed RAM.

Is it hard to add RAM? Adding RAM to a computer is fairly easy (see Figure 6.12). RAM comes with installation instructions, which you should follow carefully. RAM is also relatively inexpensive compared with other system upgrade options. Still, the cost of RAM fluctuates in the marketplace as much as 400 percent over time, so if you're considering adding RAM, you should watch the prices of memory in online or print advertisements.

Evaluating the Storage Subsystem

As you've learned, there are two ways data is saved on your computer: temporary storage and permanent storage. RAM is a form of *temporary* (or *volatile*) storage—thus, anything residing in RAM is not permanently saved. Therefore, it's critical to have means to *permanently* store data and software applications.

Fortunately, several storage options exist within every computer system. Storage devices for a typical personal computer include the hard disk drive, floppy disk drive, Zip disk drive, and CD and DVD drives. When you turn your computer off, the data stored to these devices is saved. These devices

FIGURE 6.12

Adding RAM to a personal computer is quite simple and relatively inexpensive. You simply line up the notches and push in the memory module.

are therefore referred to as *nonvolatile storage* devices. Of all the nonvolatile storage devices, the hard disk drive is used the most.

THE HARD DISK DRIVE

What makes the hard disk drive the most popular storage device? With storage capacities of up to 300 GB, **hard disk drives** (or just **hard drives**), shown in Figure 6.13, have the largest storage capacity of any storage device. The hard disk drive is also a much more economical device than floppy, Zip, or CD/DVD drives as it offers the most megabytes of storage per dollar.

Second, the hard drive's **access time**, or the time it takes a storage device to locate its stored data and make it available for processing, is also the fastest of all permanent storage devices. Hard drive access times are measured in milliseconds, or thousandths of seconds. For large-capacity disk drives, access times of approximately 9.5 milliseconds—that's less than one-thousandth of a second—are not unusual. This is much faster than the access times of other popular storage devices, like floppy and Zip disk drives.

Another reason hard drives are popular is that they transfer data to other computer components (such as RAM) much faster than the other storage devices. This speed of transfer is referred to as **data transfer rate** and depending on the manufacturer is expressed in either mega*bits* or mega*bytes* per second.

How is data stored on hard drives? As you learned in Chapter 5, a hard disk drive is composed of several coated **platters** (round thin plates of metal) stacked onto a spindle. When data is saved to a hard disk, a pattern of magnetized spots is created on the iron oxide coating each platter. Each of these spots represents a 1, while the spaces not "spotted" represent a 0. These 0s and 1s are bits (or binary digits) and are the smallest pieces of data that computers can understand. When data stored on the hard disk is retrieved (or read), your computer translates these patterns of magnetized spots into the data you have saved.

How do I know how much capacity my hard drive has? Hard drive capacity is measured in MB or GB. To check how much total capacity your hard drive has, as well as how much is being used, simply double-click on the My Computer icon, right-click on the

SOUND BYTE
HARD DRIVE ANATOMY INTERACTIVE

In this Sound Byte you'll learn about the internal construction of a hard drive and see how a hard drive reads and writes data. You'll also watch a disk being defragmented and learn how the defragmentation utility improves hard drive performance.

FIGURE 6.13

Hard disk drives are the most popular storage device for personal computers. The hard disk drive is installed permanently inside the system unit.

DIG DEEPER

HOW A HARD DISK DRIVE WORKS

The thin metal platters that make up a hard drive are covered with a special magnetic coating that enables the data to be recorded onto one or both sides of the platter. Hard disk manufacturers prepare the disks to hold data through a process called low-level formatting. In this process, **tracks** (concentric circles) and **sectors** (pie-shaped wedges) are created in the magnetized surface of each platter, setting up a grid-like pattern used to identify file locations on the hard drive. A separate process, called high-level formatting, establishes the catalog that the computer uses to keep track of where each file is located on the hard drive. As you learned in Chapter 5, this catalog is called the **File Allocation Table (FAT)**.

Hard drive platters spin at a high rate of speed, some as fast as 15,000 revolutions per minute (rpm). Sitting between each platter are special "arms" that contain **read/write heads** (see Figure 6.14). The read/write heads move from the outer edge of the spinning platters to the center, up to 50 times per second, to retrieve (read) and record (write) the magnetic data to and from the hard disk. As noted earlier, the average total time it takes for the read/write head to locate the data on the platter and return to the CPU for processing is the access time. A new hard drive should have an average access time of about 10 ms.

Access time is mostly the sum of two factors, seek time and latency. The time it takes for the read/write heads to move over the surface of the disk, between tracks, to the correct track is called the **seek time** (sometimes people incorrectly refer to this as access time). Once the read/write head locates the correct track, it may need to wait for the correct sector to spin to the read/write head. This waiting time is called **latency** (or rotational delay). The faster the platters spin (or the faster the rpm), the less time you'll have to wait for your data to be accessed. Currently, you can find new hard drives that spin between 5,400 and 7,200 rpm.

The read/write heads do not touch the platters of the hard drive; rather, they float above them on a thin cushion of air at a height of .5 microinches. As a matter of comparison, a human hair is 2,000 microinches thick and a particle of dust is larger than a human hair. Therefore, it's critical to keep your hard disk drive free from all dust and dirt, as even the smallest particle could find its way between the read/write head and the disk platter, causing a **head crash**—a stoppage of the hard disk drive that often results in data loss.

Capacities for hard disk drives in personal computers exceed 300 GB. Increasing the amount of data stored in a hard disk drive is achieved by either adding more platters and/or by increasing the amount of data stored on each platter. How tightly the tracks are placed next to each other, how tightly spaced the sectors are, and how closely the bits of data are placed affect the measurement of the amount of data that can be stored in a specific area of a hard disk. Modern technology continues to increase the standards on all three levels, enabling massive quantities of data to be stored in small places.

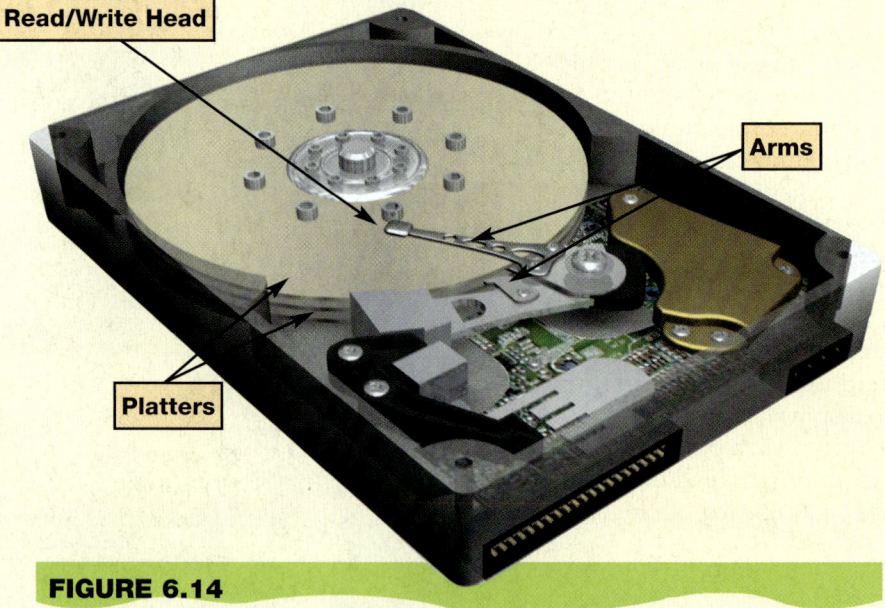

FIGURE 6.14

The hard drive is a stack of platters enclosed in a sealed case. Special arms fit in between each platter. The read/write heads at the end of each arm read from and save data to the platters.

C-drive icon, and select Properties from the shortcut menu. The Properties dialog box displays a pie chart that indicates the total capacity of your hard drive as well as the capacity being used, as shown in Figure 6.15.

How do I know how much storage capacity I need? To determine the storage capacity your system needs, calculate the amount of storage capacity basic computer programs need to reside on your computer. Since the operating system is the most critical piece of software, your hard drive needs enough space to store that program. The demands on system requirements have grown with new versions of operating systems. Windows XP, the latest Microsoft operating system, requires a whopping 1.5 GB of hard drive capacity. Five years ago, such software wouldn't have fit on most hard drives.

In addition to having space for the operating system, you need enough space to store software applications you use, such as Microsoft Office, a Web browser, and games. It's always best to check the system requirements of any software program before you purchase it to make sure your system can handle it. Storage requirements are found on the software package or on the manufacturer's Web site. Figure 6.16 shows an example of hard drive requirements for someone storing a few programs on a hard drive.

Are some hard drives faster than others? There are two basic types of hard drives: IDE (Integrated Drive Electronics) and SCSI (Small Computer System Interface). IDE drives have data transfer rates ranging from 16 to 66 MB per second. IDE drives are best if you use your computer primarily for word

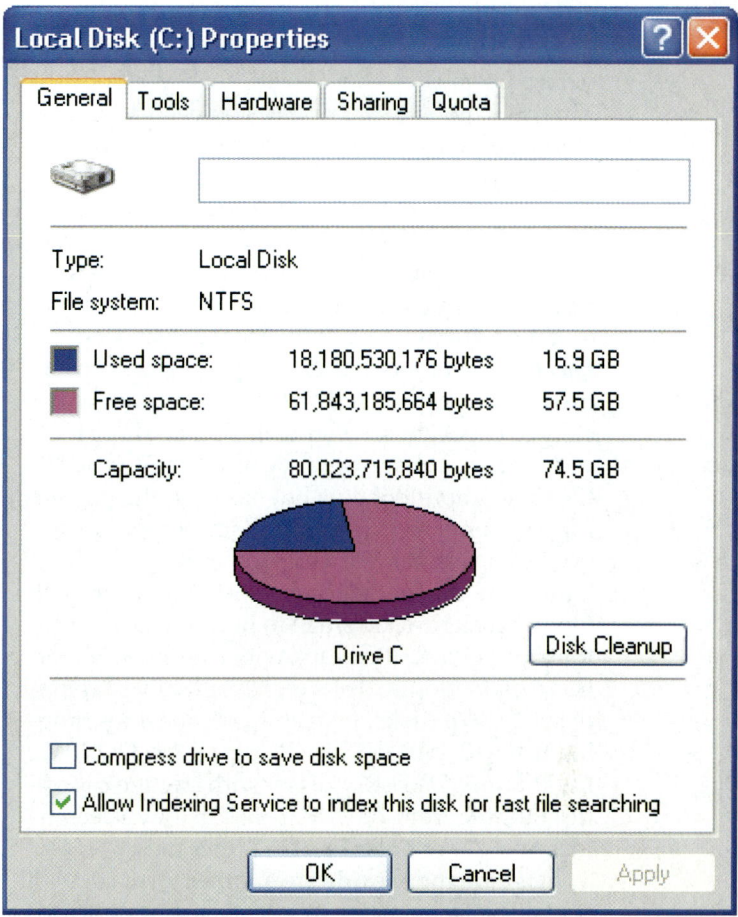

FIGURE 6.15

Using the pie chart in the General tab of the Properties dialog box, you can determine the capacity of your hard drive. This hard drive has 74.5 GB of space, with 57.5 GB of free space.

processing, spreadsheets, e-mail, and the Internet. However, "power users," such as graphic designers and software developers, may need to consider a faster SCSI hard drive. Data transfer rates for SCSI hard drives range from 40 to 80 MB per second.

Another factor affecting a hard disk's speed is access time (or the speed with which it locates data for processing). As noted above, access time is measured in milliseconds (ms). The faster the access time the better, although most hard drives will have similar access times.

PORTABLE STORAGE OPTIONS: THE FLOPPY AND BEYOND

If my hard drive is so powerful, why do I need other forms of storage? Despite all the advantages that the hard drive has as a storage device, one drawback is that data stored on it is not

FIGURE 6.16 Sample Hard Drive Requirements	
APPLICATION	**HARD DISK SPACE REQUIRED**
Windows XP	1.5 GB
Microsoft Word	265 MB
Microsoft Excel	255 MB
Internet Explorer	75 MB
Windows Media Player	521 MB
Microsoft Picture It!	150 MB
Total Required	3.44 GB

portable. To get data from one computer to another (assuming the computers aren't networked), you'll need a portable storage device, such as a floppy or Zip drive. Equally important, you need alternate storage options to back up critical data on your hard drive in case it experiences a head crash or other system problems. Finally, despite the massive storage capacity of hard drives, you should remove infrequently used files from your hard drive to maintain optimal storage capacity.

What forms of portable storage are best? Several portable storage formats (or media) are popular now, with varying ranges of storage capacity (see Figure 6.17). The **floppy disk**, with a storage capacity of 1.44 MB, holds the least amount of data but has been the primary form of convenient, portable data storage. However, today's average user is generating much larger files due to the incorporation of multimedia. Floppy disks are too small to hold even one file that has been "bulked up" with multimedia. For this reason, floppy disks will eventually become obsolete. **Zip disks**, with storage capacities ranging from 100 MB to 750 MB, and CD-R, CD-RW, DVD-R and DVD-RW disks, with storage capacities ranging from 700 MB to 9.4 GB, have become increasingly popular to store larger files.

Flash memory cards are another form of portable storage. These removable memory cards are often used in digital cameras, MP3 players, and PDAs. Some newer flash cards can hold as much as 1 GB of data. As the technology becomes more popular, capacities will continue to increase.

FIGURE 6.17 Portable Storage Capacities

STORAGE MEDIA	CAPACITY
Floppy Disk	1.44 MB
Zip Disk	100 to 750 MB
CD	700 MB
DVD	9.4 GB
Flash memory	16 MB to 1 GB

Floppy and Zip Disks

How is data stored on floppy and Zip disks? Inside the plastic cases of both floppy and Zip disks you'll find a round piece of plastic film, as shown in Figure 6.18. This film is covered with a magnetized coating of iron oxide. As was the case with hard drives, when data is saved to the disk, a pattern of magnetized spots is created on the iron oxide coating within established tracks and sectors. Each of these spots represents either a 0 or a 1, or a bit. When data stored on the disk is retrieved (read), your computer translates these patterns of magnetized spots into information. Because floppies and Zip disks use a magnetized film to store data, they are referred to as **magnetic media**.

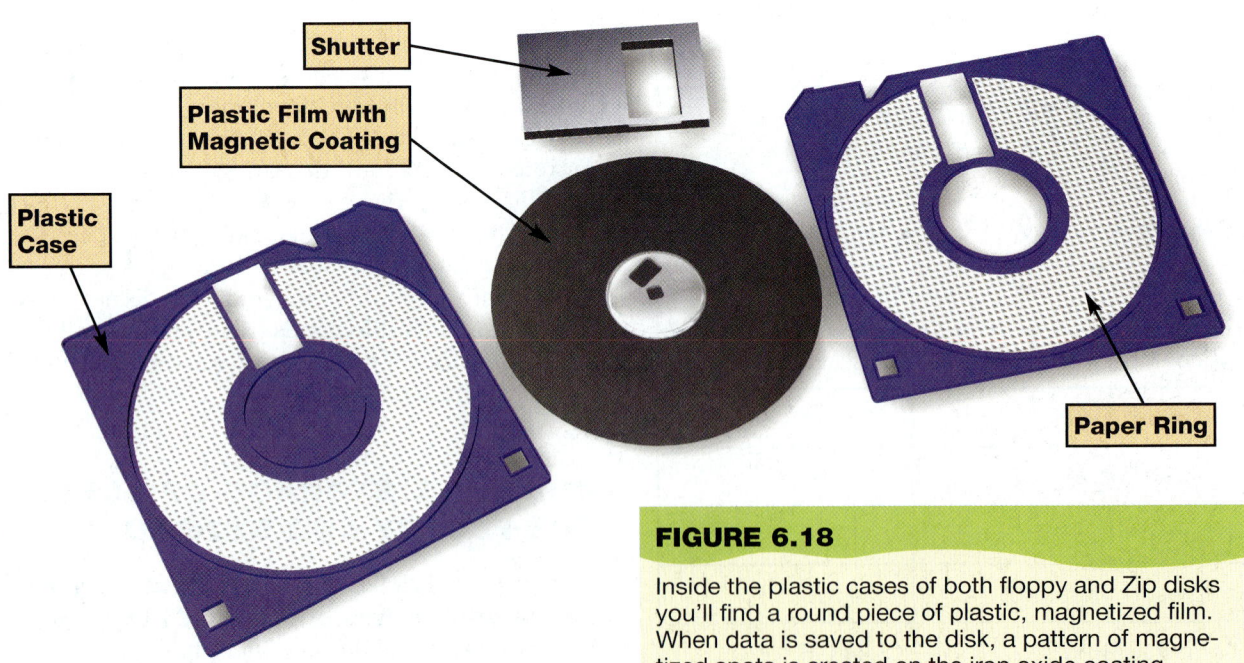

FIGURE 6.18

Inside the plastic cases of both floppy and Zip disks you'll find a round piece of plastic, magnetized film. When data is saved to the disk, a pattern of magnetized spots is created on the iron oxide coating.

BITS AND BYTES

Taking Care of Floppy and Zip Disks

The following guidelines will help you keep your floppy disks safe:

- Place disks in a protective case when carrying or mailing them.
- Keep disks away from devices that generate a magnetic field, such as speakers, televisions, and mobile phones.
- Use a felt-tip marker (not a pen or pencil) when labeling disks. Better yet, create a label and place it on the disk.
- Don't expose your disk to excessive heat or cold.
- Don't move the metal shutter that protects the disk. Dirt and dust that get into the disk can harm your data. If the metal shutter comes off the floppy, transfer the data immediately to the hard drive, as dust and dirt will corrupt the data on the exposed part of the floppy.

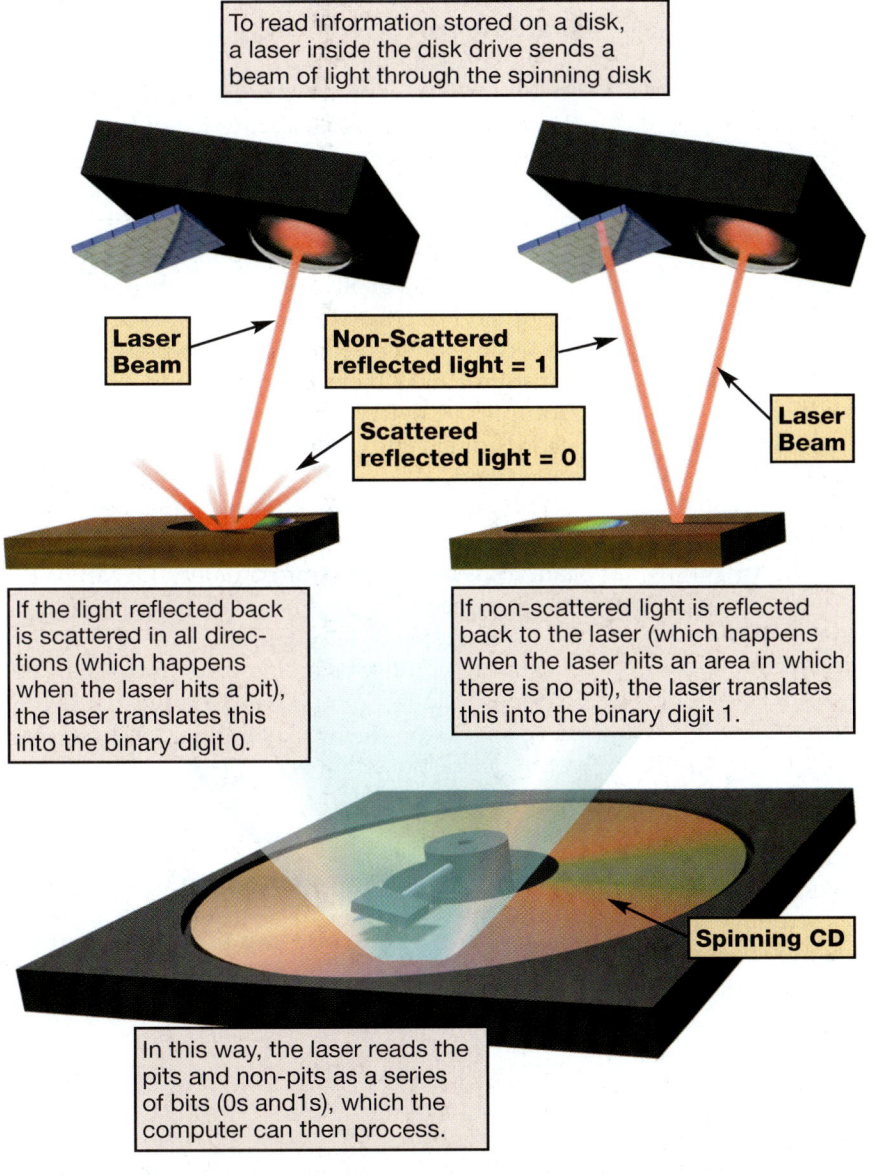

FIGURE 6.19
Data is read from a CD using focussed laser light.

SOUND BYTE
CD/DVD READING AND WRITING

In this Sound Byte you'll learn about the process of storing and retrieving data from CD-R, CD-RW, and DVD disks. You'll be amazed to see how much precision engineering is required to burn MP3 files onto a disk.

CD and DVDs

How is data saved onto a CD or DVD? Like the hard drive and floppy and Zip disks, data is saved to CDs and DVDs within established tracks and sectors. However, unlike floppy and Zip disks, which store their data on a piece of magnetized film, CDs and DVDs store data as tiny pits that are burned into a disk by a high-speed laser. These pits are extremely small, less than 1 micron in diameter, so that nearly 1,500 pits fit across the top of a pinhead. As you can see in Figure 6.19, data is read off the CD by a laser beam, with the pits and non-pits translating into the 1s and 0s of the binary code computers understand. Because CDs and DVDs use a laser to read and write data, they are referred to as **optical media**.

Why can I store data on some CDs but not others? CD-ROMs are "read-only" optical disks, meaning you can't save any data onto them. To play a CD-ROM, you use a CD-ROM drive. However, most computers today are equipped with special CD drives that allow you to save, or "burn," data onto specially designed CDs.

BITS AND BYTES
Taking Care of CDs and DVDs

The following guidelines will help you keep your CDs and DVDs safe:

- Exercise care in handling your CDs and DVDs. Dirt or oil on CDs/DVDs can keep data from being read properly, while large scratches can interrupt data completely.
- To keep CDs/DVDs from warping, avoid placing them near heat sources and store them at room temperature.
- Clean CDs/DVDs by taking a bit of rubbing alcohol on a cotton ball and wiping them from the center to the edge of the disk in long swipes. Don't rub the CD/DVD circularly, as you may cause more scratches.
- Use a felt-tip marker to label CDs/DVDs and write on the area provided for the label. Don't put stickers or labels on CDs/DVDs, unless they're specifically designed for that purpose.

CD-RW (Compact Disk–Read/Writable) disks, which use a CD-RW drive, can be written to hundreds of times. **CD-R (Compact Disk–Recordable) disks** can be written to once and can be used with either a CD-R drive or a CD-RW drive. If your computer isn't equipped with a CD-R or CD-RW drive, you can buy one at a reasonable cost, and they're quite simple to install.

What's the difference between CDs and DVDs? DVDs use the same optical technology to store data as CDs. The difference is that a DVD's storage capacity is much greater than a CD's. In order to hold more data than CDs, DVDs have less space between tracks, as well as between bits. The size of pits on the DVD are also much smaller than those on a CD. Additionally, DVD audio and video quality is superior to that of a CD. Because of their versatile nature, DVDs are quickly becoming the standard technology for audio and video files as well as graphics and data files.

SOUND BYTE
INSTALLING A CD-RW DRIVE
In this Sound Byte, you'll learn how to install a CD-RW drive in your computer.

Do I need a DVD-ROM drive *and* a CD-ROM drive? Although CDs and DVDs are based on the same technology, they are different enough to require their own devices. If your system only has a CD drive, you need to add a DVD drive to view DVDs. However, since DVD drives can read CDs, if your system has a DVD drive, you do not need to add a CD drive.

To record data to DVDs, you need **DVD-R (DVD-Recordable)** disks and a DVD-RW drive. Unfortunately, technology experts have not agreed on a standard DVD format. That's why you see both DVD-RW and DVD-R disks. These various formats are not necessarily compatible, so make sure you know what format your drive needs.

Are some CD and DVD drives faster than others? When you buy a CD or DVD drive, knowing the drive speed is important. Speeds are listed on the device's packaging. Record (write) speed is always listed first, rewrite speed is listed second (except for CD-R drives, which cannot rewrite data), and playback speed is listed last. For example, a CD-RW drive may have speeds of 40x12x48, meaning that the device can record data at 40x speed, rewrite data at 12x speed, and play back data at 48x speed. For CDs, the "x" in between each number represents the transfer of 150 KB of data per second. So, for example, a CD-RW drive with a 40x12x48 rating records data at 40x150 KB per second, or 6,000 KB per second.

DVD drives are much faster than CD drives. For example, a 1x DVD-ROM drive provides a data transfer rate of approximately 1.3 MB of data per second, which is roughly equivalent to a CD-ROM speed of 9x. CD and DVD drives are constantly getting faster. If you're in the market for a new CD-RW, you'll want to investigate the drive speeds on the market and make sure you get the fastest one you can.

UPGRADING YOUR STORAGE SUBSYSTEM

How can I upgrade my storage devices? There are several ways in which you can increase your storage capacity or add extra drives to your computer.

If you find your hard drive is running out of space, or you want a place to back up or move files to create more room on your hard drive, you have several options. You can replace the hard drive installed in your system unit with

EVALUATING THE STORAGE SUBSYSTEM

a bigger one. However, replacing your internal hard drive requires backing up your entire hard drive and reloading all the data onto your new hard drive. Instead, you may want to install an additional hard drive in your current system if you have an extra drive bay (the space reserved on the inside of your system unit for hard disk drives). In lieu of installing a bigger hard drive, there are now external hard drives you can plug directly into a free USB 2.0 or FireWire port. Figure 6.20 shows an example of such an external hard drive.

You can also upgrade your storage subsystem by adding a new Zip drive or other drive to your system. If your computer did not come with an internal Zip, CD-RW, or DVD-RW drive, and if you have an open (unused) drive bay in your system, you can easily install a drive there. Most people upgrade to a CD-RW and DVD-RW not for more storage, but because they want additional multimedia capabilities (such as the ability to burn CDs). If you don't have open bays in your system, you can still add Zip and CD/DVD drives. As is the case with hard drives, these drives are available as external units you attach to your computer through an open port.

Do you need to upgrade your storage subsystem?

	CURRENT SYSTEM	UPGRADE REQUIRED?
Hard Disk Drive Capacity		
Floppy Disk Drive		
Zip Disk Drive		
CD-ROM Drive		
CD-R Drive		
CD-RW Drive		
DVD-ROM Drive		
DVD-RW Drive		
Other Storage Devices Needed?		

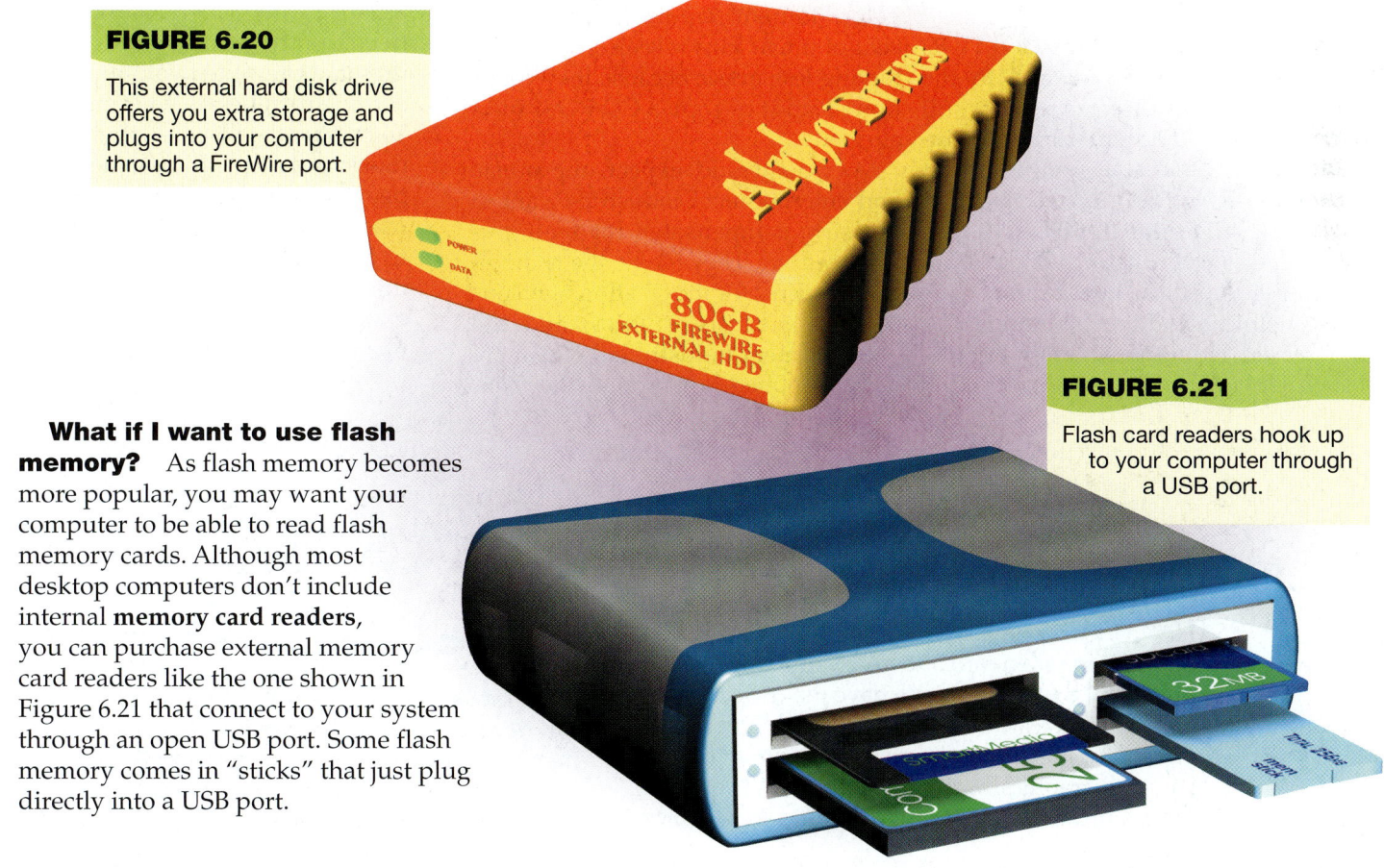

FIGURE 6.20

This external hard disk drive offers you extra storage and plugs into your computer through a FireWire port.

FIGURE 6.21

Flash card readers hook up to your computer through a USB port.

What if I want to use flash memory? As flash memory becomes more popular, you may want your computer to be able to read flash memory cards. Although most desktop computers don't include internal **memory card readers**, you can purchase external memory card readers like the one shown in Figure 6.21 that connect to your system through an open USB port. Some flash memory comes in "sticks" that just plug directly into a USB port.

TRENDS IN IT

ETHICS: CD AND DVD TECHNOLOGY: A Free Lunch—or at Least a Free Copy

Years ago, when the electronic photocopier computer made its debut, book publishers and others who distributed the printed word feared they would be put out of business. They were worried that people would no longer buy books and other printed matter if they could simply copy someone else's original. Years later, when the cassette player/recorder and VCR player/recorder arrived on the market, those who felt they would be negatively affected by these new technologies expressed similar reactions. Now, with the arrival of DVD-RW and CD-RW technology, which allows users to copy DVDs and CDs in a matter of minutes, the music and entertainment industries are up in arms.

Although copy machines and VCRs certainly didn't put an end to the industries they affected, some still say the music and entertainment industries will take a significant hit with CD-RW/DVD-RW technology. Already, CD and DVD sales are plummeting. Industry insiders are claiming that these new technologies are unethical, and they're pressing for increased federal legislation against such copying. And it's not just the CD-RW/DVD-RW technology that's causing problems—"copies" are not necessarily of the physical sort. Thanks to the Internet (and sites such as the now defunct Napster and the still up and running Kazaa), *file transferring* copyrighted works—particularly music and films—is now commonplace. Viant, a company known for its studies on Internet piracy, reported recently that, on average, between 300,000 and 500,000 films per day are being transferred over the Internet.

In a separate survey, the Recording Industry Association of America (RIAA), a trade organization that represents the interests of recording giants like Sony, Capitol Records, and other major producers of musical entertainment, reported that 23 percent of music fans revealed they were buying less music because they could download it or copy a CD-ROM from a friend.

As you would expect, the music and entertainment industries want to be fairly compensated for their creative output. They blame the technology industry for the creation of means by which artists, studios, and the entertainment industry in general are being "robbed." Although the technology exists that allows consumers to readily transfer and copy music and videos, the artists who produce these works do not want to be taken advantage of. However, others claim that the technology industry should not bear the complete burden of protecting entertainment copyrights. The RIAA sums up the future of this debate nicely: "Goals for the new millennium are to work with [the recording] industry and others to enable technologies that open up new opportunities but at the same time to protect the rights of artists and copyright owners."

What Do You Think?

1. If you were a popular recording artist, how would you feel about free file swapping of your recordings that prevented you from earning royalties on your music?
2. To combat the illegal copying of music and movie files, some people have suggested adding mandatory anticopying hardware to personal computers. Do you think such hardware poses ethical issues for computer manufacturers?

Evaluating the Video Subsystem

How video is displayed depends on two components: your video card and your monitor. It's important that your system have the correct monitor and video card to meet your needs. If you use your computer system to display files that have complex graphics, such as videos on DVD or from your camcorder, or even play graphics-rich games with a lot of fast action, you may want to consider upgrading your video subsystem.

VIDEO CARDS

What is a video card? A video card (or **video adapter**) is an expansion card that is installed inside your system unit to translate binary data (the 1s and 0s your computer uses) into the images you view on your monitor. Today,

Team Time

Problem:

In a large organization, whether it is a company or a college, the IT department often has to install several different types of computing systems. There certainly would be advantages to having every computer be identical, but because different departments have different needs, and items are purchased at different times, it is typical for there to be significant differences between two computers in the same corporation.

Process:

Split your class into groups.

STEP 1: Select a department or computer lab on campus (or within your company, at the public library, and so on). Note: If you physically cannot go to the various labs, describe the type of components that would be needed by that particular department. (For example, if you chose the computer art department, you know you would need good graphics software. You also know you would need certain levels of RAM, etc., to accommodate that graphics software.)

STEP 2: Following the worksheet in Figure 6.3, analyze the computing needs of that particular department.

STEP 3: Using the System Evaluation worksheet (found on the book's companion Web site at **www.prenhall.com/techinaction**), develop a complete systems evaluation of the computers at the lab.

STEP 4: Consider possible upgrades in hardware, software, and peripherals that would make this lab better able to meet the needs of its users.

STEP 5: Write a report that summarizes your findings. If purchasing a new system is more economical, recommend which system the lab should buy.

Conclusion:

The pace of technological change can make computer science an uncomfortable field for some. For others, it is precisely the pace of change that is exciting. Being able to evaluate a computer system and match it to the current needs of its users is an important skill.

Becoming Computer Fluent

Gerri lives across the hall from you. She heard you worry all last semester that your computer wasn't fast enough to handle your course workload. Between the simulation program for math, the reports and research you did for English, and the programming class you tried, your computer was running too slowly and you were out of storage space. She's offered to do a complete system upgrade for you but has asked you to write a letter telling her exactly what you want upgraded and why.

Instructions: Using the scenario above, write a letter using as many of the terms from the chapter as you can. Be sure your sentences are grammatically correct and technically meaningful.

Materials on the Web

In addition to the review materials presented here, you'll find extra materials on the book's companion Web site (**www.prenhall.com/techinaction**) that will help reinforce your understanding of the chapter content. These materials include:

Sound Byte Lab Guides

For each Sound Byte mentioned in the chapter, there is a corresponding lab guide located on the book's companion Web site. These guides review the material presented in the Sound Byte and direct you to various Web resources that examine the material. The Sound Byte Lab Guides for this chapter include:

- Using Windows XP to Evaluate CPU Performance
- Installing RAM
- Hard Drive Anatomy Interactive
- CD/DVD Reading and Writing
- Installing a CD-RW Drive
- Port Tour
- Letting Your Computer Clean Up After Itself

True/False and Multiple Choice Quizzes

The book's Web site includes a True/False and a Multiple Choice quiz for this chapter. You can take these quizzes, automatically check the results, and e-mail the results to your instructor.

Web Research Projects

The book's Web site also includes a number of web research projects for this chapter. These projects ask you to search the Web for information on computer-related careers, milestones in computer history, important people and companies, emerging technologies, and the applications and implications of different technologies.

OBJECTIVES

After reading this chapter, you should be able to answer the following questions:

- What is a network and what are the advantages of setting up one? (pp. 252–253)

- What is the difference between a client/server network and a peer-to-peer network? (pp. 253–254)

- What are the main components of every network? (pp. 255–256)

- What are the most common home networks? (p. 256)

- What are power line networks and how are they created? (pp. 256–258)

- What are phone line networks and how are they created? (p. 258)

- What are Ethernet networks and how are they created? (pp. 259–262)

- What are wireless networks and how are they created? (pp. 262–265)

- How can hackers attack a network and what harm can they cause? (pp. 270–275)

- What is a firewall and how does it keep my computer safe from hackers? (pp. 275–278)

- What types of viruses do I need to protect my computer from? (pp. 278–281)

- What can I do to protect my computer from viruses? (pp. 281–282)

CHAPTER 7

Networking and Security:
Connecting Computers and Keeping Them Safe from Hackers and Viruses

TECHNOLOGY IN ACTION: THE PROBLEMS OF SHARING

The Williams family is facing computer-sharing problems. Derrick and Vanessa realized that they both needed computers, and they bought their children, Stephanie and Jake, their own computers too. Still, there is trouble in this "paradise."

Scarce Resources: Jake was using his computer to scan photos for a school Web site project. Just then, his sister Stephanie burst into his room and demanded to use his scanner since she didn't have one and needed to scan images for an art project. Jake wouldn't budge. The ensuing shouting match brought their mother Vanessa to the room. Since both projects were due the next day, Vanessa told Jake he would have to let Stephanie use the scanner at some point. In exchange, Stephanie would have to let Jake use her computer so he could print from the color printer attached to it. Neither was happy, but it was the best Vanessa could do.

Internet Logjam: Derrick frowned at his e-mail inbox. He was on vacation, but his boss had e-mailed him asking him to look over a marketing plan. Meanwhile, Vanessa needed to get online to confirm the family's reservations at Disney World. Since they only had one phone line, they constantly had to take turns getting online, and with Stephanie and Jake wanting to check their e-mail too, the phone line was almost always tied up.

Virus Attack: Later that night, Derrick booted up his computer to make some changes to the marketing plan. Right away, he noticed that many of his icons had disappeared from his desktop. As he launched Microsoft Word, a message flashed on the screen that read, "The Hacker of Death Was Here!" Suddenly, his screen went black. When he rebooted his computer, it was unable to recognize the hard drive. The next day, Derrick called the computer support technician at work and learned he had caught the "Death Squad" virus, which had erased the contents of his hard drive. Derrick mused that he should have bought an antivirus program instead of Mech Warrior 5!

If this scenario doesn't reflect the situation in your home, it may in the future. As the price of computers continues to drop, more families will have multiple home computers. To avoid inconvenience and the expense of redundant equipment, computers need to be able to communicate with each other and to share peripherals (such as scanners and printers) and resources (such as Internet connections). Thus, this chapter will explore how you can network computers. In addition, you'll learn strategies for keeping unauthorized outsiders from prying into your computer when you're sharing resources with the outside world, as well as how to keep your computer safe from viruses.

Networking Fundamentals

Although you may not yet have a home network, you use and interact with networks all the time. In fact, every time you use the Internet you're interacting with the world's largest network. But what exactly *is* a network? A **computer network** is simply two or more computers that are connected together via software and hardware so they can communicate. Devices connected to a network are referred to as **nodes**. A node can be a computer, a peripheral (such as a printer), or a communications device (such as a modem). The main function for most networks is to facilitate information sharing, but networks provide other benefits as well.

What are the benefits of networks? One benefit of networks is that they allow users to share peripherals. For example, in Figure 7.1a, the computers are not networked; Computer 1 is connected to the printer but Computer 2 is not. To print files from Computer 2, users have to transfer them via floppy disk or another storage medium to Computer 1, or they have to disconnect the printer from Computer 1 and connect it to Computer 2. By networking Computer 1 and Computer 2, as shown in Figure 7.1b, both computers can print from the printer attached to Computer 1 without transferring files or moving the printer.

By networking computers, you can also transfer files from one computer to another without using external storage media such as floppy disks. And

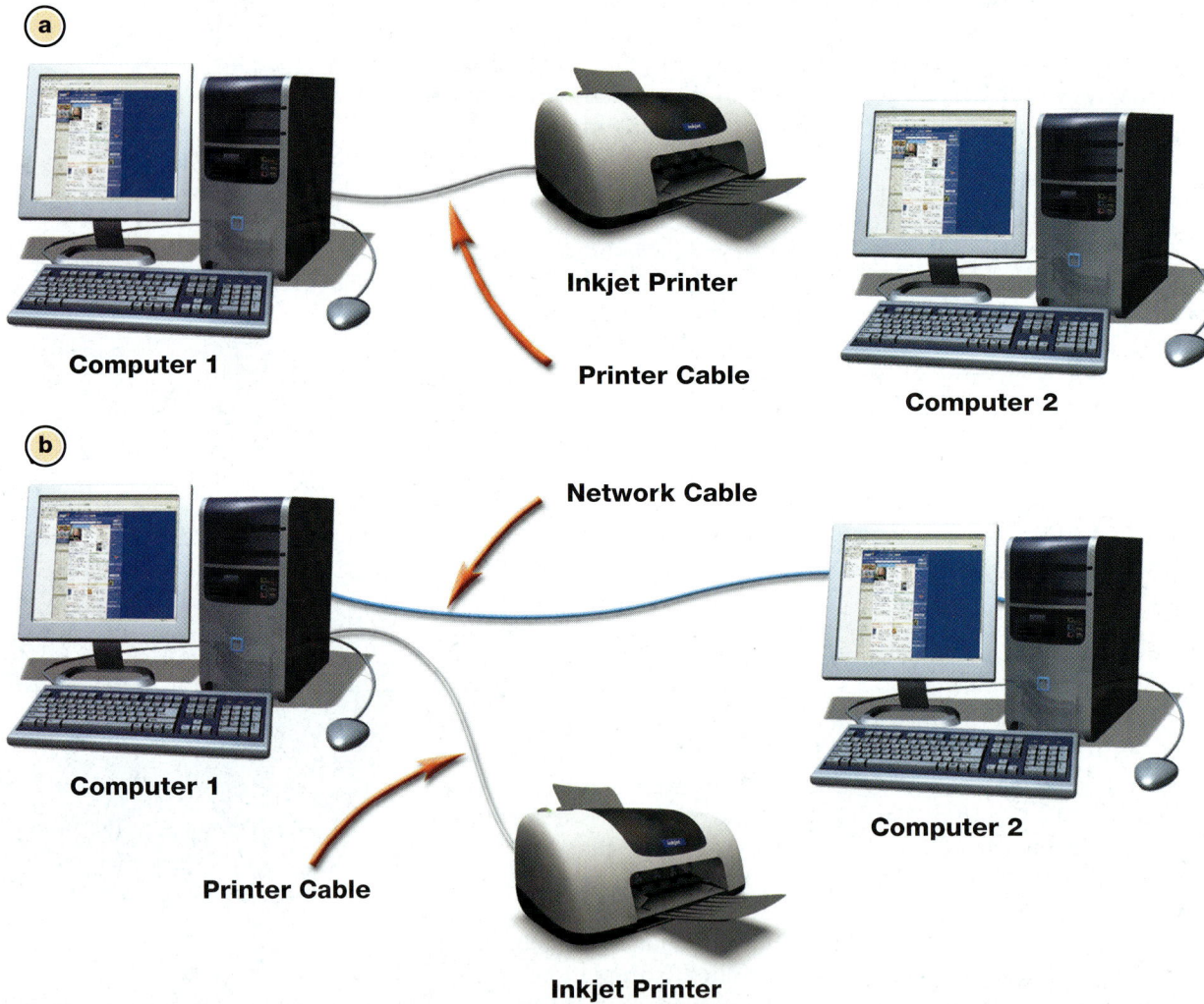

FIGURE 7.1

(a) Computer 1 and Computer 2 are not networked. Only Computer 1 can use the printer unless the printer is disconnected from Computer 1 and reconnected to Computer 2. (b) Computer 1 and Computer 2 are networked. Both computers can use the printer without having to move it.

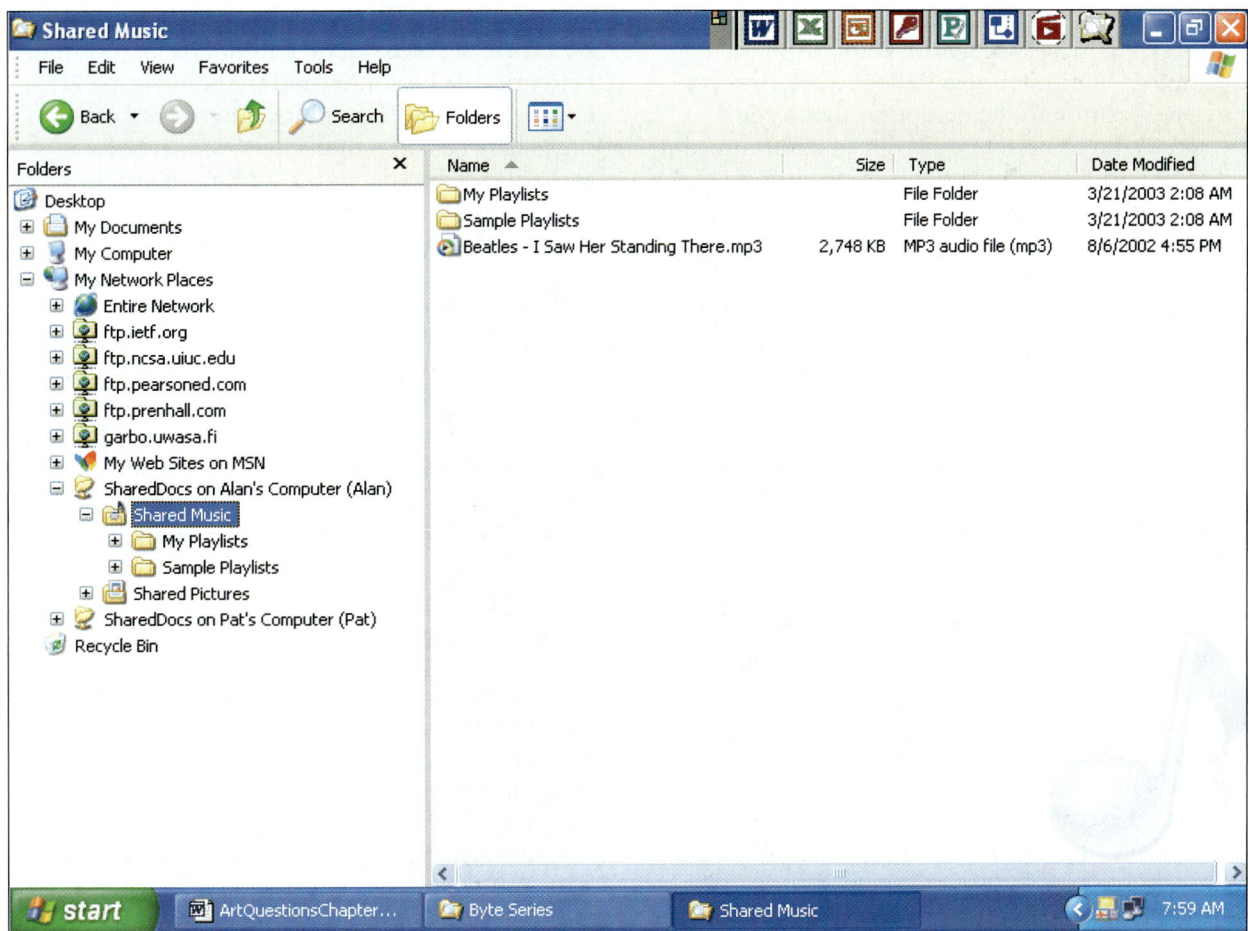

FIGURE 7.2

This Windows XP network has two computers attached to it: Alan and Pat. Shared directories (the SharedDocs folders) have been set up so that Alan and Pat can share files. When Pat is working on her computer, she can easily access the files located in the shared directory on Alan's computer, such as the folder called "Shared Music." Likewise, Alan can access the files Pat places in her shared directory.

you can set up shared directories in Windows that allow the user of each computer on the network to store files that other computers on the network may need to access, as shown in Figure 7.2.

Can I use a network to share an Internet connection? Sharing Internet connections is a significant benefit of setting up a home network. Without a network, only one user can use an Internet connection at a time. Although dial-up connections don't have sufficient bandwidth to support sharing a connection, broadband connections (such as cable and DSL) have more than enough bandwidth to allow users to share the connection.

Network Architectures

The architecture of a building refers to its design and how it fits into its surroundings. The term **network architecture** refers to the design of a network. Network architectures are classified according to the way in which they are controlled and the distance between their nodes.

DESCRIBING NETWORKS BASED ON NETWORK CONTROL

What do we mean by networks being "controlled"? There are two main ways a network can be controlled: locally or centrally. A *peer-to-peer network* is the most common example of a locally controlled network. The most common type of centrally controlled network is a *client/server network*.

What are peer-to-peer networks? In **peer-to-peer (P2P) networks**, each node connected to the network can communicate directly with every other node on the network, instead of having a separate device exercise central control over the network. Thus, all nodes on this type of network are in a sense *peers*. When printing, for example, a computer on a P2P network doesn't have to go through the computer that's connected to the printer. Instead, it can communicate directly with the printer. Figure 7.1b shows a very small peer-to-peer network.

Because they are simple to set up, P2P networks are the most common type of home network. We'll discuss different types of peer-to-peer networks that are popular in homes later in this chapter.

What are client/server networks? Very small schools and offices may have P2P networks. However, most networks that comprise 10 or more nodes are **client/server networks**. A client/server network contains two different types of computers: clients and servers. The **client** is the computer where users accomplish specific tasks (such as construct spreadsheets). The **server** is the computer that provides information or resources to the client computers on the network. The server also provides central control for functions on the network (such as printing). Figure 7.3 illustrates a client/server network in action.

As you learned in Chapter 3, the Internet is an example of a client/server network. When your computer is connected to the Internet, it is functioning as a *client computer*. When connecting to the Internet through an Internet service provider (ISP), your computer connects to a *server computer* maintained by the ISP. The server "serves up" resources to your computer so that you can interact with the Internet.

Are client/server networks ever used as home networks? Although client/server networks *can* be configured for home use, P2P networks are more often used in the home because they cost less than client/server networks and are easier to configure and maintain. To set up a client/server network in your home, you have to buy an extra computer to act as the server. (Although an existing computer could function as a server, its performance would be significantly degraded, making it impractical to use as both a client and a server.) In addition, you need extensive training to install and maintain the special software client/server networks require. Finally, the major benefits a client/server network provides (such as centralized security and administration) are not necessary in most home networks.

DESCRIBING NETWORKS BASED ON DISTANCE

How does the distance between nodes define a network? The proximity of network nodes to each other can help describe a network. **Local area networks (LANs)** are networks in which the nodes are located within a small geographic area. A network in your home or a computer lab at school is an example of a LAN. **Wide area networks (WANs)** are made up of LANs connected over long distances. Say a school has two campuses (east and west) located in different towns.

Step 1
The client computer requests a service from the server computer

Computer A
(Client)

Computer B
(Server)

Step 2
The server computer provides requested service to client computer

FIGURE 7.3

In a client/server network, a computer acts as a client, making requests for resources, or as a server, providing resources.

NETWORK COMPONENTS 255

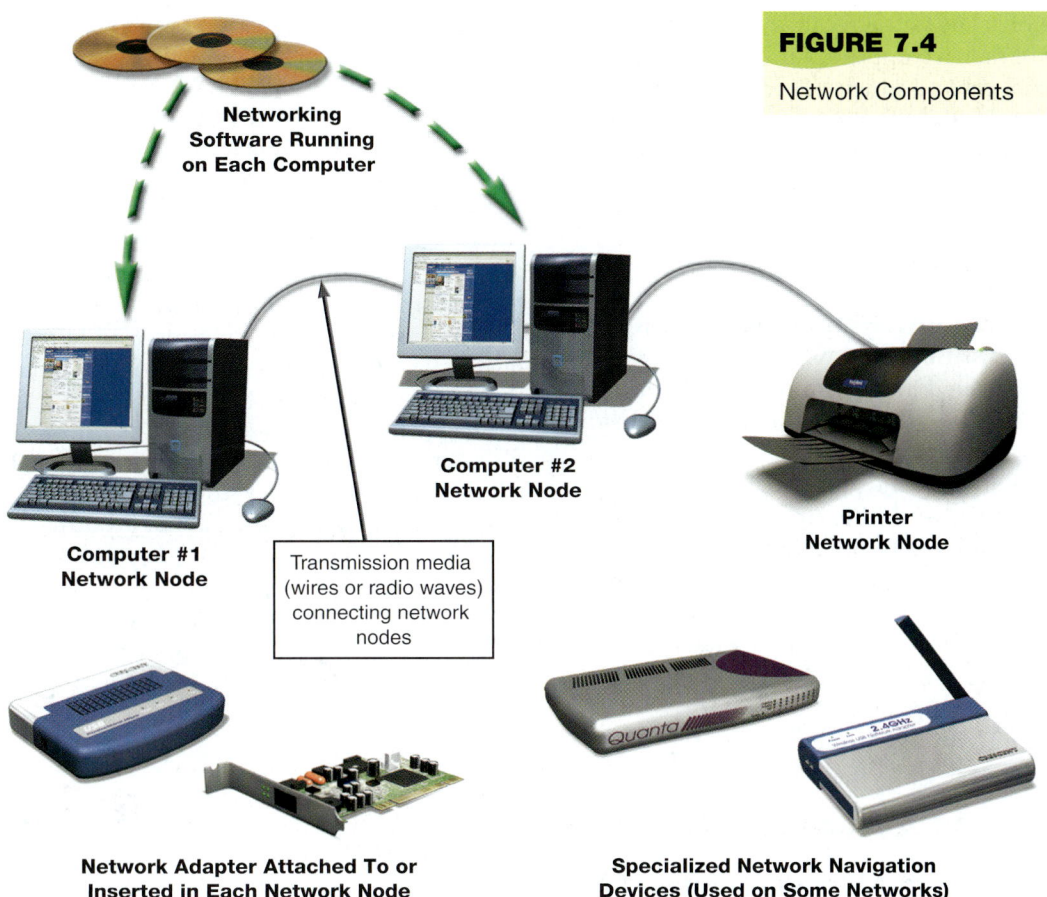

FIGURE 7.4
Network Components

Connecting the LAN at the east campus to the LAN at the west campus (via telecommunications lines) would allow the users on the two LANs to communicate with each other. The two LANs would be described as a single WAN.

Network Components

In order to function, all networks include a means of connecting the nodes on the network (via cables or wireless technology), special devices that allow the nodes to communicate with each other and to send data, and software that allows the network to run. We'll discuss each of these components, shown in Figure 7.4, next.

TRANSMISSION MEDIA

How are nodes on a network connected? All network nodes (computers and peripherals) are connected to each other and to the network by **transmission media**. A transmission medium establishes a communications channel between the nodes on a network and can take several forms:

1. Networks can use existing wiring (such as phone lines or power lines) to connect nodes.
2. Networks can use additional cable to connect nodes, such as twisted pair cable, coaxial cable, and fiber-optic cable. You have probably seen twisted pair and coaxial cable. Normal telephone wire is **twisted pair cable** and is made up of copper wires that are twisted around each other and surrounded by a plastic jacket. If you have cable TV, the cable running into your TV or cable box is **coaxial cable**. Coaxial cable consists of a single copper wire surrounded by layers of plastic. **Fiber-optic cable** is made up of plastic or glass fibers that transmit data at very fast speeds.
3. Wireless networks use radio waves instead of wires or cable to connect nodes.

Different types of transmission media transmit data at different speeds. **Data transfer rate** (also called **bandwidth** or **throughput**) is the speed at which data can be transmitted between two nodes on a network. Data transfer rate is usually measured in megabits per second (Mbps). A megabit, when applied to data transfer rates, represents

1 million bits. (As you'll recall, a *bit* is the smallest measure of data a computer can process.) Twisted pair, coaxial cable, and wireless media provide enough bandwidth for most home networks, whereas fiber-optic cable is used mostly in client/server networks.

NETWORK ADAPTERS

How do the different nodes on the network communicate? **Network adapters** are devices connected to or installed in network nodes that enable the nodes to communicate with each other and to access the network. Some network adapters take the form of external devices that plug into an available Universal Serial Bus (USB) port. Other network adapters are installed *inside* computers and peripherals as expansion cards. These adapters are referred to as **network interface cards (NICs)**. Specialized network adapters are created for the various types of networks. We'll discuss network adapters in more detail throughout the chapter.

NETWORK NAVIGATION DEVICES

How is data sent through a network? Data is sent over transmission media in bundles called **packets**. For computers to communicate, these packets of data must be able to flow freely between computers. **Network navigation devices** help to make this data flow possible. These devices, which are attached to the network, enable the transmission of data. In simple networks, navigation devices are built right into network adapters. More sophisticated networks need specialized navigation devices.

The two most common specialized navigation devices are routers and hubs. **Routers** route packets of data between two or more networks. For example, if a home network is connected to the Internet, a router is required to send data between the two networks (the home network and the Internet). **Hubs** are simple amplification devices. They receive data packets and retransmit them to all nodes on the same network (not between different networks). We'll discuss routers and hubs in more detail later in the chapter.

SOUND BYTE
INSTALLING A NETWORK

Installing a network is relatively easy if you've seen someone else do it. In this Sound Byte, you'll learn how to install the hardware and set up a wired or wireless home network.

NETWORKING SOFTWARE

What software do networks require? Home networks need operating system (OS) software that supports peer-to-peer networking. The most common versions of Windows used in the home (XP, ME, and 98) support P2P networking. You can connect computers running any of these OSs to the same network. You can also add computers that use the Windows 95 or 2000 OS to the same network, but you may need to install additional software to enable file or peripheral device sharing. The latest versions of the Mac OS also support P2P networking.

Client/server networks, on the other hand, are controlled by a central server that has specialized **network operating system (NOS)** software installed on it. This software handles requests for information, Internet access, and the use of peripherals for the rest of the network nodes. Examples of NOS software include Windows XP Professional, Windows 2000 Server, and Novell Netware.

Types of Peer-to-Peer Networks

The most common type of network you will probably encounter is a peer-to-peer network, since this is the network you would set up in your home. Therefore, we'll focus on P2P networks in this chapter. There are four main types of P2P networks:

1. Power line networks
2. Phone line networks
3. Ethernet networks
4. Wireless networks

The major differences in these networks are the transmission media by which the nodes are connected. We'll look at these networks and how each one is set up next.

POWER LINE NETWORKS

What are power line networks? **Power line networks** use the electrical wiring in your home to connect the nodes in the network. Thus, in a power line network, any electrical outlet provides a network connection. Power line networks have a maximum data transfer rate of 14 Mbps. The HomePlug Power line Alliance (www.homeplug.org) sets standards for home power line networking.

TYPES OF PEER-TO-PEER NETWORKS 257

How do I create a power line network? To create a power line network, you connect a **power line network adapter** to each computer or peripheral that you're going to attach to the network. You can buy power line network adapters in either USB or Ethernet versions, as shown in Figure 7.5. Both are easy to install, as they plug into either a USB or Ethernet port on your computer or peripheral.

After you attach a network adapter to each node on the network, you plug the adapters into an electrical outlet. Because most power line network adapters are plug-and-play compatible, your OS automatically recognizes them and knows how to interact with them. For those adapters that are not plug-and-play compatible, device drivers are provided on CD-ROMs. As you learned in Chapter 5, a **device driver** is software you install on your computer; when you attach new devices to your computer, the device driver allows your OS to interact with the device.

In addition, each computer attached to a power line network needs to have its OS configured for networking. (We'll discuss how to configure the OS later in the chapter.) Once you've set up the network, adding computers or peripherals is relatively easy. You simply attach a power line adapter to the computer or peripheral you wish to add and plug it into another electrical outlet. Figure 7.6 shows a power line network installed in a home.

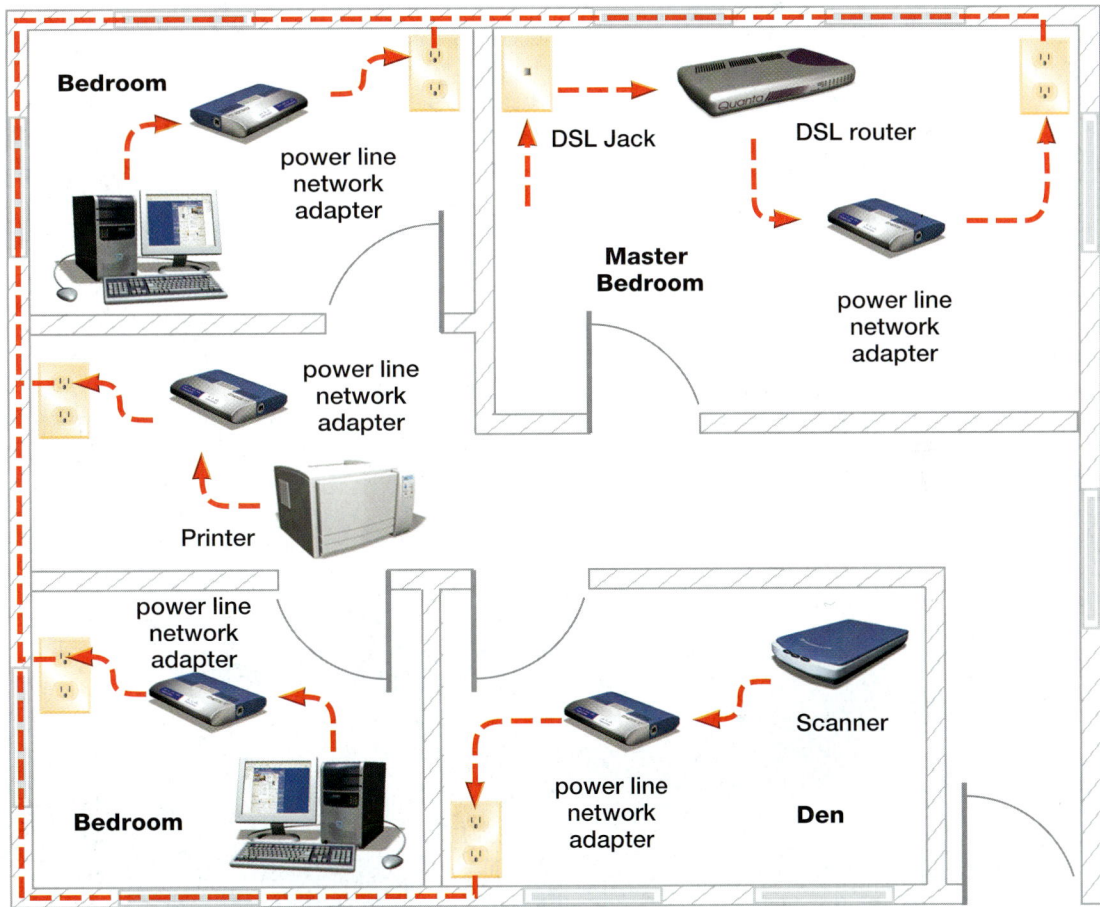

FIGURE 7.5
This USB power line network adapter is ready for installation. Ethernet power line adapters look very similar to this USB adapter.

Connect with a cable to a USB port on the system unit

FIGURE 7.6
Example of a power line Network. A power line network uses existing electrical wiring as its transmission media. Any electrical outlet in a house can therefore provide a network connection. In this power line network, the computers in the spare bedrooms can both use the scanner, printer, and DSL connection.

Are routers and hubs used on power line networks? When functioning only to allow computers to communicate with each other (not with other networks), power line networks do *not* require routers or hubs. However, if you want computers connected on a power line network to share an Internet connection, you need a router.

Will using electrical appliances interfere with a power line network? Using existing wiring doesn't interfere with the flow of electricity, so you can use electrical appliances at the same time as you're using your power line network.

PHONE LINE NETWORKS

What are phone line networks? **Phone line networks** move data through the network via conventional phone lines rather than power lines. Thus, with a phone line network, any phone jack in a house provides a network connection. Phone line networks have a maximum data transfer rate of 10 Mbps. The Home Phone line Networking Alliance (HPNA) (www.homepna.org) sets the standards for phone line networking.

How do I create a phone line network? To create a phone line network, you need to attach or install a **home phone line network adapter** (also called an **HPNA adapter**) to all computers and peripherals you want to attach to the network, as shown in Figure 7.7. You then attach the adapter to a phone jack on the wall with standard phone cord. Most phone line adapters are plug-and-play compatible, but if your OS doesn't recognize the adapter, manufacturers provide device drivers you can install on your computer.

As is the case with power line networks, each computer attached to a phone line network needs to have its OS configured for networking. To add computers or peripherals to the network, you attach a phone line adapter to the computer or peripheral you wish to add and plug it into another phone jack. Figure 7.8 shows an example of a phone line network.

Are routers and hubs used on phone line networks? As is the case with power line networks, when functioning only to allow computers to communicate with each other (not with other networks), phone line networks do *not* require routers or hubs. You need a router only if you want networked computers to share an Internet connection.

Can a telephone and a network device share a phone jack? Some phone line network adapters come with pass-through connections that allow you to plug your telephone into the adapter. For adapters *without* a pass-through connection, you can obtain a phone jack splitter, which turns one wall jack into two. This allows you to plug the network adapter and your phone into the same wall jack.

Can I hook up computers and peripherals to separate phone lines? There is no reason why you can't hook up phone line network devices to separate phone lines in the same house. However, since the phone lines are separate, devices hooked up to the first phone line can't communicate with devices hooked up to the second line. Therefore, unless you want to create two separate networks, hook up your computers to phone jacks on the same phone line.

Can I use an unused phone line (one that has no dial tone) to create a phone line network? Having a live phone line is not necessary for creating a phone line network. The only requirement for a phone line network is the phone cable itself.

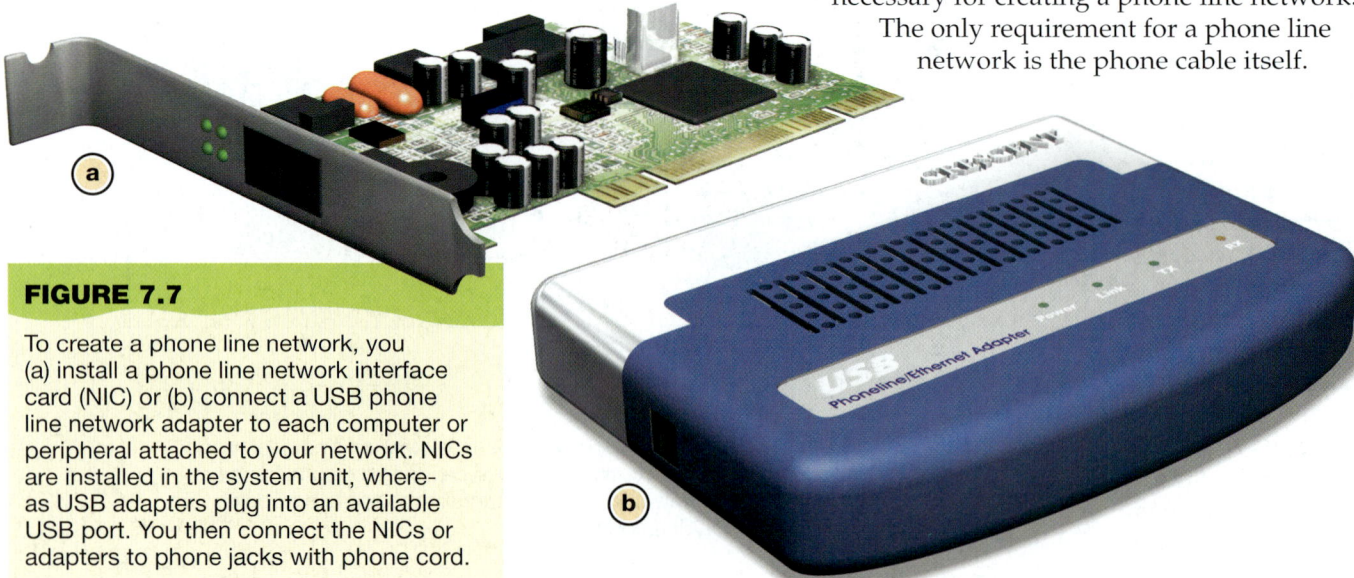

FIGURE 7.7

To create a phone line network, you (a) install a phone line network interface card (NIC) or (b) connect a USB phone line network adapter to each computer or peripheral attached to your network. NICs are installed in the system unit, whereas USB adapters plug into an available USB port. You then connect the NICs or adapters to phone jacks with phone cord.

TYPES OF PEER-TO-PEER NETWORKS 259

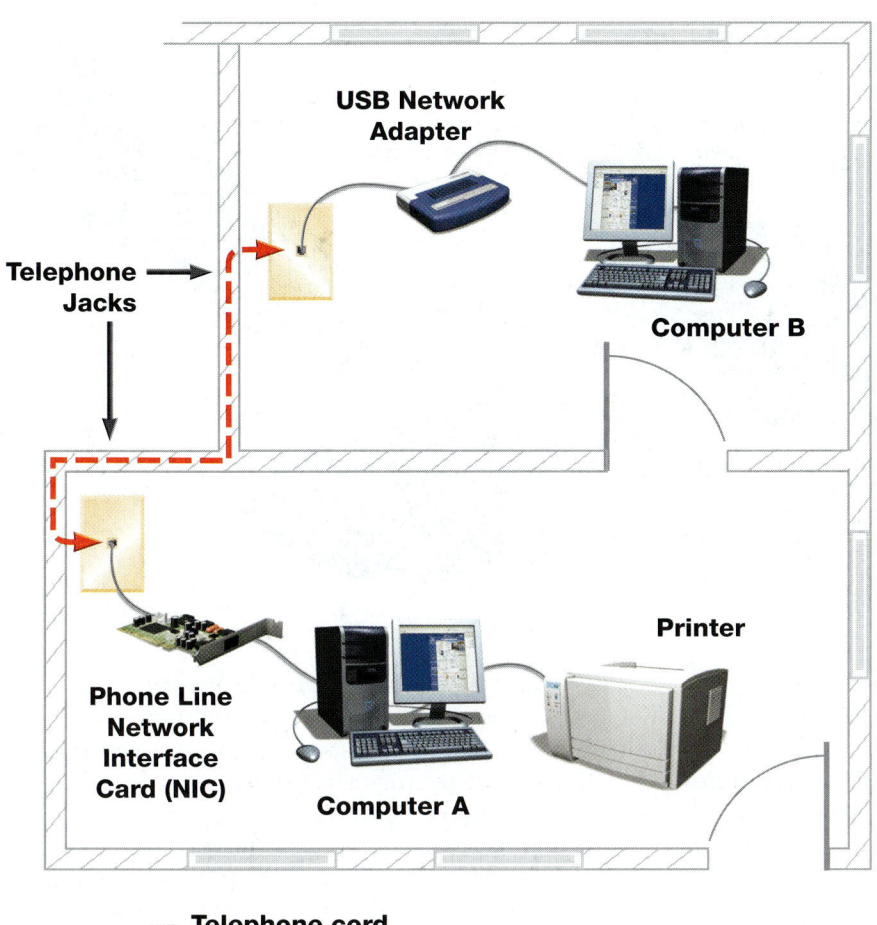

FIGURE 7.8

Example of a Phone Line Network. In this example, Computer A and Computer B are networked. Computer B can share the printer attached to Computer A. Both network interface cards (NICs) and USB network adapters can be used in phone line networks. NICs are installed inside the system unit, whereas USB network adapters plug into an available USB port.

ETHERNET NETWORKS

What are Ethernet networks? **Ethernet networks** differ from power line and phone line networks in that they use the Ethernet protocol as the means (or standard) by which the nodes on the network communicate. This protocol makes Ethernet networks extremely efficient at moving information. However, to achieve this efficiency, the algorithms for moving data through an Ethernet network are more complex than on the other peer-to-peer networks. Because of this complexity, Ethernet networks require additional devices (such as hubs and routers).

Although Ethernet networks are slightly more complicated to set up than phone line or power line networks, they are faster and more reliable. In fact, Ethernet networks are the most popular choice for home networks. Although you can install a 1-Gigabit-per-second (Gbps) Ethernet network in your home, such networks tend to be prohibitively expensive and are often unnecessarily powerful in terms of bandwidth for most home users. Thus, 100-Mbps Ethernet networks are more commonly installed in the home.

How do I create an Ethernet network? Much like phone line and power line networks, an Ethernet network requires that you install or attach network adapters to each computer or peripheral you want to connect to the network. Since Ethernet networks are so common, most computers come with Ethernet adapters preinstalled. As noted earlier, such internal network adapters are referred to as network interface cards (NICs). Modern Ethernet NICs are usually 10/100-Mbps cards. This means they can handle the old 10-Mbps Ethernet throughput traffic as well as the newer 100-Mbps throughput.

If your computer doesn't have a NIC, you can buy one and install it or you can use a USB adapter, which you plug into any open USB port on the system unit. Although you can use USB versions in laptops, PC card versions of Ethernet NICs are made especially for laptops. PC cards are about the size of credit cards and fit

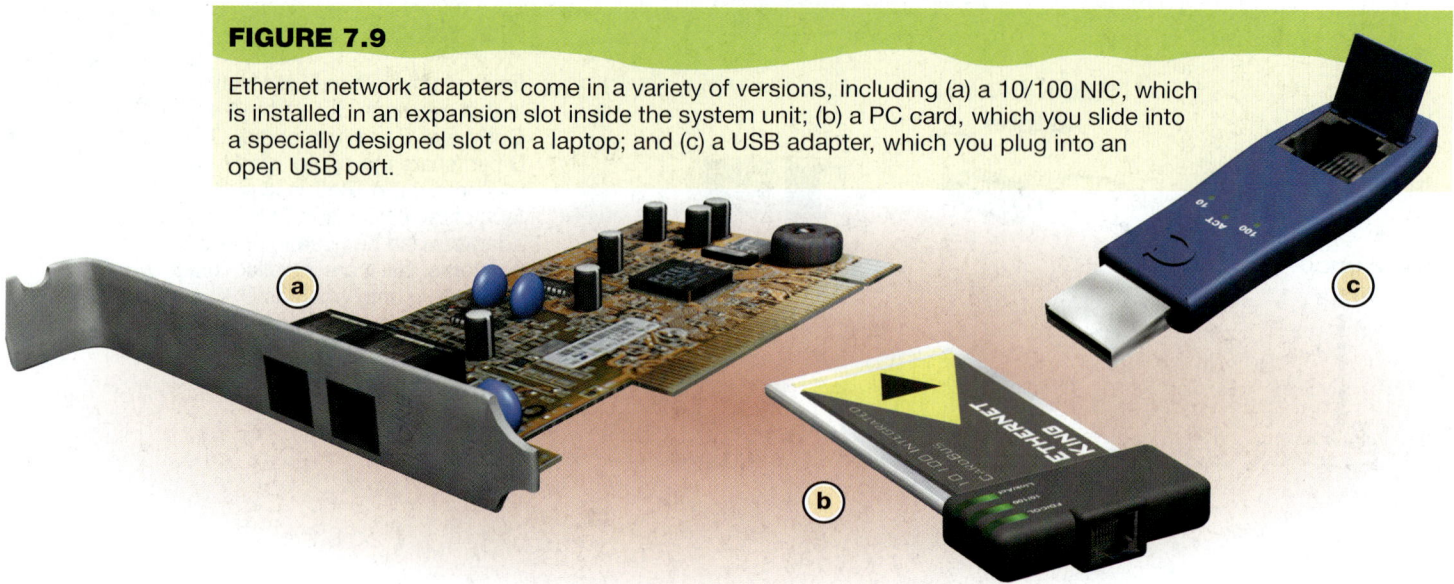

FIGURE 7.9
Ethernet network adapters come in a variety of versions, including (a) a 10/100 NIC, which is installed in an expansion slot inside the system unit; (b) a PC card, which you slide into a specially designed slot on a laptop; and (c) a USB adapter, which you plug into an open USB port.

into specially designed slots on a laptop. Figure 7.9 shows these different Ethernet network adapter options.

How are nodes connected on Ethernet networks? The most popular transmission media option for Ethernet networks is **unshielded twisted pair (UTP) cable**. UTP cable is composed of four pairs of wires that are twisted around each other to reduce electrical interference. You can buy UTP cable in varying lengths with RJ-45 jacks already attached. RJ-45 jacks resemble standard phone connectors (called RJ-11 jacks) but are slightly larger, as shown in Figure 7.10.

Do all Ethernet networks use the same kind of UTP cable? Figure 7.11 lists the three main types of UTP cable used in home Ethernet networks—CAT 5, CAT 5E, and CAT 6—and their data transfer rates. In general, it's better to install CAT 5E cable than CAT 5 since they're about the same price, and installing CAT 5E cable will enable you to take advantage of higher bandwidth Ethernet systems when they become cost effective for home use. CAT 6 cable is faster but not necessary for most home networks.

Is UTP cable difficult to install? UTP cable is no more difficult to install than normal phone cable, you just need to take a few precautions: Avoid putting sharp bends into the cable when running it around corners, as this can damage the copper wires inside and lead to breakage. Also, run the cable around the perimeter of the room (instead of under a rug, for example) to avoid damaging the wires from foot traffic.

How long can an Ethernet cable run be? Cable runs for Ethernet networks can't exceed 328 feet or else the signal starts to degrade. For cable runs over 328 feet, you can use **repeaters**, devices that are installed on long cable runs to amplify the signal. In effect, repeaters act as signal "boosters." Repeaters can extend run lengths to 600 feet, but they can add up to $400 to the cost of a network. When possible, you should use continuous lengths of cable. Although you can splice two cables together with a connecting jack, this presents a source of failure for the cable, as connectors can loosen up in the connecting jack and moisture or dust can accumulate on the contacts.

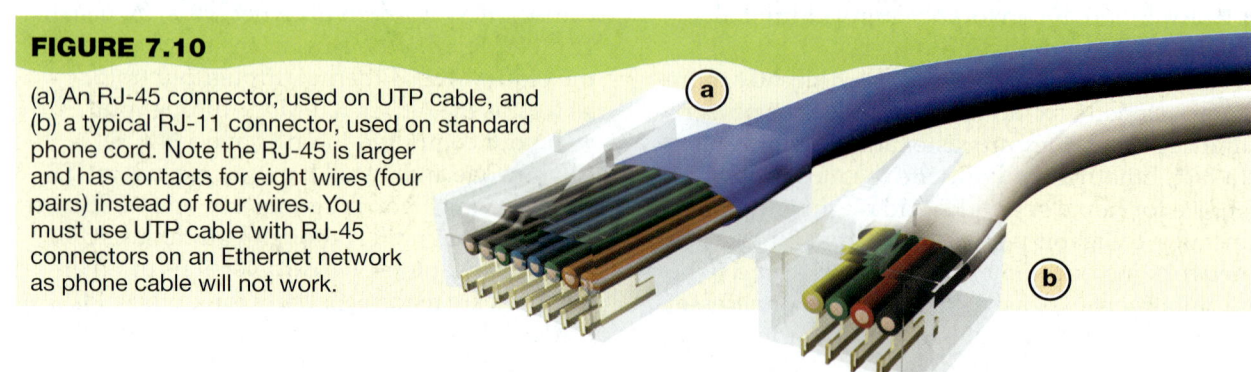

FIGURE 7.10
(a) An RJ-45 connector, used on UTP cable, and (b) a typical RJ-11 connector, used on standard phone cord. Note the RJ-45 is larger and has contacts for eight wires (four pairs) instead of four wires. You must use UTP cable with RJ-45 connectors on an Ethernet network as phone cable will not work.

TYPES OF PEER-TO-PEER NETWORKS   261

FIGURE 7.11 Data Transfer Rates for Popular Network Cable Types

CABLE CATEGORY	MAXIMUM DATA TRANSFER RATE
Category 5 (CAT 5)	100 Mbps
Category 5e (CAT 5E)	200 Mbps
Category 6 (CAT 6)	1,000 Mbps (1 Gbps)

Ethernet Hubs

How do Ethernet networks use hubs? Data is transmitted through the wires of an Ethernet network in packets. Imagine the data packets on an Ethernet network as cars on a road. If there were no traffic signals or rules of the road (such as driving on the right-hand side), we'd see a lot more collisions between vehicles, and people wouldn't get where they were going as readily (or at all). Data packets can also suffer collisions. If data packets collide, the data in them is damaged or lost. In either case, the network doesn't function efficiently.

As shown in Figure 7.12, a hub in an Ethernet network acts like a traffic signal by enforcing the rules of the data road on the transmission media. The hub keeps track of the data packets and helps them find their destination without running into each other. The hub also amplifies signals and retransmits them across the network. This keeps the network running efficiently.

How many computers and peripherals can be connected to a hub? Hubs are differentiated by the number of ports they have for connecting network devices. Four- and eight-port hubs are often used in home networks. A four-port hub can connect up to four devices to the network, while an eight-port hub can handle eight devices. Obviously, you should buy a hub that has enough ports for all the devices you want to connect to the network. Many people buy hubs with more ports than they currently need so that they can expand their network in the future.

A wonderful feature of hubs is that you can chain them together. Usually, one port on a hub is designated for plugging into a second hub. As shown in Figure 7.13, you can chain two four-port hubs together to provide connections for a total of six devices. Most hubs can be chained together to provide hundreds of ports, which would far exceed the needs of most home networks.

Ethernet Routers

How does data from an Ethernet network get shared with the Internet or another network? As we mentioned earlier, *routers* are devices that route packets of data between two or more networks. If a home network is connected to the Internet, you need a router to send data between the home network and the Internet.

Since so many people are sharing Internet access in home networks, manufacturers are making devices that combine hubs and routers and are specifically designed to connect to Digital Subscriber Line (DSL) or cable modems.

FIGURE 7.12

Hubs, like traffic signals, enforce the rules of the data road on an Ethernet network and prevent data packets from crashing into each other.

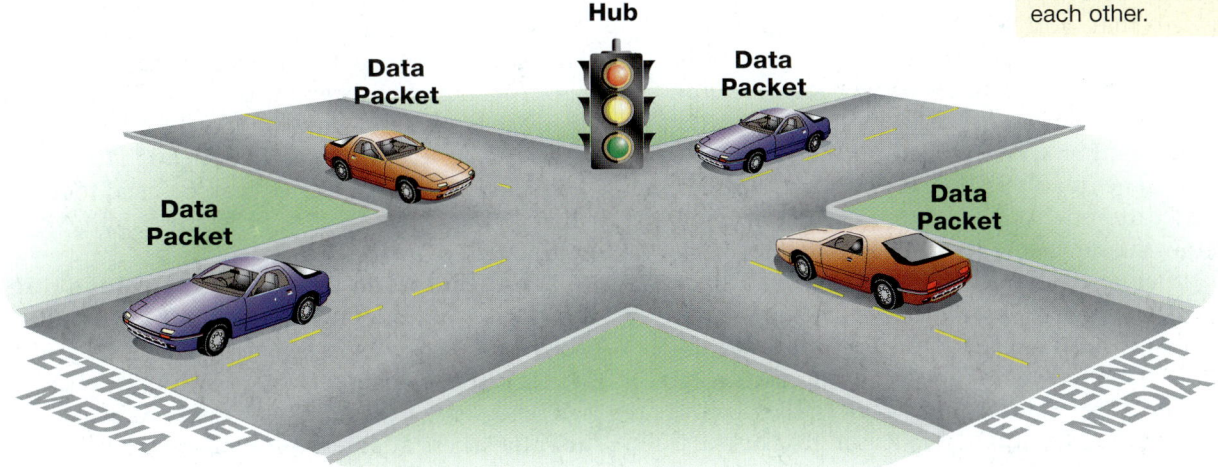

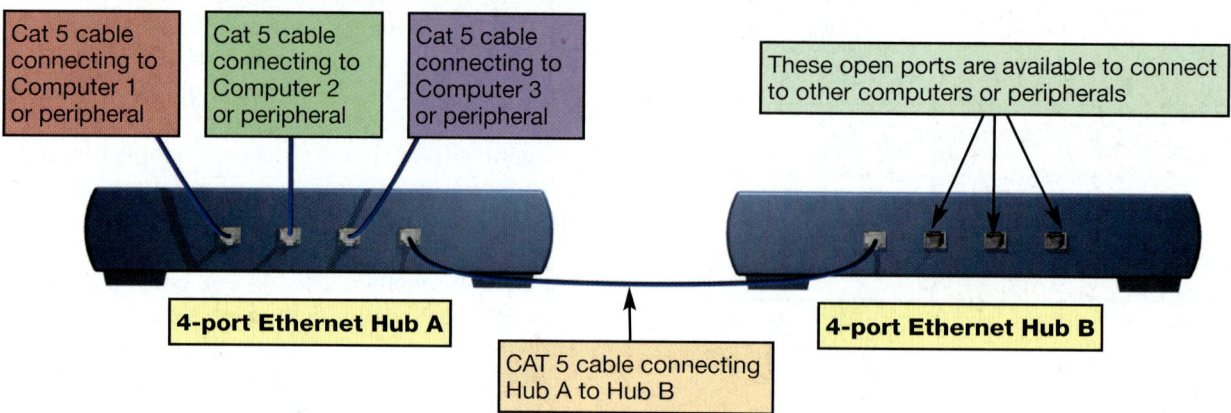

FIGURE 7.13

Adding another hub (Hub B) by chaining it to Hub A allows you to connect three more computers or peripherals to your network.

These are often referred to as **DSL/cable routers**. If you want your Ethernet network to connect to the Internet through a DSL or cable modem, obtaining a DSL/cable router is a good idea. Although you could share an Internet connection without one, a DSL/cable router provides increased throughput and is easier to configure than other methods. Figure 7.14 shows an example of an Ethernet network configured using a DSL/cable router.

BITS AND BYTES

Sticking to One Manufacturer Reduces Aggravation

Networking standards set by organizations such as the HomePlug Power Line Alliance make it easier for manufacturers to produce devices that work with a variety of computers and peripherals. In theory, such standards should benefit consumers as well since equipment from different manufacturers should work together when placed on the same network. The reality, however, is that devices from different manufacturers—even if they follow the same standards—don't always work together perfectly. This is because manufacturers sometimes introduce proprietary hardware and software that deviate from the standards. This means that an Ethernet NIC from Manufacturer A might not work with a DSL/cable router from Manufacturer B. The safe course of action is to use equipment manufactured by the same company.

WIRELESS NETWORKS

What is a wireless network? As its name implies, a **wireless network** uses radio waves instead of wires or cable as its transmission media. For wireless network devices to work with other networks and devices, standards were established. Current wireless networks in the United States are based on the **802.11 standard** established in 1997 by the Institute of Electrical and Electronics Engineers (IEEE, pronounced "I-triple-E"). The 802.11 standard is also known as **Wi-Fi** (short for Wireless Fidelity). Three standards are currently defined under 802.11: 802.11a, 802.11b, and 802.11g. The main difference between these standards is their maximum data transfer rate. For home networking, 802.11b and 802.11g are the standards used.

The 802.11b standard, which supports a maximum bandwidth of 11 Mbps, quickly became the accepted industry standard for home networks due to the low cost of implementing it. However, the newer 802.11g standard, which supports a higher throughput of 54 Mbps, is likely to be the preferred standard for home use soon, as it is much faster than 802.11b. Fortunately, 802.11g devices can be used together (that is, they have "backward compatibility") with 802.11b devices.

What do I need to set up a wireless network? Wireless networks are basically Ethernet networks configured to use radio waves instead of wires. Just like other networks, each node on a wireless network requires a **wireless network adapter**. These adapters are available as NICs that are inserted into expansion slots on the computer or as USB devices that plug into an open USB port.

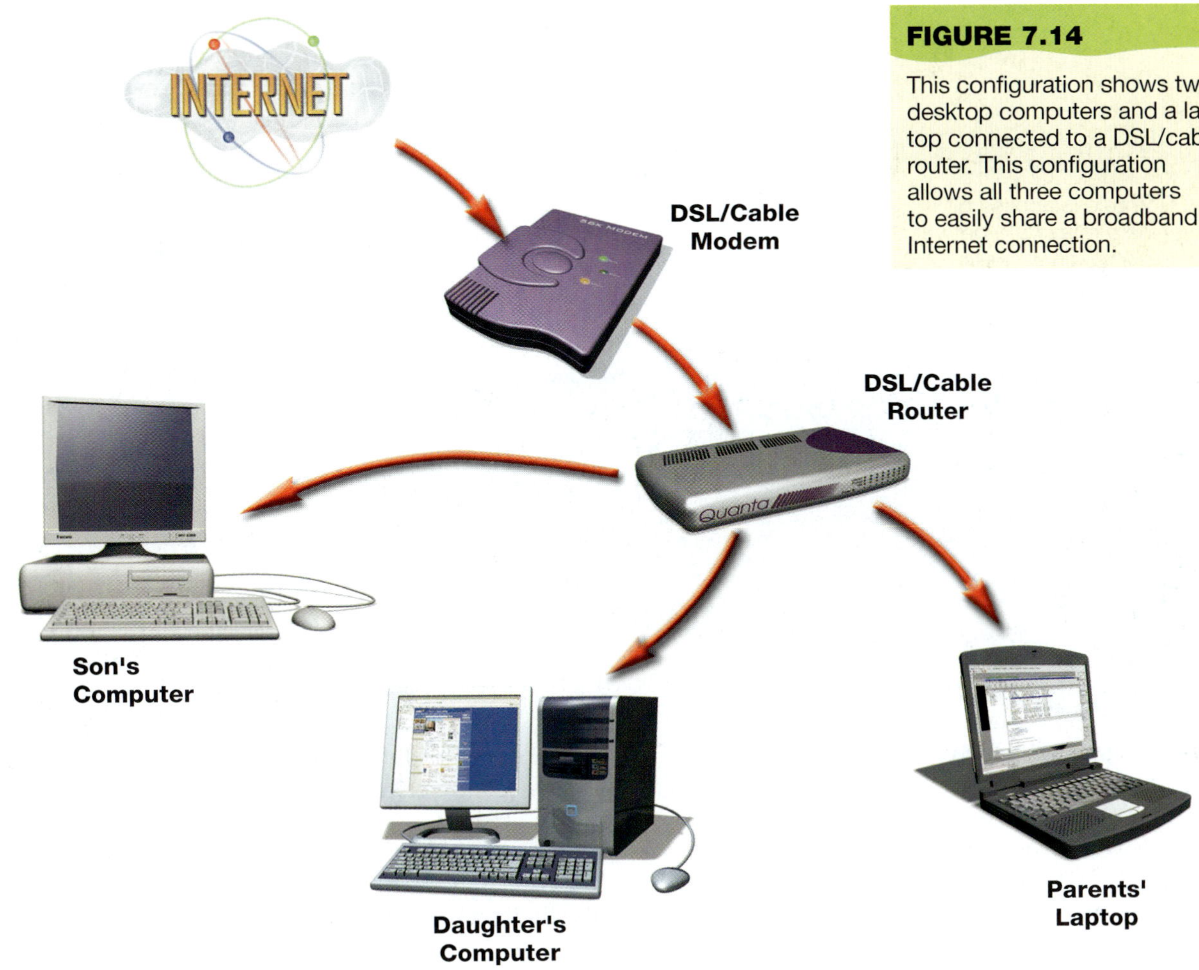

TYPES OF PEER-TO-PEER NETWORKS 263

FIGURE 7.14

This configuration shows two desktop computers and a laptop connected to a DSL/cable router. This configuration allows all three computers to easily share a broadband Internet connection.

Wireless network adapters differ from other network adapters in that they contain *transceivers*. A **transceiver** is a device that translates the electronic data that needs to be sent along the network into radio waves and then broadcasts these radio waves to other network nodes. Transceivers serve a dual function as they also receive the signals from other network nodes. As shown in Figure 7.15, wireless network adapters have antennas poking out of them, which are necessary for the transmission and reception of these radio waves.

FIGURE 7.15

Wireless network adapters have antennas poking out of them, which they use to communicate with the other devices in the network. Wireless network adapters are available as (a) NICs, which are inserted into an open expansion slot on the computer, or as (b) USB devices, which plug into an open USB port.

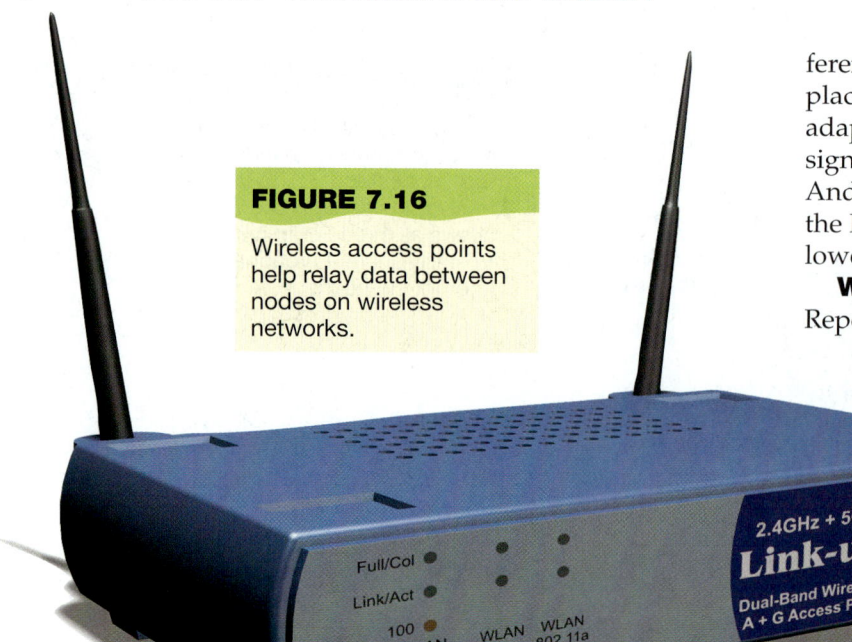

FIGURE 7.16

Wireless access points help relay data between nodes on wireless networks.

What types of problems can I run into when installing wireless networks? The maximum range of wireless devices is about 250 feet. However, as the distance between nodes increases, throughput decreases markedly. Although a computer in the same room as a cable modem may get 11 Mbps of throughput on a wireless network, a laptop connected to the network 100 feet away may get throughput of 5.5 Mbps or less. If nodes are located more than 250 feet apart, they may not be able to connect to the network at all.

Obstacles between wireless nodes also decrease throughput. Walls and large metal objects are the most common sources of interference with wireless signals. For example, placing a computer with a wireless network adapter next to a refrigerator may prevent the signals from reaching the rest of the network. And a node that has four walls between it and the Internet connection will most likely have lower than 11 Mbps of throughput.

What if the nodes can't communicate? Repositioning your computer or the wireless network adapter within the same room (sometimes even just a few inches from the original position) can often affect communication between the nodes. If this doesn't work, try moving the computers closer together or to other rooms in your house.

If these solutions don't work, you may need to add a wireless access point to your network, as shown in Figure 7.16. A **wireless access point** is a device similar to a hub in an Ethernet network. It takes the place of a wireless network adapter and helps to relay data between network nodes.

For example, as you can see in Figure 7.17, Laptop C on the back porch and Computer A in the bedroom can't connect with each other, but they can connect to Computer B in the den. By connecting a wireless access point to Computer B (instead of a wireless network adapter), all traffic from Computer A is relayed to Laptop C through the access point on Computer B.

How do I share an Internet connection on a wireless network? A *gateway* is a specialized device that enables wireless networks to share broadband connections. It is basically the router for a wireless network.

Can I have wired and wireless nodes on one network? Many users want to create a

FIGURE 7.17

With a wireless access point installed on Computer B, data can travel from Computer A (in the bedroom) to Laptop C (on the back porch).

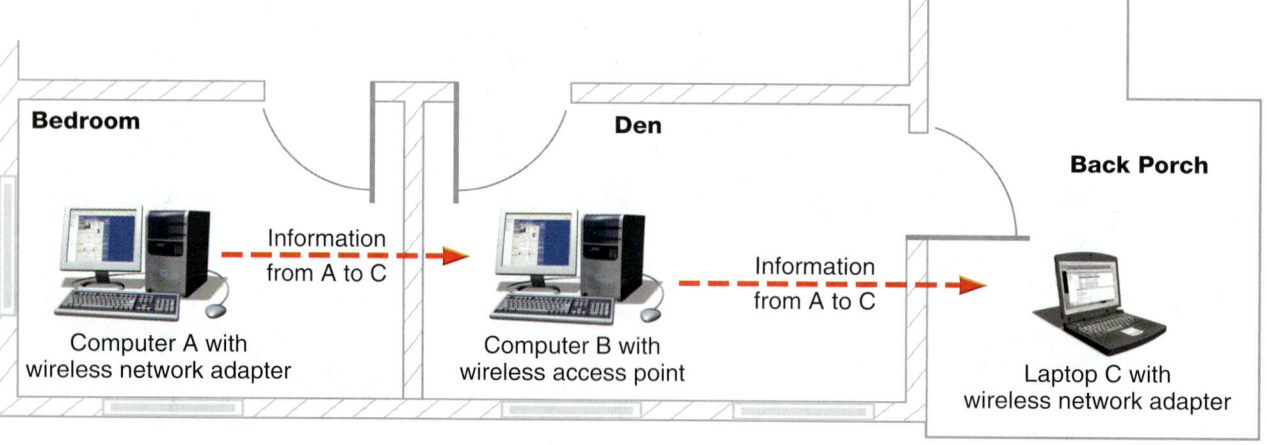

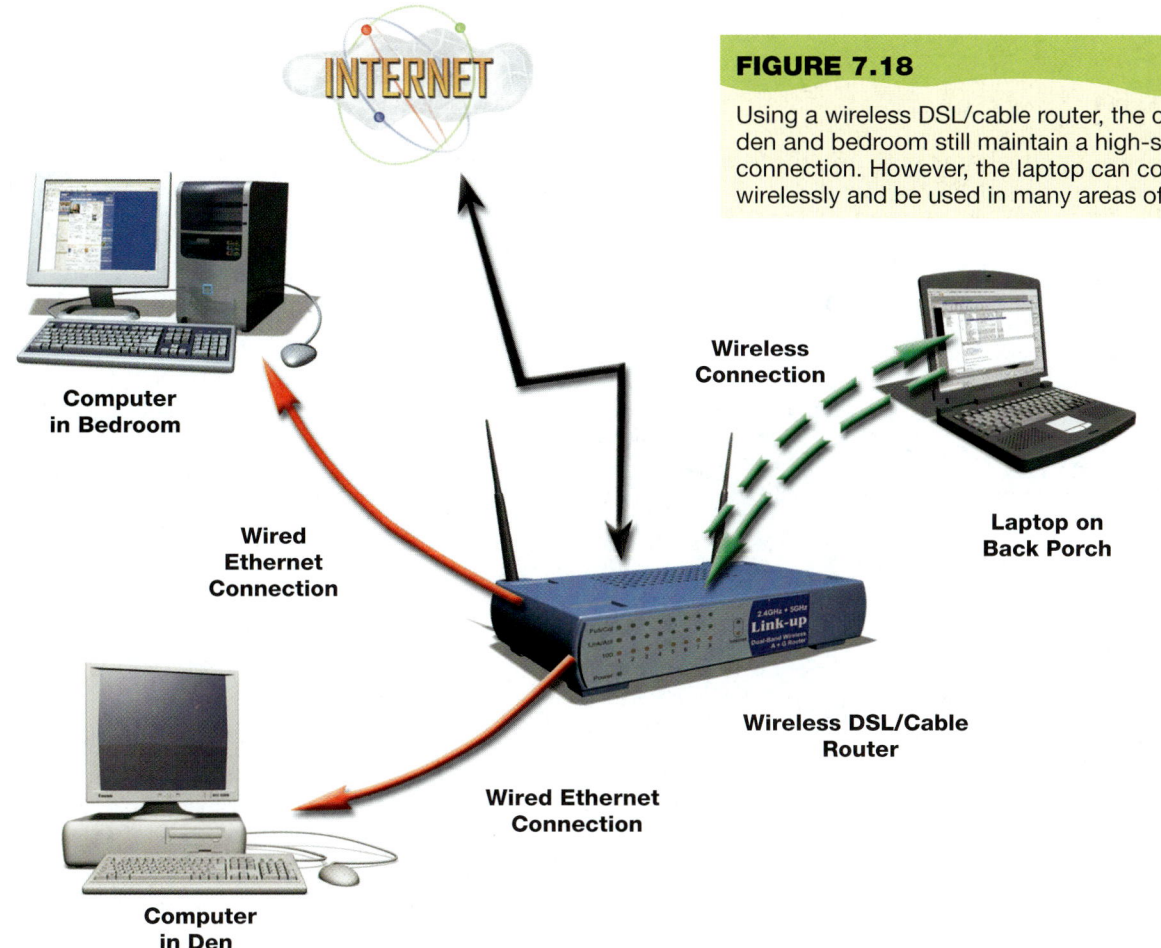

FIGURE 7.18

Using a wireless DSL/cable router, the computers in the den and bedroom still maintain a high-speed wired Ethernet connection. However, the laptop can connect to the network wirelessly and be used in many areas of the home.

network in which some computers (such as desktops) connect to the network with wires while other computers (such as laptops) connect to the network wirelessly. **Wireless DSL/cable routers** allow you to connect wireless and wired nodes to the same network. This type of router contains both a wireless access point as well as ports that allow you to connect wired nodes to the router. Figure 7.18 shows an example of a network with a wireless DSL/cable router attached to it. As you can see, the laptop maintains a wireless connection to the router while the other two computers are connected via wires. Using this type of router is a cost-effective way to have some wireless connections while still preserving the high-speed attributes of Ethernet where needed.

SOUND BYTE
SECURING WIRELESS NETWORKS

In this Sound Byte, you'll learn what "war drivers" are and why they could potentially be a threat to your wireless network. You'll also learn the simple steps to secure your wireless network against intruders.

BITS AND BYTES
Wireless Hot Spots

The rise of mobile computing is driving demand for wireless networks. To meet that demand, Starbucks, for example, offers wireless access to its customers in many of its shops. Many other businesses are also jumping on the bandwagon. Public places at which you can wirelessly connect to the Internet are known as "hot spots." Sometimes the service is free, other times there is a small charge. How can you tell? Just fire up your wireless-equipped laptop or PDA, start your browser, and try to access a Web site. If the service is not free, the store's "wireless gateway sentinel" software provides you with rates and an opportunity to pay for access. Either way, you are surfing in minutes while enjoying your latte. Going out of town and need to know where you can find a hot spot? Visit a site that offers hot spot directories, such as www.WiFinder.com or www.HotSpotList.com.

Choosing a Peer-to-Peer Network

If you're setting up a home network, the type of network you should choose depends on your particular needs. In general, consider the following factors in determining your network type:

- Whether existing wiring is available
- Whether you want wireless communications
- How fast you want your network connection to be
- How much money you can spend on your network

Figure 7.19 provides a flowchart that can help you decide which network to use in your home.

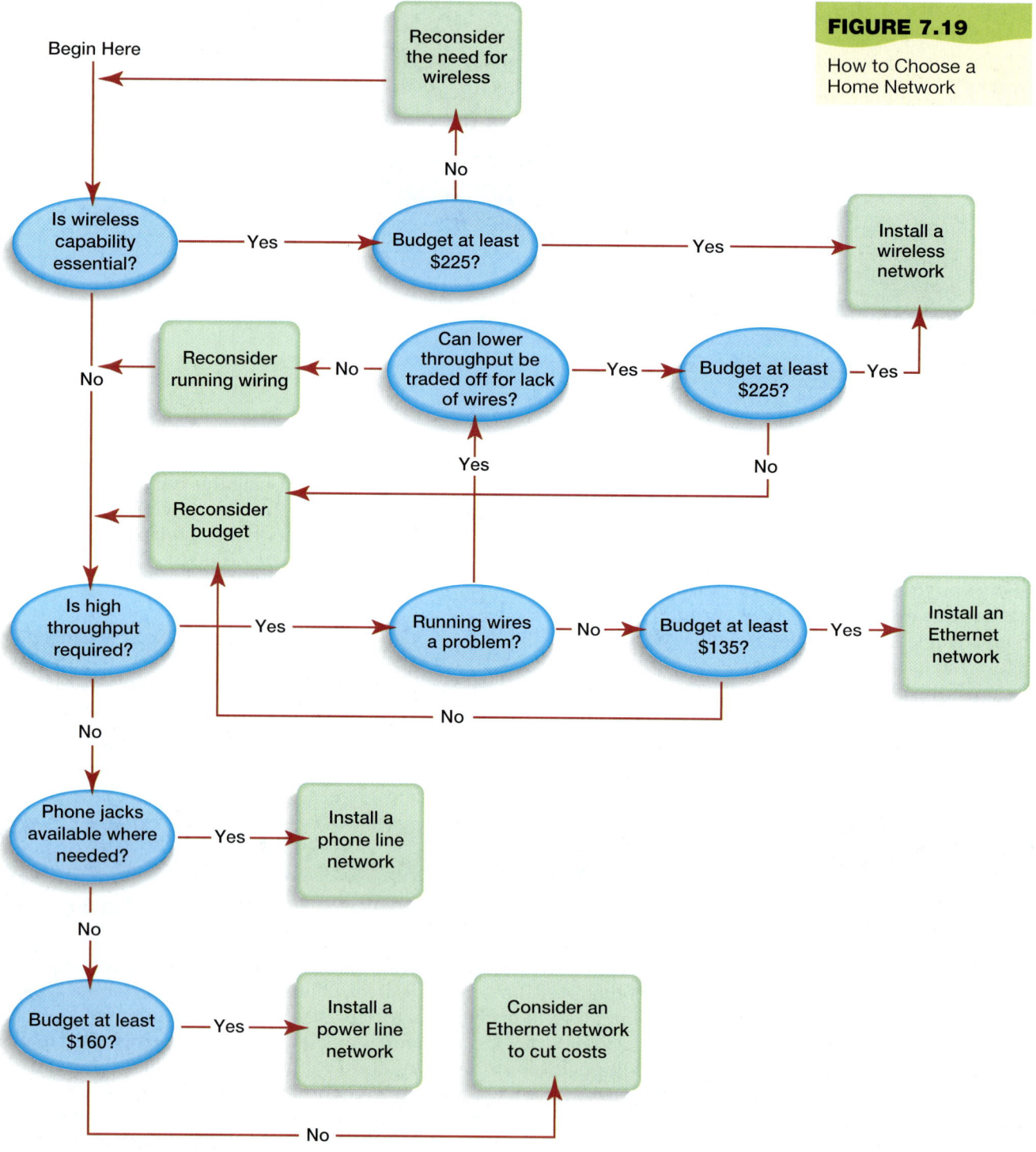

FIGURE 7.19

How to Choose a Home Network

What if I want to use existing wiring for my home network? As noted above, you can use both phone lines and power lines (electrical wiring) as media for a home network. For phone line networks, you need a phone jack in each room where you want to connect a node to the network. Likewise, for power line networks, you need an electrical outlet available in each room where you want to connect a node to the network. Since you have to plug most nodes (computers, printers, and so on) into an electrical outlet to operate them anyway, connecting a power line network is usually convenient.

What are the pros and cons of wireless networks? Wireless networks free you from having to run wires in your home. In addition, you can use any flat surface as a workstation. However, wireless networks may not work effectively in every home. Therefore, you need to install and test the wireless network to figure out if it will work. This can be expensive, so make sure you can return equipment for a refund if it doesn't work.

What network provides the fastest connection? For the majority of tasks undertaken on a home network (such as Web browsing and e-mailing), 1- or 2-Mbps throughput is sufficient. However, if high-speed data transmission is important to you (for example, if you play computer games or exchange large files), you may want a network with high throughput. Without high throughput, streaming video can appear choppy, games can respond slowly, and files can take a long time to transfer. With data transfer rates up to 1,000 Mbps, Ethernet networks are the fastest home networks.

Do I need to consider the type of broadband connection I have? Whether you connect to the Internet via DSL, cable, or satellite makes no difference in terms of the type of network you select. The differences occur with your particular network's hardware and software requirements.

What cost factors do I need to consider in choosing a network? Unfortunately, you may need to consider your budget when deciding what type of network to install. Figure 7.20 lists the approximate costs of installing the various types of peer-to-peer networks and their throughput.

FIGURE 7.20 Comparing the Major Types of Home Networks

	POWER LINE	PHONE LINE	ETHERNET	WIRELESS
Maximum data transfer rate (throughput)	14 Mbps	10 Mbps	100 to 1,000 Mbps (1 Gbps)[1]	11 to 54 Mbps
Approximate cost to network two computers (adapters and wiring only)	$160	$120	$135	$225[2]
Additional cost for adding Internet connection sharing	$80	$120	$0[3]	$40
Approximate cost to add an additional computer to network	$80	$60	$40	$60

Notes:
[1] Costs shown are for a 100-Mbps network as 1,000-Mbps networks are too expensive for most home networks.
[2] Price includes one wireless access point (a device that extends the reach of a wireless network) since this is required in many home networks.
[3] A router (a device used to route packets of data between different networks) is required for a basic Ethernet setup and is assumed to be a cable/DSL router.

Configuring Software for Your Home Network

Once you install the hardware for your network, you need to configure the software on your computers necessary for networking. In this section, you'll learn how to do just that using special Windows tools.

Is configuring software difficult? Windows XP makes configuring software relatively simple by providing the XP Network Setup Wizard, as shown in Figure 7.21. As you learned in Chapter 4, a wizard is a utility program included with Microsoft software that you can use to help you accomplish a specific task.

What if I don't have Windows XP on all my computers? Windows 98 and ME, the most common OSs found in the home other than XP, both support P2P networking. Therefore, you can network these computers with other computers using XP.

Where do I start? First, switch on all the computers and peripherals that are connected to your network. The wizard needs to check for other devices on the network, and it can't detect them if they're not turned on. If you have one computer with Windows XP but your other computers run on other versions of Windows, you should set up your Windows XP computer first with the XP Network Setup Wizard. One step of the wizard gives you an option to create an installation diskette you can use to set up non-XP computers. Create the installation diskette and then take the diskette to your Windows 98 or ME computers and follow the onscreen instructions.

What if I don't have Windows XP on any of my computers? Windows ME contains a similar wizard to Windows XP. (However, if you're mixing XP and ME computers, Microsoft recommends using the XP wizard and an installation diskette for the ME computers.) If you're networking all Windows 98 machines, there is no wizard available, so you have to set up your computers manually. Various resources on the Internet can assist you in setting up Windows 98 networks. Not surprisingly, one of the best resources is the Microsoft Web site (**www.microsoft.com**).

Why does the wizard ask me to name my computer? Each computer on the network needs a unique name so that the network can identify it. This unique name ensures that the network knows which computer is requesting services and data (such as Web page requests) so the data can be delivered to the correct computer.

Is that it? Assuming you installed and configured everything properly, your home network should be up and running and allowing you to share files, Internet connections, and peripherals. In the next section, we'll explore how you can protect your computers from intruders.

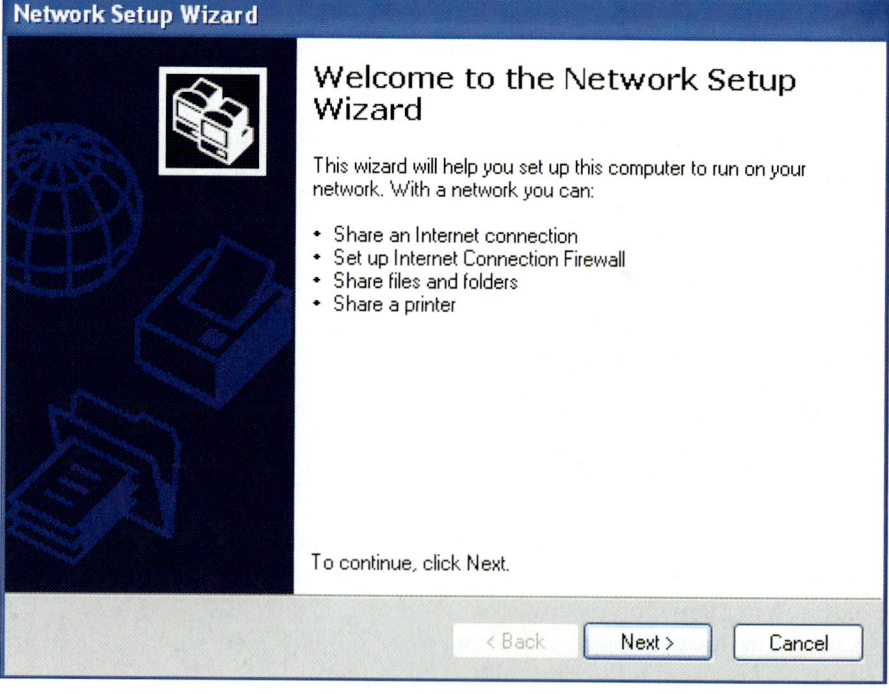

FIGURE 7.21

The XP Network Setup Wizard helps you configure your software for your network.

TRENDS IN IT

EMERGING TECHNOLOGIES: Patriot Computers: Fighting Bioterrorism with Your Computer

Want to use your computer to help wage the war against bioterrorists? Finding more effective drugs to fight diseases like smallpox doesn't require just laboratory research. Scientists also need huge amounts of computing power to analyze the data they amass in the lab. Strike a blow against terrorism and enlist your computer in the PatriotGrid, a network of computers working together to fight bioterrorism.

So when does your computer have time to work on this project? Although you may use your computer many hours each day, even at peak times, your CPU rarely exceeds 20 percent of its maximum processing capability. Computers perform calculations in bursts. In between the bursts of activity (even between keystrokes), there are periods of inactivity, which amounts to wasted computing resources. This doesn't mean your computer is inefficient, it just means that the power of today's computers is so great that you can't fully utilize it. Turning these periods of inactivity into productive time periods is the aim of **grid computing**, a technology being used to provide huge increases in computational power for a reasonable price. Using specialized software, a company can create a computer grid that can effectively link millions of computers to perform complex calculations during lulls in their processing cycles.

To become involved in the Patriot-Grid project, go to the United Devices Web site (**www.ud.com**) and download the PatriotGrid software. This software uses your computer during idle times to run chemical analyses on agents that may be effective in fighting smallpox. When you're on the Internet, the software sends the results of its calculations back to the lab and retrieves more data to analyze. Your computer is now joining millions of other computers in the PatriotGrid project. Without the PatriotGrid, the smallpox analysis project was estimated to take 45 years to complete. With the PatriotGrid, the processing time is shortened to a matter of months.

What companies use grid computing besides those fighting bioterrorism? Aircraft and spacecraft manufacturers need to perform complex design calculations, while pharmaceutical companies need to run complex tests to determine how chemical compounds will react in the human body. By using grid software, these companies can link their employees' computers to take advantage of their computing power during periods of relative inactivity (especially at night). As opposed to leasing time on a supercomputer, using their employees' computers allows these companies to cut costs. Meanwhile, online games in which massive numbers of players play simultaneously requires a lot of computing power. Companies like Butterfly.net are creating grid computing programs to harness the power of multiple computers specifically to host large gaming networks.

Grid computing today is mostly confined to grids constructed at single organizations or to volunteer organizations such as the PatriotGrid. However, as network bandwidth continues to increase, third-party companies will construct grids and lease time to corporations. These third-party companies could very well pay consumers for the use of their home computers as part of these grids. Someday, your computer could pay for itself by working during its otherwise "off" hours.

Keeping Your Home Computer Safe

The media is full of stories about computer viruses damaging computers, and attacks on corporate Web sites have brought major corporations to a standstill. These are examples of a type of criminal mischief called **cybercrime**, which is formally defined as any criminal action perpetrated primarily through the use of a computer. The existence of cybercrime doesn't mean we should fear conducting business, research, and recreation over the Internet. It does mean that computer users must maintain an awareness of computer crime and take precautions to protect themselves.

Who perpetrates computer crimes? **Cybercriminals** are individuals who use computers, networks, and the Internet to perpetrate crime. Anyone with a computer and the wherewithal to arm themselves with the appropriate knowledge can be a cybercriminal. In the next sections, we'll discuss cybercriminals and the damage they can wreak on your computer. We'll also discuss methods for protecting your computer from attacks.

Computer Threats: Hackers

Although there is a great deal of dissention (especially among hackers themselves) as to what a hacker actually is, a **hacker** (also called a **cracker**) is defined as anyone who breaks into a computer system (whether an individual computer or a network) unlawfully.

Are there different kinds of hackers? Some hackers are offended by being labeled criminals and therefore attempt to divide hackers into classes. Many hackers who break into systems just for the challenge of it (and who don't wish to steal or wreak havoc on the systems) refer to themselves as **white-hat hackers**. They tout themselves as experts who are performing a needed service for society by helping companies realize the vulnerabilities that exist in their systems.

These white-hat hackers look down on those hackers who use their knowledge to destroy information or for illegal gain. White-hat hackers refer to these other hackers as **black-hat hackers**. The terms *white hat* and *black hat* are references to old Western movies in which the heroes wore white hats and the outlaws wore black hats.

Irrespective of the opinions of the hackers themselves, the laws in the United States (and in many foreign countries) consider any unauthorized access to computer systems a crime.

What about the teenage hackers who get caught every so often? Although some of these teenagers are brilliant hackers, the majority are amateurs without sophisticated computer skills. These amateur hackers are referred to as **script kiddies**. Script kiddies don't create programs used to hack into computer systems; instead, they use tools created by skilled hackers that enable unskilled novices to wreak the same havoc as professional hackers. Unfortunately, a search on any search engine will produce links to Web sites that feature hacking tools, complete with instructions, allowing anyone to become an amateur hacker.

Fortunately, since the users of these programs are amateurs, they're usually not proficient at covering their electronic tracks. Therefore, it's relatively easy for law enforcement officials to track them down and prosecute them. Still, script kiddies can cause a lot of disruption and damage to computers, networks, and Web sites before they're caught.

TRENDS IN IT

COMPUTERS IN SOCIETY: Identity Theft: Is There More Than One You Out There?

You've no doubt heard of identity theft: a thief steals your name, address, social security number, and bank account and credit card information, and runs up debts in your name. This leaves you holding the bag as you're hounded by creditors collecting on the fraudulent debts. It sounds horrible, and it is. Many victims of identity theft spend months trying to reestablish their credit.

Stories of identity theft abound in the media, such as the Long Island, N.Y. man accused of stealing over 30,000 identities in 2002, and should serve to make the public wary. However, many media pundits would have you believe that the only way your identity can be stolen is via a computer. This is simply not true. The U.S. Federal Trade Commission (**www.ftc.gov**) has identified the following as methods thieves use to obtain others' personal information:

1. Identity thieves steal purses and wallets, where people often keep unnecessary valuable personal information (such as their ATM pin codes).
2. Identity thieves steal mail or look through trash for bank statements and credit card bills, which provide valuable personal information.
3. Identity thieves may pose as bank or credit card company representatives and trick people into revealing sensitive information over the phone.

Notice that none of these methods involve using a computer. Of course, you're at risk from online attacks too. You can give personal information to crooks by responding to bogus e-mail purportedly from your bank or ISP, for example. Once identity thieves have obtained your personal information, they can use it in a number of different ways:

Identity thieves can request a change of address for your credit card bill. By the time you realize that you aren't receiving your credit card statements, the thieves have rung up bogus charges on your account.

Identity thieves may open new credit card accounts in your name.

work navigation device such as a hub or a router. Data flows through the navigation device and the network nodes. Ethernet networks use unshielded twisted pair (UTP) cable to connect nodes and have a maximum data transfer rate of 1,000 Mbps (although 100-Mbps networks are more common in homes). Ethernet networks are the most popular type of home network.

8. **What are wireless networks and how are they created?** A wireless network uses radio waves instead of wires or cable as its transmission media. To create a wireless network, you install or attach wireless network adapters to the nodes that will make up the network. Signal interference between nodes can sometimes be a problem. If the nodes are unable to communicate, you can add a wireless access point to the network to help relay data between nodes. Wireless networks have a data transfer rate between 11 Mbps and 54 Mbps.

9. **How can hackers attack a network and what harm can they cause?** A hacker is defined as anyone who breaks into a computer system unlawfully. Hackers can use software to break into almost any computer connected to the Internet (unless proper precautions are taken). Once hackers gain access to a computer, they can (1) potentially steal personal or other important information; (2) damage and destroy data; or (3) use the computer to attack other computers.

10. **What is a firewall and how does it keep my computer safe from hackers?** Firewalls are software programs or hardware devices designed to keep computers safe from hackers. By using a personal firewall, you can close off open logical ports to invaders and potentially make your computer invisible to other computers on the Internet.

11. **What types of viruses do I need to protect my computer from?** A computer virus is a program that attaches itself to another program and attempts to spread itself to other computers when files are exchanged. Computer viruses can be grouped into five categories: (1) Boot-sector viruses replicate themselves into the master boot record of a hard drive, ensuring that the virus is loaded into memory immediately. (2) Logic bombs execute when a certain set of conditions is met, such as reaching a specific calendar date. (3) Worms are slightly different from viruses in that they attempt to travel between systems through network connections to spread their infections. (4) Scripts and macros are mini programs that are executed without your knowledge and cause destruction to your system. (5) Trojan horses, while not technically viruses, do something unintended to your computer while pretending to do something else. Once executed, they perform their malicious duties in the background, often invisible to the user.

12. **What can I do to protect my computer from viruses?** The best defense against viruses is to install antivirus software. You should update the software regularly and configure it to examine all e-mail attachments for viruses. You should periodically run a complete virus scan on your computer to ensure that no viruses have made it onto your hard drive.

Key Terms

802.11 standard (Wi-Fi)	p. 262
antivirus software	p. 281
backdoor programs	p. 272
black-hat hackers	p. 270
boot-sector viruses	p. 279
client	p. 254
client/server networks	p. 254
coaxial cable	p. 255
computer network	p. 252
computer virus	p. 278
cybercrime	p. 269
cybercriminals	p. 269
cyberterrorists	p. 272
data transfer rate (bandwidth, throughput)	p. 255
denial of service (DoS) attack	p. 273
device drivers	p. 257
distributed denial of service (DDoS) attacks	p. 273
DSL/cable routers	p. 262
dynamic addressing	p. 277
e-mail virus	p. 280
Ethernet networks	p. 259
fiber-optic cable	p. 255
firewalls	p. 275
grid computing	p. 269
hacker (cracker)	p. 270
home phone line network adapter (HPNA adapter)	p. 258
hubs	p. 256
identity theft	p. 272
Internet protocol addresses (IP addresses)	p. 276
local area networks (LANs)	p. 254
logic bombs	p. 280
logical port blocking	p. 276
logical ports	p. 275
macro	p. 280
macro viruses	p. 280
master boot record	p. 279
multipartite viruses	p. 281
network adapters	p. 256
network address translation (NAT)	p. 277
network architecture	p. 253
network interface cards (NICs)	p. 256
network navigation devices	p. 256
network operating system (NOS)	p. 256
nodes	p. 252
packet filtering	p. 276
packet sniffer	p. 272
packets	p. 256
peer-to-peer (P2P) networks	p. 253
personal firewalls	p. 275
phone line networks	p. 258
polymorphic viruses	p. 280
power line network adapter	p. 257
power line networks	p. 256
repeaters	p. 260
routers	p. 256
script kiddies	p. 270
scripts	p. 280
server	p. 254
static addressing	p. 277
stealth viruses	p. 281
transceivers	p. 263
transmission media	p. 255
Trojan horse	p. 272
twisted pair cable	p. 255
unshielded twisted pair (UTP) cable	p. 260
virus signatures	p. 281
white-hat hackers	p. 270
wide area networks (WANs)	p. 254
wireless access point	p. 264
wireless DSL/cable routers	p. 265
wireless network	p. 262
wireless network adapter	p. 262
worms	p. 280
zombies	p. 273

Buzz Words

Word Bank

- packet
- phone line
- network interface cards
- wireless
- throughput
- Ethernet
- dynamic addressing
- identity theft
- CAT 5E
- virus
- trojan horse
- firewall
- static addressing
- DDoS
- antivirus software
- cybercriminal
- peer-to-peer (P2P)
- router
- client/server
- logic bomb
- hackers
- logical ports
- zombie(s)
- power line
- electrical wiring
- phone cable
- network adapter
- wireless access point

Instructions: Fill in the blanks using the words from the Word Bank above.

Cathi needed to network three computers for her and her roommates (Sharon and Emily). After consulting her sister, a network engineer, she decided that a(n) (1)_____ network was the appropriate type to install in their apartment. Since none of them were gamers or transferred large files, they didn't need high (2)_____. Still, they decided to use the fastest type of P2P network, a(n) (3)_____ network, because it is easy to install and reliable. In order to connect the computers, they needed to buy (4)_____ cable and run it around the apartment. Since Sharon already had high-speed Internet access through the cable TV company, she needed to purchase a(n) (5)_____ for the network to ensure all the users could share the connection. Fortunately, all their computers already had (6)_____ installed, making it relatively easy to connect the computers to the network.

Cathi's roommate Emily was skeptical of being hooked up to the Internet since she had previously been the victim of (7)_____, which destroyed her credit rating. A wily (8)_____ had obtained her credit card information by posing as an employee of her bank. Cathi assured Emily that the (9)_____ could be configured as a(n) (10)_____ to repel malicious (11)_____. Turning off the unused (12)_____ would repel most attacks on their home network. With this protection, it was unlikely that a hacker would turn their PCs into (13)_____ to launch DDoS attacks with the aim of crippling major Web sites. But after the scare with the Melissa (14)_____, Cathi was careful to warn the others not to open files from untrusted sources. In addition, she made sure they all installed (15)_____ on their machines to protect them from viruses.

CHAPTER 7 NETWORKING AND SECURITY

Organizing Key Terms

Instructions: This chapter introduced many new terms and concepts. In the illustration below, fill in each of the blank boxes with keywords from the chapter in order to show how categories of ideas fit together.

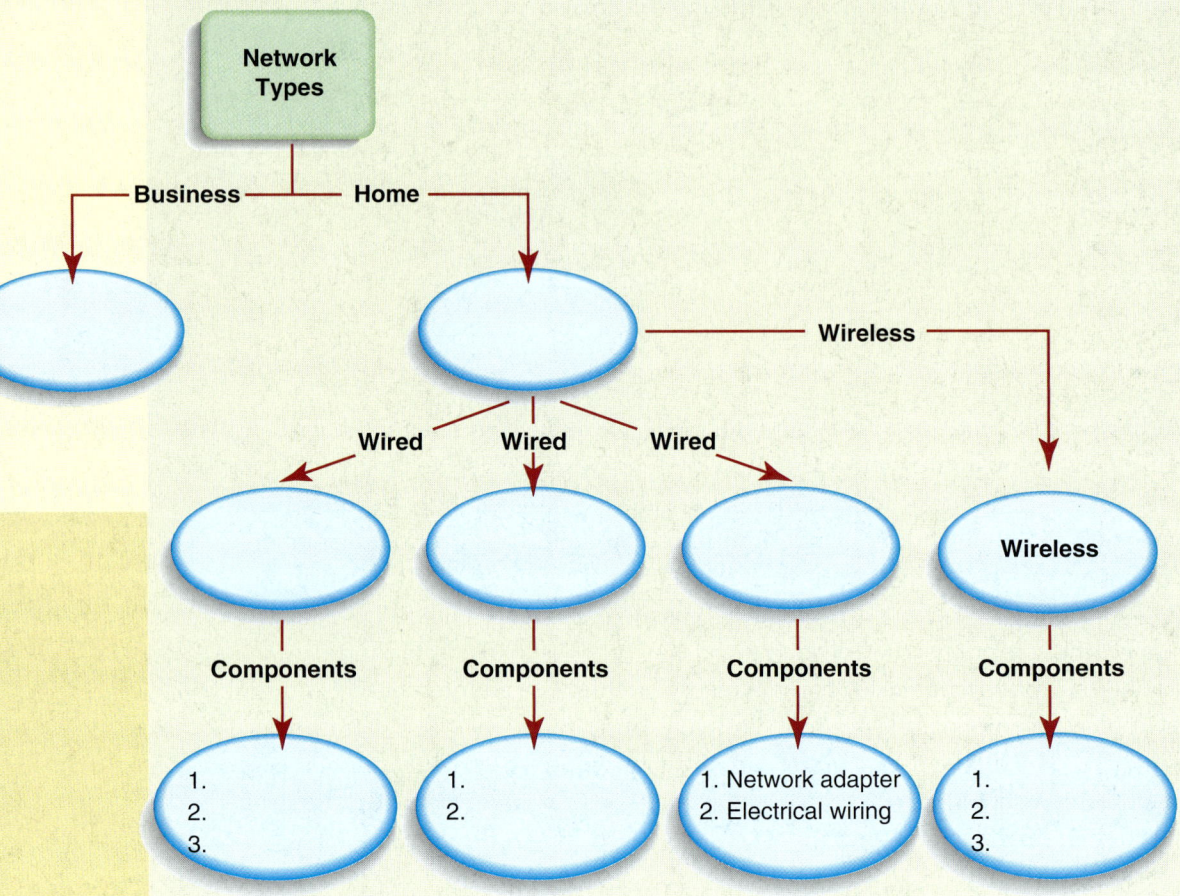

Making the Transition to . . . Next Semester

1. Dave, Jerome, and Thomas were sitting in the common room of their campus suite staring at $250 piled up on the coffee table. Selling last semester's books back to the bookstore had been a good idea. As they waited for their other roommate Phil to come home, Dave said, "Wouldn't it be cool if we could network our laptops? Then we could play Ultra Super Robot Kill-fest in team mode!" Jerome pointed out it would be even more useful if they could all have access to Dave's laser printer since he owned the only one. "And Jerome's always bugging me to use my scanner when I'm trying to sleep," remarked Thomas. "And I can't believe the only high-speed Internet connection is out here in the lounge!" The three roommates ran down the hall and rapped on your door looking for some guidance. Consider how you would answer their questions:

    a. Is $250 enough to set up a wired network for four laptops in four separate rooms? (Assume there are no phone jacks in the rooms.)
    b. Can the roommates share a printer and a scanner if they set up a wired network?
    c. How would they share the one high-speed Internet connection?
    d. Phil just returned from the campus post office with a check from his aunt for $100. Do the roommates now have enough money to set up a wireless network?

    To answer these questions, use the chapter text and the following resources:
    www.coolcomputing.com, **www.pricewatch.com**, **www.linksys.com**, **www.netgear.com**, **www.bestbuy.com**, *and* **www.compusa.com**.

2. In the Trends in IT feature, you learned about the PatriotGrid project in which computer users donate their computing power to help fight bioterrorism. Another grid computing project is being conducted by the Search for Extra Terrestrial Intelligence (SETI) Institute. The SETI@home project searches the heavens for radio waves indicating the presence of intelligent life in other solar systems. By installing free software, scientists use your computer during otherwise idle periods to process data received by the SETI project.

    a. Go to the SETI@home Web site (**http://setiathome.ssl.berkeley.edu**) and click on the Learn about SETI@home link. Then download the software and install it.
    b. Review the available information and write a short summary of how this program works.
    c. Search the Web for other grid computing projects. What other projects did you find?

3. Two students in your residence hall have recently been the victims of identity theft. You have been assigned to create a flyer telling students how they can protect themselves from identity theft in the residence hall. Using the information found in this chapter, materials you find on the U.S. federal government Web site on identity theft (**www.consumer.gov/idtheft**), as well as other Web resources, create a flyer that lists 5 to 10 ways in which students can avoid having their identities stolen.

4. Visit Gibson Research at **www.grc.com** and run the company's ShieldsUP and LeakTest programs on your computer.

    a. Did your computer get a clean report? If not, what potential vulnerabilities did the testing programs detect?
    b. How could you protect yourself from the vulnerabilities these programs can detect?

Making the Transition to . . . The Workplace

1. Your employer recently installed high-speed Internet access at the office where you work. There are 50 workstations connected to the network and the Internet. Within a week, half the computers in the office were down because of a virus that was contracted via a screensaver. In addition, network personnel from a university in England contacted the company, claiming that your employer's computer systems were being used as part of a DDoS attack on their Web site.

   a. Price out antivirus software on the Internet and determine the most cost-effective package for the company to implement on 50 workstations.
   b. Write a "virus prevention" memo to all employees suggesting strategies for avoiding virus infections.
   c. Draft a note to the CEO explaining how a firewall could prevent distributed denial of service attacks from being launched on the company network.

Critical Thinking Questions

Instructions: Albert Einstein used "Gedanken experiments," or critical thinking questions, to develop his theory of relativity. Some ideas are best understood by experimenting with them in our own minds. The following critical thinking questions are designed to demand your full attention but only require a comfortable chair—no technology.

1. Many people will be installing computer networks in their homes over the next five years.

   a. Would starting a home networking installation business be a good entry-level job for a college graduate? Could it be a good part-time job for a college student?
   b. Assuming the home networking business failed, what other careers would the technicians be prepared to assume?

2. You have just spent $325 installing a new wireless network in your home. A new wireless standard of networking will be launched a month later that is 10 times as fast as the wireless network you installed.

   a. What types of applications would you need to be using heavily to make it worth upgrading to the new standard?
   b. Can your next-door neighbor use his wireless-equipped notebook to surf off your Internet connection, thereby saving him the cost of purchasing a connection of his own? Is this ethical?

3. Hackers and virus authors cause millions of dollars worth of damage to PCs and networks annually. But hacking is a very controversial subject. Many hackers believe they are actually working for the "good of the people" or "exercising their freedom" when they engage in hacking activities. However, in most jurisdictions in the United States, hacking is punishable by stiff fines and jail terms.

   a. Hackers often argue that hacking is for the good of all people since it points out flaws in computer systems. Do you agree with this? Why or why not?
   b. What should the punishment be for convicted hackers and why?
   c. Who should be held accountable at a corporation whose network security is breached by a hacker?

4. Many of us rely on networks every day, often without realizing it. Whether using the Internet, ordering a book from Amazon.com, or accessing your college e-mail from home, you are relying on networks to relay information. But what if terrorists destroyed key components of the Internet or other networks that we depend upon?

   a. What economic problems would result from DDoS attacks launched by terrorists on major e-commerce sites?
   b. What precautions should the U.S. military take to ensure that networks involving national defense remain secure from terrorist attacks?

5. In the Trends in IT feature, we discussed an example of grid computing called the PatriotGrid. In this project, computer users run software that can help scientists find ways to fight bioterrorism. What other uses of grid computing can you think of? Would you be willing to take part in a grid computing project for free?

6. Do you have a firewall or antivirus software installed on your home computer? If not, why not? Have you ever been a victim of a hacker or a virus?

Team Time

Problem:

Wireless technology is being adopted by leaps and bounds both in the home and in the workplace. Offering easy access free of physical tethers to networks seems to be a solution to many problems. However, wireless computing also has problems, ranging from poor reception to hijackers stealing your bandwidth.

Task:

Your campus has recently undertaken a wireless computing initiative. As part of the plan, your dorm has just been outfitted with wireless access points (base stations) to provide students with connectivity to the Internet and the college network. However, since the installation, students have reported poor connectivity in certain areas and extremely low bandwidth at other times. Your group has volunteered to research the potential problems and to suggest solutions to the college IT department.

Process:

Break the class into three teams. Each team will be responsible for investigating one of the following issues:

Detecting Poor Connectivity: Research methods that can be used to find areas of poor signal strength such as signal sniffing software (**www.netstumbler.com**) and handheld scanning devices such as Wi-Fi sniffer (**www.idetect.com.sg**). Investigate maximum distances between access points and network nodes (equipment manufacturers such as **www.dlink.com** and **www.linksys.com** provide guidelines) and make appropriate recommendations.

Signal Boosters: Research alternatives that can be used to increase signal strength in access points, antennas, and wireless cards. Signal boosters are available for access points. You can purchase or construct replacement antennas or antenna enhancements. Wi-Fi cards that offer higher power than conventional cards are now available.

Security: "War drivers" (people who cruise neighborhoods looking for open wireless networks from which to steal bandwidth) may be the cause of the bandwidth issues. Research appropriate measures to keep wireless network traffic secure from eavesdropping by hackers. In your investigation, look into the new Wi-Fi Protected Access (WPA) standard developed by the Wi-Fi Alliance. Check out the security section on the Wi-Fi Alliance Web site to start (**www.weca.net**).

Present your findings to your class and discuss possible causes and preventative measures for the problems encountered at your dorm. Provide your instructor with a report suitable for eventual presentation to the college IT department.

Conclusion:

As technology improves, wireless connectivity should eventually become the standard method of communication between networks and network devices. As with any other technology, security risks exist. Understanding those risks and how to mitigate them will allow you to participate in the design and deployment of network technology and provide peace of mind for your network users.

Becoming Computer Fluent

While attending college, you are working at the Snap-Tite company, a small manufacturer of specialty fasteners. Currently, they use "sneakernet" (copying files on floppy diskettes) to transfer files between the four PCs the company owns. Only the company president has access to the Internet, although the other employees could certainly use it for doing research and placing orders for raw material and supplies. The accounts payable clerk is the only one who has a printer and is constantly being interrupted by the other employees when they want to print their files from floppy diskettes. Your boss heard that you were taking a computer course and volunteered you to create a solution.

Instructions: Using the scenario above, draft a networking plan for Snap-Tite utilizing as many of the key words from the chapter as you can. Be sure that the relatively computer-illiterate company president can understand the report.

Materials on the Web

In addition to the review materials presented here, you'll find extra materials on the book's companion Web site (**www.prenhall.com/techinaction**) that will help reinforce your understanding of the chapter content. These materials include:

Sound Byte Lab Guides

For each Sound Byte mentioned in the chapter, there is a corresponding lab guide located on the book's companion Web site. These guides review the material presented in the Sound Byte and direct you to various Web resources that examine the material. The Sound Byte Lab Guides for this chapter include:

- Installing a Network
- Securing Your Wireless Network
- Installing a Personal Firewall
- Protecting Your Computer

True/False and Multiple Choice Quizzes

The book's Web site includes a True/False and a Multiple Choice quiz for this chapter. You can take these quizzes, automatically check the results, and e-mail the results to your instructor.

Web Research Projects

The book's Web site also includes a number of Web research projects for this chapter. These projects ask you to search the Web for information on computer-related careers, milestones in computer history, important people and companies, emerging technologies, and the applications and implications of different technologies.

Protecting Your Computer and Backing Up Your Data

Technology in Focus

Just like any other valuable asset, computers and the data they contain require protection from damage, thieves, and unauthorized users. Although it's impossible to protect your computer and data completely, following the suggestions outlined in this **Technology In Focus** will provide you with peace of mind that you have done all you can to protect your computer from theft and keep it in working order.

Physically Protecting Your Computer

Your computer obviously isn't useful to you if it is damaged. Therefore, it's essential to select and ensure a safe environment for your computer. This includes protecting it from environmental factors, power surges, and power outages.

Environmental Factors

There are a number of environmental factors you need to consider in order to protect your computer.

1. Sudden movements can damage your computer or mobile device's internal components, causing them to operate erratically. Therefore, take special care in handling your computer. If you drop it, have it professionally tested by a computer repair facility to uncover any hidden damage.

2. Electronic components do not like excessive heat. Unfortunately, computers generate a lot of heat. This is why they contain a fan to cool their internal components. Make sure that you place your computer so that the fan's input vents (usually found on the rear of the system unit) are unblocked so that air can flow inside. Naturally, a fan drawing air into a computer also draws in dust and other particles, which can wreak havoc on your system. Therefore, keep the room in which your computer is located as clean as possible. Placing your computer in the workshop where you do woodworking and generate sawdust would obviously be a poor choice!

3. Since food crumbs and liquid can damage keyboards and other computer components, consume food and beverages far away from your computer to avoid food-related damage.

Power Surges

Power surges occur when electrical current is supplied in excess of normal voltage (120 volts in the United States). Old or faulty wiring, downed power lines, malfunctions at electric company substations, and lightning strikes can all cause power surges (see Figure 1). **Surge protectors** are devices that protect your computer against power surges. To use a surge protector, you simply plug all your electrical devices into the outlets of the surge protector, which in turn plugs into the wall.

TECHNOLOGY IN FOCUS

How Surge Protectors Work

During **minor** surges, the MOVs (MOV stands for metal-oxide varistor) bleed off excess current and feed it to the ground wire, where it harmlessly disappears. The MOVs can do this while still allowing normal current to pass through to the devices plugged into the surge protector. This is why it is critical to plug surge protectors into grounded power outlets.

During **major** surges that overwhelm the MOVs, the fuse blows, stopping **all** current from passing through to devices plugged into the surge protector. After a major surge, the surge protector will no longer function and must be replaced.

Over time, the MOVs lose their ability to bleed off excess current, which is why you should replace your surge protector every 2 to 3 years.

FIGURE 1

Surge protectors wear out over time (usually in less than five years), so buy a surge protector that includes indicator lights. Indicator lights illuminate when the surge protector is no longer functioning properly, letting you know you need to replace it. Note that old surge protectors can still function as multiple-outlet power strips, delivering power to your equipment without protecting it. A power surge could ruin your computer and other devices if you don't protect them. Thus, at around $30, a surge protector is an excellent investment.

It's important to protect *all* your electronic devices, not just computers, from surges. TVs, printers, and other computer peripherals all require protection. However, it can be expensive to use individual surge protectors on everything. A more practical method is to install a **whole-house surge protector**, shown in Figure 2. Whole-house surge protectors function like other surge protectors but they protect *all* electrical devices in the house. Electricians usually install whole-house surge protectors, which cost $200 to $300 installed.

Data lines (transmission media), such as the coaxial cable or phone wires that attach to your modem, can also carry surges. Installing a **data line surge suppressor** for each data line connected to your computer through another device (such as a modem) provides you with additional protection (see Figure 3). A data line surge suppressor is connected to the data line at a point before it reaches the modem or other device. In this way, it intercepts surges on the data line before they reach sensitive equipment.

Surge protectors won't necessarily guard against all surges. Lightning strikes can generate such high amounts of voltage that they can overwhelm a surge protector. As tedious as it sounds, unplugging computers and peripherals during an electrical storm is the only way to achieve absolute protection.

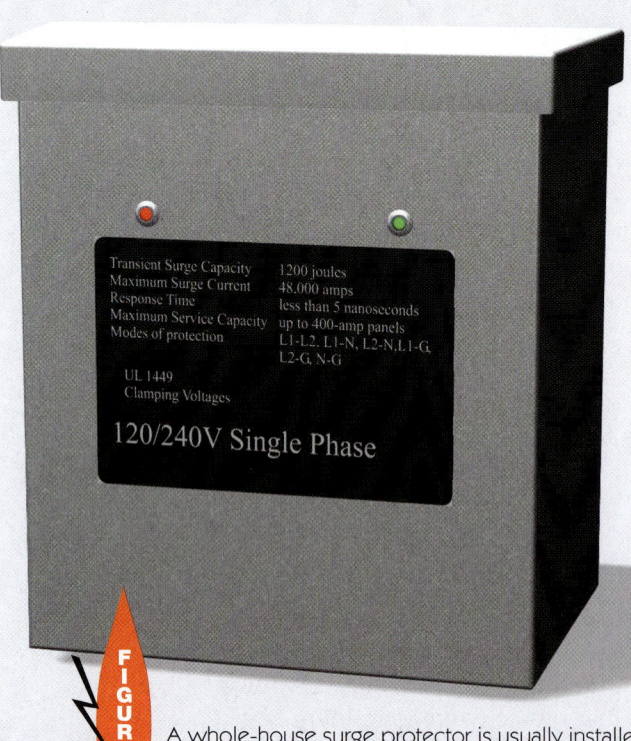

FIGURE 2 A whole-house surge protector is usually installed at the breaker panel or near the electric meter. It protects all appliances in the home from electrical surges.

DETERRING THEFT 297

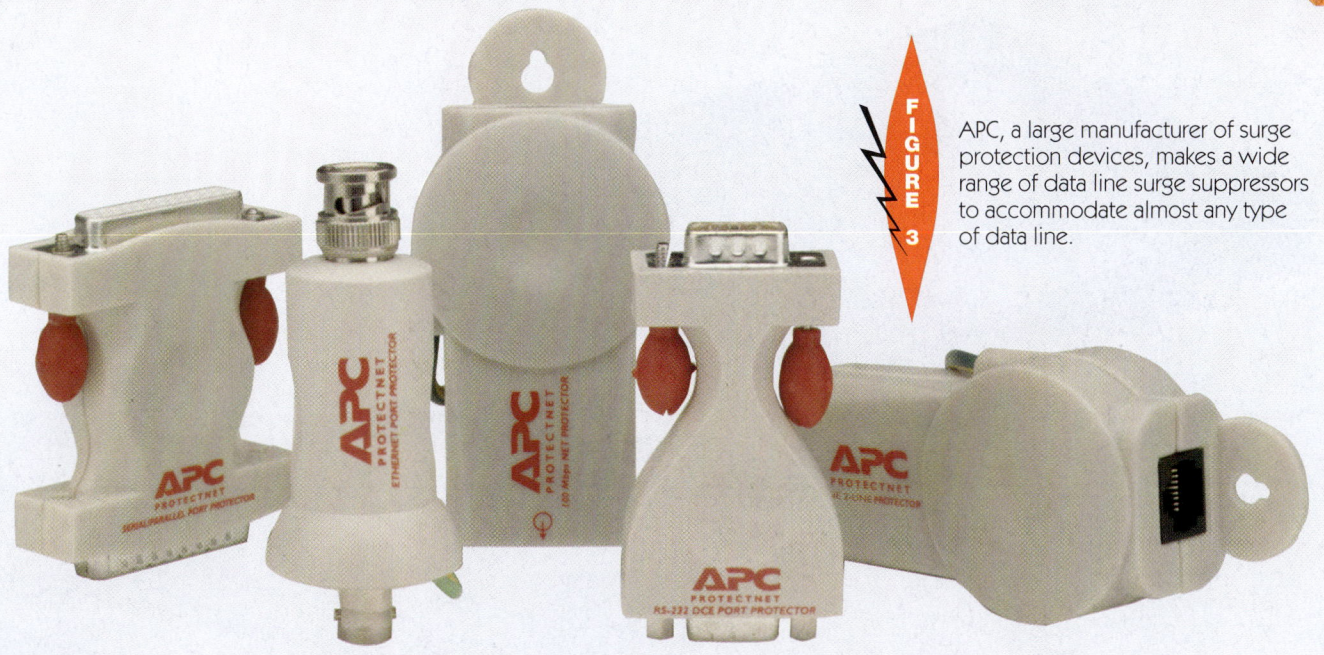

FIGURE 3 APC, a large manufacturer of surge protection devices, makes a wide range of data line surge suppressors to accommodate almost any type of data line.

Power Outages

Like power surges, power outages can wreak havoc on a system. Mission-critical computers such as web servers are often protected by **uninterruptible power supplies (UPSs)**, shown in Figure 4. A UPS is a device that contains surge protection equipment and a large battery. When power is interrupted (such as during a blackout), the UPS continues to send power to the attached computer from its battery. Depending on the battery capacity, you have between about 20 minutes and 3 hours to save your work and shut down your computer properly.

Deterring Theft

Because they are portable, laptops and personal digital assistants (PDAs) are easy targets for thieves. Common sense dictates that you don't leave your laptop or PDA unattended or in places where it can be easily stolen. Three additional approaches to deterring computer theft include alarming them, locking them down, or allowing the devices to tell you when they are stolen.

SOUND BYTE
SURGE PROTECTORS

In this Sound Byte, you'll learn about the major features of surge protectors and how they work. You'll also learn about the key factors you need to consider before buying a surge protector, and you'll see how easy it is to install one.

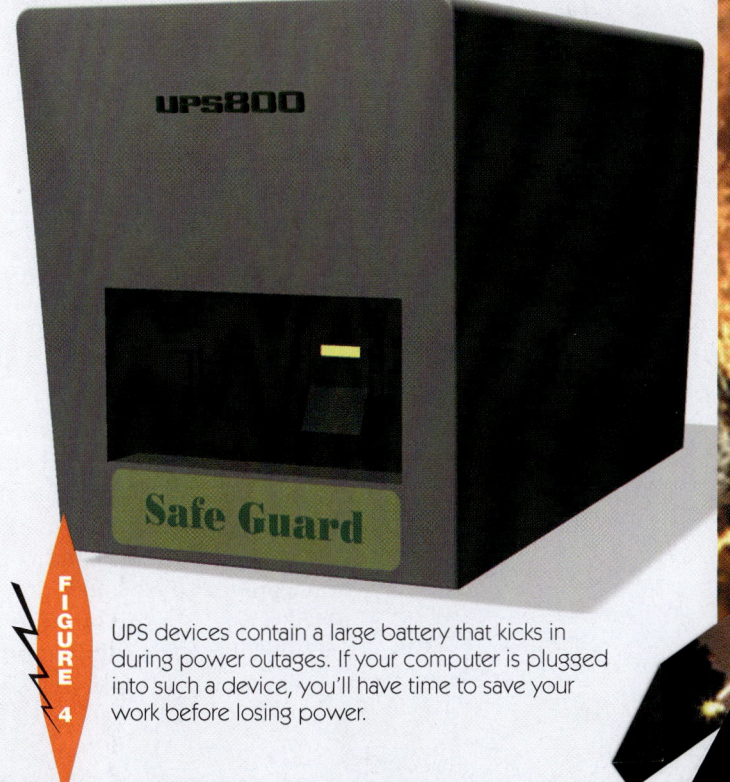

FIGURE 4 UPS devices contain a large battery that kicks in during power outages. If your computer is plugged into such a device, you'll have time to save your work before losing power.

TECHNOLOGY IN FOCUS

FIGURE 5 A laptop alarm (left) sends out an ear-piercing sound if your laptop is moved before you deactivate the alarm.

FIGURE 6 Cable locks are effective, but do require users to keep track of the keys to the locks

Alarms

To prevent your laptop from being stolen, you can attach a motion alarm to it, shown in Figure 5. When you leave your laptop, you use a small device called a "key fob activator" to activate the alarm. If your laptop is moved while the alarm is activated, it emits a wailing 85-decibel sound. The fact that the alarm is visible acts as an additional theft deterrent, just like a "beware of dog" sign in a front yard.

Locks and Surrounds

Chaining a laptop to your work surface can be an effective way to prevent theft. As shown in Figure 6, a special locking mechanism is attached to the laptop (some laptops are even manufactured with locking ports), and a hardened steel cable is connected to the locking mechanism. The other end of the cable is looped around something large and heavy, such as a desk. The cable lock obviously requires that you use a key to free the laptop from its mooring.

Many people associate computer theft only with laptops or PDAs. But desktop computers are vulnerable to theft also, especially theft of internal components such as RAM. Cable locks are available that connect through special fasteners on the back of desktop computers, but components can still be stolen because these cables often don't prevent the system unit case from being opened. A more effective theft deterrent for desktops is a **surround** (or **cage**), shown in Figure 7. A surround is a metal box that encloses the system unit and makes it impossible to remove the case, while still allowing access to ports and devices such as CD players.

PROTECTING YOUR COMPUTER FROM UNAUTHORIZED ACCESS 299

Computers that "Phone Home"

You've probably heard of LoJack, the popular theft-tracking device used in cars. Car owners install a LoJack transmitter somewhere in their vehicle. Then, if the vehicle is stolen, police activate the transmitter and use its signal to locate the car. Similar systems now exist for computers. Tracking software such as Computrace (**www.computrace.com**) and zTrace Gold (**www.ztrace.com**) enables the computer it is installed on to alert authorities as to its location if it is stolen.

To use this computer version of LoJack, you install the tracking software on your computer's hard drive. Once you install the software, it contacts a server at the software manufacturer's web site each time you connect to the Internet. If your computer is stolen, you notify the software manufacturer, who then instructs your computer to transmit tracking information (such as an IP address) that will assist authorities in locating and retrieving the stolen computer.

The files and directories holding the software are not visible to thieves looking for such software. What if the thieves reformat the hard drive in an attempt to destroy all files on the computer? The tracking software is written in such a way that it detects a reformat and hides the software code in a safe place in memory or on the hard drive (some sectors of a hard drive are not rewritten during most formats). That way, it can reinstall itself after the reformatting is completed.

Protecting Your Computer from Unauthorized Access

To protect yourself even further, you may want to restrict access to the sensitive data on your computer. Both software and hardware solutions exist to restrict others from accessing your computer, helping you keep its content safe.

Password Protection and Access Privileges

Windows XP has built-in password protection of files as well as the entire desktop. If your computer has been set up for multiple users with password protection, the Windows logon screen requires users to enter a password to gain access

FIGURE 7 Computer surrounds deter theft by making access to the internal components of the computer difficult while still allowing access to ports and drives.

TECHNOLOGY IN FOCUS

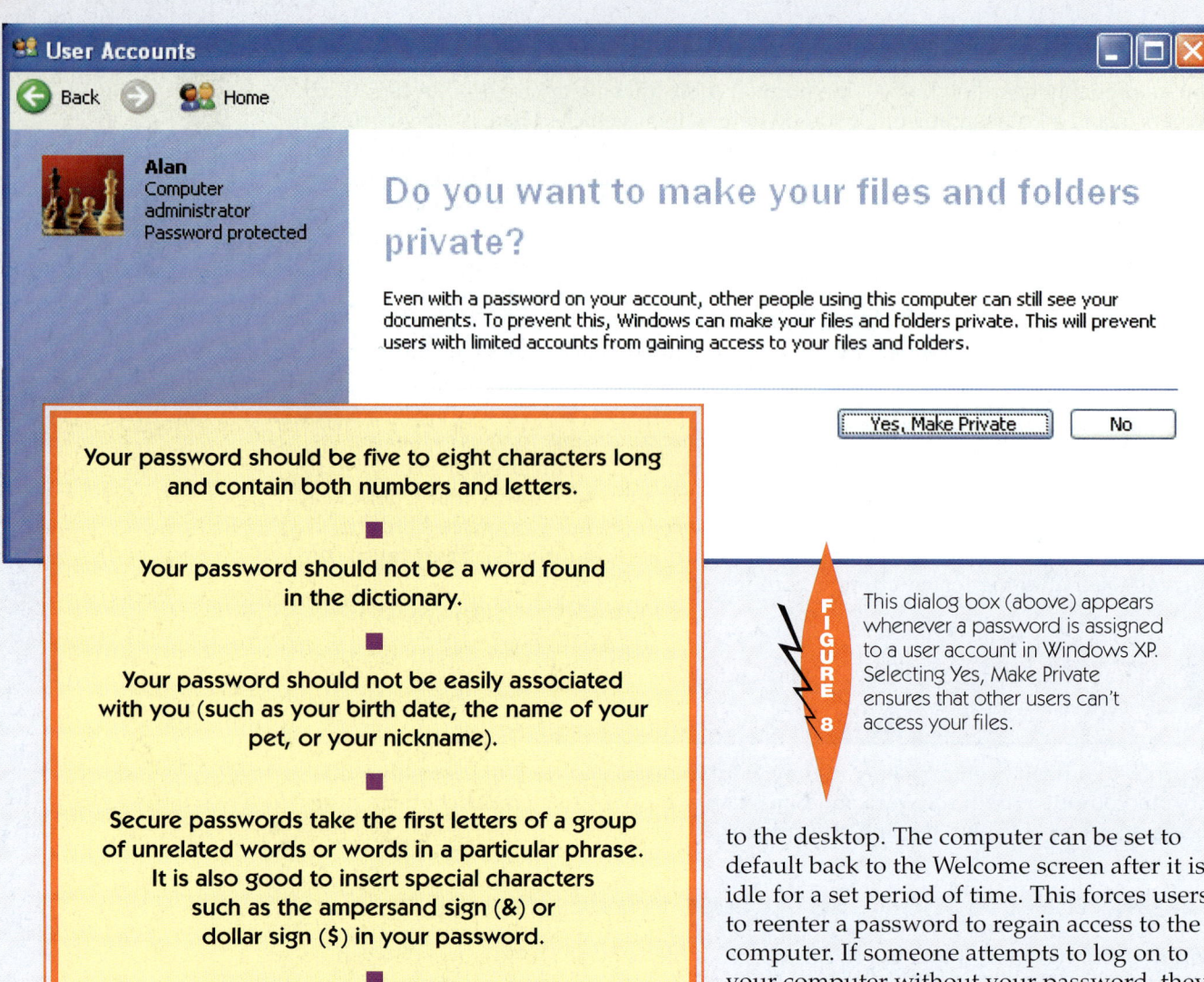

FIGURE 8 This dialog box (above) appears whenever a password is assigned to a user account in Windows XP. Selecting Yes, Make Private ensures that other users can't access your files.

- Your password should be five to eight characters long and contain both numbers and letters.
- Your password should not be a word found in the dictionary.
- Your password should not be easily associated with you (such as your birth date, the name of your pet, or your nickname).
- Secure passwords take the first letters of a group of unrelated words or words in a particular phrase. It is also good to insert special characters such as the ampersand sign (&) or dollar sign ($) in your password.
- You should never tell anyone your password or write it down in a place where others might see it.
- You should change your password if you think someone may know it.

to the desktop. The computer can be set to default back to the Welcome screen after it is idle for a set period of time. This forces users to reenter a password to regain access to the computer. If someone attempts to log on to your computer without your password, they won't be able to gain access.

Setting up a password also forces users to decide whether to share their files with other users. Figure 8 shows the dialog box that appears in Windows XP asking you to make this choice when you set up a user account password. Files not shared remain safe from the prying eyes of other users unless they know your password.

Of course, password protection only works as well as your password does. Creating a secure password is therefore very important. To do so, follow the basic guidelines shown above.

Figure 9 shows some possible passwords and explains why they make good or bad candidates.

PDA Bomb Software

PDAs can be vulnerable to unauthorized access if they are left unattended or are stolen. **PDA bomb software** features data and password protection for your PDA in an attempt to combat this problem. If you have PDA bomb software, a thief who steals your PDA is forced to crack your password to gain access. When a thief launches a brute force attack (repetitive tries to guess a password) on the PDA, the software's bomb feature kicks in

PROTECTING YOUR COMPUTER FROM UNAUTHORIZED ACCESS

FIGURE 9

Good and Bad Password Candidates

GOOD PASSWORD	REASON	BAD PASSWORD	REASON
L8t2me	Uses letters and numbers to come up with memorable phrase "Late to me"	jsmith	Combination of first initial, last name
YDKHYMM	First initials of first line of your favorite Uncle Cracker song, "You Don't Know How You Met Me"	4smithkids	Even though this has alphanumeric combination, it is too descriptive of a family
P1zzA	Easily remembered word with mix of alphanumeric characters and upper-/lowercase letters	Brown5512	Last name and last four digits of phone number is easily decoded
S0da&IC	Mix of numbers, characters and letters. Stands for "Soda and Ice Cream"	123Main	Using your street address is an easily decoded password

after a certain number of failed password attempts. The "bomb" erases all data contained on the PDA, thereby protecting your sensitive information. PDA Defense, found at **www.pdadefense.com**, is a popular example of PDA bomb software.

Biometric Authentication Devices

Biometric authentication devices are devices you can attach to your computer or PDA that read a unique personal characteristic, such as a fingerprint or the iris pattern in your eye, and convert that pattern to a digital code. When you use the biometric device, your pattern is read and compared to the pattern stored on the computer. Only users having an exact fingerprint or iris pattern match are allowed to access the computer.

Since no two people have the same biometric characteristics (fingerprints and iris patterns are unique), these devices provide a high level of security. They also eliminate the human error that can occur in password protection. You might forget your password, but you won't forget to bring your fingerprint to the computer! Some newer PDAs feature built-in fingerprint readers (Figure 10 shows a mouse that includes a fingerprint reader). Other biometric devices include voice authentication and facepattern recognition systems.

Firewalls

As we noted in Chapter 7, unauthorized access often occurs when your computer is connected to the Internet. You can best prevent such cases of unauthorized access by using either hardware or software **personal firewalls**. Hardware firewalls are often built into a router. Figure 11 lists popular software firewall programs. Setting up either a hardware or software firewall should adequately protect you from unauthorized access related to being connected to the Internet.

BITS AND BYTES
Are Klingonese Passwords Safe?

Many computer users are diehard science fiction fans. Growing up watching Star Trek, Babylon 5, and Battlestar Galactica has provided computer users with loads of planet names, alien races, alien vocabulary (Klingon words from the Star Trek series are very popular), and starship names to use as passwords. Unfortunately, hackers are on to this ploy. Recently developed hacking programs use dictionaries of "geek-speak" to attempt to break passwords. While "Qapla" (Klingonese for "success" and also used as "goodbye") might seem like an unbreakable password, don't bet your data on it!

FIGURE 10 The BioLink U-Match Mouse is a conventional two-button mouse that includes a digital fingerprint reader.

Fingerprint reader

Popular Personal Firewall Software

Firewall	URL
Norton Personal Firewall	www.norton.com
McAfee Firewall	www.mcafee.com
Zone Alarm	www.zonelabs.com
BlackICE PC Protection	http://blackice.iss.net

FIGURE 11

Dealing with Online Annoyances

As you learned in Chapter 7, **computer viruses** can damage files and data on your computer. This type of destruction obviously impacts your productivity. As we noted in Chapter 7, you can eliminate this threat fairly easily by installing **antivirus software**, such as Norton AntiVirus or McAfee VirusScan. However, other problems can affect your ability to complete your work and can violate your privacy. In this section, we'll discuss how to deal with such problems as spam, pop-ups, and spyware.

Spam

As we noted in Chapter 3, **spam** is unwanted e-mail that is sent out by companies in an attempt to solicit business. While not damaging to your computer, it can be annoying as your inbox fills up with useless messages. It was recently estimated that 45 percent of e-mail is now spam. That's a lot of junk mail!

Most Internet-based mail services, such as Yahoo! Mail, offer **spam filters** that filter spam out of your e-mail (see Figure 12). A spam filter is an option you can select in your e-mail account that places known spam messages into a folder other than your inbox. But what if you use Outlook to manage your e-mail? Then you need to obtain **spam filtering software** and install it on your computer. Programs that provide you with some control over spam include

SOUND BYTE
SECURING WIRELESS NETWORKS

In this Sound Byte, you'll learn what "war drivers" are and why they could potentially be a threat to your wireless network. You'll also learn simple steps to take to secure your wireless network against intruders.

DEALING WITH ONLINE ANNOYANCES   **303**

FIGURE 12

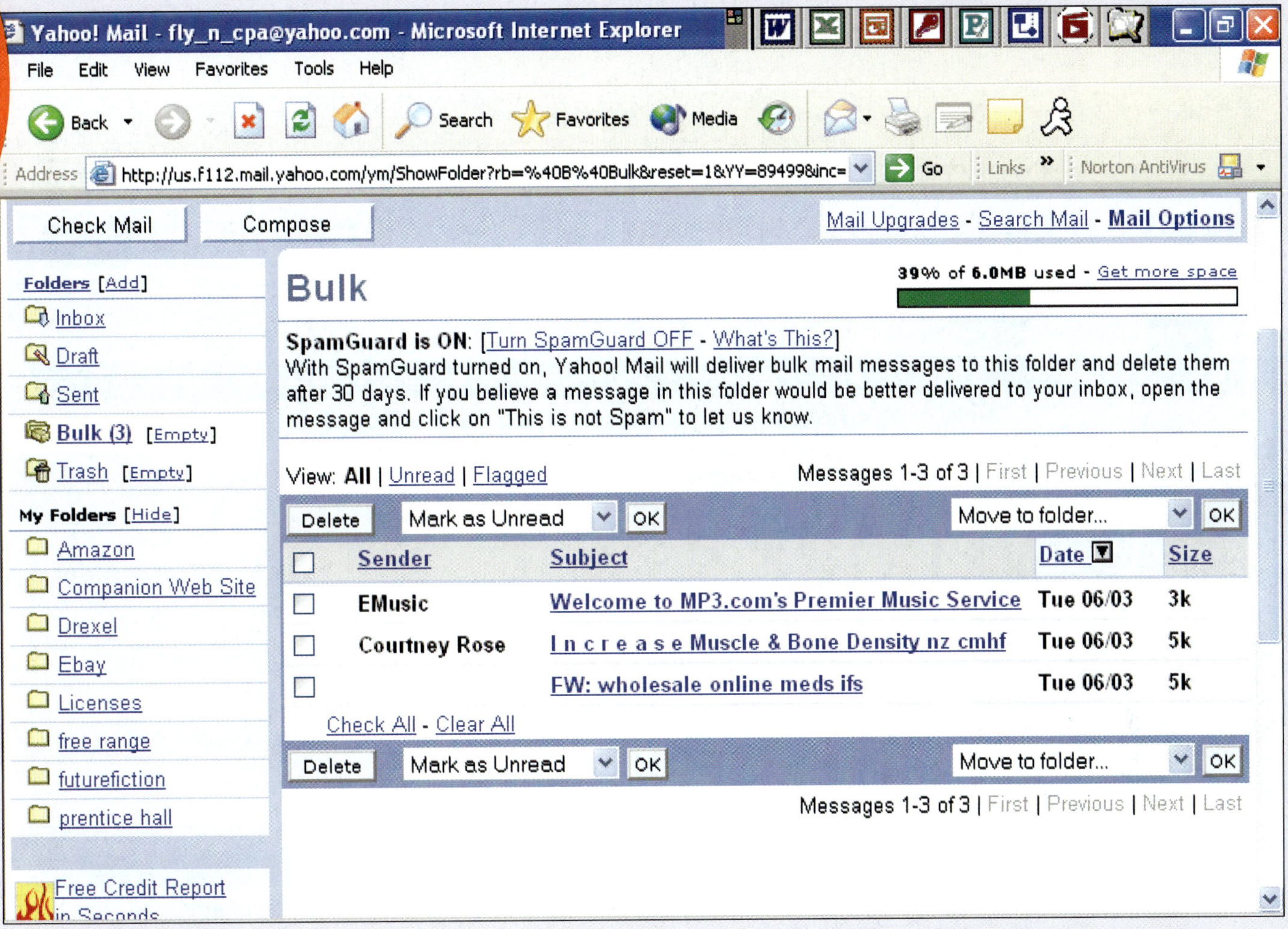

[Screenshot of Yahoo! Mail showing the Bulk folder with SpamGuard turned on.]

In Yahoo! Mail, turning on the SpamGuard feature alerts the Yahoo! Mail server to screen your incoming mail for obvious or suspected spam. This mail is then directed into a folder called "bulk" where you can easily delete all the spam at one time. This keeps your inbox relatively free from annoying spam.

MailWasher Pro, Spam Alarm, and SpamButcher, all of which can be obtained at **www.download.com**.

Spam filters and filtering software can catch up to 95 percent of spam. They work by checking incoming e-mail subject headers and sending addresses against databases of known spam. Spam filters also check your e-mail for frequently used spam patterns and keywords (such as "Viagra" and "free"). E-mail that the filter identifies as spam does not go into your inbox but rather to a folder called Spam. Since spam filters aren't perfect, you should check the Spam folder rather than just deleting its contents, as legitimate e-mail might occasionally end up there.

Pop-Ups

Pop-up windows are the unsightly billboards of the Internet (see Figure 13). These windows pop up when you enter web sites, sporting "useful" information or touting

BITS AND BYTES

Free Protection Software

Protecting your computer often requires that you purchase software. The good news is that you can download plenty of protection programs for free off the Internet. Many companies offer free versions of their software that either expire after a certain number of days or offer fewer features than commercial versions. In many cases, these free versions are sufficient for home use. One site for downloading free software is www.download.com. This should be your first stop when looking for free software you can use to protect your computer.

TECHNOLOGY IN FOCUS

FIGURE 13 Pop-ups like the Father's Day Gift Guide box do draw your attention to useful information at times. And the boxes on Barnes and Noble merely pop up the first time you are on the main page that day. But if you surf a lot of sites that constantly bombard you with pop-ups, you may wish to install anti-pop-up software to thwart their appearance altogether.

products. Although some sites use pop-ups to increase the functionality of their site (your account balance may pop up at your bank's web site, for example), most pop-ups are just plain annoying.

To stop pop-ups from appearing, you need **anti-pop-up software**. Two free anti-pop-up programs are Pop-Up Stopper and Pop-Up Defender, both of which are available at **www.download.com**. When a site attempts to launch a pop-up, these programs either prevent it from being launched or close it immediately. The software also includes options that enable you to allow pop-ups that you do want to appear.

Spyware

If you ever install software from a web site, you may be installing something on your computer together with the software. Called **spyware** (or **adware**), these unwanted piggyback programs run in the background of your system and gather information about you, usually your Internet surfing habits, without your knowledge. They then periodically transmit this information to the owner of the spyware program so that it can be used for marketing purposes.

Many spyware programs use **cookies** to collect information. As you learned in Chapter 3, cookies are small pieces of text that are stored on your hard drive that have many legitimate uses. For instance, when you customize the look and feel of a web site such as Amazon.com, that customization information is stored in a cookie file on your hard drive. However, cookies can also be used to help spyware track your movements.

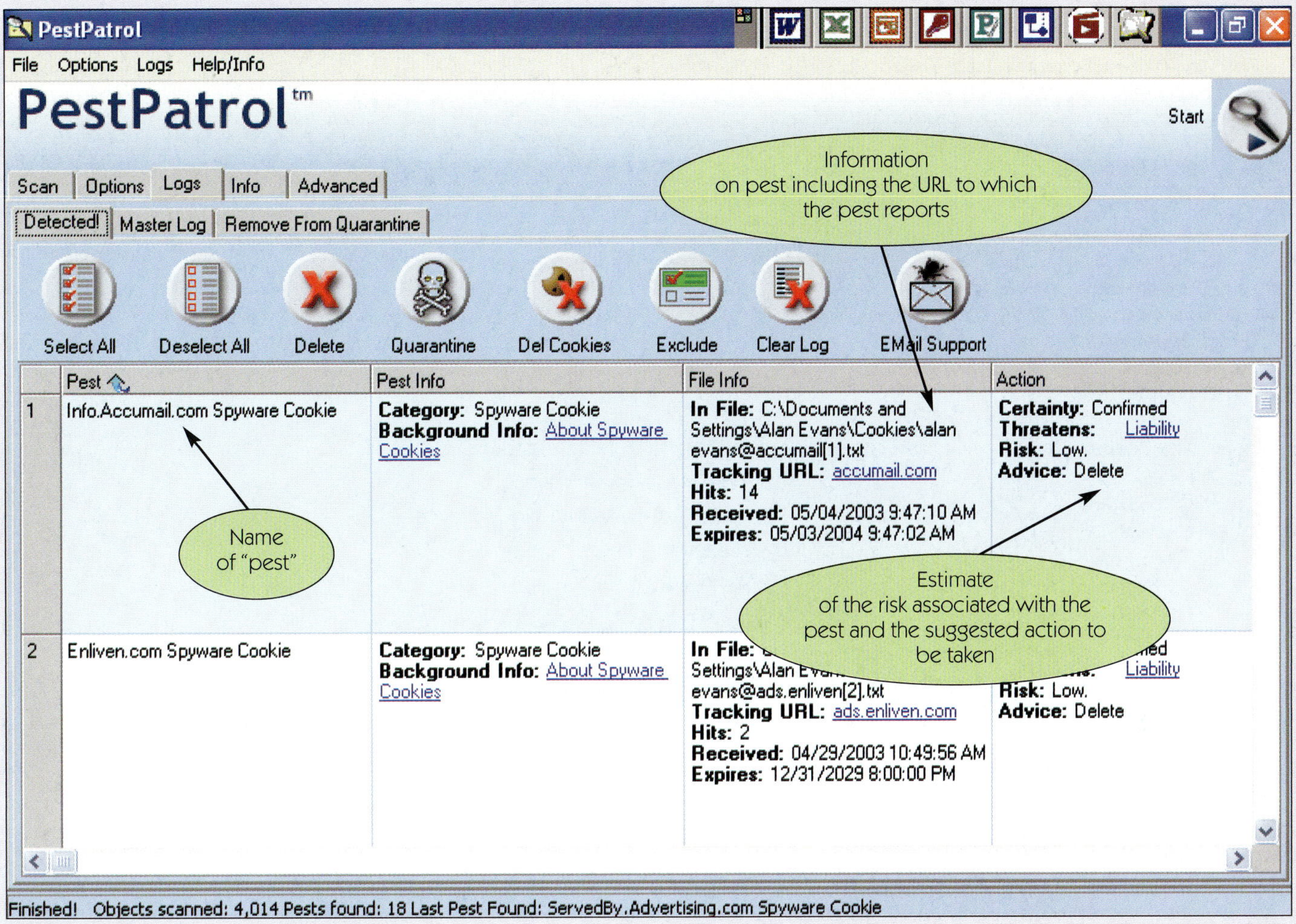

BACKING UP YOUR DATA

A PestPatrol log screen showing pests detected on a computer after a routine scan. PestPatrol allows you to easily detect and delete unwanted "pests" from your computer.

Most antivirus software doesn't detect spyware or prevent spyware cookies from being placed on your hard drive. However, you can obtain **spyware removal software** and run it on your computer to delete unwanted spyware. Since new spyware is created all the time, you should update your spyware removal software regularly. Ad-aware (which is available for free at www.download.com) and PestPatrol (available at www.pestpatrol.com) are programs that are easy to install and update. Figure 14 shows an example of PestPatrol in action.

 ## Backing Up Your Data

The data on your computer faces three major threats: unauthorized access, tampering, and destruction. As we noted in Chapter 7, a hacker can gain access to your computer and steal or alter your data. However, a more likely scenario is that you will lose your data unintentionally. You may accidentally delete files; your hard drive may break down, resulting in complete data loss; a virus may destroy your original file; or a fire may destroy the room that houses your computer. Since many of these factors are beyond your control, you should have a strategy for backing up your files.

A file **backup** is merely a copy of that file that you can use to replace the original if it is lost or damaged. When you make a backup of your files, it's important that you store the copy in a different place than the original. Having two copies of the same file on the same hard drive does you little good when the hard drive breaks down.

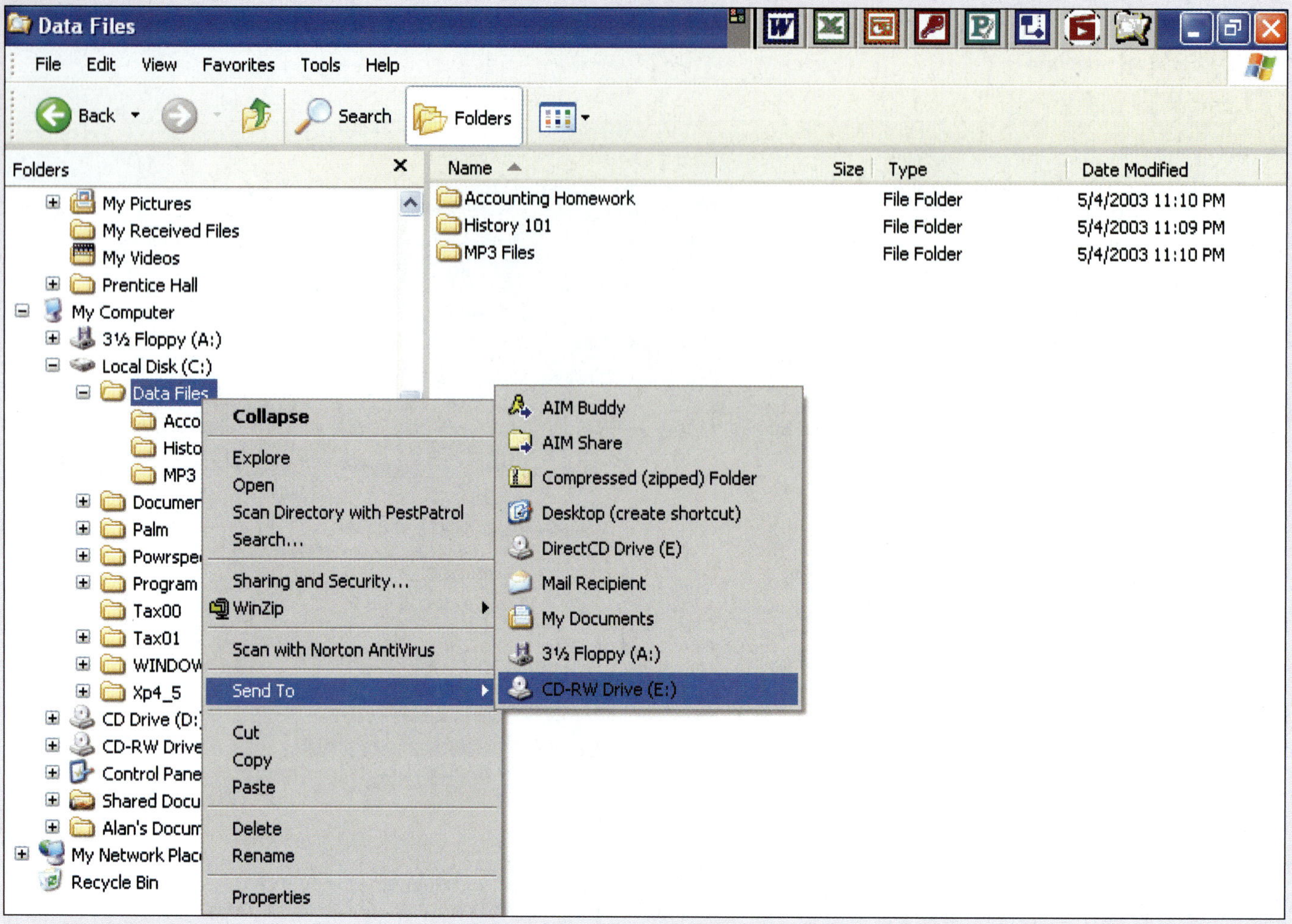

TECHNOLOGY IN FOCUS

FIGURE 15 All of the data files on this computer are located in subfolders under the main Data Files folder. Right-clicking on the Data Files folder brings up a menu which allows the user to send (or copy) the files to the CD-RW drive installed on this computer. Assuming a blank CD was inserted in the drive, this creates a backup CD containing a copy of all the data files.

Likewise, storing your backup copies in the same room as your computer would not work well if there were a fire in the room. Removable storage media such as Zip disks, DVDs, and CDs are popular choices for backing up files because they hold a lot of data and can be easily transported.

Two types of files need backups, program files and data files:

- **Program files** are those files you use to install software. They should be on the CDs or DVDs that they originally came on. If any programs came preinstalled in your computer, you should still have received a CD or DVD that contains them. As long as you have the original media in a safe place, you shouldn't need to back up these files.

- **Data files** are those files you create (such as spreadsheets, letters, and so on), as well as contact lists, address books, e-mail archives, and your Favorites list from your browser.

You should back up your files (especially important ones) frequently, depending on how much work you can afford to lose. You should always back up data files when you make changes to them, especially if those changes involve hours of work. It may not seem important to back up your history term paper file when you finish it, but do you really want to do all that work again if your computer crashes before you have a chance to turn your paper in?

To make backups easier, store all your data files in one folder on your hard drive. For example, you can create a folder on your hard drive called Data Files. You can then create subfolders (such as History Homework, Music Files, and so on) within the Data Files folder. If you store all your data files in one place, to back up your files, you simply copy the Data Files folder and all of its subfolders onto an alternate storage media. If you have a DVD/CD-RW drive in your computer, open Windows Explorer (as shown in Figure 15) and right-click on the Data Files folder. You can then copy the contents of the folder to the appropriate media.

Backup Software

Of course, the backup method described above assumes that you will take the initiative to back up your files on a regular basis. However, many people forget to do so, and only learn when it's too late that they should have been performing a systematic backup routine. The good news is that **backup software**, such as Norton Ghost, allows you to schedule regular backups that occur automatically, with no intervention on your part. These products can back up individual files, folders, or an entire hard drive to another hard drive, such as an external drive connected to your computer via a USB port.

Backing up your entire hard drive greatly speeds up the recovery process should you experience a hard drive failure. With a backup of your entire hard drive, you won't need to reinstall all of the program software from the original CDs. Instead, you just replace the broken hard drive with the backup hard drive (or copy the entire contents of the backup drive to a new drive). This can save you time and may be worth the expense of an additional hard drive.

Online Backups

A final backup solution is to store backups of your files online. For a fee, companies such as NetMass (**www.systemrestore.com**) can provide you with such online storage. If you store all of your backups on the Internet, you don't need to buy an additional hard drive. This method also takes the worry out of keeping your backups in a safe place since they're always stored in an area far away from your computer (such as on the NetMass sever). However, if you would like to store your backups online, make sure you have high-speed Internet access such as cable or DSL; otherwise, your computer could be tied up as you transfer files.

Additional Resources

Hackers, spammers, and advertisers are constantly developing new methods for circumventing the protection that security software provides. Although the manufacturers of such software are constantly updating and improving it, you should keep abreast of new techniques being employed that could threaten your privacy and security. *SC Magazine* is a security magazine available in a free online version at **www.scmagazine.com**. Take a few minutes each month and scan the articles to make sure you have taken the appropriate protective measures on your computer to keep it safe and secure.

SOUND BYTE

PROTECTING YOUR COMPUTER

In this Sound Byte, you'll learn how to use a variety of tools to protect your computer, including antivirus software and Windows utilities.

OBJECTIVES

After reading this chapter, you should be able to answer the following questions:

- What are the advantages and limitations of mobile computing? (pp. 310–311)

- What are the various mobile computing devices? (p. 312)

- What can pagers do and who uses them? (pp. 312–313)

- How do cell phone components resemble a traditional computer and how do they work? (pp. 313–317)

- What can I carry in an MP3 player and how does it store data? (pp. 317–319)

- What can I use a PDA for and what internal components and features does it have? (pp. 321–327)

- How can I synchronize my mobile devices with my desktop computer? (pp. 325–326)

- What is a tablet PC and why would I want to use one? (pp. 329–331)

- How powerful are laptops and how do they compare to desktop computers? (pp. 331–335)

CELLULAR PHONES

FIGURE 8.4

The Motorola Talkabout T900 is an example of a two-way pager. It displays four large lines of text and supports preprogrammed replies and an address book.

Are there different kinds of pagers? **Numeric pagers** display only numbers on their screens, telling you that you have received a page and providing you with the number you should call. Numeric pagers do not allow you to send a response. A somewhat more sophisticated style of pager is the **voice pager**, which offers all the features of a numeric pager but also allows you to receive voice messages. **Alphanumeric pagers** are much like numeric pagers but they also can display text messages. Like numeric pagers, alphanumeric pagers do not allow you to send messages.

More useful are **two-way pagers**, which support both receiving and sending text messages. Two-way pagers have a small built-in keyboard so you can compose text messages, as shown in Figure 8.4. More advanced two-way pagers support preprogrammed replies. Instead of having to type out an entire reply, you can insert a standard phrase (such as "I'll meet you at") from a list, using just one keystroke. Advanced pagers also include an address book that stores the phone numbers and e-mail addresses of your contacts. In addition, two-way pagers can notify you of e-mail and allow you to check it and send replies.

How do pagers work? Pagers are radio devices. This means that all pagers have receivers that receive radio waves and then translate those waves into the signals needed to display numbers or characters on the display screen. In addition, every pager is assigned a unique code. When a message is sent to a pager, it is sent together with the unique code on a special channel, or frequency. This unique code ensures that the correct pager receives the message. Two-way pagers also have a transmitter that allows you to send messages.

With all the new devices on the market, does anyone still use pagers? People who need to be reachable but want an inexpensive and lightweight device are the primary market for pagers. Staff who are on call (such as doctors), expectant fathers, and teenagers are typical pager owners. Pagers cost less than a cell phone (they currently sell for about $50), and with monthly fees of around $10, they provide an inexpensive alternative to cell phones. However, as other devices like cell phones continue to drop in price and shrink in size, pagers will have a difficult time finding a market. We may see pagers phased out completely within the next few years.

Cellular Phones

Cellular phones have evolved from their early days as large, clunky, box-like devices to become the compact, full-featured communication and information storage devices they are today. Cell phones offer all of the features available on a home telephone system, including auto-redial, call timers, and voice-mail capabilities. Some cell phones also feature voice-activated dialing, which is important for hands-free operation. In addition to these services, cell phones offer Internet access, text messaging, personal information management (PIM) features, and more, all within the palm of your hand.

CELL PHONE HARDWARE

Is a cell phone considered a computer? Cell phones are so advanced that they have the same components as a computer: a processor (or CPU), memory, and input and output devices, as shown in Figure 8.5. A cell phone also requires software and has its own operating system (OS). One popular operating system for full-featured cell phones is the **Symbian OS**. As cell phones begin to handle e-mail, images, and even video,

FIGURE 8.5

Inside your cell phone, you'll find some familiar components, including a processor (CPU), a memory chip, input devices such as a microphone and a keypad, and output devices such as a display screen and a speaker.

Are Cell Phones Bad for Your Health?

Cell phones work by sending electromagnetic waves into the air from an antenna. Depending on the phone's design—that is, whether there is a shield between the antenna and the user's head—up to 60 percent of the radiation emitted penetrates the area around the head. It is not yet clear whether there are long-term health consequences of using cell phones due to the electromagnetic waves they emit. What is clear is that using a cell phone while driving is dangerous. Studies show that motorists are four to nine times more likely to crash when talking on a cell phone while driving, a risk similar to the effects of driving drunk. In many states, "hands-free" cell phone use is mandatory. For more information on this topic as well as the legislative status in your state, visit the *Car Talk* Web site, "Drive Now, Talk Later," at **http://cartalk.cars.com/About/Drive-Now**.

more complex operating systems like Symbian are required to translate the user's commands into instructions for the processor.

What does the processor inside a cell phone do? Although the processor inside a cell phone is obviously not as fast or as high powered as a processor in a desktop computer, it is still responsible for a great number of tasks. The processor coordinates sending all of the data between the other electronic components inside the phone. It also runs the cell phone's operating system, which provides a user interface so that you can change phone settings, store information, play games, and so on.

What does the memory chip inside a cell phone do? The operating system and the information you save into your phone (such as phone numbers and addresses) need to be stored in memory. The operating system is stored in read-only memory (ROM) since the phone would be useless without that key piece of software. As you learned earlier in this text, there are two kinds of memory used in computers: *volatile memory*, which requires power to save data, and *nonvolatile memory*, which can store data even when the power is turned off. ROM is nonvolatile, or permanent, memory. This means that when you turn off your cell phone, the data you have saved to ROM (including the operating system) does not get lost.

The phone contact data is stored in separate internal memory chips. Full-featured phones have as much as 750 KB of memory that can be used to store contact data, ring tones, images, and even small software applications such as currency converters or a world clock.

What input and output devices do cell phones use? The input devices for a cell phone are primarily the microphone, which converts your voice into electronic signals that the processor can understand, and a keypad, which is used for numeric or text entry. Newer phones like the Kyocera 7135 Smartphone (shown in Figure 8.6a) feature the ubiquitous Palm Graffiti pad as well as touch-sensitve screens that allow you to input data. In addition, more and more cell phones include digital cameras. The Nokia 3650 (shown in Figure 8.6b) offers a built-in camera that can input photographs or even eight seconds of video to your phone.

Cell phone output devices include a speaker and an LCD display. Higher-end models include full-color, high-resolution plasma displays. Such displays are becoming increasingly popular as more people are using their cell phones to send and receive the digital images included in multimedia text messages and e-mail. Certain cell

CELLULAR PHONES

phones (such as the Motorola T720, shown in Figure 8.6c) include two displays: an outside LCD display you can see when the phone is folded and a separate color display inside.

HOW CELL PHONES WORK

How do cell phones work? When you speak into a cell phone, the sound enters the microphone as a sound wave. Because analog sound waves need to be digitized (that is, converted into a sequence of 1s and 0s that the cell phone's processor can understand), an **analog-to-digital converter chip** converts your voice's sound waves into digital signals. Next, the digital data must be compressed, or squeezed, into the smallest possible space so that it will transmit more quickly to another phone. The processor cannot perform the mathematical operations required at this stage quickly enough, so a specialized chip, called the **digital signal processor**, is included in a cell phone to handle the compression work. Finally, the digital data is transmitted as a radio wave through the cellular network to the destination phone.

When you receive an incoming call, the digital signal processor *decompresses* the incoming message. An amplifier boosts the signal to make it loud enough, and it is then passed on to the speaker, from which you hear the sound.

What's "cellular" about a cell phone? A set of connected "cells" makes up a cellular network. Each cell is a geographic area centered on a **base transceiver station**, which is a large communications tower with antennas, amplifiers, and receivers/transmitters. When you place a call on a cell phone, a base station picks up the request for service. The station then passes the request on to a central location, called a **mobile switching center.** (The reverse process occurs when you receive an incoming call on a cell phone.) A telecommunications company builds its network by constructing a series of cells that overlap, in an attempt to guarantee that its cell phone customers have coverage no matter where they are.

As you move during your phone call, the mobile switching center monitors the strength of the signal between your cell phone and the closest base station. When the signal is no longer strong enough between your cell phone and the base station, the mobile switching center orders the next base station to take charge of your call. When your cell phone "drops out," it sometimes does so because the distance between base stations was too great to provide an adequate signal.

FIGURE 8.6

(a) The Kyocera Smartphone includes a built-in Graffiti pad and touch-sensitive color display. (b) The Nokia 3650 weighs less than 4.5 ounces including the battery, and is just 5 inches x 2 inches. Its camera can input both still photographs and video to your phone. (c) The Motorola T720 includes an outside LCD display as well as an inside color display.

CELL PHONE FEATURES: TEXT MESSAGING

What is text messaging? Short Message Service (SMS) (often just called **text messaging**) is a technology that allows you to send short text messages (up to 160 characters) over mobile networks. To send SMS messages from your cell phone, you simply use the numeric keypad or a presaved template and type in your message. You can send SMS messages to other mobile devices (such as cell phones or pagers) or to any e-mail address. You can also use SMS to send short text messages from your home computer to mobile devices, such as your friend's phone.

How does SMS work? Unlike the text messages you send with a pager, which use radio waves, SMS uses the cell phone network to transmit messages. When you send an SMS message, an SMS calling center receives the message and delivers it to the appropriate mobile device using something called "store-and-forward" technology. This technology allows users to send SMS messages to any other SMS device in the world.

Many SMS fans like text messaging because it can be cheaper than a phone call and it allows the receivers to read messages when it is convenient for them. In fact in some countries, such as Japan, text messaging is more popular than voice messaging. However, entering text using your cell phone keypad can be time-consuming and hard on your thumbs. Frequent SMS users save typing time by using a number of abbreviations, some of which are shown in Figure 8.7.

Can I send and receive multimedia files over a cell phone? SMS technology allows you to send only text messages. However, an extension of SMS called **Multimedia Message Service (MMS)** allows you to send messages that include text, sound, images, and video clips to other phones or e-mail addresses. MMS messages actually arrive as a series of messages; you view the text and then the image and then the sound, and so on. You can then choose to save just one part of the message (such as the image), all of it, or none of it. MMS users can subscribe to financial, sports, and weather services that will "push" information to them, sending it automatically to their phones in MMS format.

CELL PHONE INTERNET CONNECTIVITY

How do I get Internet service for my phone? Just as you pay an Internet service provider (ISP) for Internet access for your desktop computer, connecting your cell phone to the Internet requires that you have a **wireless Internet service provider**. Phone companies that provide cell phone calling plans (such as Verizon or T-Mobile) usually double as wireless ISPs. As noted above, accessing the Internet on a mobile device comes with limitations. For one, the connection is often very slow. While you may be able to connect to the Internet at speeds ranging from 44 to 800 kilobits per second (Kbps) at home, your cell phone will connect at a maximum speed of just 14.4 Kbps.

In addition, because cell phones have a very limited amount of screen space and low screen resolution, the Internet experience is quite different. To make it possible for you to access the Internet, special **microbrowser** software runs on your cell phone. Figure 8.8 shows a typical microbrowser screen on

FIGURE 8.7 Popular Text Messaging Abbreviations

AFAIK	As far as I know	IDK	I don't know
B4N	Bye for now	JAS	Just a sec
BRB	Be right back	LOL	Laughing out loud
CUL	See you later	QPSA	¿Qué pasa?
FBM	Fine by me	T+	Think positive
F2T	Free to talk	TTYL	Talk to you later
G2G	Got to go	WUWH	Wish you were here
HRU	How are you?	YBS	You'll be sorry

FIGURE 8.8

Microbrowser software helps you access the Internet from your cell phone.

BITS AND BYTES

Your Phone Called . . . It Needs a Tequila Shooter!

Someday soon, you may be sitting in a bar and instead of finishing that last shot of Jack Daniels, you may instead pour it into your cell phone! Researchers are currently working on a technology called a "biofuel cell" that could power mobile devices using alcohol. A standard fuel cell (that is, a battery) derives its power from chemical reactions between its elements (such as zinc and manganese-oxide). A biofuel cell would use biological molecules (in this case, alcohol enzymes) to generate power. The advantage would be that the batteries are easily rechargeable—just add beer! Once the battery is charged, you would probably only need to add a few drops of alcohol to keep it running for a month or longer. More research is needed to develop these fuel cells before they can become commercially viable, but in the future you may hear someone say, "Bartender, another round for me and my cell phone!"

---

a cell phone. As you can see, the Internet experience is not as enjoyable as at your home desktop computer.

Web sites are beginning to create content specifically designed for wireless devices to make such Internet connections worthwhile. This specially designed content, which is text-based and contains no graphics, is written in a format called **Wireless Mark-up Language (WML)**. These sites are designed so that they fit the tiny display screens of cell phones and PDAs.

MP3 Players

MP3 is a format for efficiently storing music as digital files, or a series of bits. An **MP3 player** is a small portable device that enables you to carry your MP3 files around with you. Depending on the player, you can carry several hours of music or your entire CD collection in an incredibly small device. For example, the Apple iPod is 4 inches x 2.4 inches and can hold over 7,500 songs. The most compact players are the size of a cigarette lighter (although they hold far less music than the iPod). Figure 8.9 shows several popular models of MP3 players.

MP3 HARDWARE

How do I know how much music an MP3 player can hold? The number of songs an MP3 player can hold depends on how much storage space it has. MP3 players use memory chips and hard disk drives to store music files. Inexpensive players use memory chips (ranging from 64 MB to 256 MB), whereas expensive models use a built-in hard drive, which provides up to 40 GB of storage. Some MP3 players allow you to add storage capacity by purchasing removable memory cards.

Another factor that determines how much music a player can hold is the quality of the MP3 music files. The size of an MP3 file depends on the digital sampling of the song. The **sampling rate** is the number of times per second the music is measured and converted to a digital value. Sampling rates are measured in kilobits per second (Kbps). The same song could be sampled at 192 Kbps or 64 Kbps. The size of the song file will be three times larger if it is sampled at 192 Kbps instead of the lower sampling rate of 64 Kbps. The higher the sampling rate, the better quality the sound, but the larger the file size.

If you are "ripping," or converting, a song from a CD into a digital MP3 file, you can select the sampling rate yourself. You decide by considering what quality sound you want as well as how many songs you want to fit onto your MP3 player. For example, if your player has 64 MB of storage and you have ripped songs at 192 Kbps, you can fit about 45 minutes of music onto the player. The same 64 MB could store 133 minutes of music if it were sampled at 64 Kbps. Whenever you are near your computer, you can connect your player and download a different set of songs, but you are always limited by the amount of storage your player has.

MP3 FLASH MEMORY AND FILE TRANSFER

What if I want to store more music than what my MP3 player's memory allows? Some MP3 players allow you to add additional, removable memory. The removable memory

FIGURE 8.9 Popular MP3 Players and Their Characteristics

MP3 PLAYER	APPROXIMATE NUMBER OF MP3 SONGS	EXPANDABLE MEMORY	HARD DISK DRIVE CAPACITY	CONNECTION TO COMPUTER	OTHER FEATURES
CREATIVE LABS' NOMAD MuVo	About 30 songs	128 MB	None	USB 1.0 port	Size of a cigarette lighter
SAMSUNG YEPP 55-v	About 60 songs	256 MB	None	USB 1.0 port	Includes voice recorder, FM tuner, and backlit display
APPLE iPOD	About 2,500 (for 10 GB) to 10,000 songs (for 40 GB)	None	10 GB, 15 GB, or 30 GB	FireWire or USB 2.0 port	Latest version includes a calendar, contact database, and can store images, corporate logos, and live URLs
ARCHOS MULTIMEDIA ENTERTAINMENT CENTER	About 5,000 songs or 1,000 hours of voice recordings or 40 hours of video.	None	20 GB	USB 2.0 port	MPEG4 Player/Recorder, Photo Wallet & Viewer and acts as a 20GB Hard Drive to store data and files.

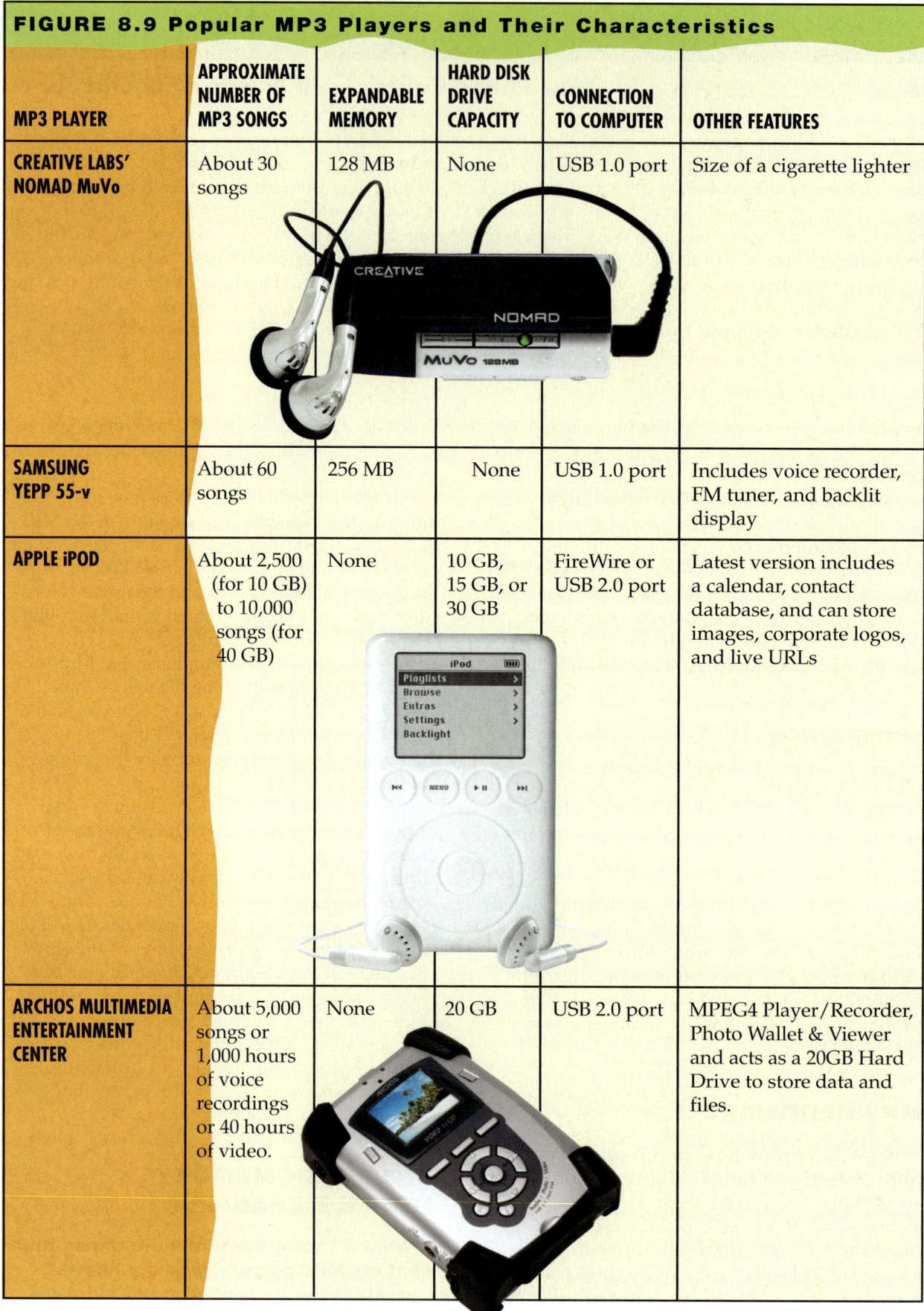

MP3 players use is a type of portable, nonvolatile memory called **flash memory**. Flash memory cards are noiseless, very light, use very little power, and slide into a special slot in the player. If you've ever played a video game on Playstation 2 or Nintendo and saved your progress to a memory card, you have used flash memory. Because flash memory is nonvolatile, when you store data on a flash memory card you won't lose it when you turn off the player. In addition, flash memory can be erased and re-written with new data.

What types of flash memory cards do MP3 players use? You can use several different types of flash cards with MP3 players, as shown in Figure 8.10. One popular type is **CompactFlash** cards. These are about the size of a matchbook and can hold between 64 MB and 1 GB of data. They are very durable, so the data you store on them is safer than it would be on a floppy disk, for example. Multimedia cards (**MMCs**) and **SmartMedia** cards are about the same size as CompactFlash cards but are thinner and less rugged. They can hold up to 128 MB of data. A newer type of memory card called **Secure Digital** is faster and offers encryption capabilities so your data is secure even if you lose the card. The stamp-sized Secure Digital cards can hold up to 256 MB of data.

Sony has its own brand of flash memory called the **Memory Stick**. These tiny cards—measuring just 2 inches x 1 inch and weighing a fraction of an ounce—are currently found only on Sony devices but are becoming more widely accepted. Particular models of MP3 players can only support certain types of flash cards, so check your manual to be sure you buy compatible memory cards.

How do I transfer MP3 files to my MP3 player? All MP3 players come with software that enables you to transfer your MP3 files from your computer onto the player. As noted above, players that hold thousands of songs use internal hard drives to store music. For example, the Apple iPod is available with a 10-GB, 20-GB, or 40-GB hard drive. To move that volume of data between your computer and MP3 player, you need a high-speed port. The iPod uses a FireWire port or the even faster Universal Serial Bus (USB) 2.0 port, whereas some other MP3 players use the slower USB 1.0 port.

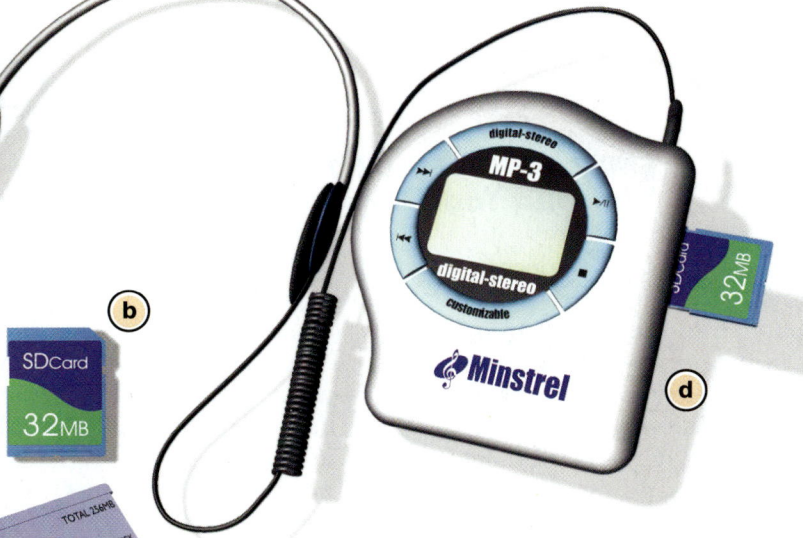

FIGURE 8.10

(a) CompactFlash memory cards offer rugged, portable storage. They can be dropped and exposed to hot and cold weather and still protect your data. (b) Secure Digital cards are faster and offer security protection for your data. (c) Memory Sticks are currently used only in Sony devices, but are becoming more widely accepted. (d) To use flash memory with an MP3 player, you simply slide the card into the player's memory card slot.

Using a FireWire port, you can transfer a complete CD to the iPod in less than 10 seconds.

Can MP3 players carry more than just music? MP3 players have become so popular that some manufacturers are redesigning their software and operating systems to support the transfer and storage of nonmusical data as well. The latest version of the iPod sports a calendar and a contact database, and can store images, corporate logos, and live URLs in addition to music files. Other high-end MP3 devices, such as the Archos Multimedia Entertainment Center, shown in Figure 8.9, offer you the ability to transfer and store music, video, and image files.

MP3 ETHICAL ISSUES: NAPSTER AND BEYOND

Is it illegal to download MP3 files? The initial MP3 craze was fueled by sites such as MP3.com. MP3.com differs from controversial sites such as the former Napster because its files are on a public server, the songs are placed on

TRENDS IN IT

EMERGING TECHNOLOGIES:
Wearing Your Applications on Your Sleeve

By now, most of us expect to have Internet access in our wristwatches and radio transmitters in our coat buttons in the not-so-distant future. But how about having computers woven into your clothes?

The idea may not be so farfetched. Flexible synthetic "yarns" that can transmit electrical signals may soon be common components of cotton and polyester fabrics. Called *electrotextiles*, these textiles are already being woven into soldiers' vests to serve as radio antennas. Researchers see almost no limits to their possibilities, and are working to make electrotextiles not only wearable but also washable, customizable, and even programmable. Experts predict that nearly every type of clothing will have some sort of electronic functions in 10 years, ranging from Global Positioning Systems (GPSs) to medical sensors to fabrics that change colors and patterns at your command.

The new fibers are woven into fabric or sewn on in ribbon-like strips around the neck or sleeves of a garment. They are then connected to chips and batteries so that, for instance, blankets and clothing can adapt to body temperature and car seat fabric can tell a car's air bags to adjust their force to match the passenger's weight. Researchers at the Georgia Institute of Technology are already working on a T-shirt for firefighters that tracks the heart rate, body temperature, and other vital signs of the person wearing it, and transmits this data to a wearable pager. And scientists at Germany's Infineon Technologies have made a prototype of a hooded jacket (in stylish charcoal gray) that includes the electronic elements of an MP3 player in the pockets. The hood's drawstrings are used as the MP3 player's headphones, and controls are found on the sleeve.

And just think: once solar cells can be woven into fabric, you might be able to wear all your applications *and* your power source—in next year's hottest colors.

Someday you may be wearing your computer on your sleeve.

TechTV

For more information on new portable gear see 2003 Consumer Electronics Show, a TechTV clip found at www.prenhall.com/techinaction

the server with the permission of the original artist or recording company, and you often pay a fee for the right to download the song. Therefore, you are not infringing on a copyright by downloading songs from *legal* MP3 sites.

What was Napster all about? Napster was a file-exchange site created to correct some of the small annoyances found by users of MP3.com and similar sites. One such annoyance was the limited availability of popular music in MP3 format. With the MP3 sites, if you found a song you wanted to download, often the links to the sites the file was found on no longer worked. Napster differed from MP3.com because songs or locations of songs were not stored in a central public server but instead were "borrowed" directly from other users' computers. This process of users transferring files between computers is referred to as **peer-to-peer (P2P) sharing**. Napster also provided a search engine dedicated to finding MP3 files. This direct search and sharing eliminated the inconvenience of searching links only to find them unavailable.

Why did Napster disappear? The problem with Napster was that it was so good at what it did. Napster's convenient and reliable mechanism to find and download popular songs in MP3 format became a huge success. The rapid acceptance and use of Napster—at one point, it had nearly 60 million users—led the music industry to claim it was losing money. Recording companies and artists sued for copyright infringement, and Napster was closed in June 2002.

If Napster is illegal, why are there still Napster clones? While Napster was going through its legal turmoil, other P2P Web sites were quick to take advantage of a huge opportunity. Napster was "easy" to shut down because it used a central index server that queried other Napster computers for requested songs. Current P2P Web sites (such as Gnutella and Kazaa) differ from Napster in that they do not limit themselves to sharing only MP3 files. More importantly, these sites don't have a central index server. Instead, they operate in a true P2P sharing environment in which computers connect directly to other computers.

The argument these P2P networks make to defend their legality is that they do not run a central server like Napster but only facilitate connections between users. Therefore, they have no control over what the users trade, and not all P2P file sharing is illegal. Those against the file-sharing sites contend that the sites know their users are distributing illegal files. This argument has not yet been definitively settled.

Personal Digital Assistants (PDAs)

A **personal digital assistant (PDA)** is a small device that allows you to carry digital information. Often called *palm computers* or *handhelds*, PDAs are about the size of your hand and usually weigh less than 5 ounces. Although small, PDAs are quite powerful and can carry all sorts of information, from calendars to contact lists to specially designed personal productivity software programs (such as Excel and Word), to songs, photos, and games. And you can easily "synchronize" your PDA and your home computer so that the changes you make to your schedules and files on your PDA are made on your home or office computer files as well.

PDA HARDWARE

What hardware is inside a PDA? Like any computer, a PDA includes a processor (CPU), operating system software, storage capabilities, input and output devices, and ports. Because of their small size, PDAs (like cell phones) must use specially designed processors and operating system software. They store their operating system software in ROM and their data and application programs (which are specially designed to run on the PDA) in random access memory (RAM).

What kinds of input devices do PDAs use? All PDAs feature touch-sensitive screens that allow you to enter data directly with a penlike device called a **stylus**. To make selections, you simply tap or write on the screen with the stylus. Other PDAs include integrated keyboards or support small, portable, folding keyboards. Figure 8.12 shows all of these input options.

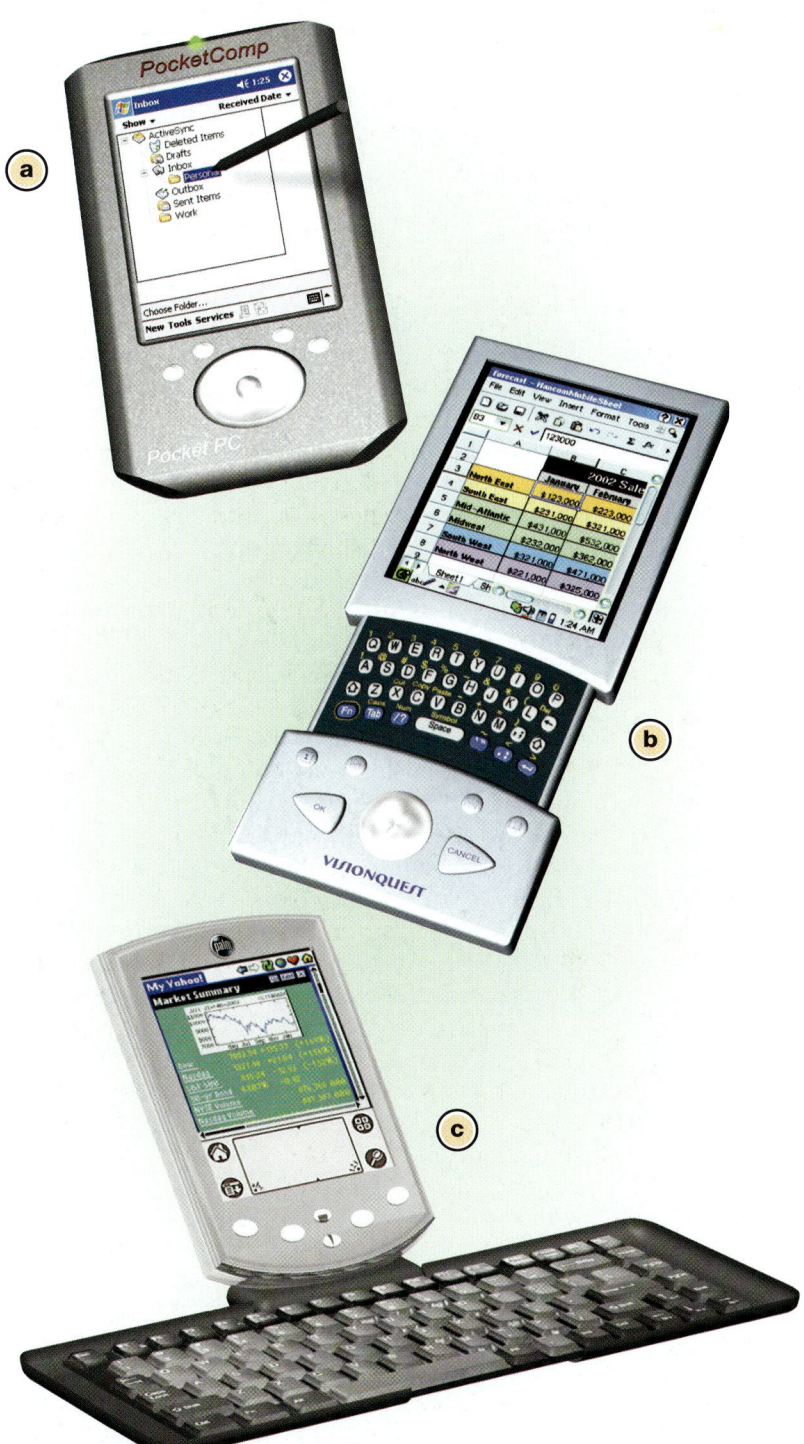

FIGURE 8.12

To enter information, PDAs offer different options: (a) making selections using a stylus (b) an integrated keyboard (shown here on a Sharp Zaurus SL-5500), and (c) a folding keyboard accessory.

With a touch screen and stylus, you can use either handwritten text or special notation systems to enter data into your PDA. One of the more popular notation systems is the **Graffiti** text system. As shown in Figure 8.13, with Graffiti, you must learn special strokes that represent each letter, such as an upside-down V for the letter A. Another popular system, **Microsoft Transcriber**, doesn't require special strokes and can recognize both printed and cursive writing with fairly decent accuracy. PDAs also support an onscreen keyboard so that you can use your stylus and "type" (tap out) messages.

Can I take photos with my PDA? Some newer PDAs include digital cameras that can take photos and record videos. High-end models include digital cameras with flash, auto-focus, and red-eye-reduction features. They can record videos in the MPEG4 (Moving Picture Experts Group) format, a high quality standard for creating compressed video files, and play them back right on the PDA display or on a television through an audio/video output port.

What kinds of displays do PDAs have? PDAs come with LCD screens in a variety of resolutions. The more inexpensive models use 16 levels of gray (grayscale). For appointment schedules and to-do lists, this is fine. However, if you plan on using your PDA to display photos and play video clips, you should consider buying a PDA with a color display. High-end color displays can have resolutions as high as 320 x 480 and support for 65,000 colors.

How do I compare processors for PDAs? Popular PDA processors (CPUs) on the market today include the Motorola DragonBall, the Texas Instruments OMAP, and the Intel XScale processor. When comparing PDA processors, one consideration to keep in mind is **processor speed**. Processor speed, which is measured in hertz (Hz), is the number of operations (or cycles) the processor completes each second. For example, the Dell Axim X5 PDA uses an Intel XScale processor running at 400 megahertz (MHz), or 400 million cycles per second. The Palm Zire uses a Motorola DragonBall EZ processor running at 16 MHz.

If you're interested in running demanding software on your PDA, like games and image-editing applications, getting a fast processor is important. For basic PDA functions like a to-do list and a calendar, a slower processor works fine. The type of application software you plan on running should help you determine whether the additional processing power is worth the cost.

However, processor speed is not the only aspect of the processor that affects performance. The internal design of the processor, both in the software commands it speaks and its internal hardware, are other factors. To measure performance, PDA reviewers often run the same task on competing PDAs and then compare the time it takes to complete the task. This process is called **benchmarking** and gives a good indication of the unit's overall system performance. When comparing PDAs, look for benchmarks in magazine and online reviews.

As well as having different speeds, each processor uses different amounts of power, affecting how long the PDA can run on a single battery. When comparing PDAs, look for the expected operating time on one battery charge.

PDA OPERATING SYSTEMS

How do I compare PDA operating systems? The two main operating system competitors on the PDA market today are the **Palm OS** from Palm and the **Pocket PC** system from Microsoft (formerly called Windows CE). Which operating system is better depends on your personal needs. Palm OS is found on PDAs made by Palm and Sony. Pocket PC is used by Compaq, Hewlett-Packard, and Toshiba on their PDA models. As you can see in Figure 8.14, both operating systems offer you graphical user interfaces.

Palm OS is in some ways a better match to the PDA environment. It requires less memory, is easy to use, and focuses on supporting only the features most commonly used by PDA owners, such as a calendar, to-do list, and contact information. PDAs using the Palm OS can recognize and support Microsoft Word, Excel, and PowerPoint files, although this requires that you buy a software application named Documents To Go. Similarly, you can use the Palm OS to view movies or listen to MP3 files, but only with the purchase of separate application software.

FIGURE 8.13

The Graffiti system uses special strokes to represent all the letters of the English alphabet.

FIGURE 8.28

Laptop computers offer larger displays and more powerful processors than desktops could offer just a few years ago. Here you see (a) a PC laptop and (b) an Apple laptop.

Recently, the Pentium M was combined with two additional components, the Intel 855 chipset and the Intel PRO/Wireless network connection, in a package called the Intel Centrino. The Centrino is optimized to work in a highly mobile setting as it uses less power, is more stable, and provides integrated wireless capability.

LAPTOP OPERATING SYSTEMS AND PORTS

Are there special operating systems for laptops? Laptops use the same operating systems that run on desktop systems. However, laptop operating systems do have some special settings, such as power management profiles. A power management profile contains recommended power saving settings, such as turning off your hard drive after 15 minutes of no use, shutting down the display after 20 minutes of no movement, and switching the machine to standby or "hibernation" modes after a certain length of time.

Can laptops connect to other devices easily? As you can see in Figure 8.29, laptops include a full set of ports, including FireWire, USB 1.0 and 2.0, serial, parallel, infrared (IrDA); RJ-11 jacks for a modem connection; and Ethernet ports for wired networking connections. A newer

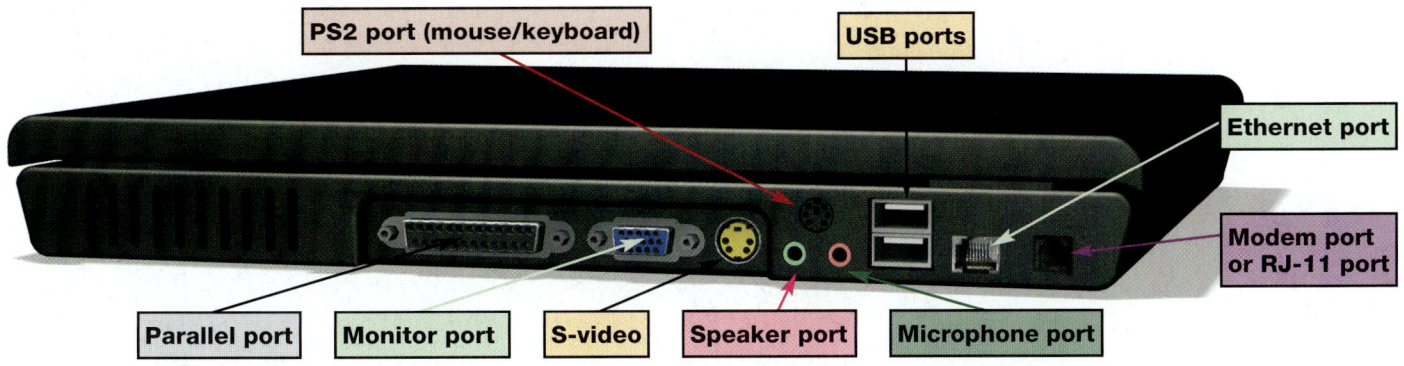

FIGURE 8.29

Laptops have a number of ports like the ones shown above. This laptop has ports along the back for audio and video input and output as well as network connections. Often, additional ports like IrDA and FireWire are also available on either the back or sides of the system.

BITS AND BYTES

More RAM Means Longer Battery Life for Your Laptop

Your laptop stores data both in RAM and on the hard disk drive. Increasing your RAM capacity makes your laptop perform more quickly since data can be read from RAM much faster than from a hard drive. Adding more RAM to your laptop will also make your battery last longer. This is because if the data needed is not in RAM, the hard drive must be powered up, requiring about 30 times as much battery power as simply reading directly from RAM.

version of FireWire, FireWire 800, is beginning to appear on laptop units as well. This port allows transfer speeds of twice the original FireWire connection, up to 800 Mbps.

Often, the size limitations of laptops mean that they don't offer as many of each type of port as in a desktop system. Therefore, if you buy a laptop you should consider how you will use your machine and whether you will need an expansion hub. An expansion hub is a device that connects to a USB port, creating three or four USB ports from one. If you will be connecting a USB mouse, printer, and scanner to your laptop, you may be short one USB port, in which case a hub would come in handy.

How do laptops connect to wireless networks? Most laptops have integrated support for wireless connectivity. As we discussed in Chapter 7, the 802.11b Wi-Fi wireless standard is the standard used in most wireless networks, while the faster 802.11g Wi-Fi standard is also becoming available on laptops. The 802.11g standard allows wireless connections to operate at up to 54 Mbps instead of the 11 Mbps offered in the 802.11b standard. Several laptops also offer built-in Bluetooth chips that allow you to connect to other Bluetooth-enabled devices.

LAPTOP BATTERIES AND ACCESSORIES

What types of batteries are there for laptops? Rechargeable batteries are lithium based (Li-ion batteries) or nickel-based (Ni-Cad). Lithium-based batteries are lighter

FIGURE 8.30

(a) The Brother MPrint micro printer is equipped with Bluetooth and USB interfaces, so you can print with or without connecting cables. It fits in your jacket pocket and can print on plain paper, address labels, and even carbon paper. The Brother MPrint is compatibale with Windows based tablets, and notebooks; Pocket PC based PDAs and Palm based PDAs. (b) The Dell 3200MP projector weighs about 3 pounds and can project to an auditorium or conference room.

than nickel-based batteries and do not show the "memory effect" that nickel-based batteries do. **Memory effect** means that the battery must be completely used up before it is recharged. If not, the battery won't hold as much charge as it originally did.

How long does a laptop battery last? The capacity of a battery is measured in ampere-hours (A-hrs). Ampere is a measure of current flow, so a battery rated at 5 A-hrs can provide 5 amps of current for an hour. Battery power depends on the device you're using and the work you're doing. A high-performance battery can operate a laptop for up to five hours if fully charged. However, using the laptop's DVD drive can consume a high-performance battery in as little as 90 minutes. Some laptop systems allow you to install two batteries at the same time, doubling the battery life but forcing you to purchase a second battery and have both fully charged.

Do I need to use my battery everywhere I go? Some environments support easy access to power for laptops. For example, you can use an AC/DC or DC/DC converter to enable a laptop to run in a car without using battery power. In addition, many airplanes offer laptop power connections at each seat, although to use such a connection you have to buy a power converter and adapter.

What special purchases might a laptop require? Because a laptop is so easily stolen, purchasing a security lock is a wise investment. And as is the case with regular desktops, power surges can adversely affect laptops, so investing in a portable surge protector is also a good idea. (Be sure to read the Technology in Focus feature "Protecting Your Computer and Backing Up Your Data" for more information about protecting your laptop.)

If you frequently make presentations to large groups, adding a lightweight projector to your laptop might prove useful. Printers have also become travel-sized. Figure 8.30 shows some of these special laptop accessories.

LAPTOP OR DESKTOP?

How does a laptop compare to a desktop? Desktop systems are invariably a better value than laptops. Because of the laptop's small **footprint** (the amount of space on the desk it takes up), you pay more for each component. Each piece has had extra engineering time invested to make sure it fits in the smallest space.

FIGURE 8.31

PC cards add functionality to your laptop.

In addition, a desktop system offers you more expandability options. It's easier to add new ports and devices because of the amount of room available in the desktop computer's design.

Desktop systems are also more reliable. Because of the amount of vibration that a laptop experiences, as well as the added exposure to dust, water, and temperature fluctuations, laptops do not last as long as desktop computers. Manufacturers offer extended warranty plans that cover accidental damage and unexpected drops, although at a price.

How long should I be able to keep my laptop? The answer to that question depends on how easy it is to upgrade your system. Take note of the maximum amount of memory you can install in your laptop. Internal hard disks are not easy to upgrade in a laptop, but if you have a FireWire or USB 2.0 port, you can add an external hard drive for more storage space. Most laptops are equipped with PC card slots, which are slots on the side of the laptop that accept special credit card–sized devices called **PC cards**, shown in Figure 8.31. PC cards can add fax modems, network connections, wireless adapters, USB 2.0 and FireWire ports, and other capabilities to your laptop.

You can also add a device that allows you to read flash memory cards, such as Compact Flash, Memory Sticks, and Secure Digital cards. As new types of ports and devices are introduced, many will be manufactured in PC card formats so that you can make sure your laptop is not obsolete before its time.

SOUND BYTE
TABLET AND LAPTOP TOUR

In this Sound Byte, you'll take a tour of a tablet PC and a laptop computer, learning about the unique features and ports available on each.

TRENDS IN IT

EMERGING TECHNOLOGIES:
Nanotubes: The Next Big Thing Is Pretty Darn Small!

In the classic 1967 film *The Graduate*, Dustin Hoffman is a young man uncertain about which career he should embark upon. At a cocktail party, an older gentleman provides him with some career advice, telling him "I've got just one word for you . . . plastics!" This made sense at the time since plastics were coming on strong as a replacement for metal. If *The Graduate* were remade today, the advice would be, "I've got just one word for you . . . nanotubes!"

As you learned in Chapter 1, *nanoscience* involves the study of molecules and structures (called *nanostructures*) whose size ranges from 1 to 100 nanometers (or one-billionth of a meter). Using nanotechnology, scientists are hoping to one day build resources from the molecular level by manipulating individual atoms instead of using raw materials already found in nature (such as wood or iron ore). This would allow us to create microscopic computers, the ultimate in portable devices. Imagine nano-sized robotic computers swimming through your arteries clearing them of plaque. Consider carrying a supercomputer with you the size of a pencil eraser, or even better, having the power of your desktop computer implanted in your body as a nano-sized chip.

The possibilities of miniaturization are endless, but what would the computer circuits for these devices be constructed from? Carbon nanotubes are poised to be the building blocks of the future. You're familiar with carbon from pencils. The graphite core in a pencil is composed of sheets of carbon atoms laid out in a honeycomb pattern. Individual sheets of graphite are very strong, but don't bond well to other sheets. This makes them ideal for use in a pencil since as you write, the graphite flakes off and leaves marks on the paper. Unfortunately, graphite doesn't conduct electricity very well. This, coupled with the lack of strong bonding principles, makes graphite unsuitable as a material to manufacture circuits.

In 1991, carbon nanotubes were discovered. Nanotubes are essentially a sheet of carbon atoms (much like graphite) laid out in a honeycomb pattern but rolled into a spherical tube, as shown in Figure 8.32. Arranging the carbon in a tube increases its strength astronomically. It is estimated that carbon nanotubes are 10 to 100 times stronger per unit of weight than steel. This should make them ideal for constructing many types of devices and building materials. Some day we may have earthquake-proof buildings constructed from nanotubes or virtually indestructible clothing woven from nanotube fibers.

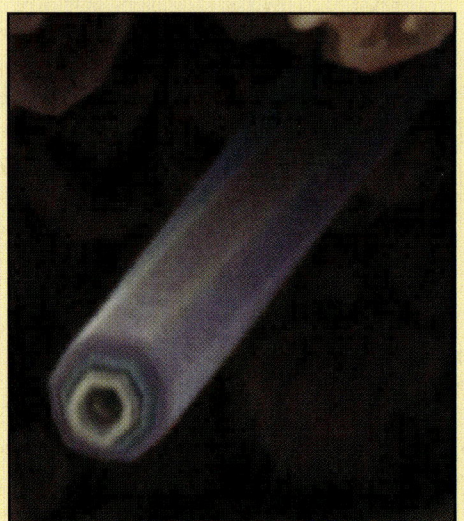

FIGURE 8.32

Here is a highly magnified close-up of a carbon nanotube. Rolling the sheets of carbon atoms into a tube shape gives them incredible strength.

But how does this help us build a computer? Aside from strength, the most interesting property of nanotubes is that they are good conductors of electricity. Nanotubes are actually classified as semimetal, meaning they can have properties that are a cross between semiconductors (such as silicon, which is used to create computer chips) and metals. In fact, depending on how a nanotube is constructed, it can change from a semiconductor to a metal along the length of the tube. These properties make it vastly superior to silicon for the construction of transistor pathways in computer chips, as it provides engineers with more versatility.

In addition, while the smallest silicon transistors that are likely to be produced in the future will be millions of atoms wide, scientists believe that transistors constructed of nanotubes would be only 100 to 1,000 atoms wide. This represents a quantum leap in miniaturization even surpassing the original invention of the transistor. Just imagine what can be done when nanotube transistors replace silicon transistors!

So when can you buy that pencil eraser–sized computer? Not for quite a while. At this point, researchers can manufacture nanotubes only in extremely small quantities at a large cost. But the U.S. government and many multinational corporations are expected to pour billions of dollars into nanoscience research over the next five years. The ongoing research will hopefully lead to breakthroughs in manufacturing technology that will result in nano-scale computers within your lifetime.

Summary

1. **What are the advantages and limitations of mobile computing?** Mobile computing allows you to communicate with others, remain productive, and have access to your information no matter where you are. However, mobile devices are more expensive and less rugged than desktop equipment. In addition, battery life limits the usefulness of mobile devices, the screen is typically small, the Internet connection is often slow, and wireless Internet coverage is limited.

2. **What are the various mobile computing devices?** There is a range of mobile computing devices on the market including pagers, cell phones, MP3 players, PDAs, tablet PCs, and laptop computers.

3. **What can pagers do and who uses them?** A pager allows you to receive messages on a small display screen. Two-way pagers support both receiving and sending messages. Pagers are compact and are the most inexpensive mobile computing device. People who need to be reachable but want an inexpensive device are the primary market for pagers.

4. **How do cell phone components resemble a traditional computer and how do cell phones work?** Cell phones include a processor, memory, input and output devices, software, and an operating system. When you speak into a cell phone, the sound enters as a sound wave. An analog-to-digital converter converts these waves into digital signals. The digital data is then compressed so that it transmits more quickly to another phone. Finally, the digital data is transmitted as a radio wave through the cellular network to the destination phone.

5. **What can I carry in an MP3 player and how does it store data?** MP3 players store mainly digital music files, but some players also allow you to carry video and image files. Inexpensive players use only memory chips to store data, whereas more expensive models use a built-in hard drive, which provides more storage. Some MP3 players allow you to add removable memory called flash memory.

6. **What can I use a PDA for and what internal components and features does it have?** PDAs can carry calendars, contact lists, software, songs, and more. PDAs include a processor, operating system, input and output devices, and ports. The two main PDA operating systems are the Palm OS and the Pocket PC system. PDAs do not come with built-in hard drives, but for memory needs beyond their built-in RAM and ROM, PDAs use flash memory.

7. **How can I synchronize my mobile devices with my desktop computer?** To synchronize your desktop and PDA, you can place the PDA in a cradle and touch a "hot sync" button. This begins the process of data transfer that updates both sets of files to the most current version.

8. **What is a tablet PC and why would I want to use one?** A tablet PC is a portable computer that includes advanced handwriting and speech recognition. Tablet PCs can be used either in a laptop mode or in "tablet mode." The most innovative input technology on the tablet PC is digital ink. Supporting digital ink, the tablet's screen is pressure-sensitive and reacts to a digital pen. A tablet PC can be the ideal solution when you require a lightweight, portable computer with full desktop power.

9. **How powerful are laptops and how do they compare to desktop computers?** The most powerful mobile solution is a laptop (or notebook). Laptops offer large displays and can be equipped with DVD/CD-RW drives, hard drives, and up to 1 GB of RAM. Many models feature hot-swappable bays and a full set of ports. Still, desktop systems are more reliable and cost-effective than laptops and are easier to upgrade. How-ever, many users feel the mobility laptops offer is worth the added expense.

CHAPTER 8 MOBILE COMPUTING: KEEPING YOUR DATA ON HAND

Key Terms

alphanumeric pagers	p. 313	Napster	p. 320
analog-to-digital converter chip	p. 315	numeric pagers	p. 313
base transceiver station	p. 315	paging device (pager)	p. 312
benchmarking	p. 322	Palm OS	p. 322
BlueBoard	p. 326	PC cards	p. 335
Bluetooth	p. 326	peer-to-peer (P2P) sharing	p. 320
cellular phones	p. 313	personal digital assistant (PDA)	p. 321
Compact Flash	p. 319	Pocket PC	p. 322
cradle	p. 325	processor speed	p. 322
digital ink	p. 329	sampling rate	p. 317
digital pen	p. 329	Secure Digital	p. 319
digital signal processor	p. 315	Short Message Service (SMS) (text messaging)	p. 316
docking station	p. 330	smart battery	p. 324
flash memory	p. 319	SmartMedia	p. 319
footprint	p. 335	stylus	p. 321
Global Positioning System (GPS)	p. 328	Symbian OS	p. 313
Graffiti	p. 322	synchronizing	p. 325
hot-swappable bays	p. 332	tablet PC	p. 329
laptop computer (notebook computer)	p. 332	two-way pagers	p. 313
memory effect	p. 335	voice pager	p. 313
Memory Stick	p. 319	WAP (Wireless Application Protocol)	p. 326
microbrowser	p. 316	Web clipping	p. 326
MMCs (multimedia cards)	p. 319	Web-enabled	p. 310
mobile computing devices	p. 310	wireless Internet service provider	p. 316
mobile switching center	p. 315	Wireless Mark-up Language (WML)	p. 317
MP3 player	p. 317		
MS Transcriber	p. 322		
Multimedia Message Service (MMS)	p. 316		

Buzz Words

Word Bank

- mobile switching center
- WML
- processor speed
- microbrowser
- tablet PC
- PDA
- memory effect
- flash memory card
- SMS
- synchronize
- stylus
- MP3 player
- laptop
- GPS
- mobile device(s)
- Analog-to-digital converter chip
- Pocket PC
- MMS
- cradle
- Bluetooth
- cell phone
- Graffiti
- pager
- Palm OS
- digital ink

Instructions: Fill in the blanks using the words from the Word Bank above.

Kathleen's new job as a sales rep is going to mean a lot of travel. She'll need to start thinking about using (1)_____ to stay productive when she's out of the office. Because she needs voice communication and not just text exchange, she'll be selecting a (2)_____ rather than a (3)_____. With her cell phone, she will be able to exchange quick text messages with her coworkers using (4)_____. When she accesses the Internet from her cell phone, she will use (5)_____ software to check the latest stock prices.

Because she travels a lot and loves music, she has been considering investing in a digital (6)_____. However, she has instead decided to purchase a more expensive, more powerful (7)_____ that includes MP3 capabilities. That way, she can also use the device as more than just an MP3 player. Seeing that she needs more storage room, she also invests in a removable (8)_____ on which she'll store her MP3 files. Because she's a hiker, she wants to use her PDA as a navigation device, so she has purchased a (9)_____ accessory to go with it.

To make sure she is getting the best device with the most powerful processor, Kathleen has been comparing benchmarks that measure (10)_____. She knows this is an important part of getting the most powerful device. In addition, she wants to make sure she can (11)_____ her PDA with her desktop computer, so the files on both always match. Thus, she bought a PDA that includes the wireless (12)_____ technology, as well as an external (13)_____ that connects her PDA to her computer through a USB port.

Because she still needs to run powerful software packages when she's out of the office, she bought a (14)_____ as well. It was a better choice than a full-sized (15)_____ because she carries it with her all day, taking notes while standing on the production floor.

CHAPTER 8 MOBILE COMPUTING: KEEPING YOUR DATA ON HAND

Organizing Key Terms

Instructions: This chapter introduced many new terms and concepts. In the illustration below, fill in each of the blanks with a key term from the chapter in order to show how categories of ideas fit together.

Mobile Devices

- **Pager**
  - Types of Pagers:
    1.
    2.
    3.

- **Cell Phone**
  - Cell Phone:
    1.
    2.
  - Output Devices:
    1. Speaker
    2.
  - Popular Operating System:
    1.

- **MP3 Player**
  - Types of Storage:
    1. Hard Drive
    2.
    3.
  - Types of Data It Can Hold:
    1. MP3 files
    2. Photos
    3.
    4.

- **PDA**
  - Input Devices:
    1. Stylus (Graffiti)
    2.
    3.
  - Popular Operating Systems:
    1. Palm OS
    2.
  - Storage:
    1. ROM
    2.
  - Popular CPUs:
    1. Xscale
    2.

- **Tablet PC**
  - Input Technologies:
    1. Digital Pen/Ink (Handwriting Recognition)
    2.
  - Popular CPUs:
    1. Transmeta Caruso
    2.

- **Laptop**
  - Hardware:
    1. Hard drive
    2. RAM
    3.
  - Popular CPUs:
    1. Pentium M
    2.
  - Ports:
    1. Firewire
    2. USB 1.0 and 2.0
    3.
    4.
    5.
    6.

Making the Transition to . . . Next Semester

1. As a student, which devices discussed in this chapter would have the most immediate impact on the work you do each day? Which would provide the best value (that is, the most increase in productivity and organization per dollar spent)?
2. Compare the Apple PowerBook series of laptop computers with the Dell Inspiron 8500 series. Consider price, performance, expandability, and portability. Explain which would be the better investment for your needs next semester.
3. Fill out the table below to determine how many minutes of MP3 files you could store depending on the sampling rate of the MP3 files. Use the following to help you fill in the table:
   - Say you sample music at 192 kilobits per second. There are 8 bits in one byte, so 192 kilobits per second = 192/8 = 24 KB per second.
   - There are 60 seconds in one minute, so the number of kilobytes per minute is equal to 24KB per second times 60 seconds = 1,440 KB per minute.
   - With 256 MB of space, which is 256,000 KB, there would be room for 256,000 KB divided by 1,440 KB per minute = approximately 177 minutes of songs that can be stored

SAMPLING RATE KILOBITS (Kb) PER SECOND	NUMBER OF KILOBYTES PER SECOND (KB/SEC)	NUMBER OF KILOBYTES PER MINUTE (KB/MINUTE)	FLASH CARD MEMORY	MINUTES OF SONGS THAT CAN BE STORED
192 Kbps	192/8 = 24 KBps	24 KBps * 60 seconds = 1,440 KB/minute	256 MB	256,000 KB/ 1,440 KB/ minute = 177 minutes
128 Kbps			256 MB	
96 Kbps			256 MB	
64 Kbps			256 MB	

4. How wireless is your campus? Find out if there are areas of your campus where wireless support is in place right now—perhaps the library, the student center, or certain buildings. Is the entire campus wireless? Are there plans for that in the near future?
5. Major national cellular providers include AT&T, Verizon, T-Mobile, and Sprint. Visit their Web sites and compare the prices and features of their popular cellular plans for both minimal users and "power" users. Based on your research, which cell phone plan would be best for your needs?
6. Visit the Electronics store at Amazon.com and locate the PDA section. Compare three different PDA models and list their price, input and output devices, operating systems, built-in memory, processor speed, and other special features.
   a. Which of the three models you compared is the best value for your needs?
   b. What special features does that PDA have? What accessories would you buy for your PDA to make it more useful?
   c. Would you consider buying a used PDA? Why or why not?
   d. Next, investigate PDA software on sites such as www.download.com. What software would you buy for a PDA and which operating system does that software require?

Making the Transition to . . . The Workplace

1. Imagine your company is boosting its sales force and looking to the future of mobile technology. Your boss has asked you to research the following issues surrounding mobile computing for the company:

   a. Do mobile computing devices present increased security risks? What would happen if you left a flash memory card at a meeting and a competitor picked it up? Are there ways to protect your data on mobile devices?
   b. Can viruses attack mobile devices? Is there any special software on the market to protect mobile devices from viruses?
   c. Is there a role for mobile computing devices even if employees don't leave the building? Which devices would be important for a company to consider for use within corporate offices?

2. The next generation of telecommunications (nicknamed "3G" for "third generation") allows the speed of cellular network transmissions to rise from 144 Kbps to 2 Mbps. How does that compare to dial-up and cable modem access for wired networks? What implications does it have on information access and e–commerce?

3. Research the Internet and determine what speed limitations are expected to exist in the future with regards to mobile devices.

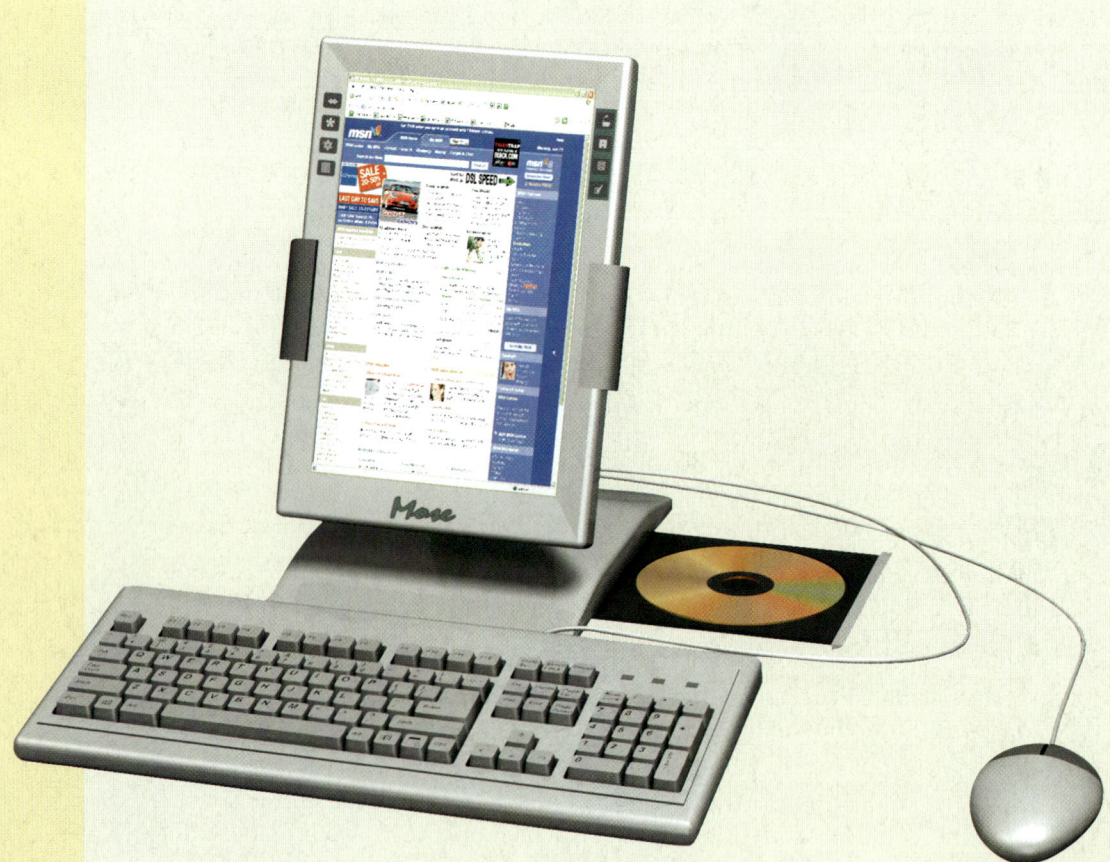

Critical Thinking Questions

Instructions: *Albert Einstein used "Gedanken experiments," or critical thinking questions, to develop his theory of relativity. Some ideas are best understood by experimenting with them in our own minds. The following critical thinking questions are designed to demand your full attention but require only a comfortable chair—no technology.*

1. What social changes do you predict from increased mobile access to information? Will there be more social interaction? Less? Will mobile computing promote increased understanding between people? More isolation?

2. As devices become lighter and smaller we are seeing a combining of multiple functions into one device. What would the ultimate convergent mobile device be for you? Is there a limit in weight, size, or complexity?

3. The recording industry, recording artists, and consumers find themselves in a complex discussion when the topic of peer-to-peer sharing systems is brought up. What kind of solution would you propose to safeguard the business interests of the industry, the intellectual property rights of the musicians, and the freedoms of the consumers?

4. Have you ever downloaded music off the Web? If so, did you download the music from a legal site? Do you think sites like Napster should be allowed to exist?

5. Consider the implications of the "bat" tracking device we discussed in the Trends in IT feature. Would you agree to be "tracked" at work if it meant a more convenient way to use communication tools? What sorts of privacy risks do such devices pose?

6. What applications can you think of for the "wearable computer" technology discussed in the Trends in IT on page 000 feature? Do you currently own anything that you would consider a wearable computer? What sorts of wearable computers do you think you would be likely to purchase?

7. Do you think we will ever become a completely wireless society? Will there always be a need for some land lines (physical wired connections)?

8. Would you agree to insert a GPS-enabled tracking device into your pet? Your child? What legislation would be required if tracking data were available on you? Would you be willing to sell that information to marketing agencies? Should that data be available to the government if you were suspected of a crime?

9. America Online Instant Messenger (AIM) service is now available on some cellular phone systems (for example, T-Mobile provides this service). Would the ability to be alerted to IM buddies on your cell phone be useful to you? What would you be willing to pay for this feature?

10. The Trends in IT on page 000 feature on nanotechnology describes a number of ways in which nano-sized computers may one day be used. What other applications for such powerful and tiny computers can you think of? Can you think of any security or privacy risks associated with such nano-sized computers?

Team Time

Problem:

You have formed a consulting group and that advises clients on how to move their businesses into the new mobile computing age.

Task:

Each group will be defined as an "expert" resource in one of the mobile devices presented in this chapter: cell phones, PDAs, tablet PCs, or laptops. For each of the scenarios described by a client, the group should assess how strong a fit their device is to that client's needs.

Process:

Divide the class into three or four groups and assign each group a different mobile device (cell phone, PDA, tablet PC, or laptop).

STEP 1: Research the current features and prices for the mobile device your group has been assigned.

STEP 2: Consider the following three clients:
- An elementary classroom that wants to have students carry mobile devices to the nearby creek to do a science project on water quality
- A manufacturing plant that wants managers to be able to report back hourly on the production line's performance and problems with a mobile device
- A pharmaceutical company that wants to outfit their sales reps with the devices they need to be prepared to promote their products when they visit physicians

Discuss the advantages and disadvantages of your device for each of these clients. Consider value, reliability, computing needs, and communication needs as well as expandability for the future.

STEP 3: Prepare a final report for the group that considers the costs, availability, and unique features of the device that led you to recommend or not recommend it for each client.

STEP 4: Bring the research materials from the individual team meetings to class. Looking at the clients' needs, make final decisions as to which mobile device is best suited for each client.

Conclusion:

There are a number of mobile computing devices on the market today. Finding the best mobile device to use in any given situation depends on factors such as value, reliability, expandability, and the computing and communication needs of the client.

Becoming Computer Fluent

You have a job as a sales representative at a large publishing company. Your boss is considering investing in some kind of mobile device as well as Internet access for it to help you perform your duties. However, first she requires a justification. What mobile device(s) would be best suited for your position? Why would you need (or not need) Internet access for your mobile device? Does it depend on which type of device you are using? Does it depend on your job responsibilities? What advantages would there be to the company? What hardware would be required?

Instructions: Using the scenario above, write a report using as many of the key words from the chapter as you can. Be sure the sentences are grammatically and technically correct.

Materials on the Web

In addition to the review materials presented here, you'll find extra materials on the book's companion Web site (**www.prenhall.com/techinaction**) that will help reinforce your understanding of the chapter content. These materials include:

Sound Byte Lab Guides

For each Sound Byte mentioned in the chapter, there is a corresponding lab guide located on the book's companion Web site. These guides review the material presented in the Sound Byte and direct you to various Web resources that examine the material. The Sound Byte Lab Guides for this chapter include:

- PDAs on the Road and at Home
- Tablet and Laptop Tour

True/False and Multiple Choice Quizzes

The book's Web site includes a True/False and a Multiple Choice quiz for this chapter. You can take these quizzes, automatically check the results, and e-mail the results to your instructor.

Web Research Projects

The book's Web site also includes a number of Web research projects for this chapter. These projects ask you to search the Web for information on computer-related careers, milestones in computer history, important people and companies, emerging technologies, and the applications and implications of different technologies.

OBJECTIVES

After reading this chapter, you should be able to answer the following questions:

- What is a switch and how does it work in a computer? (pp. 348–349)

- What is the binary number system and what role does it play in a computer system? (pp. 349–353)

- What is inside the CPU and how do these components operate? (pp. 353–359)

- How does a CPU process data and instructions? (pp. 353–355)

- What is cache memory? (pp. 356–357)

- What types of RAM are there? (pp. 360–361)

- What is a bus and how does it function in a computer system? (pp. 361–362)

- How do manufacturers make CPUs so that they run faster? (pp. 363–365)

CHAPTER 9

Behind the Scenes:
Inside the System Unit

TECHNOLOGY IN ACTION: TAKING A CLOSER LOOK

Although twins, Jim and Joe are completely different in certain ways. Joe checks the oil in his car regularly, knows the air pressure in his tires, and can tell when the fan belt should be replaced. Jim, on the other hand, asks, "Oil level? What oil?" and drives from place to place relying on service departments to keep his car running. Although Jim's lack of maintenance hasn't resulted in any catastrophic problems, Joe warns his brother that by not learning a few things about cars, he'll end up paying more to keep up his car, if it even lasts that long.

Similarly, after taking a class in college, Joe has a strong but basic understanding of his computer, and keeps his PC running well through periodic maintenance and upgrades. Not surprisingly, Jim is hands-off when it comes to his computer. He's happy to just turn it on and open the files he needs. He can't be bothered with all the acronyms: CPU, RAM, and all the rest. When there's a problem, Jim just calls his brother. He hates waiting and paying for technical service and is afraid he'll mess up his system if he tries to fix things himself. But after getting another 2:00 A.M. call from Jim after his computer crashed, Joe told him, "Take a class or pay a technician!"

When it comes to your computer, are you most like Jim or Joe? Joe found out how easy it is to understand his computer and isn't dependent on anyone else to keep it running. But if you use a computer without understanding the hardware inside, you'll have to pay a technician to fix or upgrade it. Meanwhile, it won't be as efficient as if you were fine-tuning it yourself, and you may find yourself buying a new computer earlier than necessary.

There are other advantages to having a deeper understanding of computer hardware. If you're preparing for a career in programming, for example, understanding computer hardware will affect the speed and efficiency of the programs you design. In addition, if you're interested in computers, you're no doubt excited by advances you hear about. How do you evaluate the impact of a new type of memory or a new processor? A basic appreciation of how a computer system is built and designed is a good start.

In this chapter, we'll build on what you've learned about computer hardware in other chapters and go behind the scenes, looking at your system unit's components in more detail. First we'll examine how computers translate the commands you input into the digits they can understand: 1s and 0s. Next, we'll analyze the internal workings of the CPU and memory. We'll then look at *buses*, the highways that transport data between the CPU, memory, and other devices connected to the computer. But first, let's look at the building blocks of computers: switches.

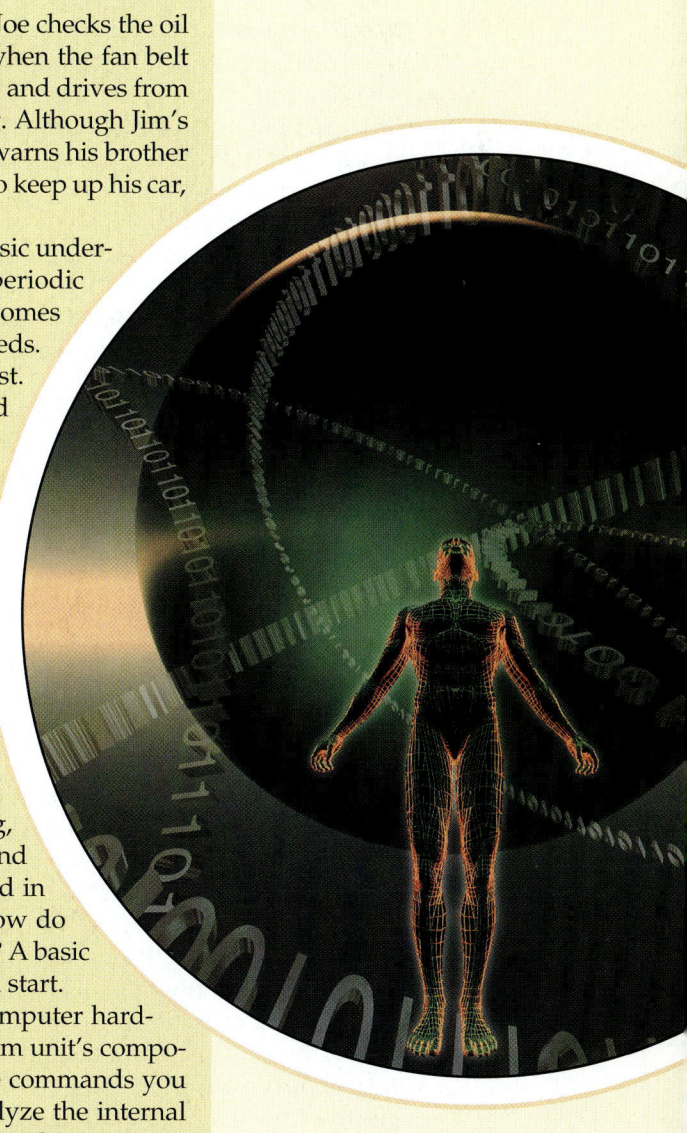

Digital Data: Switches and Bits

In earlier chapters, you learned that the **system unit** is the box that contains the central electronic components of the computer, including the central processing unit (CPU), memory, motherboard, and many other circuit boards that help the computer to function. But how exactly does the computer perform all of its tasks? How does it process the data you input? In this section, we'll discuss how the CPU performs its functions—adding, subtracting, moving data around the system, and so on—using nothing but a large number of on/off switches. In fact, as you'll learn, a computer system can be viewed as just an enormous collection of on/off switches.

ELECTRONIC SWITCHES

What are switches and what do they do? You learned earlier that unlike humans, computers work exclusively with numbers (not words). In order to process data into information, computers need to work in a language they understand. This language, called **binary language**, consists of just two numbers: 0 and 1. Everything a computer does (such as process data or print a report) is broken down into a series of 0s and 1s.

Why do computers use 0s and 1s to process data? Because modern computers are electronic, digital machines, they understand only two states of existence: on and off. Computers represent these two possibilities, or states, using the numbers (or digits) 1 and 0. **Electronic switches** are devices inside the computer that can be flipped between these two states: 1 or 0, on or off.

Although the notion of switches may seem complex, you use various forms of switches every day. For example, a button is a mechanical switch: pushed in, it could represent the value 1, while popped out, it could represent the value 0. Another switch you use each day is a water faucet. As shown in Figure 9.1, shutting it off so no water flows could represent the value 0, while turning it on could represent the value 1.

Since computers are built from a huge collection of switches, using buttons or water faucets obviously would limit the amount of data computers could store. It would also make computers very large, and cause them to run at very slow speeds. Thus, the history of computers is really a story about creating smaller and faster sets of electronic switches so that more data can be stored and manipulated quickly.

FIGURE 9.1

Water faucets can be used as binary switches. Turning the faucet on could represent the value 1, while shutting the faucet off so no water flows could represent the value 0.

What were the first switches used in computers? The earliest generation of electronic computers used devices called **vacuum tubes** as switches, as shown in Figure 9.2a. Vacuum tubes act as computer switches by allowing or blocking the flow of electrical current. The problem with vacuum tubes is that they take up a lot of space. The first high-speed digital computer, the Electronic Numerical Integrator and Computer (referred to as the ENIAC), was deployed in 1945 and used nearly 18,000 vacuum tubes as switches, which filled approximately 1,800 square feet of floor space. In addition to being very large, vacuum tubes produce a lot of heat. Thus, although they are still used as switches in some high-end audio equipment, they make for impractical switching devices in personal computers.

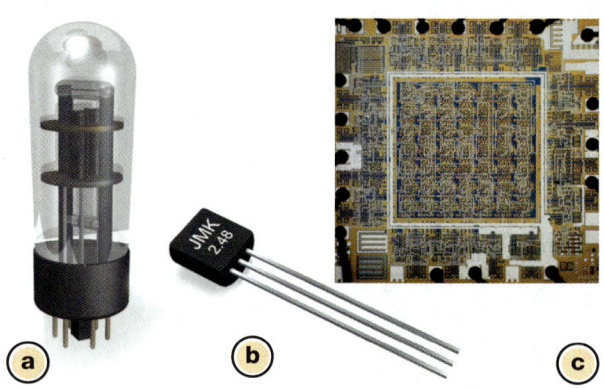

FIGURE 9.2

Electronic switches have become smaller and faster over time, from (a) vacuum tubes to (b) discrete single transistors to (c) integrated circuits that can hold more than 220 million transistors.

What do personal computers use as switching devices? Since the vacuum tubes of the ENIAC, two major revolutions have occurred in the design of switches, and consequently computers, to make them smaller and faster: the invention of the *transistor* and the fabrication of *integrated circuits*.

What are transistors? Transistors are electrical switches that are built out of layers of a special type of material called a **semiconductor**. A semiconductor is any material that can be controlled to either conduct electricity or act as an insulator (not allow electricity to pass through). Silicon, which is found in common sand, is the semiconductor material used to make transistors.

By itself, silicon does not conduct electricity particularly well, but if specific chemicals are added in a controlled way to the silicon, it begins to behave like a switch. It allows electrical current to flow easily when a certain voltage is applied, and prevents electrical current from flowing otherwise, thus behaving as an on/off switch. This kind of behavior is exactly what is needed to store digital information, the 1s and 0s in binary language.

Early transistors were built in separate units as small metal cans, each can acting as a single on/off switch, as shown in Figure 9.2b. These first transistors were much smaller than vacuum tubes, produced very little heat, and could be switched from on to off (allowing or blocking electrical current) very quickly. They were also less expensive than vacuum tubes.

However, it wasn't long before transistors reached their limits. Continuing advances in technology began to require more transistors than circuit boards at the time could reasonably handle. Something was needed to pack more transistor capability into a smaller space. Thus, integrated circuits, the next technical revolution in switches, developed.

What are integrated circuits? **Integrated circuits** (or chips) are very small regions of semiconductor material, such as silicon, that support a huge number of transistors, as shown in Figure 9.2c. Along with all the many transistors, other components critical to a circuit board (such as resistors, capacitors, and diodes) are also located on the integrated circuit. Most integrated circuits are no more than a quarter inch in size.

Why are integrated circuits important? Because so many transistors can fit into such a small area, integrated circuits have enabled computer designers to create small yet powerful **microprocessors,** which are chips that contain a CPU. In 1971, the Intel 4004 was the first complete microprocessor to be located on a single integrated circuit chip, marking the beginning of true miniaturization of computers. The Intel 4004 contained slightly more than 2,300 transistors. Today more than 220 *million* transistors can be manufactured in a space as tiny as the nail of your pinky finger!

This incredible feat has fueled an industry like none other. In 1951, the Univac I computer was 10 feet high by 10 feet wide by 10 feet long (or 1,000 cubic feet) and cost $1 million. Thanks to advances in integrated circuits, the IBM PC released just 30 years later took up just 1 cubic foot of space, cost $3,000, and performed 155,000 times more quickly. (For more information about computer history, see Technology in Focus, "The History of the PC.")

But how can computers store information in a set of on/off switches? You understand that computers use on/off switches to perform their functions. But how can these simple switches be organized so that they enable us to use a computer to pay our bills online or write an essay? How could a set of switches describe a number or a word, or give a computer the command to perform addition? Recall that to manipulate the on/off switches, the computer works in binary language, which uses only two digits, 0 and 1. Therefore, in order to understand how a computer works, we must first look at how the computer uses a special numbering system called the *binary number system* to represent all of its programs and data.

THE BINARY NUMBER SYSTEM

What is a number system? A **number system** is an organized plan for representing a number. Although you may not realize it, you are already familiar with one number system. The **base 10 number system**, also known as **decimal notation**, is the system you use to represent all of the numeric values you use each day. It's called "base 10" because it uses 10 digits, 0 through 9, to represent any value.

To represent a number in base 10, you break the number down into groups of ones, tens, hundreds, thousands, and so on. For example, using base 10, in the whole number 6,954, there are 6 sets of thousands, 9 sets of hundreds, 5 sets of tens, and 4 sets of ones. Working from right to left, each place in a number represents an increasing power of ten, as shown below:

$$6{,}954 = 6*(1{,}000) + 9*(100) + 5*(10) + 4*(1)$$
$$= 6*10^3 + 9*10^2 + 5*10^1 + 4*10^0$$

Anthropologists theorize that humans developed a base 10 number system because we have 10 fingers. But computer systems, with their huge collections of on/off switches, are not well suited to thinking about numbers in groups of 10. Instead, computers describe a number as powers of 2 since each switch can be in one of two positions: on or off. This numbering system is referred to as the **binary number system**. It is the number system used by computers to represent all data.

How does the binary number system work? Because it only includes two digits (0 and 1), the binary number system is also referred to as the **base 2 number system**. However, even with just two digits, the binary number system can still represent all the same values that a base 10 number system can. Instead of breaking the number down into sets of ones, tens, hundreds, and thousands, as is done in base 10 notation, the binary number system describes a number as the sum of powers of 2. Binary numbers therefore are used to represent *every* piece of data stored in a computer: all of the numbers, all of the letters, and all of the instructions that the computer uses to execute work.

Representing Numbers in the Binary Number System

How does the binary number system represent a whole number? As noted above, in the base 10 number system, a whole number is represented as the sum of ones, tens, hundreds, thousands, or the sums of powers of ten. The binary system works in the same way but describes a value as the sum of powers of 2: 1, 2, 4, 8, 16, 32, 64, and so on. Let's look at the number 67. In base 10, the number 67 would be 6 sets of 10 and 7 sets of 1s, as follows:

Base 10: $67 = 6 * 10^1 + 7 * 10^0$

In base 2, we work from the largest possible power of 2 that could be in the number 67. Therefore,

67 has	1	group of	64 (leaving 3) and
3 has	0	groups of	32
		0 groups of	16
		0 groups of	8
		0 groups of	4
		1 group of	2 (leaving 1) and
1 has	1	group of	1 leaving 0

Therefore, the binary number for 67 is 1000011 as shown below:

Base 2

$$67 = 64 + 0 + 0 + 0 + 0 + 2 + 1$$
$$= (1*2^6) + (0*2^5) + (0*2^4) + (0*2^3) + (0*2^2) + (1*2^1) + (1*2^0)$$
$$= (1\ 0\ 0\ 0\ 0\ 1\ 1)\ \text{base 2}$$

Is there an easier way to convert a base 10 number to a base 2 number? You can convert base 10 numbers to binary manually by repeatedly dividing the number by 2 and examining the remainder at each stage. An example will make this clearer. Let's convert the base 10 number 67 into binary:

$$67 \div 2 = 33 \text{ remainder } 1$$
$$33 \div 2 = 16 \text{ remainder } 1$$
$$16 \div 2 = 8 \text{ remainder } 0$$
$$8 \div 2 = 4 \text{ remainder } 0$$
$$4 \div 2 = 2 \text{ remainder } 0$$
$$2 \div 2 = 1 \text{ remainder } 0$$
$$1 \div 2 = 0 \text{ remainder } 1$$
$$1\ 0\ 0\ 0\ 0\ 1\ 1$$

The binary number is then read from left to right. Therefore, 100011 is the binary (base 2) equivalent of the base 10 number 67.

Is there a faster way to convert between base 10 and binary? Programmers and engineers who work with binary codes daily learn to convert between decimal and binary mentally. However, if you use binary notation less often, it is easier to use a calculator. Some calculators identify this operation with a button labeled DEC (for decimal) and one labeled BIN (for binary). In Windows, you can access a scientific calculator that supports base conversion between decimal (base 10) and binary (base 2) by choosing Start, Programs, Accessories, then choosing Calculator, then clicking on the View menu to select the Scientific Calculator.

DIGITAL DATA: SWITCHES AND BITS

DIG DEEPER
ADVANCED BINARY AND HEXADECIMAL NOTATIONS

You understand how the binary number system represents a positive number, but how can it represent a negative number? In the decimal (base 10) system, a negative value is represented with a special symbol, the minus sign (–). In the binary (base 2) system, one way to represent a negative value is to place an extra bit (or digit) in front of the binary number. This extra bit is referred to as a sign bit. The sign bit is set to 1 if the binary number has a negative value and is set to 0 if the binary number has a positive value. Therefore, using a sign bit, the base 10 number +13 is written as 01101 in binary while the number –13 is written as 11101 in binary. This is referred to as signed integer notation and is one way to represent negative numbers in binary.

But how does the computer know that what it is looking at is a negative number and not just a longer binary number? The binary pattern 11101 can represent more than one number. If we know it is a binary number using a sign bit, we read the first bit (the sign bit) as 1 and therefore know that the number is a negative number. Following the sign bit are the digits that represent the value of the number itself. Since 1101 in binary (base 2) has the value 13 in base 10, the final interpretation of the bits 11101 would be –13.

But what if we were told in advance that 11101 is definitely a positive number? We would then read this number differently and compute 1*16 + 1*8 + 1*4 + 0*2 + 1*1 and get the base 10 value of 29. The bits themselves are exactly the same. The only thing that has changed is our agreement on what the same five digits mean: the first time they represented a negative number, and the second time they represented a positive number.

The binary number system can also represent a decimal number. How can a string of 1s and 0s capture the information in a value like 99.368? Because every computer must store such numbers in the same way, the Institute of Electrical and Electronic Engineers (IEEE) has established a standard called the "floating-point" standard that describes how numbers with fractional parts should be represented in the binary number system. Using a 32 bit system, an incredibly wide range of numbers can be represented. The method dictated by the standard works the same for any number with a decimal point, such as the number –0.75. The first digit, or bit (the sign bit), is used to indicate whether the number is positive or negative. The next eight bits store the *magnitude* of the number, indicating whether the number is in the hundreds or millions, for example. The standard says to use the next 23 bits to store the *value* of the number.

As you can imagine, some numbers in binary result in quite a long string of 0s and 1s. For example, the number 123,456 is a 17-digit sequence of 1s and 0s in binary code, 11110001001000000. When working with these long strings of 0s and 1s it is easy for a human to make a mistake. Thus, many computer scientists use **hexadecimal notation**, another commonly used number system, as a form of shorthand.

Hexadecimal notation is a base 16 number system, meaning it uses 16 digits to represent numbers instead of the 10 digits used in base 10 or the 2 digits used in base 2. The 16 digits it uses are the 10 numeric digits, 0 to 9, plus six extra symbols: A, B, C, D, E, F, with each of the letters, A through F, corresponding to a numeric value. So A equals 10, B equals 11, and so on. Looking back at the number we started with: 123,456 is represented as 1E240 in hexadecimal notation. This is much easier for computer scientists to use than the long string of binary code. The scientific calculator in Windows XP (mentioned earlier) can also perform conversions to hexadecimal notation.

When will you ever use hexadecimal notation? Unless you write your own Web pages (where hexadecimal notation is used to represent colors), your only likely encounter with hexadecimal notation will be when you see an error code on your computer. Generally, the location of the error will be represented in hexadecimal notation.

Representing Letters and Symbols: ASCII and Unicode

How can the binary number system represent letters and punctuation symbols? We have just been converting numbers from base 10, that we understand, to base 2, or binary state that the computer understands. Similarly, we need a system that converts letters and other symbols that we understand to a binary state that the

SOUND BYTE
BINARY

This Sound Byte helps remove the mystery surrounding binary numbers. Base conversion between decimal, binary and hexadecimal can be learned interactively using colors, sounds and images.

SOUND BYTE
WHERE DOES BINARY SHOW UP?

In this Sound Byte, you'll learn to use tools that come with the Windows operating system to work with binary, decimal, and hexadecimal numbers. You'll also learn where you might see binary and hexadecimal values showing up as you use a computer.

computer understands. In order to provide a consistent means for representing letters and other characters, there are codes that dictate how to represent characters in binary format. Older mainframe computers use EBCDIC (Extended Binary-Coded Decimal Interchange Code, pronounced "Eb sih dik"). However, most of today's personal computers use the American National Standards Institute (ANSI) standard code, called the **ASCII (American Standard Code for Information Interchange) code**, to represent each letter or character as an 8-bit (or 1-byte) binary code.

As you know by now, binary digits correspond to the on and off states of your computer's switches. Each of these digits is called a **binary digit**, or **bit** for short. Eight binary digits (or bits) combine to create one **byte**. In the above discussions, we have been converting base 10 numbers to a binary format. In such cases, the binary format has no standard length. For example, the binary format for the number 2 is two digits (10) while the binary format for the number 10 is four digits (1010). Although binary numbers can have more or less than 8 bits, each single alphabetic or special character is 1 byte (or 8 bits) of data and consists of a unique combination of a total of eight 0s and 1s. Eight bits is the standard length upon which computers are built.

The ASCII code represents the 26 uppercase letters and 26 lowercase letters used in the English language, along with a number of punctuation symbols and other special characters, using 8 bits. Figure 9.3 shows a number of examples of ASCII code representation of letters and characters.

Can ASCII represent the alphabets of different languages? Because it represents letters and characters using only 8 bits, the ASCII code can assign only 256 (or 2^8) different codes for unique characters and letters. While this is enough to represent English and many other characters found in the world's languages, ASCII code cannot represent *all* languages and symbols. Thus, a new encoding scheme, called **Unicode**, was created. By using 16 bits instead of the 8 bits used in ASCII, Unicode can represent more than 65,000 unique character symbols, enabling it to represent the alphabets of all modern languages and all historic languages and notational systems, including such languages as Tibetan, Tagalog, Japanese, and Canadian-Aboriginal syllabics. As we continue to become a more global society, it is anticipated that Unicode will replace ASCII as the standard character formatting code.

So *all* data inside the computer is stored as bits? Yes! As noted in the Dig Deeper, positive and negative numbers can be stored using signed integer notation, with the first bit (the sign bit) indicating the sign and the rest of the bits indicating the value of the number. Decimal numbers are stored according to the IEEE floating-point standard, while letters and symbols are stored according to the ASCII code or Unicode. All of these different number systems and codes exist so that computers can store different types of information in their on/off switches. No matter what kind of data you input in a computer—a color, a musical note, or a street address—that information will be stored as a string of 1s and 0s. The important lesson is that the *interpretation* of 1s and 0s is what matters. The same binary pattern could represent a postive number, a negative number, a fraction, or a letter.

How does the computer know which interpretation to use for the 1s and 0s? When your brain processes language it takes sounds you hear and uses the rules of English along with other clues to build an interpretation of the sound as a word. If you are in New York City and hear someone shout "Hey Lori!"

FIGURE 9.3 ASCII Standard Code for a Sample of Letters and Characters

ASCII CODE	REPRESENTS THIS SYMBOL	ASCII CODE	REPRESENTS THIS SYMBOL
01000001	A	01100001	a
01000010	B	01100010	b
01000011	C	01100011	c
01011010	Z	00100011	#
00100001	!	00100100	$
00100010	"	00100101	%

you expect someone is saying hello to a friend. If you are in London and hear the same sound—"Hey! Lorry!"—you jump out of the way because a truck is coming at you! You knew which interpretation to apply to the same sound because you had some other information—that you were in England.

Likewise, the CPU is designed to understand a specific language, a set of instructions. But certain instructions tell the CPU to expect a negative number next or to interpret the following bit pattern as a character. Because of this extra information, the CPU always knows which interpretation to use for a series of bits. This all happens without any extra work on the part of the user or the programmer because of the work the CPU design team has done.

The CPU: Processing Digital Information

The **central processing unit** (**CPU** or **processor**), the "brains" of the computer, executes every instruction given to your computer. As you learned earlier, the entire CPU fits on a tiny chip, called the microprocessor, which contains all of the hardware responsible for processing information, including millions of transistors (the switches we discussed earlier).

The CPU is located in the system unit on the computer's **motherboard**, the main circuit board that connects all of the electronic components of the system: the CPU, memory, the expansion slots where you can insert expansion (or adapter)

BITS AND BYTES

Do I Ever See Binary Numbers on My Computer?

Internally, the computer "thinks" in binary numbers and stores all of your data and all the commands it is given in binary code. However, because computers interface with human users, who don't think in terms of binary code, messages and user interfaces are always presented in a style that is more comfortable to us. The only time most computer users ever encounter binary or hexadecimal code is when certain error messages appear, describing what the internal machine settings look like when an error occurred. Although confusing to most users, these strings of code can be useful to service technicians working to understand and correct computer problems. If you ever encounter such an error code, write down the complete error message so that you can better work with technicians to solve your problem.

cards, and all of the electrical paths that connect these components together. Figure 9.4 shows a typical motherboard and the location of each of these components.

Looking at a CPU chip itself gives you very little information about how exactly it accomplishes its work. However, understanding more about

FIGURE 9.4

The motherboard is the home of all the most essential computer hardware, including the CPU, memory, and expansion slots where you can insert expansion (or adapter) cards.

FIGURE 9.5

Externally a CPU chip itself reveals nothing about its internal architecture. (a) The Pentium 4 chip is used in many Windows-based PCs. (b) The Motorola PowerPC G4 chip is used in most Apple Macintosh computers.

how the CPU is designed and how it operates will give you greater insight into how computers work, what their limitations are, and what technological advances may be possible in the future.

What CPUs are used in desktop computers? Only a few major companies manufacture CPUs for desktop computers. Intel manufactures the Xeon, Celeron and Pentium processors (including the Pentium II, III, and 4), shown in Figure 9.5a. Advanced Micro Devices (AMD) produces the AMD-K6 and the Athlon XP processors. Both Intel and AMD chips are used in the majority of Windows-based PCs.

Apple computer systems (such as the iMac and PowerBook series of computers) use a very different CPU design. Until recently, the Motorola PowerPC G4 chip, shown in Figure 9.5b, was the only CPU chip for Apple computers. However, IBM now manufactures the PowerPC G5 chip installed in the newest Apple computers. As you learned in earlier chapters, the processor used on a desktop computer also determines the operating system used. The combination of operating system and processor is referred to as a computer's *platform*.

What makes CPUs different? The primary distinction between CPUs is processing power, which is determined by the number of transistors on each CPU. In addition, as you'll learn in the next section, other factors differentiate CPUs, but the greatest differentiators are how quickly the processor can work (called its *clock speed*) and the amount of immediate access memory the CPU has (called its *cache memory*). Figure 9.6 shows the basic specifications of several of the major processors on the market today.

FIGURE 9.6 Processors on the Market Today

PROCESSOR	MANUFACTURER	NUMBER OF TRANSISTORS	CLOCK SPEED	LEVELS OF CACHE STORAGE	NOTES
Athlon XP	AMD	54.3 million	2.20 GHz	2	AMD's premiere chip. It competes with the Pentium 4.
Centrino	Intel	77 million	1.70 GHz	2	Designed specifically for mobile computers; has built-in wireless local network capabilities.
Itanium 2	Intel	410 million	1.50 GHz	3	The Itanium 2 chip is seen in high-end server computers.
Pentium 4	Intel	55 million	3.20 GHz	2	The latest version of the Pentium chip. It has the largest market in Windows home user PCs.
Pentium 4 Processor-M	Intel	55 million	2.60 GHz	2	The M is for mobile. This chip uses less power so it can run longer on a battery charge.
PowerPC G4	Motorola	57 million	1.00 GHz	3	The processor that powers the Apple line of computers (iMacs, PowerBooks, and so on).
PowerPC G5	IBM	58 million	2.00 GHz	2	The most powerful processor yet for the home computer.

THE CPU MACHINE CYCLE

What exactly does the CPU do? Any program you run on your computer is actually a long series of binary code, 1s and 0s, describing a specific set of commands the CPU must perform. Each CPU is a bit different in the exact steps it follows to perform its tasks, but all CPUs must perform a series of similar general steps. These steps are referred to as a CPU **machine cycle** (or **processing cycle**) and are shown in Figure 9.7:

- **Fetch**: When any program begins to run, the 1s and 0s that make up the program's binary code must be "fetched" from their temporary storage location in random access memory (RAM) and moved to the CPU before they can be executed.
- **Decode**: Once the program's binary code is in the CPU, it is "decoded" into the commands the CPU understands.
- **Execute**: Next, the CPU actually performs the work described in the command. Specialized hardware on the CPU performs addition, subtraction, multiplication, division, and other mathematical and logical operations at incredible speeds.
- **Store:** The result is stored in **registers**, special memory storage areas built into the CPU, which are the most expensive, fastest memory in your computer. The CPU is then ready to fetch the next set of bits encoding the next instruction.

No matter what program you are running, be it an Internet browser or a word processing program, and no matter how many programs you are using at one time, the CPU performs these four steps over and over at incredibly high speeds. Shortly, we'll look at each stage in more detail so you can understand the complexity of the CPU's design, how to compare different CPUs on the market, and what enhancements to expect in CPU designs of the future. But first, let's examine a few other components of the CPU that help it perform its tasks.

The System Clock

How does the CPU know when to begin the next stage in the machine cycle? In order to move from one stage of the machine cycle to the next, the motherboard contains a built-in **system clock**. This internal clock is actually a special crystal that shouts out "Next! . . . Next! . . . Next! . . ." over and over, thereby controlling when the CPU moves to the next stage of processing.

These "ticks" of the system clock, known as the **clock cycle**, set the pace by which the computer moves from process to process. The pace, known as **clock speed**, is measured in hertz (Hz), a unit of measure that describes how many times something happens per second. Today's system clocks are measured in megahertz (MHz), or one million clock ticks per second, and gigahertz (GHz), or one billion clock ticks per second. Therefore, in a 3-GHz system, there are three billion clock ticks each second. Computers with older processors would sometimes need one or more cycles to process one instruction. Today, however, CPUs are designed to handle more

FIGURE 9.7

The CPU Machine Cycle

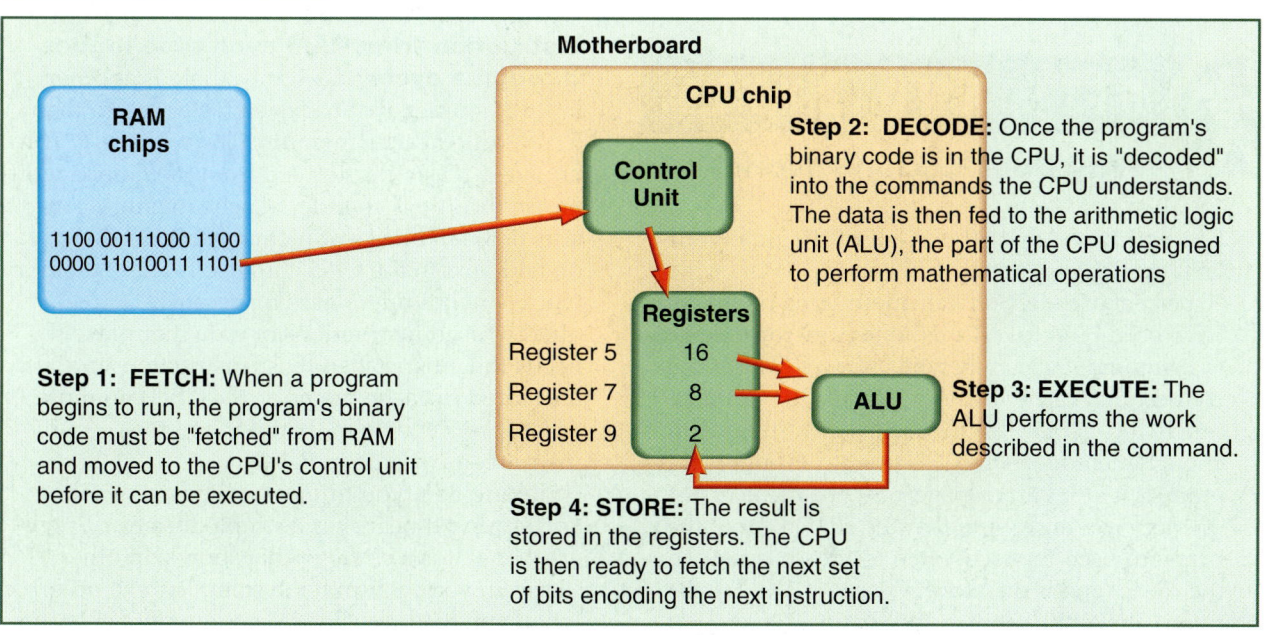

SOUND BYTE
MEMORY HIERARCHY
There are so many different types of memory in a computer—cache and registers and RAM and hard drives. This Sound byte focuses on the differences between each of these types of memory and how they can be upgraded for improved performance.

instructions more efficiently, therefore executing more than one instruction per cycle.

The Control Unit

How does the CPU know which stage in the machine cycle is next? The CPU, like any part of the computer system, is designed from a collection of switches. How can simple on/off switches "remember" the fetch-decode-execute-store sequence of the CPU machine cycle? How can they perform the work required in each of these stages?

The **control unit** of the CPU controls the switches inside the CPU. It is programmed by CPU designers to remember the sequence of processing stages for that CPU and how each switch in the CPU should be set, on or off, for each stage. As soon as the system clock shouts "Next!" the control unit moves each switch to its correct setting (on or off) and then performs the work of that stage.

Let's now look at each of the stages in the machine cycle in a bit more depth.

STAGE 1: THE FETCH STAGE

Where does the CPU find the necessary information? There are different areas in the computer system where the data and program instructions the CPU needs are stored. Data and program instructions move between these areas as needed or not needed by the CPU for processing. Programs (like Microsoft Word) are permanently stored on the hard disk because the hard disk offers nonvolatile storage, meaning the programs remain stored there even when you turn the power off. However, when you launch a program (that is, when you double-click on an icon to execute the program), the program, or sometimes only the essential parts of a program, is transferred from the hard disk into RAM.

The program moves to RAM because the CPU can access the data and program instructions stored in RAM more than one million times faster than if they are left on the hard drive. This is because RAM is much closer to the CPU than is the hard drive. As specific instructions from the program are needed, they are moved from RAM into *registers* (the special storage areas located on the CPU itself) where they wait to be executed.

Why doesn't the CPU chip just contain enough memory to store an entire program? The CPU's storage area is not big enough to hold everything it needs to process at the same time. If enough memory was located on the CPU chip itself, an entire program could be copied to the CPU from RAM before it was executed. This certainly would add to the computer's speed and efficiency since there would not be any delay to stop and fetch instructions from RAM to the CPU. However, including so much memory on a CPU chip would make these chips very expensive. CPU design is so complex that there is only a limited amount of storage space available on the CPU itself.

Cache Memory

So the CPU needs to fetch every instruction from RAM each time it goes through a cycle? Actually, there is another layer of storage that has even faster access than RAM, called **cache memory**. The word *cache* (pronounced "cash") is derived from the French word *cacher*, meaning "to hide." Cache memory consists of small blocks of memory located directly on and next to the CPU chip itself. These memory blocks are holding places for recently or frequently used instructions or data that the CPU needs the most. When these instructions or data are stored in cache memory, the CPU can more quickly retrieve them than if it had to access the instructions or data in RAM.

Taking data you think you'll be using soon and storing it nearby is a simple idea but a powerful one. It is a strategy that shows up other places in your computer system. For example, when you are browsing Web pages, images take

Why Does Caching Work?

Did you know that 80 percent of the time your CPU spends processing, it is working on the same 20 percent of code? Software monitoring programs have been built to test this conjecture for specific systems and it generally holds up well. This is the concept that cache memory exploits. It would be much too expensive to design a system with enough memory on the CPU to store an entire program. But if careful management of a cache of fast memory can make sure that the 20 percent of the program used the most often is already sitting in the cache (and is therefore closer to the CPU), the improvement in overall performance is great.

a long time to download. Your browser software automatically stores images on your hard drive so you don't have to wait to download them again if you want to go back and view a page you've already visited. Although this "cache" of files is not related to the cache storage space designed into the CPU chip, the idea is the same.

How does cache memory work? Modern CPU designs include a number of types of cache memory. If the next instruction to be fetched is not already located in a CPU register, instead of looking directly to RAM to find it, the CPU first searches Level 1 cache. **Level 1 cache** is a block of memory that is built onto the CPU chip for the storage of data or commands that have just been used.

If the command is not located in Level 1 cache, the CPU searches Level 2 cache. Depending on the design of the CPU, **Level 2 cache** is located on the CPU chip itself but is slightly farther away from the CPU, or it's on a separate chip next to the CPU and therefore takes somewhat longer to access. Level 2 cache contains more storage area than does Level 1 cache. For the Intel Pentium 4, for example, the Level 1 cache is 8 KB and the Level 2 cache is 512 KB.

Only if the CPU doesn't find the next instruction to be fetched in either Level 1 or Level 2 cache will it make the long journey to RAM to access it, as shown in Figure 9.8.

Are there any other types of cache memory? The current direction of design in processors is toward larger and larger multilevel CPU cache structures. Therefore, some newer CPUs have an additional third level of cache memory storage, called **Level 3 cache.** On computers with Level 3 cache, the CPU checks this area for instructions and data after it looks in Level 1 and Level 2 cache, but before it makes the longer trip to RAM. The Level 3 cache holds between 2 MB and 4 MB of data. With 4 MB of Level 3 cache, there is almost enough storage for an entire program to be transferred to the CPU for its execution.

How do I use cache memory? As an end user of computer programs, you do nothing special to use cache memory. In fact, you will not even be able to notice that caching is being used—nothing special lights up on your system unit or keyboard. However, the advantage of having more cache memory is that you'll experience better performance as the CPU won't have to make the longer trip to RAM to get data and instructions as often. Unfortunately, because it

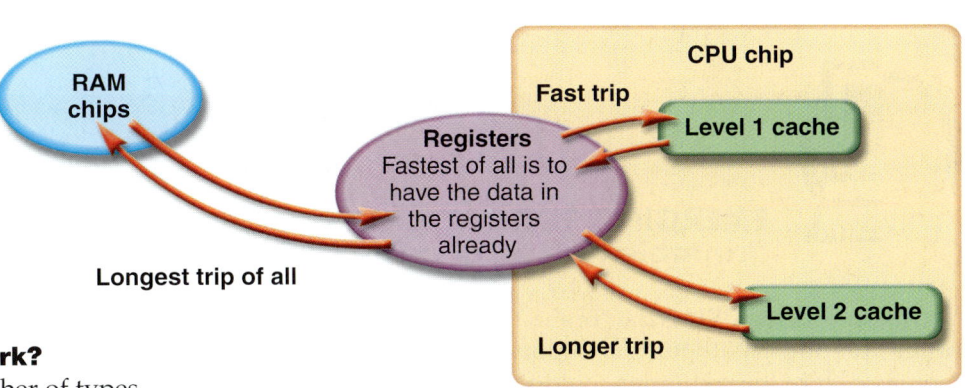

FIGURE 9.8

Modern CPUs have two or more levels of cache memory, which leads to faster CPU processing.

is built into the CPU chip or motherboard itself, you can't upgrade cache: it is part of the original design of the computer system. Therefore, like RAM, it's important when buying a computer to consider buying the one, if everything else is equal, with the most cache memory.

STAGE 2: THE DECODE STAGE

What happens during the decode stage? The main goal of the decode stage is for the CPU's control unit to translate (or decode) the program's instructions (or its lines of binary code) into the commands the CPU can understand. A CPU can only understand a very small set of commands. The collection of commands a specific CPU can execute is called the **instruction set** for that system. Each CPU has its own unique instruction set. For example, the G4 processor used in an Apple PowerBook has a different instruction set than does the Intel Pentium 4 used in a Dell Inspiron laptop. The control unit interprets the code's bits according to the instruction set the CPU designers laid out for that particular CPU. Based on this process of translation, the control unit then knows how to set up all the switches on the CPU so that the proper operation will occur.

What does the instruction set look like? Since humans are the ones to write the instructions initially, all of the commands in an instruction set are written in a language that is easier for humans to work with, called **assembly language.** However, since the CPU only knows and recognizes patterns of 0s and 1s, it cannot understand assembly language, so these human-readable instructions are translated in long strings of binary code. These long strings of binary code, called **machine language,** are used by the control unit to set up the hardware in the CPU for

TRENDS IN IT

EMERGING TECHNOLOGIES:
Printable Processors: The Ultimate in Flexibility

You know that the CPU is the "brains" of the computer. Without this important little chip, the computer couldn't process information. The innovations of the transistor and then the integrated circuit have shrunk the processor to a size so small that even a pen-like instrument can house computer processing capabilities. Miniaturization has made technology very much a part of our lives.

Manufacturing tiny bits of electronic circuitry on silicon is a time-consuming and costly process. But imagine if making microprocessors were as easy as printing them out on your inkjet printer. Or for larger projects, imagine printing computer components out on rolls similar to those that are fed through newspaper presses. Sound crazy? Not to Michael Sauvante and Jim Sheats, founders of Rolltronics Corporation. Their visions of what will one day be possible with computer technology make even the Jetsons seem old-fashioned.

According to Rolltronics, if computer processors could be printed out on common materials, such as flexible plastic or even paper, rather than manufactured on silicon, computers could be cheaper, smaller, and more completely incorporated into objects we use every day. In fact, the computer would be nearly invisible. You might, for example, download a processor from the Internet, then print this processor directly onto a plastic-type substance using your desktop printer. You could then incorporate these plastic-based processors into everything—even wallpaper that could change images or provide lighting for a room.

Printable processors might even have uses you would expect to see in a James Bond movie, such as wearable computers. Your jacket might have a processor with a built-in thermostat that "reads" your body temperature, and your sunglasses could include processors in the lenses that display visual information.

Sauvante and Sheats anticipate their technology will lead to other innovations, such as lightweight medical devices like programmable heart and blood pressure monitors, food cans that could tell you when they are out of date, high-capacity memory devices, ultrathin batteries that are safe and inexpensive, and flexible information devices (like today's PDAs) that could roll up and fit inside your purse or pocket. They also anticipate that printable processors will become extremely cheap, lowering the price as well as the size of most computing devices and helping to close the so-called "digital divide."

Possibilities like these represent only the tip of the iceberg in terms of printable electronics. Another company, Plastic Logic, is also working on printing electronic circuitry onto plastic. In addition to the new and creative applications that plastic microprocessors would produce, Plastic Logic touts an added environmental benefit of printable processors. The technology to make them doesn't use toxic or environmentally damaging materials that are currently used in manufacturing silicon chips. So, in a few years, when the coat you're wearing senses that you're still cold and turns on a built-in heater, you may be able to thank the innovative processes of printable processors!

FIGURE 9.9 Representations of Sample CPU Commands

HUMAN LANGUAGE FOR COMMAND	CPU COMMAND IN ASSEMBLY LANGUAGE (LANGUAGE USED BY PROGRAMMERS)	CPU COMMAND IN MACHINE LANGUAGE (LANGUAGE USED IN THE CPU'S INSTRUCTION SET)
Add	ADD	1110 1010
Subtract	SUB	0001 0101
Multiply	MUL	1111 0000
Divide	DIV	0000 1111

the rest of the operations it needs to perform. Machine language is a binary code for computer *instructions* much like the ASCII code is a binary code for letters and characters. Similar to each letter or character having its own unique combination of 0s and 1s assigned to it, a CPU has a table of codes consisting of combinations of 0s and 1s for each of its commands. If the CPU sees that pattern of bits arrive, it knows the work it must do. Figure 9.9 shows a few commands in both assembly language and machine language.

Many CPUs have similar commands in their instruction sets, including ADD (add), SUB (subtract), MUL (multiply), DIV (divide), MOVE (move data to RAM), STORE (move data to a

Summary

1. **What is a switch and how does it work in a computer?** Electronic switches are devices inside the computer that flip between two states: 1 or 0, on or off. Transistors are switches built out of layers of semiconductor. Integrated circuits (or chips) are very small regions of semiconductor material that support a huge number of transistors. Integrated circuits enable computer designers to fit millions of transistors into a very small area.

2. **What is the binary number system and what role does it play in a computer system?** The binary number system uses only two digits, 0 and 1. It is used instead of the base 10 number system to manipulate the on/off switches that control the computer's actions. Even with just two digits, the binary number system can still represent all the same values that a base 10 number system can. In order to provide a consistent means for representing letters and other characters, there are codes that dictate how to represent characters in binary format. The ASCII code uses 8 bits (0s and 1s) to represent 255 characters. Unicode uses 16 bits of data for each character and can represent more than 65,000 character symbols.

3. **What is inside the CPU and how do these components operate?** The CPU executes every instruction given to your computer. CPUs are differentiated by their processing power (how many transistors are on the microprocessor chip), how quickly the processor can work (called clock speed), and the amount of immediate access memory the CPU has (called cache memory). The CPU consists of two primary units: the control unit controls the switches inside the CPU and the arithmetic logic unit (ALU) performs logical and arithmetic calculations.

4. **How does a CPU process data and instructions?** All CPUs must perform a series of similar general steps. These steps, referred to as a CPU machine cycle, include: fetch (loading program and data binary code into the CPU), decode (translating the binary code into commands the CPU can understand), execute (carrying out the commands), and store (placing the results in special memory storage areas, called registers, before the process starts again).

5. **What is cache memory?** Cache memory consists of small blocks of memory located directly on and next to the CPU chip itself that hold recently or frequently used instructions or data that the CPU needs the most. The CPU can more quickly retrieve data and instructions from cache than from RAM.

6. **What types of RAM are there?** RAM is volatile storage, meaning that when you turn off your computer, the data stored there is erased. The cheapest and most basic type of RAM is DRAM (dynamic RAM). Other types of RAM include SDRAM, DDR RAM, RAMBUS, and RDRAM. All of these forms of RAM store data that the CPU can access quickly.

7. **What is a bus and how does it function in a computer system?** A bus is an electrical wire in the computer's circuitry through which data (or bits) travel between the computer's various components. Local buses are on the motherboard and run between the CPU and the main system memory. Expansion buses expand the capabilities of your computer by allowing a range of different expansion cards to connect to the motherboard. The width of the bus (or the bus width) determines how many bits of data can be sent along a given bus at any one time.

8. **How do manufacturers make CPUs so that they run faster?** Pipelining is a technique that allows the CPU to work on more than one instruction (or stage of processing) at a time, thereby boosting CPU performance. A dual-processor design has two separate CPU chips installed on the same system. Dual- or multiple-processor systems are often used when intensive computational problems need to be solved. In parallel processing, computers in a large network each work on a portion of the same problem at the same time.

Key Terms

Accelerated Graphics Port (AGP) buses	p. 362	Industry Standard Architecture (ISA) bus	p. 361
access time	p. 359	instruction set	p. 357
arithmetic logic unit (ALU)	p. 359	integrated circuits	p. 349
ASCII (American Standard Code for Information Interchange) code	p. 352	Level 1 cache	p. 357
		Level 2 cache	p. 357
		Level 3 cache	p. 357
assembly language	p. 357	local buses	p. 361
base 10 number system (decimal notation)	p. 349	machine cycle (processing cycle)	p. 355
binary digit (bit)	p. 352	machine language	p. 357
binary language	p. 348	microprocessors	p. 349
binary number system (base 2 number system)	p. 350	motherboard	p. 353
		number system	p. 349
bus	p. 361	parallel processing	p. 365
bus width	p. 361	Peripheral Component Interconnect (PCI) bus	p. 362
byte	p. 352		
cache memory	p. 356	pipelining	p. 364
central processing unit (CPU, processor)	p. 353	random access memory (RAM)	p. 359
		read only memory (ROM)	p. 361
clock cycle	p. 355	registers	p. 355
clock speed	p. 355	semiconductor	p. 348
control unit	p. 356	static RAM (SRAM)	p. 361
DRAM (dynamic RAM)	p. 360	system clock	p. 355
dual-processor	p. 365	system unit	p. 348
electronic switches	p. 348	transistors	p. 349
expansion bus	p. 361	Unicode	p. 352
Extended Industry Standard Architecture (EISA) bus	p. 361	vacuum tubes	p. 348
		word size	p. 359
hexadecimal notation	p. 351		

Buzz Words

Word Bank

- binary
- ALU
- AGP
- PCI
- registers
- bytes
- instruction set
- DDR SDRAM
- expansion bus
- ASCII
- control unit
- decoded
- DRAM
- data transfer rates
- buses
- hexadecimal
- fetch
- Level 1 cache
- dual processors
- supercomputer
- number system
- cache
- RDRAM
- pipelining
- Level 2 cache
- Centrino
- PowerPC G5

Instructions: Fill in the blanks using the words from the Word Bank above.

Computers are based on a system of switches, which can be either on or off. The (1)_____ number system, which only has two digits, models this well. A (2)_____ is a set of rules for the representation of numbers. Bits are organized into groups of eight, or (3)_____, so they are easier to work with. The (4)_____ code organizes bytes in unique combinations of 0s and 1s to represent characters, letters, and numerals.

The CPU organizes switches to execute the basic commands of the system. No matter what command is being executed, the CPU steps through the same four processing stages. First it needs to (5)_____ the instruction from RAM. Next the instruction is (6)_____ and the (7)_____ sets up all of the CPU hardware to perform that particular command. The actual execution takes place in the (8)_____. The result is then stored by storing it in the (9)_____ on the CPU. Another form of memory the CPU uses is (10)_____ memory. (11)_____ is the form of this type of memory located closest to the CPU. (12)_____ is located a bit farther from the CPU.

RAM comes in several different types. (13)_____ must be refreshed each cycle to keep the data it stores valid. The pathways connecting the CPU to memory are known as (14)_____. The speeds at which they can move data, or the data transfer rates, vary. (15)_____ is a bus designed primarily to move three-dimensional graphics data quickly.

Organizing Key Terms

Instructions: This chapter introduced many new terms and concepts. In the illustration below, fill in each of the blanks with words from the Word Bank in order to show how categories of ideas fit together.

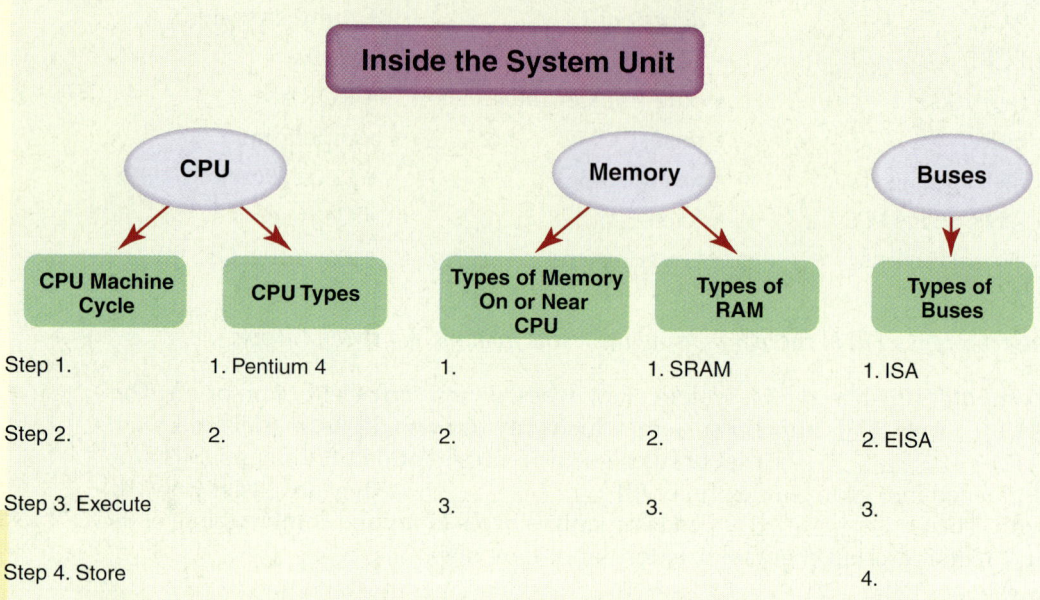

Making the Transition to . . . Next Semester

1. As your collegiate career continues, are you finding your computer needs to do more and more? Upgrading the RAM in your machine can greatly improve performance. What kind of RAM is installed in the computer you use for schoolwork (your own or the lab system you use)? How much would an upgrade to 512 MB of RAM cost for that type of RAM? An upgrade to 1 GB of RAM? Use the supplier Crucial Technology (**www.crucial.com**) to get information about the type of RAM in your system.

2. It is always challenging for administrators to keep computer laboratories up to date. Investigate the type of processor installed in the computer systems you use in your lab. Visit the manufacturer's Web site to get detailed specifications about the design of that processor—how many levels of cache it has, how much total cache memory it has, its speed, and the number of pipeline stages in the CPU. How does that processor compare with the latest model available from the manufacturer?

3. Do some comparative shopping. Pick three relatively comparable, moderately priced computer systems. Create a spreadsheet that outlines all the specific features of each machine. What kind of processor does each machine include? How fast is it? How many levels of cache and how much storage capacity does each cache level have? Look at the RAM: what kind and how much of RAM does each machine have? What is the bus architecture of each machine?

Making the Transition to . . . The Workplace

1. Almost every business today uses networks to connect the computer systems they own. Each machine is assigned its own identifying number, called a network adapter address. This value is a long binary number that uniquely labels each adapter card in the business. On the networked machine, type "winipcfg" in the Run command from the Start menu to find your own network adapter address. Is it presented in binary? hexadecimal? decimal? Why?

2. In an effort to stay technologically current, you have been asked to research the most recently released CPUs to determine whether it's worth buying new machines with the new CPUs or waiting for perhaps the next generation. Investigate the newest CPUs released by Intel, Motorola, and IBM. Compare these new CPUs with the current "best" CPUs. What technological advancements are present in the latest CPUs? From a cost perspective, does it make sense to replace the old systems for these new CPUs? What is the buzz on the next-generation CPUs? Would it be better to wait for these future CPUs to come out?

3. You work in a car assembly plant, so production lines are a familiar concept to your boss. However, he still doesn't understand the concept of pipelining and how it expedites a computer's processing cycle. Create a presentation for your boss that describes pipelining in enough detail so your boss will understand it. In doing so, compare it with the automobile assembly process.

Critical Thinking Questions

Instructions: *Albert Einstein used "Gedanken experiments," or critical thinking questions, to develop his theory of relativity. Some ideas are best understood by experimenting with them in our own minds. The following critical thinking questions are designed to demand your full attention but require only a comfortable chair—no technology.*

1. Consider the current limitations of the design of memory, how it is organized, and how a CPU operates. Think radically—what extreme ideas can you propose for the future of processor design? What do you think the limit of clock speed for a processor will be? How could a CPU communicate more quickly with memory? What could future cache designs look like?

2. SIMD and 3DNow! (used by AMD in its processors) are two approaches to modifying the instruction set to speed up graphics operations. What do you think will be the next important type of processing users will expect from computers? How could you customize the commands the CPU understands so that processing occurs faster on the CPU you are designing?

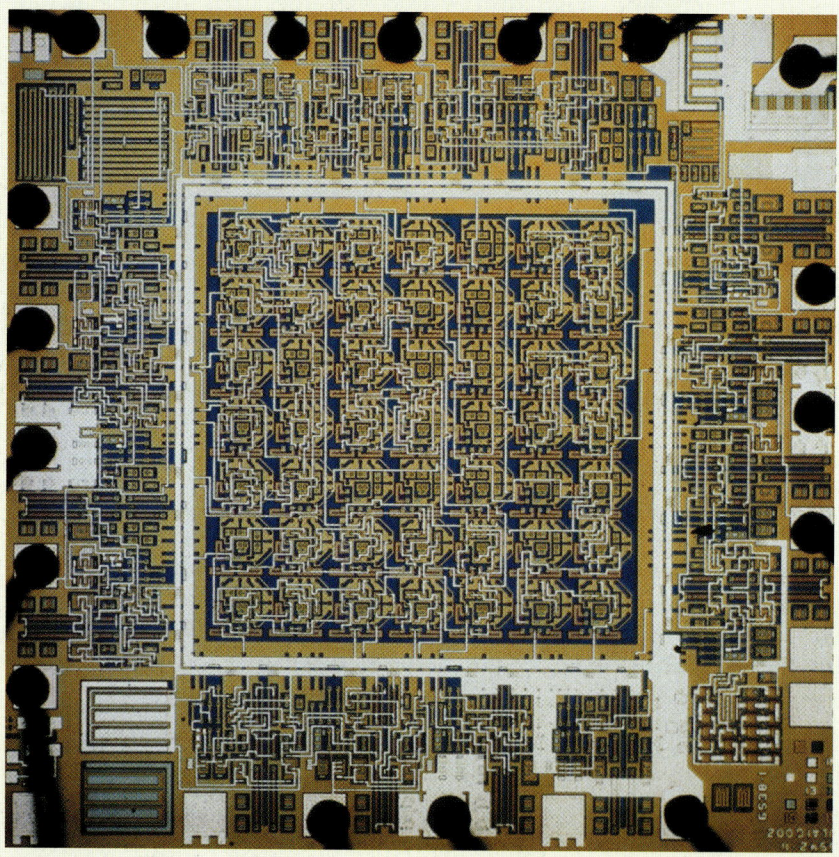

Team Time

Problem:

For a system to be effective, it must be balanced—that is, the performance of each subsystem must be well matched so that there are no bottlenecks in the overall performance. In this exercise, teams will develop balanced hardware designs for specific systems within several different price ranges.

Task:

Each group will select one part of a computer—either the CPU, the memory, or the bus architecture. The group will be responsible for researching the available options and collecting information on both price and performance specifications. The group will write a report that recommends a specific product for each of three price ranges—entry level, mid-range and high performance. Finally the three groups will combine their reports so that the team has developed a specification for the entire system.

Process:

STEP 1: Divide into three groups: processor, memory, and bus design.

STEP 2: Consider the following three price ranges:

- $500 to $1000
- $1001 to $2500
- Unlimited

For each price range, try to specify at least two components that would keep the system cost in range and would provide the best performance. Keep track of all performance information so you can later meet with the other groups and make sure each subsystem is well matched.

STEP 3: Bring the research materials from the group meetings to one final team meeting. Looking at the system level, make final decisions on the system design for each of the three price ranges. Each range is the sum of total cost that can be expended for hardware for the system unit (monitor and other peripherals not included). The system case, power supply, motherboard, RAM, video card, and storage must be included. Research vendors that supply parts to home developers, such as TigerDirect.com.

STEP 4: Produce a report that documents the decisions and trade-offs evaluated en route to your final selections.

Conclusion:

A performance increase in one subsystem contributes to the overall performance, but only in proportion to how often it is used. This affects system design and how limited financial resources can be spent to provide the most balanced, best performing system.

374 CHAPTER 9 BEHIND THE SCENES: INSIDE THE SYSTEM UNIT

Becoming Computer Fluent

Some Apple Macintosh users develop a "religious" attachment to their machines. Apple has a reputation for elegant design but has made other business decisions that have limited its penetration into the business and home marketplace. Several of the revolutionary features of operating system and hardware design that have begun at Apple are now available from many PC clone vendors. Your new boss is unsure what the differences between high-end systems are and would like you to compile a report on two high-end systems, a G5 Macintosh and a Windows-based PC. She has asked you to compare the price/performance ratio, the hardware features including CPU design, and memory capacities.

Instructions: Using the scenario above, write a report using as many of the words from the Word Bank as you can. Be sure the sentences are grammatically correct and technically meaningful.

Materials on the Web

In addition to the review materials presented here, you'll find extra materials on the book's companion Web site (**www.prenhall.com/techinaction**) that will help reinforce your understanding of the chapter content. These materials include:

Sound Byte Lab Guides

For each Sound Byte mentioned in the chapter, there is a corresponding lab guide located on the book's companion Web site. These guides review the material presented in the Sound Byte and direct you to various Web resources that examine the material. The Sound Byte Lab Guides for this chapter include:

- Binary
- Where Does Binary Show Up?
- Memory Hierarchy
- Computer Architecture

True/False and Multiple Choice Quizzes

The book's Web site includes a True/False and a Multiple Choice quiz for this chapter. You can take these quizzes, automatically check the results, and e-mail the results to your instructor.

Web Research Projects

The book's Web site also includes a number of Web research projects for this chapter. These projects ask you to search the Web for information on computer-related careers, milestones in computer history, important people and companies, emerging technologies, and the applications and implications of different technologies.

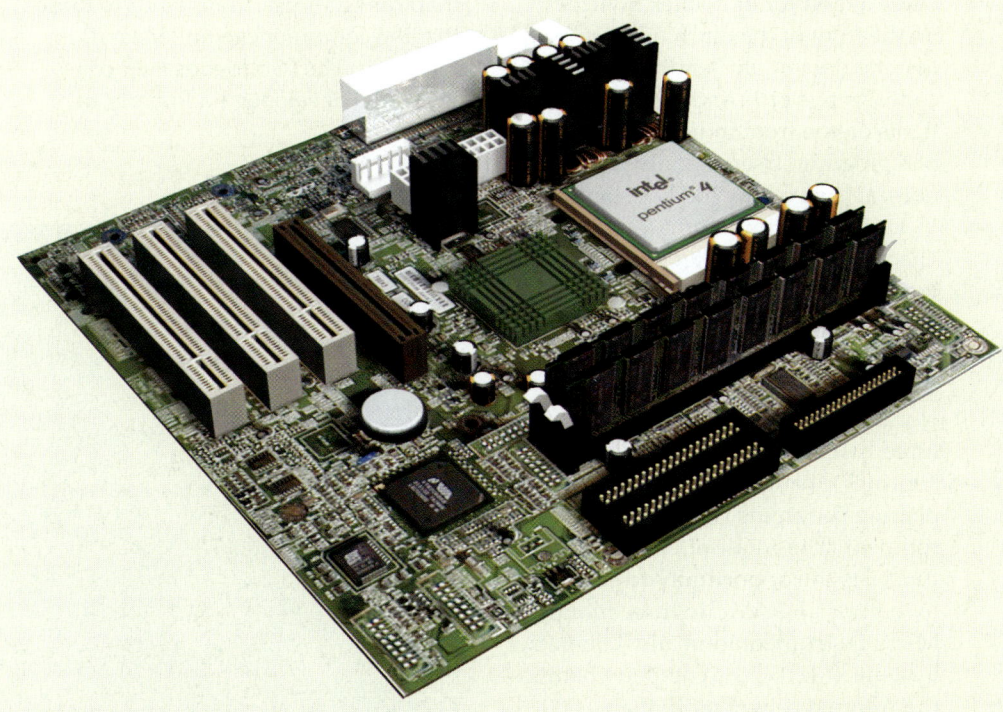

Technology in Focus: THE HISTORY

DO YOU EVER WONDER how big the first personal computer was, or how much the first portable computer weighed? Computers are such an integral part of our lives that we don't often stop to think about how far they've come or where they got their start. But in just 30 years, computers have evolved from expensive, huge machines that only corporations owned, to small, powerful devices found in millions of homes. In this Special Feature, we'll look at the history of the computer. Along the way, we'll discuss some developments that helped make the computer powerful and portable, as well as the people who contributed to its development. But first, we'll start with the story of the personal computer and how it grew to be as integral to our lives as the automobile.

FIGURE 1 In 1975, the Altair was touted as the "world's first minicomputer" in the January issue of *Popular Electronics*.

The First Personal Computer: The Altair

Our journey through the history of the personal computer starts in 1975. At that time, most people were unfamiliar with the "mainframes" and "supercomputers" that large corporations and the government owned. With price tags exceeding the cost of buildings, and with few if any practical home uses, these monster machines were not appealing or attainable to the vast majority of Americans. But the January 1975 cover of *Popular Electronics* announced a change to that with the debut of the **Altair 8800**, touted as the first "personal computer" (see Figure 1). For just $395 for a do-it-yourself kit or $498 for a fully assembled unit (about $1,000 in today's dollars), the price was reasonable enough so that computer fanatics could finally own their own computer.

The Altair was a very primitive computer, with just 256 bytes (not *kilo*bytes, just bytes) of memory. It didn't come with a keyboard, nor did it include a monitor or printer. Switches on the front of the machine were used to enter data in unfriendly machine code (strings of 1s and 0s). Flashing lights on the front indicated the results of a program. User-friendly it was not—at least not by today's standards.

Despite its limitations, computer "hackers" (as computer enthusiasts were called then) flocked to the machine. Many who bought the Altair had been taught to program, but until that point had access only to big, clumsy computers. They were often hired by corporations to program "boring" financial, statistical, or engineering programs in a workplace environment. The Altair offered these enthusiasts the opportunity to create their own programs. Within three months, Micro Instrumentation and Telemetry Systems (MITS), the company behind the Altair, received more than 4,000 orders for the machine.

The release of the Altair marked the start of the personal computer (PC) boom. In fact, two men who would play large roles in the development of the PC were among the first Altair owners. Recent high school grads Bill Gates and Paul Allen were so enamored by this "minicomputer" as these personal computers were called at the time, that they wrote a compiling program (a program that translates user commands to those that the computer can understand) for the Altair. The two friends later convinced its developer, Ed Roberts, to

BITS AND BYTES

Why Was It Called the "Altair"?

For lack of a better name, the Altair's developers originally called the computer the PE-8, short for Popular Electronics 8-bit. However, Les Soloman, the *Popular Electronics* writer who introduced the Altair, wanted the machine to have a catchier name. The author's daughter, who was watching *Star Trek* at the time, suggested the name Altair (that's where the *Star Trek* crew was traveling that week). The first star of the PC industry was born.

buy their program. This marked the start of a small company called Microsoft. But we'll get to that story later. First, let's see what their future archrivals were up to.

The Apple I and II

Around the time the Altair was released, **Steve Wozniak,** an employee at Hewlett-Packard, was becoming fascinated with the burgeoning personal computer industry and was dabbling with his own computer design. He would bring his computer prototypes to meetings of the "Homebrew Computing Club," a group of young computer fans who met to discuss computer ideas in Palo Alto, California. **Steve Jobs,** who was working for computer game manufacturer Atari at the time, liked Wozniak's prototypes and made a few suggestions. Together, the two built a personal computer, later known as the **Apple I,** in Wozniak's garage (see Figures 2 and 3).

OF THE PC

FIGURE 2 Steve Jobs and Steve Wozniak were two computer hobbyists who worked together to form the Apple Computer Company.

FIGURE 3 The first Apple computer, the Apple I, looked like a typewriter in a box. It was one of the first computers to incorporate a keyboard.

FIGURE 4 The Apple II came with the addition of a monitor and an external floppy disk drive.

Why Is It Called "Apple"?

Steve Jobs wanted Apple Computer to be the "perfect" computer company. Having recently worked at an apple orchard, Jobs thought of the apple as the "perfect" fruit—it was high in nutrients, came in a nice package, and was not easily damaged. Thus he and Wozniak decided to name their new computer company Apple.

BITS AND BYTES

In that same year, on April 1, 1976, Jobs and Wozniak officially formed the **Apple Computer Company**.

No sooner had the Apple I hit the market, Wozniak was working to improve it. A year later, in 1977, the **Apple II** was born (see Figure 4). The Apple II included a color monitor, sound, and "game paddles." Priced around $1,300 (quite a bit of money in those days), it included 4 KB of random access memory (RAM) as well as an optional floppy disk drive that enabled users to run additional programs. Most of these programs were games. However, for many users, there was a special appeal to the Apple II: the program that made the computer function when the power was first turned on was stored in read only memory (ROM). Previously, such "routine task" programs had to be rewritten every time the computer was turned on. This "automation" made it possible for the least technical computer enthusiast to write programs.

An instant success, the Apple II would be the most successful in the company's line, outshining even its successor, the **Apple III,** released in 1980. Eventually, the Apple II included a spreadsheet program and word processing and desktop publishing software. These programs gave personal computers like the Apple functions beyond just gaming and special programming, leading to their increased popularity. We'll talk more about these advances later. For now, other players were entering the market.

TECHNOLOGY IN FOCUS
Enter the Competition

Around the time Apple was experiencing success with its computers, a number of competitors entered the market. The largest among them were Commodore, Radio Shack, and IBM. As Figure 5 shows, just years after the introduction of the Altair, the market was filled with personal computers from a variety of manufacturers.

FIGURE 5 Personal Computer Development

YEAR	APPLE	IBM	OTHERS
1975			MITS Altair
1976	Apple I		
1977	Apple II		Tandy Radio Shack's TRS-80 Commodore PET
1980	Apple III		
1981		IBM PC	Osborne
1983	Lisa		
1984	Macintosh	286-AT	IBM PC clones

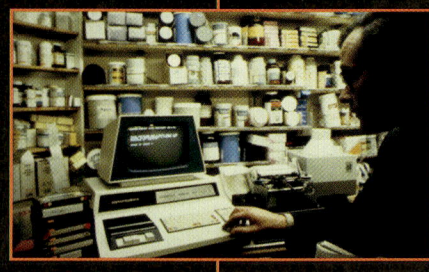

FIGURE 6 The Commodore PET was well received because of its "all-in-one" design.

FIGURE 7 The TRS-80 hid its circuitry under the keyboard. The computer was nicknamed "trash-80," which was more a play on its initials than a reflection of its capabilities.

FIGURE 8 The Osborne was introduced as the first "portable" personal computer. It weighed a whopping 24.5 pounds and contained just 64 KB of memory.

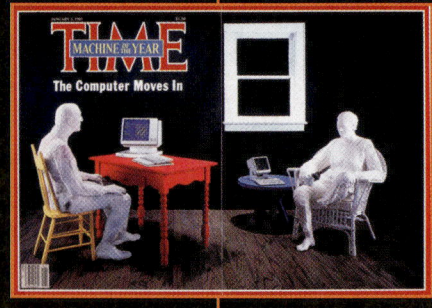

FIGURE 9 The IBM PC was the first (and only) non-human object chosen as "man of the year" (actually, "machine of the year") by *Time Magazine* in its January 1983 issue. This designation indicated the impact the PC was having on the general public.

THE COMMODORE PET AND TRS-80

Among Apple's strongest competitors were the **Commodore PET 2001,** shown in Figure 6, and Tandy Radio Shack's **TRS-80,** shown in Figure 7. Commodore introduced the PET in January 1977. It was featured on the cover of *Popular Science* in October 1977 as the "new $595 home computer." Tandy Radio Shack's home computer also garnered immediate popularity. Just one month after its release in 1977, the TRS-80 sold approximately 10,000 units. Priced at $599.95, the easy-to-use machine included a color display and 4 KB of memory. Many other manufacturers followed suit over the next decade, launching new desktop products, but none were as successful as the TRS-80 and the Commodore.

THE OSBORNE

The Osborne Company introduced the **Osborne** in April 1981 as the industry's first "portable" computer (see Figure 8). Although portable, the computer weighed 24.5 pounds, and its screen was just 5-inches wide. In addition to its hefty weight, it came with a hefty price tag of $1,795. Still, the Osborne included 64 KB of memory, two floppy disk drives, and software programs installed (such as word processing and spreadsheet software). The Osborne was an overnight success, with sales quickly reaching 10,000 units per month. However, despite the Osborne's popularity, the release of a successor machine, called the **Executive,** reduced sales of the Osborne significantly, and the Osborne Company eventually closed. Compaq bought the Osborne design and later produced its first portable in 1983.

IBM PCS

By 1980, IBM recognized it needed to get its feet into the personal computer market. Up until that point, the company had been a player in the computer industry, but primarily with mainframe computers, which it sold only to large corporations. It had not taken the smaller, personal computer seriously. In August 1981, however, IBM released its first personal computer, appropriately named the **IBM PC.** Because many companies were already familiar with IBM mainframes, they readily adopted the IBM PC. The term "PC" soon became the term used to describe personal computers.

The IBM PC came with up to 256 KB of memory and started at $1,565. IBM marketed its PC through retail outlets such as Sears and Computerland in order to reach the home market, and it quickly dominated the playing field. In January 1983, *Time Magazine,* playing on its annual "man of the year" issue, named the computer "1982 machine of the year" (see Figure 9).

Other Important Advancements

It was not just the *hardware* of the personal computer that was developing during the 1970s and 1980s. At the same time, advances in programming languages and operating systems and the influx of application software were leading to more useful and powerful machines.

The Importance of BASIC

The software industry began in the 1950s with the development of programming languages such as FORTRAN, ALGOL, and COBOL. These languages were used mainly by businesses to create financial, statistical, and engineering programs for corporate enterprises. But the 1964 introduction of **BASIC (Beginners All-purpose Symbolic Instruction Code)** revolutionized the software industry. BASIC was a programming language that the beginning programming student could easily learn. It thus became enormously popular—and the key language of the PC. In fact, **Bill Gates** and **Paul Allen** (see Figure 10) used BASIC to write the program for the Altair. As we noted earlier, this program led to the creation of **Microsoft,** a company that produced software for the "micro" computer.

The Advent of Operating Systems

Since data on the earliest personal computers was stored on audiocassettes (not floppies), many programs were not saved or reused. Rather, programs were rewritten as needed. Then Steve Wozniak developed a floppy disk drive called the **Disk II,** which he introduced in July 1978. With the introduction of the floppy drive, programs could be saved with more efficiency, and operating systems (OSs) developed.

OSs were (and still are) written to coordinate with the specific processor chip that controlled the computer. Apples run on a Motorola chip, while PCs (IBMs and so on) run on an Intel chip. **DOS (Disk Operating System),** developed by Wozniak and introduced in December 1977, was the OS that controlled the first Apple computers. The **CP/M (Control Program for Microcomputers),** developed by Gary Kildall, was the first OS designed for the Intel 8080 chip (the processor for PCs). Intel hired Kildall to write a compiling program for the 8080 chip, but Kildall quickly saw the need for a program that could store computer operating instructions on a floppy disk rather than on a cassette. Intel wasn't interested in buying the CP/M program, but Kildall saw a future for the program and thus founded his own company, Digital Research.

In 1980, when IBM was considering entering the personal computer market, it approached Bill Gates at Microsoft to write an OS program for the IBM PC. Although Gates had written versions of BASIC for different computer systems, he had never written an OS. He therefore recommended IBM investigate the CP/M OS, but no one from Digital Research returned IBM's call. Microsoft then jumped on the opportunity and developed **MS-DOS** for IBM computers. (This was one phone call Digital Research certainly regrets not returning!)

MS-DOS was based on an OS developed by the Seattle Computer Products company called **QDOS (Quick and Dirty Operating System).** Microsoft bought the nonexclusive rights to QDOS and distributed it to IBM. Eventually, virtually all personal computers running on the Intel chip used MS-DOS as their OS, and Microsoft's reign as one of the dominant players in the PC landscape had begun. Meanwhile, many other software programs were being developed, taking personal computers to the next level of user acceptance.

FIGURE 10 Paul Allen and Bill Gates are the founders of Microsoft.

The Software Application Explosion: VisiCalc and Beyond

Inclusion of floppy disk drives in personal computers not only facilitated the storage of operating systems, but also set off a software application explosion, since the floppy disk was a convenient way to distribute software. Around that same time, in 1978, Harvard Business School student Dan Bricklin recognized the potential for a spreadsheet program that could be used on PCs. He and his friend Bob Frankston (see Figure 11) thus created the program **VisiCalc**. VisiCalc not only became an instant success, it was also one of the main reasons for the rapid increase in PC sales. Finally, ordinary home users could see how owning a personal computer could benefit their lives. More than 100,000 copies of VisiCalc were sold in its first year.

After VisiCalc, other electronic spreadsheet programs entered the market. **Lotus 1-2-3** came on the market in 1982, and **Microsoft Excel** entered the scene in 1985. These two products became so popular that they eventually put VisiCalc out of business.

Meanwhile, word processing software was gaining a foothold in the PC industry. Up to this point, there were separate, dedicated word processing machines, and the thought hadn't occurred to enable the personal computer to do word processing. Personal computers, it was believed, were for "computation" and "data management." However, once **WordStar**, the first word processing application, came out in disk form in 1979 and was available on personal computers, word processing became another important use for the PC. In fact, word processing is now one of the most common PC applications. Competitors such as **Word for MS-DOS** (the precursor to Microsoft Word) and **WordPerfect** soon entered the market. Figure 12 lists some of the important dates in software application development.

FIGURE 11 Dan Bricklin and Bob Frankston created VisiCalc, the first business application developed for the personal computer.

FIGURE 12 Software Application Development

YEAR	APPLICATION
1979	**VisiCalc:** First electronic spreadsheet application **WordStar:** First word processing application
1980	**WordPerfect:** Thought to be the best word processing software for the PC. WordPerfect was eventually sold to Novell, then later acquired by Corel
1982	**Lotus 1-2-3:** Added integrated charting, plotting, and database capabilities to spreadsheet software
1983	**Word for MS-DOS:** Introduced in *PC World* magazine with first magazine-inserted demo disk
1985	**Excel:** One of the first spreadsheets to use a graphical user interface **PageMaker:** First desktop publishing software

The Graphical User Interface

Another important advancement in personal computers was the introduction of the **graphical user interface (GUI),** which allowed users to more easily interact with the computer. Until that time, users had to use complicated command- or menu-driven interfaces to interact with the computer. Apple was the first company to take full commercial advantage of the GUI, but competitors were fast on its heels, and soon the GUI became synonymous with personal computers. But who developed the idea of the GUI? You'll probably be surprised to learn that a company known for its photocopiers was the real innovator.

Xerox

In 1972, a few years before Apple had launched its first PC, photocopier manufacturer **Xerox** was hard at work in its Palo Alto Research Center (PARC) designing a personal computer of its own. Named the **Alto** (shown in Figure 13), the computer included a word processor, based on the "What You See Is What You Get" (WYSIWYG) principle, that was a file management system with direc-

FIGURE 13 The Alto was the first computer to use a graphical user interface, and it provided the basis for the GUI Apple used. However, due to marketing problems, the Alto never was sold.

THE GRAPHICAL USER INTERFACE

tories and folders. It also had a mouse and could connect to a network. None of the other personal computers of the time had any of these features. Still, for a variety of reasons, Xerox never sold the Alto commercially. Several years later, it developed the Star Office System, which was based on the Alto. Despite its convenient features, the Star never became popular, as no one was willing to pay the $17,000 asking price.

The Lisa and the Macintosh

Xerox's ideas were ahead of their time. But many of the ideas of the Alto and Star would soon catch on. In 1983, Apple introduced the **Lisa,** shown in Figure 14. Named after Apple founder Steve Job's daughter, the Lisa was the first successful PC brought to market to use a GUI. Legend has it that Jobs had seen the Alto during a visit to PARC in 1979 and was influenced by its GUI. He therefore incorporated a similar user interface into the Lisa, providing features such as windows, drop-down menus, icons, a hierarchical file system with folders and files, and a point-and-click device called a mouse. The only problem with the Lisa was its price. At $9,995 ($20,000 in today's dollars), few buyers were willing to take the plunge.

THE INTERNET BOOM

The GUI made it easier for users to work on the computer. The Internet provided another reason for consumers to buy computers. Now they could conduct research and communicate with each other in a new and convenient way. In 1993, the web browser **Mosaic** was introduced. This browser allowed users to view multimedia on the web, causing Internet traffic to increase by nearly 350 percent. Meanwhile, companies discovered the Internet as a means to do business, and computer sales took off. IBM-compatible PCs became the personal computer system of choice when, in 1995, Microsoft (the predominant software provider to PCs) introduced **Internet Explorer,** a web browser that integrated web functionality into Microsoft Office applications, and **Windows 95,** the first Microsoft OS not based on MS-DOS.

A year later, in 1984, Apple introduced the **Macintosh,** shown in Figure 15. The Macintosh was everything the Lisa was and then some, and at about a third of the cost. The Macintosh was also the first personal computer to introduce 3.5-inch floppy disks with a hard cover, which were smaller and sturdier than the previous 5.25-inch floppies.

FIGURE 14 The Lisa was the first computer to introduce a GUI to the market. Priced too high, it never gained the popularity it deserved.

FIGURE 15 The Macintosh became one of Apple's best-selling computers, incorporating a graphical user interface along with other innovations such as the 3.5-inch floppy disk drive.

TECHNOLOGY IN FOCUS
Making the PC Possible: Early Computers

Since the first Altair was introduced in the 1970s, more than a billion personal computers have been distributed around the globe. Because of the declining prices of computers and the growth of the Internet, it's estimated that a billion more computers will be sold within the next decade. But what made all this possible? The computer is a compilation of parts, all of which are the result of individual inventions. From the earliest days of humankind, we have been looking for a more systematic way to count and calculate. Thus, the evolution of counting machines has led to the development of the computer we know today.

THE PASCALENE AND THE JACQUARD LOOM

The **Pascalene** was the first accurate mechanical calculator. This machine, created by the French mathematician **Blaise Pascal** in 1642, used revolutions of gears to count by tens, similar to odometers in cars. The Pascalene could be used to add, subtract, multiply, and divide. The basic design of the Pascalene was so sound that it lived on in mechanical calculators for more than 300 years.

Nearly 200 years later, **Joseph Jacquard** revolutionized the fabric industry by creating a machine that automated the weaving of complex patterns. While not a counting or calculating machine, the **Jacquard Loom** (shown in Figure 16) was significant because it relied on stiff cards with punched holes to automate the process. Much later this process would be adopted as a means to record and read data via punch cards in computers.

16

17

BABBAGE'S ENGINES

Decades later, in 1834, **Charles Babbage** designed the first automatic calculator, called the **Analytical Engine**. The machine was actually based on another machine, called the **Difference Engine**, which was a huge steam-powered mechanical calculator Babbage designed to print astronomical tables. Babbage stopped working on the Difference Engine to build the Analytical

FIGURE 16 The Jacquard Loom used holes punched in stiff cards to make complex designs. This technique would later be used in the form of punch cards to control the input and output of data in computers.
FIGURE 17 The Analytical Engine, designed by Charles Babbage, was never fully developed, but included components similar to those found in today's computers. **FIGURE 18** The Atanasoff-Berry Computer laid the design groundwork for many computers to come.

18

Engine. Although it was never developed, Babbage's detailed drawings and descriptions of the machine include components similar to those found in today's computers, including the store (RAM), the mill (central processing unit), as well as input and output devices. This invention gave Charles Babbage the title of the "father of computing."

THE HOLLERITH TABULATING MACHINE

In 1890, **Herman Hollerith**, while working for the U.S. Census Bureau, was the first to take Jacquard's punch card concept and apply it to computing. Hollerith developed a machine called the **Hollerith Tabulating Machine** that used punch cards to tabulate census data. Up until that time, census data had been tabulated in a long, laborious process. Hollerith's tabulating machine automatically read data that had been punched onto small punch cards, speeding up the tabulation process. Hollerith's machine became so successful that he left the Census Bureau in 1896 to start the Tabulating Machine Company. His company later changed its name to International Business Machines, or IBM.

THE Z1 AND ATANASOFF-BERRY COMPUTER

German inventor **Konrad Zuse** is credited with a number of computing inventions. His first, in 1936, was a mechanical calculator called the **Z1**. The Z1 is thought to be the first computer to include features that are integral to today's systems, including a control unit and separate memory functions, noted as important breakthroughs for future computer design.

In late 1939, John Atanasoff, a professor at Iowa State University, and his student, Clifford Berry, built the first electrically powered digital computer, called the **Atanasoff-Berry Computer (ABC)**, shown in Figure 18. The computer was the first to use vacuum tubes to store data instead of the mechanical switches used in older computers. Although revolutionary at its time, the machine weighed 700 pounds, contained a mile of wire, and took about 15 seconds for each calculation. (In comparison, today's personal computers can calculate more than 300 billion operations in 15 seconds.) Most importantly, the ABC computer was the first to use the binary system. It was also the first to have memory that repowered itself upon booting. The design of the ABC would end up being central to that of future computers.

THE HARVARD MARK I

From the late 1930s to the early 1950s, **Howard Aiken** and **Grace Hopper** designed the Mark series of computers at Harvard University. The U.S. Navy used these computers for ballistic and gunnery calculations. Aiken, an electrical engineer and physicist, designed the computer, while Hopper did the programming. The **Harvard Mark I**, finished in 1944, could perform all four arithmetic operations (addition, subtraction, multiplication, and division).

However, many believe Hopper's greatest contribution to computing was the invention of the **compiler**, a program that translates English language instructions into computer language. The team was also responsible for a common computer-related expression. Hopper was the first to "debug" a computer when she removed a moth that had flown into the Harvard Mark I, causing it to break down (see Figure 19). Thus she coined the term "computer bug."

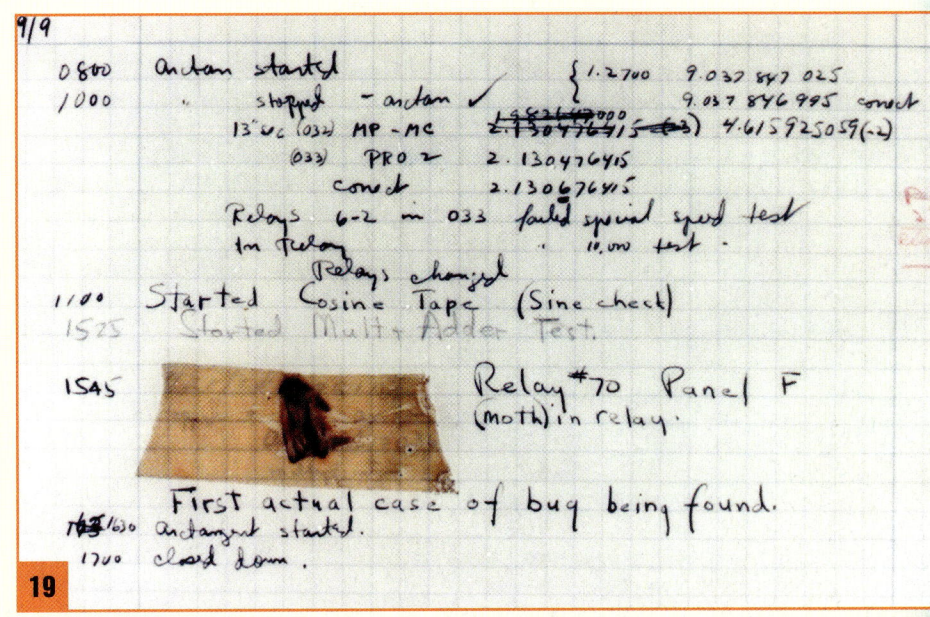

FIGURE 19 Grace Hopper coined the term "computer bug" when she removed a moth that had flown into the Harvard Mark I, causing it to break down.

THE TURING MACHINE

Meanwhile, in 1936, the British mathematician **Alan Turing** created an abstract computer model that could perform logical operations. The **Turing Machine** was not a real machine but rather a hypothetical model that mathematically defined a mechanical procedure (or algorithm). Additionally, Turing's concept described a process by which the machine could read, write, or erase symbols written on squares of an infinite paper tape. This concept of an infinite tape that could be read, written to, and erased was the precursor to today's RAM.

THE ENIAC

The **ENIAC (Electronic Numerical Integrator and Computer)**, shown in Figure 20, was another U.S. government-sponsored machine developed to calculate the settings used for weapons. Created by

TECHNOLOGY IN FOCUS

John W. Mauchly and **J. Presper Eckert** at the University of Pennsylvania, it was placed in operation in June 1944. Although the ENIAC is generally thought of as the first successful high-speed electronic digital computer, it was big and clumsy. The ENIAC used nearly 18,000 vacuum tubes and filled approximately 1,800 square feet of floor space. Although inconvenient, the ENIAC served its purpose and remained in use until 1955.

THE UNIVAC

The **Universal Automatic Computer, or UNIVAC**, was the first commercially successful electronic digital computer. Completed in June 1951 and owned by the company Remington Rand, the UNIVAC operated on magnetic tape (as opposed to its competitors, which ran on punch cards). The UNIVAC gained notoriety when, in a 1951 publicity stunt, it was used to predict the outcome of the Stevenson-Eisenhower presidential race. By analyzing only 5 percent of the popular vote, the UNIVAC correctly identified Dwight D. Eisenhower as the victor. After that, the UNIVAC soon became a household name. The UNIVAC and computers like it were considered **first-generation computers**, and were the last to use vacuum tubes to store data.

TRANSISTORS AND BEYOND

Only a year after the ENIAC was completed, scientists at the Bell Telephone Laboratories in New Jersey invented the **transistor** as a means to store data. The transistor replaced the bulky vacuum tubes of earlier computers, and was smaller and more powerful. It was used in almost everything, from radios to phones. Computers that used transistors were referred to as **second-generation computers**. Still, transistors were limited as to how small they could be made.

A few years later, in 1958, **Jack Kilby**, while working at Texas Instruments, invented the world's first **integrated circuit**, a small chip capable of containing thousands of transistors. This consolidation in design enabled computers to become smaller and lighter. The computers in this early integrated circuit generation were considered **third-generation computers**.

Other innovations in the computer industry further refined the computer's speed, accuracy, and efficiency. However, none were as significant as the 1971 introduction by the Intel Corporation of the **microprocessor chip**, a small chip containing millions of transistors. The microprocessor functions as the CPU, or brains, of the computer. Computers that used a microprocessor chip were called **fourth-generation computers**. Over time, Intel and Motorola became the leading manufacturers of microprocessors. Today, the Intel Pentium 4 chip, shown in Figure 21, contains more than 42 million transistors.

FIGURE 20 The ENIAC took up an entire room and required several people to manipulate it.
FIGURE 21 The Pentium 4 chip contains more than 42 million transistors.

AS YOU CAN SEE, personal computers have come a long way since the Altair, and have a number of inventions and people to thank for their amazing popularity. What will the future bring? If current trends continue, computers will be smaller, lighter, and more powerful. The advancement of wireless technology will also certainly play a big roll in the development of the personal computer.

What Are the Computers of Tomorrow?

So now you know where today's computers came from. But what will the computers of tomorrow look like? They will most likely continue two trends that led to the development of today's computers: 1) Computers will become smaller and even more portable, and 2) Different materials will be employed in their construction. Let's look at two interesting possibilities.

Shrinking Computers: How Small Will Computers Eventually Be?

With the rise of nanoscience, we may one day have nearly invisible computers. Already, "smart dust"—tiny machines that can detect and send data—is being developed. These machines are made up of small particles of silicon embedded with microscopic detection equipment. Although the current size of the particles is about 100 cubic millimeters (the size of a vitamin pill), scientists hope to create dust smaller than 1 cubic millimeter, making it fairly undetectable.

So what will smart dust do? Sensors can be embedded in the dust so that it can detect metal objects or toxins. The military is interested in spreading the dust on battlefields to track enemy troop movements, and the government could spread the dust in public areas to act as a warning system for nerve or germ warfare agents.

To gather and transmit data, the dust requires a power source and an operating system. Solar cell–powered dust is being tested, but so far the cells are too big and low powered. Eventually, nuclear power may provide a solution. Dust Inc., a company developing smart dust, has designed an OS that uses little memory, making it suitable for small objects. No word yet on a competing Microsoft product (Dusty Windows?)!

Once smart dust can be manufactured in large quantities and made affordable for the public, new products—and privacy problems—could emerge. Suspect that an employee is launching a competing business and stealing your customers? Sprinkle smart dust in his or her office and let the data roll in!

Forget Megahertz: How Many Carats of CPU Would You Like?

In the future, computers will not just be smaller, they'll also be made of different materials. As noted earlier, current CPUs are manufactured from silicon wafers. Processing data through the circuits embedded on these CPUs produces a large amount of heat. Unfortunately, today's chips cannot be made much more powerful without the heat generated by them rendering the chips inoperable (you don't want a puddle of silicon in your computer!).

It may surprise you, but the ideal substance to replace silicon in computer chips is diamond. While best known for being extremely hard, diamonds also have an extremely high degree of thermal conductivity. Massive amounts of heat can be passed through a diamond without damaging it. But diamonds are too expensive to use as computer chips . . . at least for now.

Gemesis and Apollo Diamond are two U.S. companies that are currently pioneering processes to manufacture diamonds in the laboratory. The companies use two different approaches. Gemesis uses a manufacturing process featuring tremendous pressure and temperature, which mimics the geological conditions that form diamonds within the Earth. Apollo's system uses intense heat to generate a plasma cloud of carbon precipitates, and the carbon literally rains onto a platform where diamond wafers are formed.

Although neither process is yet producing diamonds at costs that would make them attractive for use in computing, they may reach this point (perhaps as low as five dollars per carat) within the next decade. So one day you may just have a computer worth its weight in diamonds!

OBJECTIVES

After reading this chapter, you should be able to answer the following questions:

- What is a system development life cycle and what are the phases in the cycle? (pp. 388–391)

- What is the life cycle of a program? (p. 391)

- What role does a problem statement play in programming? (pp. 392–393)

- How do programmers create algorithms? (pp. 394–398)

- How do programmers move from algorithm to code and what categories of language might they code in? (pp. 399–401)

- How does a programmer move from code in a programming language to the 1s and 0s the CPU can understand? (pp. 404–407)

- How is a program tested? (pp. 408–409)

- What steps are involved in completing the program? (p. 409)

- How do programmers select the right programming language for a specific task? (pp. 410–413)

- What are the most popular programming languages for Windows and Web applications? (pp. 410–415)

CHAPTER 10

Behind the Scenes: Software Programming

TECHNOLOGY IN ACTION: UNDERSTANDING SOFTWARE PROGRAMMING

Every day we face a wide array of tasks. Some tasks are complex and need a human touch, some require creative thought and higher-level organization. However, some tasks are routine, such as alphabetizing a huge collection of invoices. Tasks that are repetitive, work with electronic information, and follow a series of clear steps are candidates for computerization.

For many tasks, there already exist well-designed computer programs. For example, if you want to write a research paper and include footnotes, Microsoft Word allows you to do just that. The program has already been designed to translate the tasks you want to accomplish into computer instructions. In order to do your work, you need only be familiar with the interface of Word; you do not have to create a program yourself.

However, for users who cannot find an existing software product to accomplish a task, programming is mandatory. Say a medical company comes up with a new smart bandage that is designed to transmit medical information about a wound directly to a diagnostic computer (these are under development). There is no software product on the market that is designed to accumulate and relay information in this manner. Therefore, the company will need to deploy a team of software developers to generate the appropriate software for this bandage to function as designed.

Even if you'll never create a program of your own, knowing the basics of computer programming is still helpful. For example, most modern software applications allow you to customize and automate various features using small custom-built "miniprograms" called *macros*. By creating macros, you can ask the computer to execute a complicated sequence of steps with a single command. Understanding how to program macros enables you to add custom commands to Word or Excel, for example, and automate frequently performed tasks, providing a huge boost to your productivity.

Understanding programming is therefore an important piece of working well with a computer system. If you plan to use only "off-the-shelf" software, having a basic knowledge of programming enables you to understand how application software is constructed and to add features that support your personal needs. If you plan to create custom applications from scratch, a detailed knowledge of programming is key to the successful completion of any project. Thus, in this behind-the-scenes chapter, we'll explore the stages of program development and survey the most popular programming languages.

The Life Cycle of an Information System

Generally speaking, a *system* is a collection of pieces working together to achieve a common goal. Your body, for example, is a system of body parts, muscles, organs, and other components working together. The college you attend is a system with administrators, faculty, students, and maintenance personnel working together. An **information system** includes data, people, procedures, hardware, and software. You interact with information systems all the time whether you are at a grocery store, bank, or restaurant. In any of these instances, the parts of the system work together toward a similar goal. Since teams of individuals are required to develop such systems, an organized process (or set of steps) needs to be followed to ensure that development proceeds in an orderly fashion.

This set of steps is usually referred to as the *system development life cycle (SDLC)*. In this section, we'll provide you with an overview of systems development and show you how programming fits into the cycle.

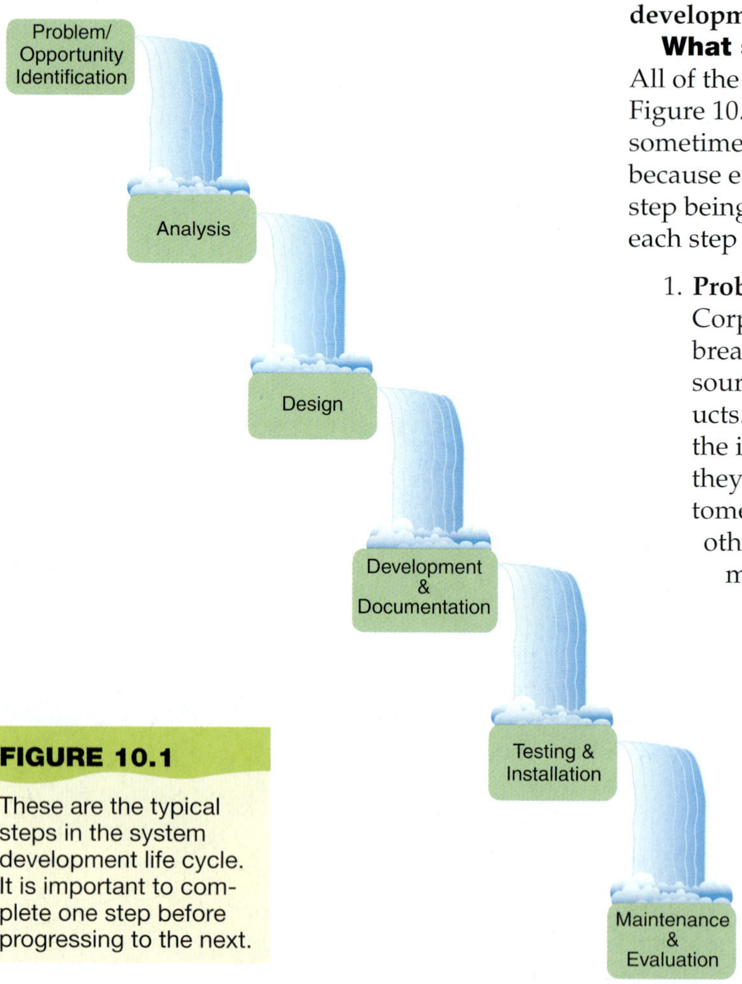

SYSTEM DEVELOPMENT LIFE CYCLE

Why do you need a process to develop a system? If you have programming skills, you can sit down in a day, perhaps, to write a little program to balance your checkbook or to organize your CD collection. If you don't have such skills, or the time or inclination to write a program, you can run out and buy one. However, one person does not develop in a day the programs you buy in a store. Those programs are generally far more complex than we would write ourselves, and they require many phases to make the product complete and saleable. Therefore an entire team of people and a systematic approach are necessary. Since teams of individuals are required to develop systems, an organized process (or set of steps) needs to be followed to ensure that development proceeds in an orderly fashion. As noted above, this set of steps is usually referred to as the **system development life cycle (SDLC)**.

What steps constitute the SDLC? All of the six steps of the SDLC are shown in Figure 10.1. As the figure shows, this system is sometimes referred to as a "waterfall" system because each step is dependent on the previous step being completed first. A brief synopsis of each step follows:

1. **Problem/Opportunity Identification.** Corporations are always attempting to break into new markets, develop new sources of customers, or launch new products. When the founders of eBay developed the idea of an online auction community, they needed a system that could serve customers and allow them to interact with each other. At other times, systems development is driven by serving existing customers more efficiently or responding to problems. When traditional brick-and-mortar businesses wanted to launch e-commerce sites they needed to develop systems for customers to purchase products. Whether solving an existing problem or exploiting an opportunity, corporations usually generate more ideas for systems than they have the time and money to implement. Large corporations typically form

FIGURE 10.1

These are the typical steps in the system development life cycle. It is important to complete one step before progressing to the next.

a development steering committee to evaluate systems development proposals. The committee reviews ideas and decides which projects to take forward based on available resources (personnel and funding).

2. **Analysis.** In this phase, analysts explore in depth the problem to be solved and develop a *program specification*. The program specification is a clear statement of the goals and objectives of the project at hand. It is also at this stage that the first feasibility assessment is performed. The feasibility assessment determines whether the project should go forward or not. You might have a great idea, but that doesn't mean that the company has the technical expertise or the financial or operational resources to develop it. Similarly, there may not be enough time to fully develop the product.

   Assuming the project is feasible, the analysis team studies the current system (if any) and defines the user requirements of the proposed system. Finally, the analysts recommend a solution or plan of action, and the process moves to the design phase.

3. **Design.** Before a house is built, blueprints are developed so that the workers have a plan to follow. The design phase of the SDLC has the same objective—generating a detailed plan for programmers to follow. The current and proposed systems are documented using flowcharts and data-flow diagrams. Data-flow diagrams trace all data in an information system from the point at which data enters the system to its final resting place (storage or output). Figure 10.2 shows a high-level data-flow diagram illustrating the flow of concert ticket information.

   The ultimate goal of the design phase with respect to software development is to design documents that programmers can follow in developing the actual system. It is also in this phase where the "make or buy" decision is made. Once the system plan is designed, existing software packages (off-the-shelf software) can be evaluated to determine whether purchasing a solution is feasible.

   For instance, if you wanted to start an online auction site to compete with eBay, you might not have to build your own system. There are numerous online auction software packages for sale now that might meet your needs. If an existing package cannot be found that meets your needs, then you would have to develop your own system. Or, perhaps, you can *outsource*, or hire someone outside the corporation, to develop the program you need.

4. **Development and Documentation.** This is the phase where actual programming takes place. This phase is also the first part of the program development life cycle (PDLC), described in detail in the rest of the chapter.

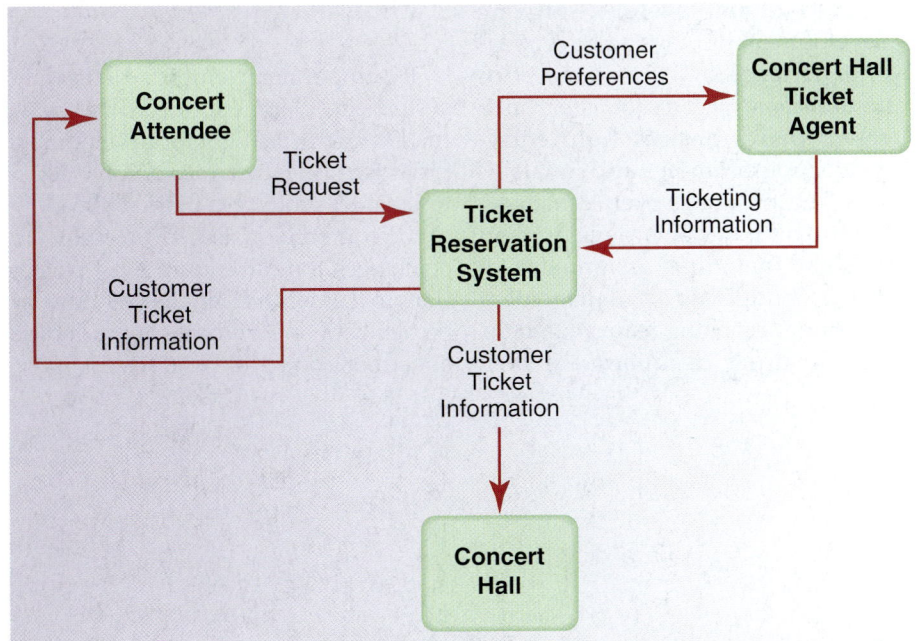

FIGURE 10.2

Data-flow diagrams trace the flow of data such as ticketing information. This is a top-level overview diagram. More detailed diagrams would be prepared for each piece of the system, showing specific pieces of information being tracked (such as customer name, payment information, and so on).

TRENDS IN IT

CAREERS:
Considering a Career in Systems Development?

Since large projects involve many people to complete them, there are many opportunities for jobs in systems development. Although there are a number of people involved in the process, the key players in systems development are systems analysts, programmers, and project managers:

- **Systems analysts** spend most of their time in the beginning stages of the SDLC. They talk with end users to gather information about problems and existing information systems. They document systems and propose solutions to problems. Having good people skills is essential to success. In addition, an analyst works with programmers during the development phase to design appropriate programs to solve the problem at hand. Therefore, many organizations insist on hiring systems analysts with both a solid business background and prior programming experience (at least at a basic level). For entry-level jobs, a four-year degree is usually required. Many colleges and universities offer degrees in Management Information Systems (MIS) that include a mixture of systems development, programming, and business courses.

- **Programmers** participate in the SDLC, attending meetings to document user needs and working closely with systems analysts during the design phase. In addition, programmers need excellent written communication skills since they often generate detailed systems documentation for end-user training purposes. Since programming languages are mathematically based, it is essential for programmers to have strong math skills and an ability to think logically. Programmers should also be proficient at more than one programming language. As you'll learn later, developers use different languages for different problems. While Java, C++, or Visual Basic are popular languages today, programmers need to be prepared to learn new languages (such as C#) as well.

  A four-year degree is required for entry-level programming positions. Computer science is the major of choice, but some employers will hire entry-level employees with degrees in MIS with mathematics and additional programming courses as electives.

- **Project managers** are usually not entry-level employees in software development but instead have years of experience as a programmer or systems analyst. This job is part of a career path upwards from entry-level programming and analysis jobs. The job entails overall management of the systems development process, including assignment of staff, budgeting, management reporting, coaching team members, and ensuring deadlines are met. Project managers need excellent time-management skills since they are pulled in several directions at once. Many project managers obtain master's degrees to supplement their undergraduate degrees in computer science or MIS.

In addition to these people, the following are also involved in the systems development process:

- **Technical writers** provide documentation for the new system.
- **Network engineers** help the programmers and analysts design compatible systems, since many systems are required to run in certain environments (UNIX or Windows, for instance) and must work well in conjunction with other programs.
- **Database analysts and database administrators (DAs and DBAs)** design and implement database structures, since most systems interface with or populate one type of database or another.
- **Graphic designers and interface designers** create attractive and effective interface screens

It is important to emphasize that all systems development careers are stressful. Deadlines are tight for development projects, especially if they involve getting a new product to market ahead of the competition. But if you enjoy challenges and can endure a fast-paced, dynamic environment, there should be plenty of opportunities in the decade ahead for good systems developers.

controlling security. While people who are new to a language might not immediately understand how these keywords are used, examining the keywords of a language you are learning will often tell you what special features the language has and will help you ask important questions about it.

Each time programmers want to store data in their program, they must ask the operating system for storage space at a RAM location. **Data types** describe the *kind* of data that is being stored at the memory location. Each programming language has its own unique data types (although there is some degree of overlap among languages). For example, C++ includes data types representing integers, real numbers, characters, and Boolean (true/false) values. These C++ data types show up in code statements as *int* for integer, *float* for real numbers, *char* for characters, and *bool* for Boolean values.

Since it takes more room to store a real number like 18,743.23 than it does to store the integer 1, programmers use data types in their code to indicate to the operating system how much memory it needs to allocate. Programmers must be familiar with all of the data types available in the language so that they can assign the most appropriate data type for each input and output value so as not to waste memory space.

Operators are the coding symbols that represent the fundamental actions of the language. Each programming language has its own set of operators. Many languages include common algebraic operators such as + , - , * , / to represent the mathematical operations of addition, subtraction, multiplication, and division, respectively. The language APL (A Programmer's Language) was specifically designed to solve mathematics problems. APL therefore includes the common mathematical operators, but it also includes the operators rho, sigma, and iota, each representing complex mathematical operations. Because it contains many unique operators, APL requires that programmers use a special keyboard when they input code, as shown in Figure 10.15.

Programming languages sometimes include other unique operators. For example, the C++ operator << is used to tell the computer to read data from the keyboard or from a file. The C++ operator && is used to tell the computer to check whether two statements are both true. "Is your weight greater than 120 AND less than 150?" is a question that requires the use of the && operator.

In the following C++ code, several operators are being used. The > operator checks whether the number of hours worked is larger than 0. The && operator checks that the number of hours worked is both positive AND less than or equal to 8 at the same time. If that happens, then the = operator sets the output Pay equal to the number of hours paid at $7.32 per hour:

```
if (Hours > 0 && Hours <= 8 )
```

FIGURE 10.15

APL requires programmers to use a unique APL keyboard that includes the many specialized operators in the language.

As long as the program is running, these RAM cells will be saved for the Day variable; no other program can use that memory until the program ends. From that point on, when the program encounters the expression Day, it will mean access the memory cells it reserved as Day and find the integer stored there.

The following line of C++ code asks that a real number (represented by the keyword "float") be stored in RAM:

```
float TotalPay;
```

This line asks the operating system to find enough storage space for one real number.

Can programmers leave notes to themselves inside a program? Programmers often insert **comments** (or **remarks**) into program code to explain the purpose of sections of code, to indicate the date they wrote the program, and to include other important information about the code so that fellow programmers can more easily understand and update it should the original programmer no longer be available. Comments are written into the code in plain English. Languages provide a special symbol or keyword to indicate that what follows is a comment, not part of the executable program. In C++, the symbol // at the beginning of a line indicates that the rest of the line is a comment. In Visual Basic, the keyword "REM," short for "REMark," does the same thing.

What would completed code for a program look like? Figure 10.16 presents a completed C++ program for our example parking garage problem. Each statement in a program is executed sequentially (that is, in order from the first statement to the last) unless the program encounters a keyword that changes the flow. In the figure, the program begins (Step 1) by declaring the variables needed to store the program's inputs and outputs in RAM. Next, the "for" keyword begins a looping pattern (Step 2). All of the steps between the very first bracket "{"and the last bracket "}" (shown in red for better identification) will be repeated seven times in order to gather the total pay for each day of the week.

The next section (Step 3) collects the input data from the user. It then (Step 4) checks that the user entered a reasonable value (in this case, a positive number for hours worked) and if needed, reads another input value. Now (Step 5) the program processes the data. If the user worked eight hours or less, he or she is paid at the rate of $7.32, while hours exceeding eight are billed at $11.73.

The final statement (Step 6) updates the value of the TotalPay variable. The last bracket "}" indicates the program has reached the end of a loop. The program will repeat the loop to collect and process the information for the next day. When the seventh day of data has been processed, the Day variable will be bumped up to the next value, 8. The program then fails the test (Day <= 7 ?). At that point, the program exits the loop, prints out the results, and quits.

Are there ways in which programmers can make their code more useful for the future? One aspect of converting an algorithm into good code is the programmer's ability to design general code that can adapt easily to new settings. Sections of code that will be used repeatedly, with only slight modification, can be packaged together into reusable "containers" or components. These reusable components, depending on the language, are referred to as *functions, procedures, subroutines, modules, or packages.*

In our program, we could create a function that implements the overtime pay rule. As it stands in Figure 10.16, the code will work only in situations where the hourly pay is exactly $7.32 and the bonus pay is exactly $11.73. However, if we rewrote this part of the processing rules as a function, we could have code that would work for any base pay rate and any overtime rate. If the base pay rate or overtime rate changed, the function would use whatever values it was given as input to compute the output pay variable. Such a function, as shown in Figure 10.17 on the next page, could be reused in many settings, without changing any of the code.

COMPILATION

How does a programmer move from code in a programming language to the 1s and 0s the CPU can understand? **Compilation** is the process by which code is converted into machine language, the language the CPU can understand. The **compiler** is the program that understands both the syntax of the programming language and the exact structure of the CPU and its machine language. It can "read" the **source code**—the instructions programmers have written in the higher-level language—and translate the source code directly into machine language—the binary patterns that will execute commands on the CPU.

Each programming language has its own compiler. In addition, separate versions of the com-

TechTV For information on creating computer games without knowing how to program, see "Computer Game WYSIWYG"— a TechTV clip found at www.prenhall.com/techinaction

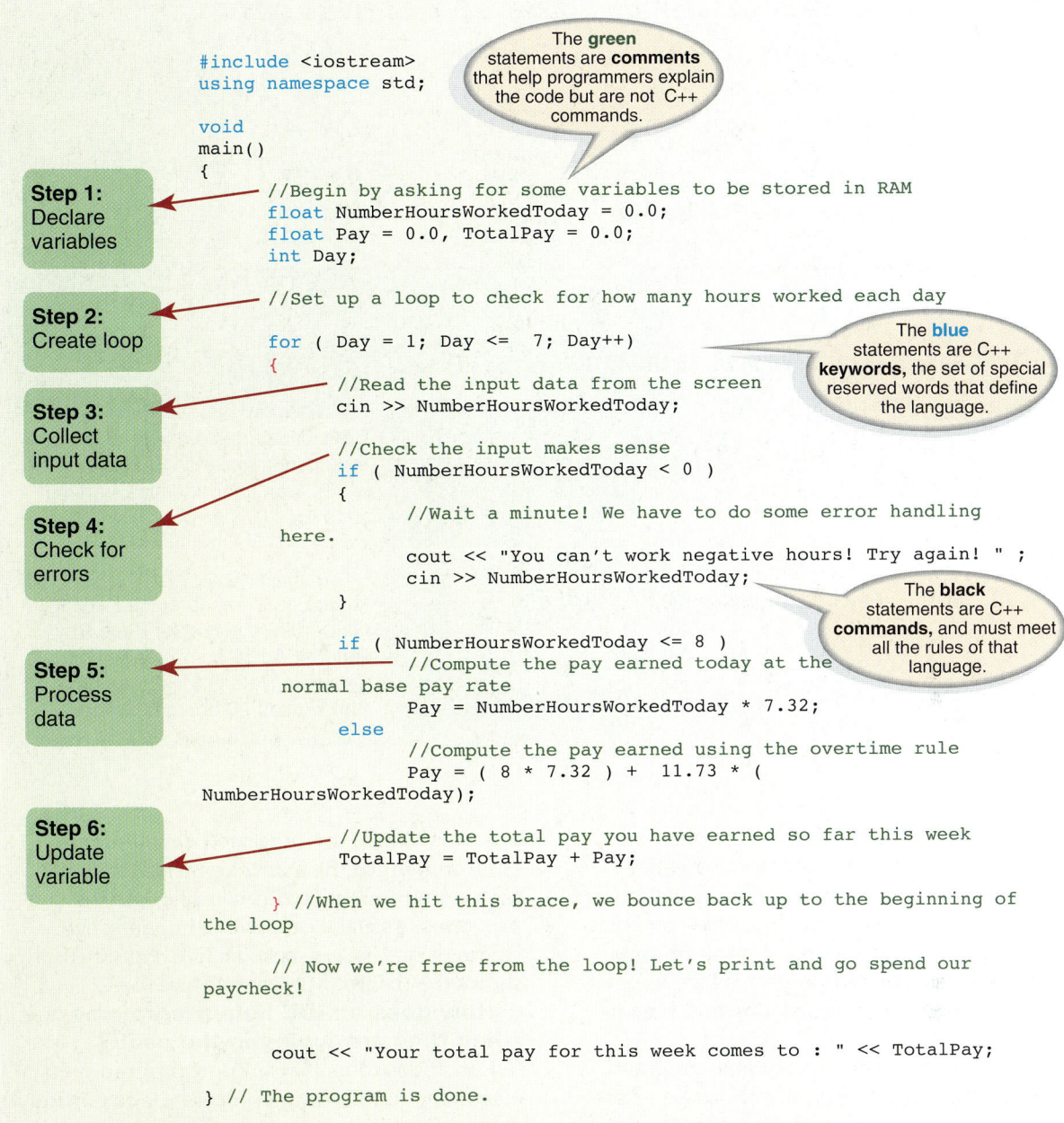

FIGURE 10.16

This is an example of a complete C++ program that solves the parking garage pay problem.

piler are required to compile code that will run on each different type of processor. One version of the compiler would create finished programs for a Motorola G4 processor and another version of the compiler would create programs for an Intel Pentium CPU.

At this stage, programmers finally have produced an **executable program,** the binary sequence that instructs the CPU to run their code. Executable programs cannot be read by human eyes since they are pure binary codes. They are stored as *.exe or *.com files on Windows systems.

Does every programming language have a compiler? Some programming languages do not have a compiler but use an *interpreter* instead. An **interpreter** translates the source code into an intermediate form, line by line. Each line is then executed as it is translated.

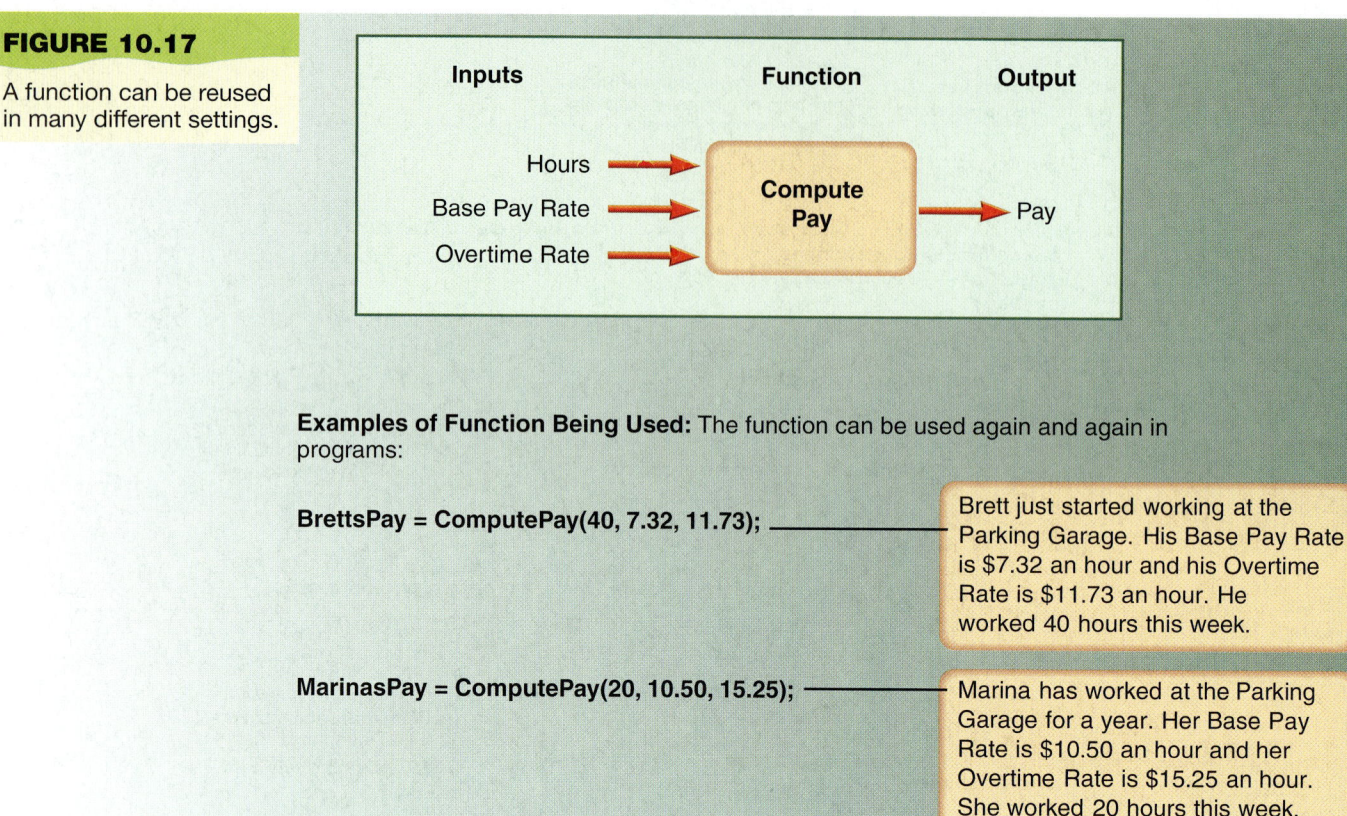

FIGURE 10.17

A function can be reused in many different settings.

The compilation process takes longer than the interpretation process because in compilation, *all* of the lines of source code are translated into machine language before any lines are executed. However, the finished compiled program runs faster than an interpreted program since the interpreter is constantly translating and executing as it goes.

If producing the fastest executable program is important, programmers would choose a language that uses a compiler instead of an interpreter. For developmental environments where a lot of changes are still being made to the code, interpreters have an advantage: programmers do not have to wait for the entire program to be recompiled each time they make a change. With interpreters, programmers can immediately see the results of their program changes as they are making them in the code.

CODING TOOLS: INTEGRATED DEVELOPMENT ENVIRONMENTS

Are there any tools that make the coding process easier? Modern programming is supported with a collection of tools to make the writing and testing of software easier. Compiler products feature an **integrated development environment (IDE)**, a developmental tool that helps programmers write, compile, and test their programs. As is the case with compilers, every language has its own specific IDE. Figure 10.18 shows the IDE for Microsoft Visual C++.

How does an IDE help programmers when they are typing in the code? The IDE includes tools that support programmers at every step of the coding process. **Code editing** is the step in which programmers actually type the code into the computer. IDEs include an **editor**, a special tool that helps programmers as they enter the code, highlighting keywords and alerting them to typos. Modern IDE editors also automatically indent the code correctly, aligning sections of code appropriately, and color-code comments to remind programmers that these lines will not be executed as code. In addition, IDEs provide help files that document and provide examples of the proper use of keywords and operators.

How does the IDE help programmers after code editing is finished? Editing is complete when the entire program has been keyed into the editor. At that time, the programmer clicks on a button in the IDE and the com-

grams often have a number of common features—scroll bars, title bars, text boxes, buttons, and expanding/collapsing menus to name a few. Several languages include customized controls that allow programmers to easily include these features in their programs. In these languages, programmers can simply use the mouse to lay out on the screen where the scroll bars and buttons will be in their application. The code needed to explain this to the computer is then written automatically when the programmer says the layout is complete. This is referred to as **visual programming**, and it helps programmers produce a final application much more quickly. Figure 10.19 shows an example of visual programming in use.

Visual Basic

What if programmers want to have a model of their program before it's fully developed? Earlier in the chapter, you read

BITS AND BYTES

Cup of Java, Anyone?

The word *Java* is slang for coffee (a substance programmers use heavily to meet project deadlines), and there are many Java puns based on this slang. *JavaBeans*, for example, are tiny prewritten modules of code programmers can use to quickly add functionality to their programs. Java is also an island in Indonesia. One of the most popular sites for the exchange of ideas and code samples for Java programmers is **www.gamelan.com**. Gamelan is the Javanese word for an orchestra comprised mainly of instruments like gongs, metal chimes, and xylophones. Computer programmers seem to love odd connections!

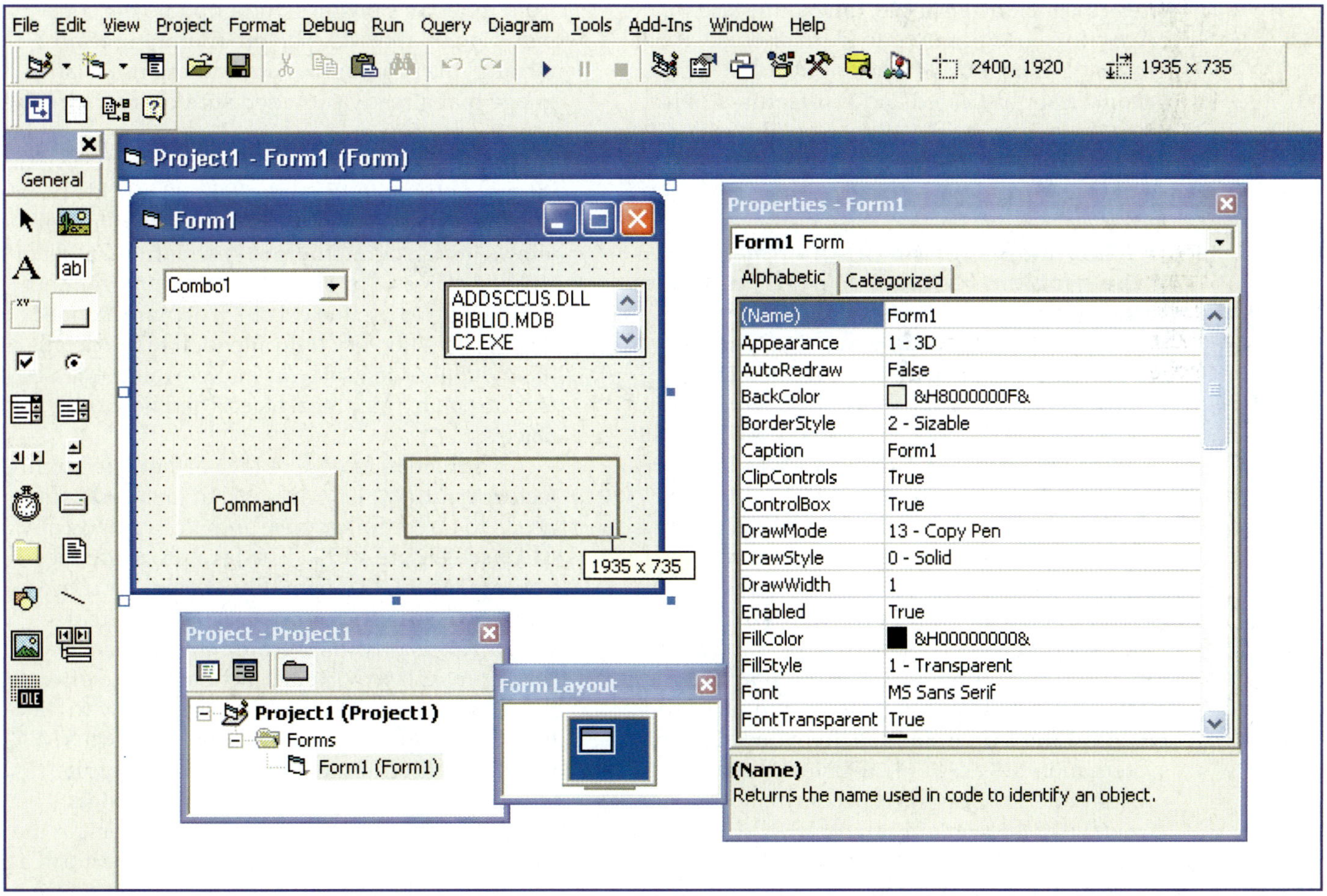

FIGURE 10.19

Microsoft Visual Basic supports visual programming by allowing the "drag and drop" of objects directly to build the application.

about how information systems are developed through the system development life cycle. Although the SDLC has been around for quite some time, it doesn't necessarily work for all environments and instances. Programmers often like to build a **prototype**, or small model, of their program at the beginning of a large project.

While the entire project won't be finished for several months, it can be useful to have a simple shell of what the final program will look like to help with design. Prototyping is a form of **rapid application development (RAD)**, which is an alternative to the "waterfall" approach of systems development described at the beginning of the chapter. Instead of developing detailed system documents prior to the production of the system, developers create a prototype first and they generate system documents as they use and remodel the product.

Prototypes for Windows applications are often coded in **Microsoft Visual Basic (VB)**, which is a powerful programming language used to build a wide range of Windows applications. VB's strengths include a simple, quick interface that is easy for a programmer to learn and use. It has grown from its roots in the language BASIC to become a sophisticated and full-featured object-oriented language. It is often used in the creation of graphical user interfaces for Windows.

C and C++

What languages do programmers use if the problem requires a lot of "number crunching"? A Windows application that demands raw processing power because there are difficult repetitive numerical calculations is most often a candidate for C or C++. Several companies sell C/C++ compilers equipped with a design environment that makes Windows programming as visual as with VB.

C, the predecessor of C++, was developed originally for system programmers. It was defined by Brian Kernighan and Dennis Ritchie of AT&T Bell Laboratories in 1978 as a language that would make accessing the operating system easier. It provides higher-level programming language features (like if statements and for loops) but still allows programmers to manipulate the system memory and CPU registers directly. This mix of high- and low-level access makes C very attractive to "power" programmers. Most modern operating systems (Windows XP, Mac OS X, and Linux) have been written in C.

C++ takes C to the next level. Bjarne Stroustrup, the developer of C++, used all of the same symbols and keywords as C, but extended the language with additional keywords, better security, and more support for the reuse of existing code through object-oriented design.

Neither C nor C++ was intended as a teaching language. The notation and compactness of the languages make them relatively difficult to master. They are in demand in industry, however, because C/C++ can produce fast-running code that uses a small amount of memory. Programmers often choose to learn C/C++ because their basic components (operators, data types, and keywords) are common to many other languages.

Java

What language do programmers use for applications that need to collect information from networked computers? Say a program that an insurance company runs each night needs to communicate with networked computers in many offices around the country, collect the policy changes and updates from that day's business, and update the company's main records. The programming team writing this program would want to use a language that already provided support for network communications.

Java would be a good choice. Sun Microsystems introduced Java in the early 1990s. It quickly became popular because its object-oriented model allows Java programmers to benefit from its set of existing classes. For example, a Java programmer could begin to use the existing "network connection" class with very little attention to the details of how that code itself was implemented. Classes exist for many graphical objects and network objects.

An attractive feature of Java is that it is "architecture neutral." This means that Java code only needs to be compiled once and it can run on any CPU (see Figure 10.20). The Java program does not care what CPU, operating system, or user interface is running on the machine where it lands. This is possible because the target computer runs a Java Virtual Machine (VM), software that can explain to the Java program how to function on any specific system. There is a Java VM installed with Internet Explorer, for example, which allows IE to execute any **Java applets** (small Java-based programs) it encounters on the Internet. Although Java code does not perform as fast as C++, the advantage of only needing to compile once before it can be distributed to any system is very important.

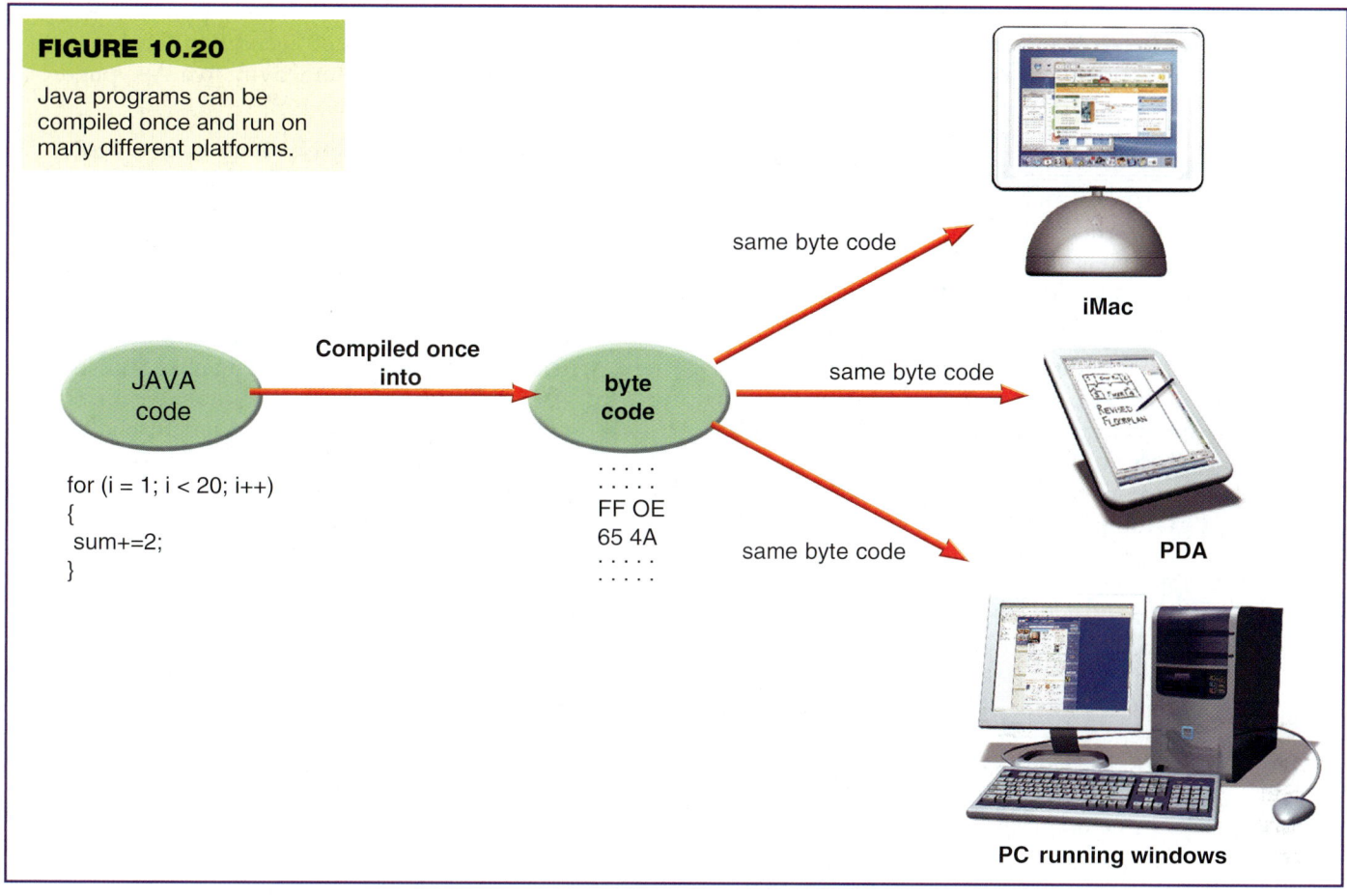

FIGURE 10.20

Java programs can be compiled once and run on many different platforms.

WEB APPLICATIONS: HTML AND BEYOND

What is the most basic language for developing Web applications? A document that will be presented on the Web must be written using special symbols called *tags*. These markers control how a browser will display the text. Figure 10.21 shows several examples of these tags and their effect on the display of text. The tags are examples of **HTML**, the **Hypertext Markup Language**. While HTML knowledge is required to program for the Web, it is not itself a programming language. HTML is just a series of tags that modify the display of text.

Many good HTML tutorials are available on the Web at sites such as **www.learnthenet.com** and **www.webmonkey.com**. These sites include lists of the major HTML tags that can be used to create HTML documents.

FIGURE 10.21 HTML Tags and Displays		
HTML TAG	SAMPLE HTML CODE	DISPLAYED TEXT
Bold	<bold> READ ME</bold>	**READ ME**
Small	<font size="1">tiny words</font>	tiny words
Link (or anchor)	<a href="http://www.amazon.com"> Read more books! </a>	Read more books!
Color	<font color = "red"> Red </font>	Red
Font	<font = "Arial> new font </font>	new font

Are there tools that help programmers write in HTML? Several different programs are available to assist in the generation of HTML. Macromedia Dreamweaver and Microsoft FrontPage present Web page designers with an interface that is similar to a word processor. Web designers can place text, images, and hyperlinks freely, and the corresponding HTML tags are inserted automatically, as shown in Figure 10.22. For simple, static (nonchanging) Web pages, no programming is required.

JavaScript and VBScript

What programming languages do programmers use to make complex Web pages? To make their Web pages more visually appealing and interactive, programmers include *JavaScript* code. **JavaScript** is a programming language often used to add interactivity to Web pages. JavaScript is not as full-featured as Java, but its syntax, keywords, data types, and operators are a subset of Java's. In addition, JavaScript has a set of classes that represent the objects often used on Web pages: buttons, check boxes, and drop-down lists.

The JavaScript button class, for example, describes a button with a name and a type (whether it is a regular button or a Submit or Reset button). The language includes behaviors like click() and can respond to user actions. For example, when a user moves his or her mouse over a button and pushes down to select it, the button "knows" the user is there and jumps in and performs a special action (such as playing a sound).

Often, programmers more familiar with Visual Basic than Java or C++ will use *VBScript* to introduce dynamic decision making into Web pages. **VBScript** is a subset of Visual Basic and is also used to introduce interactivity to a Web page.

ASP and JSP

How are interactive Web pages built? To build Web sites with interactive abilities, programmers use **ASP** (**Active Server Pages**) or **JSP** (**Java Server Pages**) to adapt the HTML page to the user's selections. ASP code is used to translate the user's information into a request for information from the company's main computer, often using a database query language like SQL. More ASP or JSP code controls the automatic writing of the custom HTML page that is returned to the user's computer.

What does additional programming bring to my Web page? The most advanced Web pages interact with the user, collecting information and then customizing what they present based on the user's feedback. For example, as shown in Figure 10.23, an online store's page will collect a customer's bicycle order and then ask the company's main server computer if that particular type of bicycle is in stock. An ASP program then creates a new HTML page and delivers that to the user's browser, telling the customer what color bicycles for that model and size are currently in stock.

Thus, ASP programs can have HTML code as their output. They use what the user has told them from the list boxes, check boxes, and buttons on the page to make decisions. Based on those results, the ASP decides what HTML to write. A small example of ASP writing its own HTML code is shown in Figure 10.24.

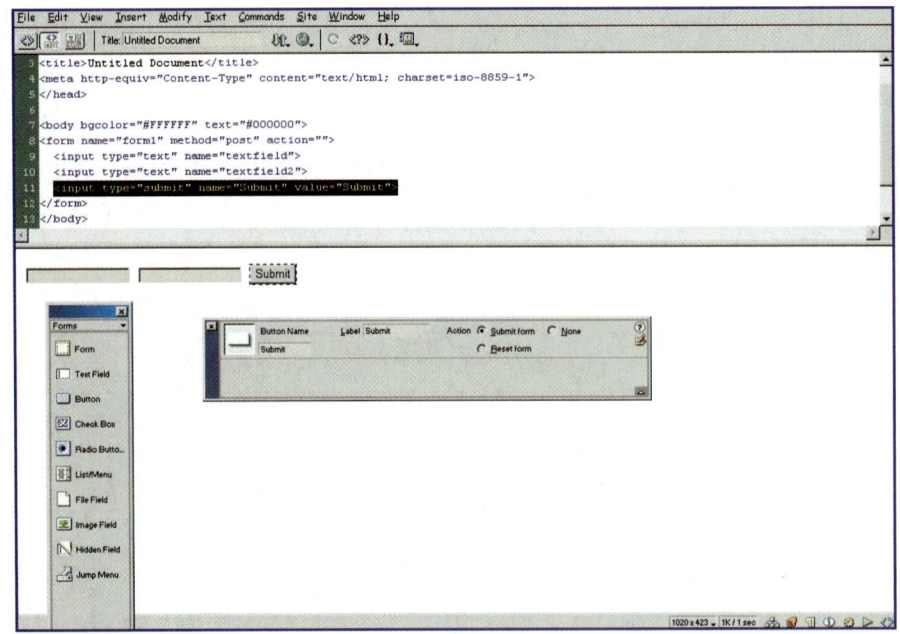

FIGURE 10.22

Macromedia Dreamweaver is a popular tool for creating Web pages.

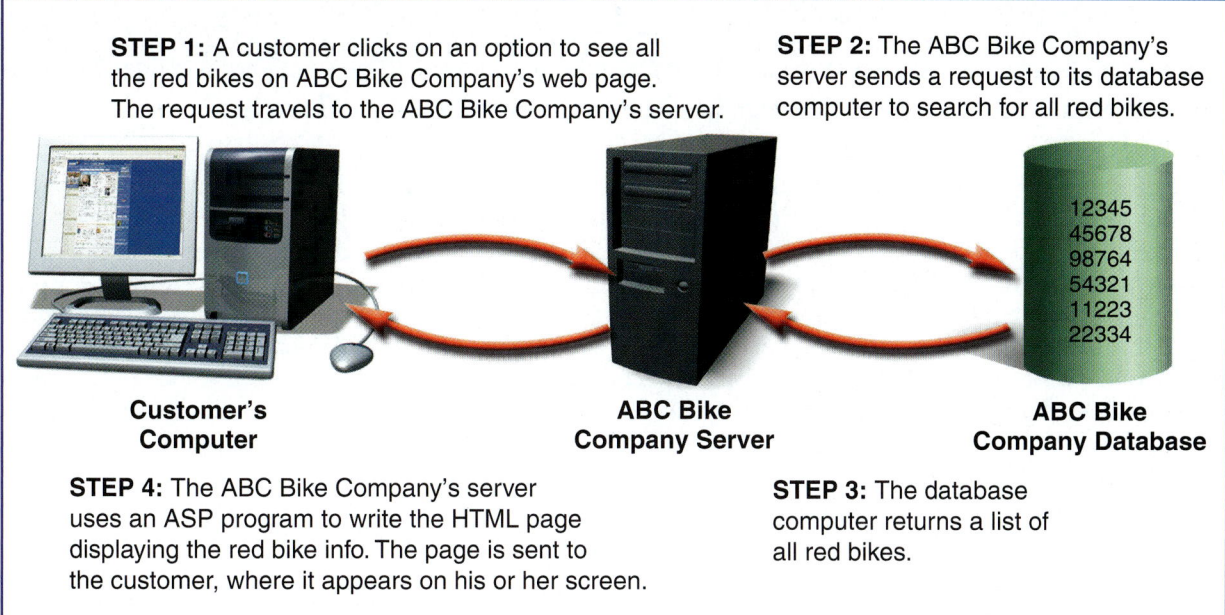

FIGURE 10.23
An online store is an example of the three-tier client/server type of Internet application.

Flash and XML

What if a programmer wants to create a Web page that includes sophisticated animation? Many Web sites feature elaborate animations that interact with visitors. These sites include buttons and hyperlinks along with animation effects. These components can be designed with **Macromedia Flash**, a software product for developing Web-based multimedia. Flash includes its own programming language named ActionScript, which is very similar to JavaScript in its selection of keywords, operators, and classes.

Is HTML the only markup language for the Web? When Web sites communicate with humans, HTML works well because the formatting it controls is important. People respond immediately to the visual styling of textual information—its layout, color, size, and font design all help to transfer the message off the page to the reader. When computers want to communicate with each other, however, all of these qualities just interfere. **XML (Extensible Markup Language)** allows designers to define their own data-based tags, making it much easier for a Web site to transfer the key information on its page to another site.

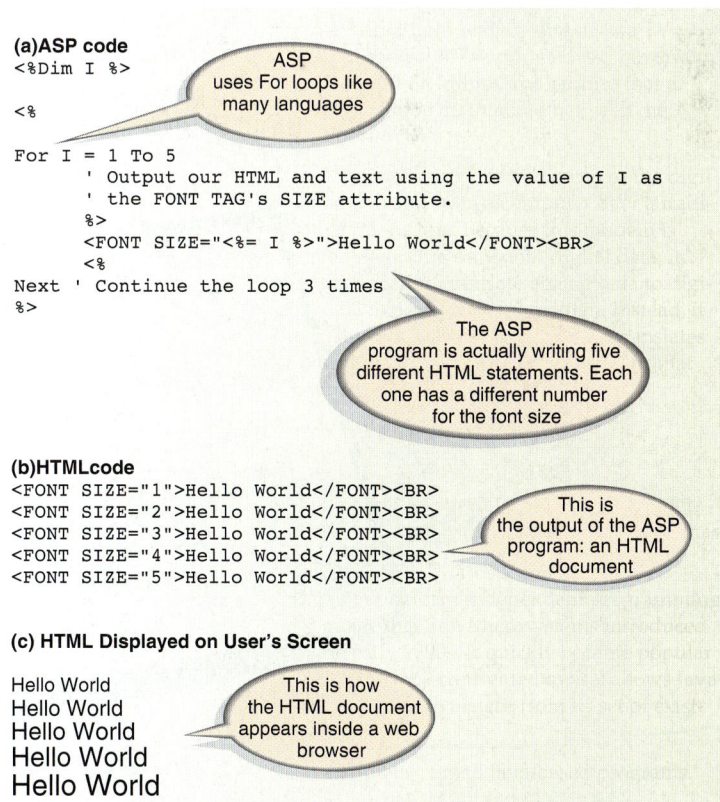

FIGURE 10.24
(a) An ASP program can write
(b) HTML code as its output.
(c) The HTML page it writes would show up in a browser.

BITS AND BYTES

How Does a Programmer Write Code for Everyday Appliances?

Programming code exists inside many of the appliances and devices in modern life. Washing machines make decisions on temperature and agitation time based on the types of stains in the laundry. Toasters include a microchip to decide when the toast is perfectly done. These are examples of embedded applications, very small software programs that are tucked away inside an appliance, automobile, or other device. The types of programs that are used in toasters or washing machines need to be very efficient and compact. They have no need for graphics systems or for networking. Embedded code is often written in C/C++ and then compiled into assembly language. Using assembly language, a programmer can fine-tune the use of registers and memory so that the final program is very efficient.

Without XML, a Web site that wanted to look up current stock pricing information at another site would have to retrieve the HTML page, then sort through the formatting information, and try to recognize which text on the page identified the data needed. With XML, groups can agree on standard systems of tags that represent important data elements. For example, the XML tag <stock></stock> might delimit key stock quote information. Mathematicians have created a standardized set of XML tags for their work named MathML, while biometrics groups are developing an XML standard to describe and exchange biometric data like DNA, fingerprints, and iris and face scans. We'll discuss both HTML and XML in more detail in Chapter 13.

THE NEXT GREAT LANGUAGE

What will be the next great language?
It is never easy to predict which language will become the next "great" language, but software experts predict that as software projects continue to grow in size, the amount of time for a completed project to compile will also grow. It is not uncommon for a large project to require 30 minutes or more to recompile. Interpreted languages, however, have virtually zero compile time as compilation occurs while the code is being edited. As projects get larger, this ability to be instantaneously compiled will become even more important. Thus, interpreted languages like Python, Ruby, and Smalltalk could become more important in the coming years.

Will all languages someday converge to one? Certain characteristics of modern programming languages correspond well to how programmers actually think. These traits support good programming practices and are therefore emerging as common features of most modern programming languages. The object-oriented paradigm is one example. Both Visual Basic and COBOL have moved toward a support of objects.

There will always be a variety of programming languages, however. Forcing a language to be so general that it can work for any task also forces it to include components that make it slower to compile, as well as produce larger final executables and require more memory to run. Having a variety of languages and mapping a problem to the best language creates the most efficient software solutions.

So what path of learning should I follow if I want to learn languages that will be relevant in the future? There is no absolute set of languages that is best to learn and no one best sequence in which to learn them. The Association for Computing Machinery (ACM) encourages educators to teach a core set of mathematical and programming skills and concepts, but departments are free to offer a variety of languages.

Some geographical considerations come into play when selecting the sequence of programming language courses you should study. For example, in an area where there is a large number of pharmaceutical companies, there may be a demand for Massachusetts General Hospital Utility Multi-programming System, or MUMPS. MUMPSs is a language often used to build clinical databases, an important preoccupation of the pharmaceutical industry. Review the advertisements for programmers in area newspapers and investigate resources like ComputerJobs (**www.computerjobs.com**) to identify languages in particular demand in your area. Figure 10.25 presents a possible two-year college sequence of programming courses for students with different career objectives.

FIGURE 10.25 Computer Science Course Sequences for a Variety of Goals			
SEQUENCE	**NETWORKING**	**SOFTWARE ENGINEERING**	**WEB PROGRAMMING**
SEMESTER 1	Intro to C++ DOS	Intro to C++ HTML	Intro to C++ HTML
SEMESTER 2	Any language	Visual Basic Data Structures in C++	Unix JavaScript
SEMESTER 3	Data Structures in C++	Advanced VB Advanced C++	ASP Java
SEMESTER 4	SQL	SQL Java Assembly language	SQL Advanced C++

Summary

1. **What is a system development life cycle (SDLC) and what are the phases in the cycle?** There are six steps in the SDLC. First, a problem or opportunity is *identified*. Next, the problem is *analyzed* and a program specification document is created. Then a detailed *design* or plan for programmers to follow is created, from which the *development* and *documentation* of the program occur. The program is then *tested* and *installed*. *Maintenance* and *evaluation* ensure a working product is maintained.

2. **What is the life cycle of a program?** Each programming project follows a number of stages from conception to deployment. The *problem statement* identifies the task to be computerized and describes how the program will behave. An *algorithm* specifies the steps the program must take and then is *translated* into *code*. The code goes through *debugging*, in which programmers find and repair errors. The results of the project are *documented* and users are *trained*.

3. **What role does a problem statement play in programming?** The problem statement is a description of what tasks the program must accomplish and how the program will execute these tasks. It describes the *input* data that users will have at the start of the job, the *output* that the program will produce, and the *processing* that converts these inputs to outputs.

4. **How do programmers create algorithms?** For simple problems, programmers create an *algorithm* by converting the problem statement into a list of steps the program will take. Algorithms are documented in a flowchart or in pseudocode. Programmers use a top-down or object-oriented analysis to produce the algorithm.

5. **How do programmers move from algorithm to code and what categories of language might they code in?** Programming languages are classified in groupings, referred to as "generations," with the first generation being machine language, the 1s and 0s that the computer understands. Assembly language is the next generation that uses English-like commands that give the programmer direct control of hardware resources. Each successive generation becomes more matched to how humans think.

6. **How does a programmer move from code in a programming language to the 1s and 0s the CPU can understand?** Compilation is the process of converting code into machine language. The compiler understands the syntax of the programming language and the structure of the CPU and its machine language.

7. **How is a program tested?** Programmers debug the program by running it to find errors and to make sure it behaves properly. Once debugging has detected all the errors, users test the program both as it was intended to be used and in ways only new users may think up.

8. **What steps are involved in completing the program?** Once testing is completed, technical writers create documentation and user manuals. Once the software is distributed, training teaches users how to use the software.

9. **How do programmers select the right programming language for a specific task?** Certain languages are best used with specific problems. The target language should be well matched to the amount of space available for the final program. Some projects require a language that can produce code that executes in the fastest possible time. Choosing a language that the programmers are familiar with is also helpful.

10. **What are the most popular languages for developing Windows and Web applications?** Visual Basic, C, C++, and Java are languages that allow programmers to include Windows control features such as scroll bars and menus. Programmers use HTML for basic Web design, or for more complex Web designs, programs such as JavaScript and VBScript. Web page animations are done with ASP, JSP, Flash, and XML.

Key Terms

algorithm	p. 393
ASP (Active Server Pages)	p. 414
assembly languages	p. 400
base class	p. 399
beta version	p. 409
binary decisions	p. 395
C	p. 412
C++	p. 412
classes	p. 398
code editing	p. 406
coding	p. 399
comments (or remarks)	p. 404
compilation	p. 404
compiler	p. 404
data types	p. 403
data	p. 398
debugger	p. 408
debugging	p. 408
decision points	p. 394
derived class	p. 399
documentation	p. 409
editor	p. 406
error handling	p. 392
executable program	p. 405
fifth-generation languages (5GLs)	p. 401
first-generation languages (1GLs)	p. 400
flowcharts	p. 396
for and next	p. 402
fourth-generation languages (4GLs)	p. 401
higher-level languages	p. 401
HTML (Hypertext Markup Language)	p. 413
if else	p. 402
information system	p. 388
inheritance	p. 398
initial value	p. 396
integrated development environment (IDE)	p. 406
interpreter	p. 405
Java applets	p. 412
Java	p. 412
JavaScript	p. 414
JSP (Java Server Pages)	p. 414
keywords	p. 402
loop	p. 395
machine languages	p. 400
Macromedia Flash	p. 415
methods or behaviors	p. 398
Microsoft Visual Basic (VB)	p. 412
object	p. 398
object-oriented analysis	p. 398
operators	p. 403
Pascal	p. 410
portability	p. 401
problem statement	p. 392
program development life cycle (PDLC)	p. 391
programming language	p. 399
programming	p. 391
prototype	p. 412
pseudocode	p. 396
rapid application development (RAD)	p. 412
reusability	p. 398
runtime (or logic) errors	p. 408
second-generation languages (2GLs)	p. 400
software updates (service packs)	p. 409
source code	p. 404
statements	p. 402
syntax	p. 402
syntax errors	p. 408
system development life cycle (SDLC)	p. 388
test condition	p. 396
testing plan	p. 392
third-generation languages (3GLs)	p. 401
top-down design	p. 397
variable declaration	p. 401
variables	p. 401
VBScript	p. 414
visual programming	p. 411
XML (Extensible Markup Language)	p. 415

Buzz Words

Word Bank

- algorithm
- C/C++
- XML
- HTML
- JavaScript
- embedded application
- runtime error
- top-down design
- object-oriented analysis
- documentation
- beta test
- interpreter
- compilation
- classes
- inheritance
- service pack
- variable
- testing plan
- data type
- operator
- machine language
- syntax error
- problem statement
- debugger
- Visual Basic

Instructions: Fill in the blanks using the words from the Word Bank above.

Things are not running smoothly at the Whizgig factory. We need to keep track of how many Whizgigs are made every hour, and how many are defective. We begin by calling in the programming team to work with users. Together they begin to build a software solution by creating a (1)_____. All of the input and output information required is identified as well as the (2)_____, which lists specific examples of what outputs the program will produce for certain inputs. The team then begins to design the (3)_____ by listing all the tasks and subtasks required to complete the job. This approach is called (4)_____.

An alternative to this design method is the (5)_____ design, which develops the program based on objects. Objects that have similar attributes and behaviors can be grouped into (6)_____. The benefit of using this type of design approach is that objects and classes can be reused in other programs. If necessary, new classes can be made by first "borrowing" the attributes of an existing class and then adding differentiating attributes. This concept is known as (7)_____.

To select the best language for this problem, the team considers the resources at Whizgig. Although a lot of the programmers know the visual programming language of (8)_____, the most important factors for this application will be how fast it runs, so the language (9)_____ is selected. Because Whizgig has no Web presence the programmers will not be using (10)_____. Once the program has been written in programming language, the (11)_____ translates it to (12)_____ or the binary code that the CPU understands. Now, the program is ready to be tested.

The (13)_____ looks for errors in the program code. Then the programmers put together the necessary (14)_____ that explains the program and how to use it. However, since this is a program that is to be used internally, the program does not need to be (15)_____ by a group of potential outside users.

Organizing Key Terms

Instructions: This chapter introduced many new terms and concepts. In the illustration below, fill in each of the blanks in order to show how categories of ideas fit together.

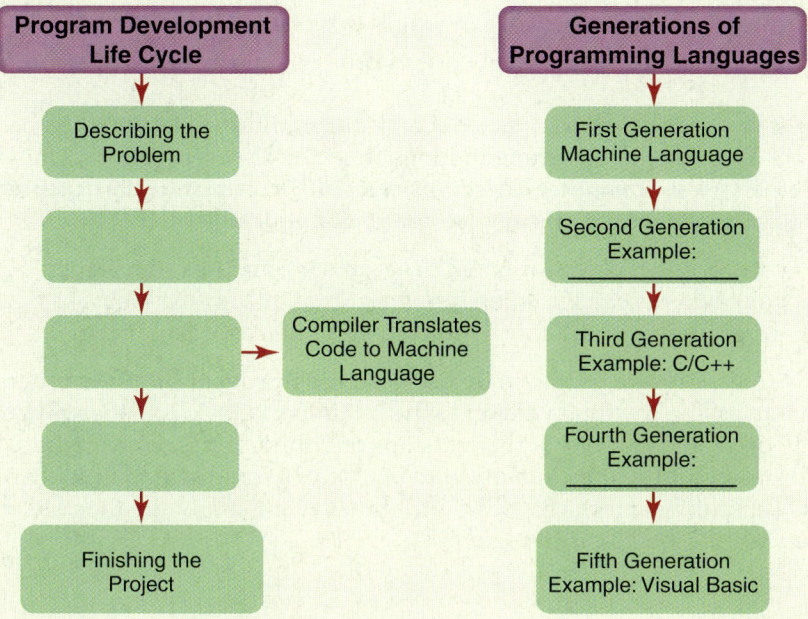

Making the Transition to . . . Next Semester

1. Research the core programming sequence at the college you are attending.

   a. How many courses are in the core sequence?
   b. How many languages does the sequence cover?
   c. How many object-oriented languages does your school offer?
   d. What is the Web-design sequence of courses?

2. How will you follow up an introductory programming class at your current school? Examine the recommendations from the Association for Computing Machinery (**www.computer.org/education/cc2001/steelman/cc2001/chapter7.htm**) and compare them with the course content at your school.

3. One of the original object-oriented programs is Smalltalk, developed by Xerox in its Palo Alto Research Center. Research Smalltalk on the Web and determine what programming situations would be best suited for Smalltalk.

4. Companies use beta testing to detect remaining problems in their programs before releasing the final version to the retail market. Go to **www.xbetas.com** or to any other site that offers beta versions of your favorite software title. What programs would you be interested to beta test? Why do you think it would make sense to beta test a software program? What, if any, are the risks involved in beta testing a software program?

APL Keyboard

Making the Transition to . . . The Workplace

1. If you work in a place that employs computer programmers, which languages do they work with?

    a. Which IDEs do they use?
    b. Why are these particular languages well matched to the goals of the department?

2. Check whether your office typically uses macros for automating common software tasks.

    a. Do users share macros or simply develop them individually?
    b. Are there other ways your office tries to automate routine tasks?

3. Using resources from the Web, determine which programming languages would be best to learn if you were going to program for the following industries:

    a. Animated movies
    b. Computer games
    c. Database management
    d. Robotics

Critical Thinking Questions

Instructions: Albert Einstein used "Gedanken experiments," or critical thinking questions, to develop his theory of relativity. Some ideas are best understood by experimenting with them in our own minds. The following critical thinking questions are designed to demand your full attention but require only a comfortable chair—no technology.

1. Think about what classes would be important in modeling a major league baseball team. What data and methods would each class need? How are the classes related to each other?

2. A common test for deciding the structure of a class hierarchy is the "is a" vs. "has a " test. For example, a motorcycle "has a" sidecar, so Sidecar would be a data field of a Motorcycle object. However, a motorcycle "is a" kind of vehicle, so Motorcycle would be a subclass of the base class Vehicle. Use the "is-a has-a" tests to decide how a class structure could be created for computer peripherals. Work to separate the unique features into objects and to extract the most common features into higher-level classes.

3. What do you think the computer programming language of the future should be able to do? How simple would it be? Would it use a graphical interface? Would it combine voice-recognition software with the coding process? Do you think computers will ever be able to create their own programs, or will humans always have to play a role in computer programming?

4. Some companies make their programmers sign agreements to prevent them from using code developed for one company with a new employer or competitor. When might this be an appropriate requirement? When might this not be necessary?

5. You learned about how programs go through a testing process to rid the program of errors. However, there are some people who are paid to create problems in the code. Why do you think this is necessary? What kinds of problems are these people trying to simulate?

Team Time

Problem:

You and your team have just been selected to write a software program that tells a vending machine how to make proper change from the bills or coins the customer inserts. The program needs to deliver the smallest possible amount of coins for each transaction.

Task:

Divide the class into three teams: Algorithm Design, Coding, and Testing. The responsibilities of each team are outlined below.

Process:

STEP 1: The Algorithm Design team is required to develop two documents. The first document should present the problem as a top-down design sequence of steps. The second document should use object-oriented analysis to identify the key objects in the problem. Each object needs to be represented as data and behaviors. Inheritance relationships between objects should be noted as well. You can use flowcharts to document your results.

STEP 2: The Coding team needs to decide which programming language would be the most appropriate for the project. This program needs to be fast and only take up a small amount of memory. Make sure your team defends its position by finding information about the language you select on the Web.

STEP 3: The Testing team must create a testing plan for the program. What set of inputs would you test with to be sure the program is completely accurate? Develop a table listing combinations of inputs and correct outputs.

STEP 4: As a group, discuss how each team would communicate their results to the other teams. Once one team has completed its work, are the team members finished or do they need to interact with the other teams?

Conclusion:

Any modern programming project requires programming teams to produce an accurate and efficient solution to the problem. The interaction of the team members within the team as well as with the other teams is vital to successful programming.

Becoming Computer Fluent

Your new manager wants to design and deploy an Internet application to collect marketing information about potential customers. She wants to gather data and analyze it in one report, which can be shipped to the marketing department.

Instructions: Write a memo to the manager describing what will have to be considered in the creation of this project. Write the letter using as many of the key terms from the chapter as you can.

Materials on the Web

In addition to the review materials presented here, you'll find extra materials on the book's companion Web site (**www.prenhall.com/techinaction**) that will help reinforce your understanding of the chapter content. These materials include:

Sound Byte Lab Guides

For each Sound Byte mentioned in the chapter, there is a corresponding lab guide located on the book's companion Web site. These guides review the material presented in the Sound Byte and direct you to various Web resources that examine the material. The Sound Byte Lab Guides for this chapter include:

- Programming for End Users: Macros
- Looping Around in the IDE

True/False and Multiple Choice Quizzes

The book's Web site includes a True/False and a Multiple Choice quiz for this chapter. You can take these quizzes, automatically check the results, and e-mail the results to your instructor.

Web Research Projects

The book's Web site also includes a number of Web research projects for this chapter. These projects ask you to search the Web for information on computer-related careers, milestones in computer history, important people and companies, emerging technologies, and the applications and implications of different technologies.

OBJECTIVES

After reading this chapter, you should be able to answer the following questions:

- What is a database and why is it beneficial to use databases? (pp. 430–433)

- What components make up a database? (pp. 432–435)

- What types of databases are there? (p. 436)

- What do database management systems do? (p. 437)

- How do relational databases organize and manipulate data? (pp. 442–451)

- What are data warehouses and data marts and how are they used? (pp. 452–456)

- What is an information system and what types of information systems are used in business? (pp. 457–461)

- What is data mining and how does it work? (pp. 461–462)

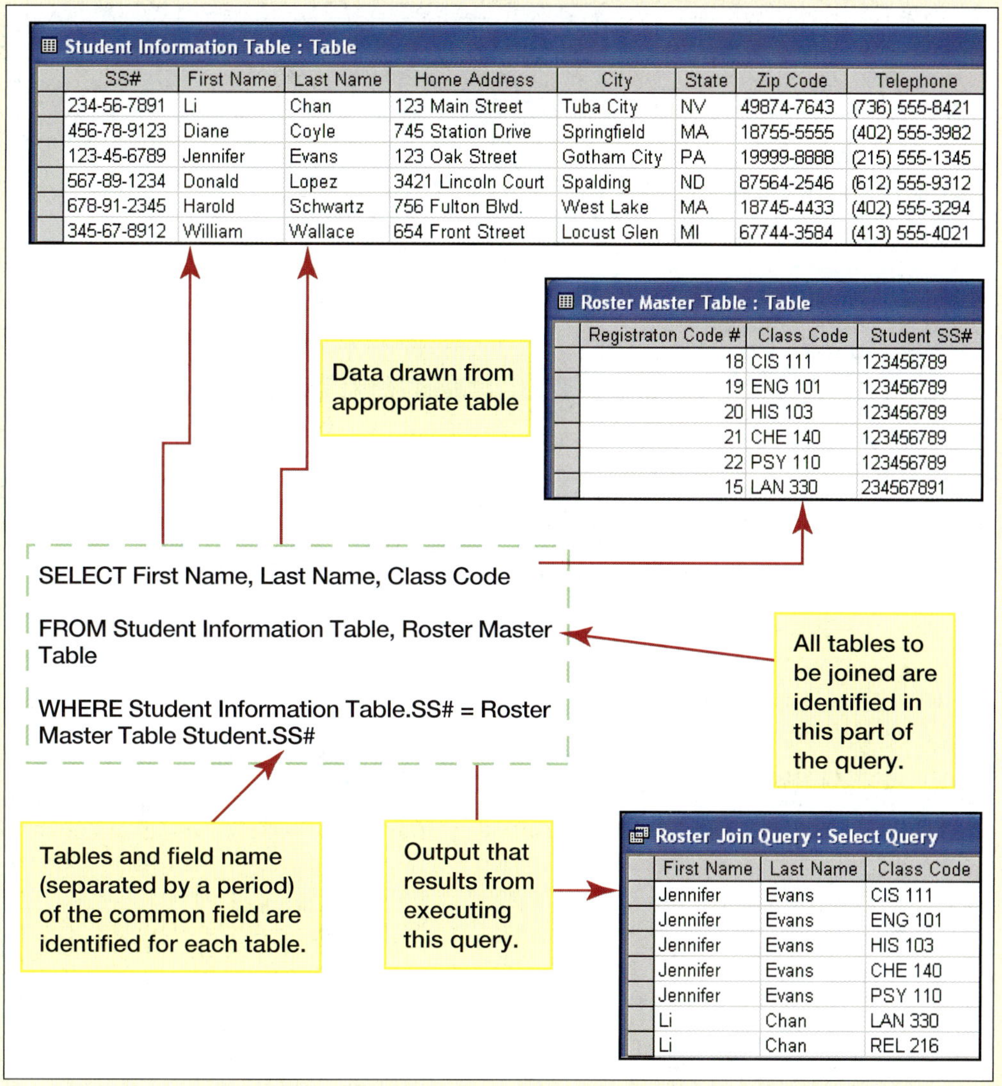

FIGURE 11.15

This join query will display a student roster for each student in the Student Information Table. Notice that the WHERE statement creates the join by defining the common fields (in this case SS# and Student SS#) in each table.

Student Information Table : Table

SS#	First Name	Last Name	Home Address	City	State	Zip Code	Telephone
234-56-7891	Li	Chan	123 Main Street	Tuba City	NV	49874-7643	(736) 555-8421
456-78-9123	Diane	Coyle	745 Station Drive	Springfield	MA	18755-5555	(402) 555-3982
123-45-6789	Jennifer	Evans	123 Oak Street	Gotham City	PA	19999-8888	(215) 555-1345
567-89-1234	Donald	Lopez	3421 Lincoln Court	Spalding	ND	87564-2546	(612) 555-9312
678-91-2345	Harold	Schwartz	756 Fulton Blvd.	West Lake	MA	18745-4433	(402) 555-3294
345-67-8912	William	Wallace	654 Front Street	Locust Glen	MI	67744-3584	(413) 555-4021

Roster Master Table : Table

Registraton Code #	Class Code	Student SS#
18	CIS 111	123456789
19	ENG 101	123456789
20	HIS 103	123456789
21	CHE 140	123456789
22	PSY 110	123456789
15	LAN 330	234567891

Data drawn from appropriate table

SELECT First Name, Last Name, Class Code
FROM Student Information Table, Roster Master Table
WHERE Student Information Table.SS# = Roster Master Table Student.SS#

All tables to be joined are identified in this part of the query.

Tables and field name (separated by a period) of the common field are identified for each table.

Output that results from executing this query.

Roster Join Query : Select Query

First Name	Last Name	Class Code
Jennifer	Evans	CIS 111
Jennifer	Evans	ENG 101
Jennifer	Evans	HIS 103
Jennifer	Evans	CHE 140
Jennifer	Evans	PSY 110
Li	Chan	LAN 330
Li	Chan	REL 216

The first line of the query contains variables for the field names you want to display. The FROM statement allows you to specify the table name from which the data will be retrieved. The last line (the WHERE statement) is used only when you wish to specify which records need to be displayed (such as all students with a GPA greater than 3.2). If you wish to display all the rows (records) in the table, you do not use the WHERE statement.

Suppose you want to create a telephone list from the Student Information Table, shown in Figure 11.14a, that includes all students. The SQL query you would send to the database would look like this:

SELECT (First Name, Last Name, Telephone)

FROM (Student Information Table)

Figure 11.14b shows the output from this query.

But what if you want a phone list only of students from Massachusetts? In that case, you would add a WHERE statement to the query as follows:

SELECT (First Name, Last Name, Telephone)

FROM (Student Information Table)

WHERE State = MA

This would restrict the output to students that live in Massachusetts, as shown in Figure 11.14c. Notice that the State field in the Student Information Table can be used by the query (in this case as a limiting criteria) but the contents of the State field are not required to be displayed in the query results. This explains why the output shown in Figure 11.14c doesn't show the State field.

When you want to extract data that is in two or more tables you use a **join query**. The query actually links (or joins) the two tables using the common field in both tables and extracts the relevant data from each. The format for a simple join query for two tables is as follows:

SELECT (Field Name 1, Field Name 2)

FROM (Table 1 Name, Table 2 Name)

WHERE (Table 1 Name.Common Field Name, Table 2 Name.Common Field Name)

AND (Selection Criteria)

Notice how similar this is to a select query, although the FROM statement must now contain two table names. Also, in a join query, the WHERE statement is split into two parts. In the first part (right after WHERE), the relation between the two tables is defined by identifying the common fields between the tables. The second part of the statement (after AND) is where the selection criteria are defined.

The AND means that both parts of the statement must be true for the query to produce results (that is, the two related fields must exist and the selection criteria must be valid). Figure 11.15 illustrates a join query for the Student Information Table and the Roster Master Table to produce a class roster for students.

FIGURE 11.16

(a) The Simple Query Wizard in Microsoft Access makes creating queries very easy. The wizard displays all fields available in your table so that you can select the ones you need to see.
(b) In this example, name and address information is selected and the output is sorted alphabetically by last name.

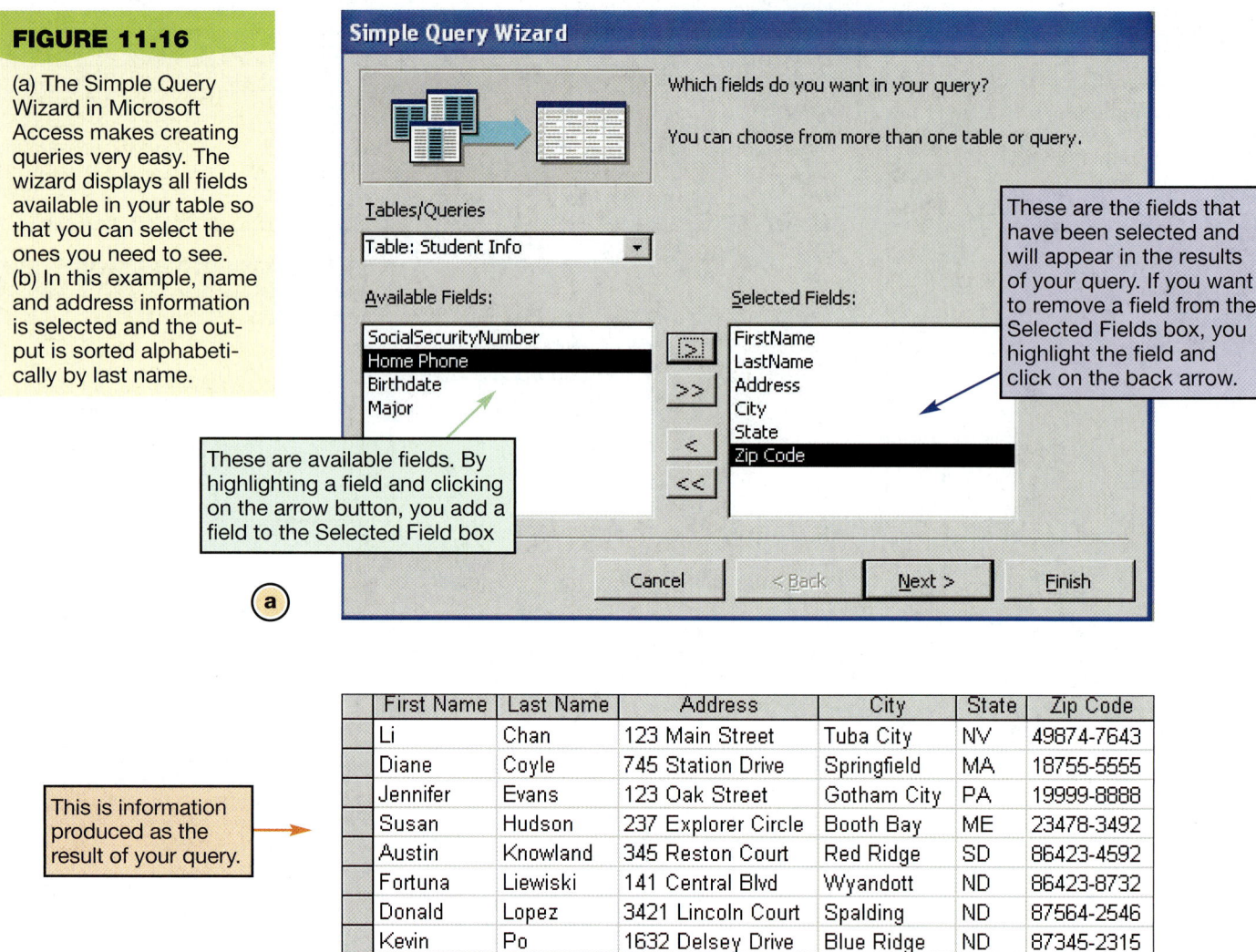

Do I have to learn a query language to develop queries for my database? Fortunately, modern database systems provide wizards to guide you through the process of creating queries. Figure 11.16 shows an example of an Access wizard being used to create a query. Not only does this speed up the process of creating queries, but you don't have to learn a query language.

Did the Simple Query Wizard use SQL to create the query? When you use Access, you're actually using SQL commands without realizing it. The Simple Query Wizard takes the criteria you specify and creates the appropriate SQL commands behind the scenes.

However, you may want to create your own SQL queries in Access, modify existing queries at the SQL language level, or view the SQL code that the wizard created. To do so, with a query open, select SQL View from the View menu. This displays the SQL code that comprises the query. Figure 11.17 shows the SQL code that the query in Figure 11.16 created.

FIGURE 11.17

The SQL View window shows the SQL code that the wizard created for the query in Figure 11.16. Although it is a relatively simple SELECT statement, it is much easier to create with the wizard.

TRENDS IN IT

COMPUTERS IN SOCIETY:
Databases Aid in Apprehending Criminals

When you watch police dramas on TV, detectives often use fingerprints to catch criminals. But until recently, the process of identifying suspects by their fingerprints was time consuming and ineffective in catching criminals who moved from state to state.

The use of fingerprints as a method of identification dates back to the beginning of the 20th century when the "Henry" method of fingerprint classification (still in use today) was developed in England. Fingerprints are unique to each individual and do not change during the course of a lifetime. Therefore, they are ideally suited for identification purposes in criminal investigations.

Originally, investigators collected fingerprints by placing ink on a person's fingertips and rolling his or her fingers on a piece of card stock. Fingerprint identification experts then manually compared new suspects' fingerprints with existing paper files of fingerprints. This was a time-consuming process, and the pool of potential suspects was limited to the fingerprints that the particular law enforcement agency had collected.

Lack of a large pool of fingerprints was less of a problem in the United States during the early 1900s since the population was not very mobile. However, in the 1930s, criminals became more mobile. The Federal Bureau of Investigation (FBI) thus undertook the maintenance of a large centralized manual database of fingerprint cards and photos of these cards, and provided fingerprint identification services to local law enforcement agencies. However, identification was still performed manually, and many local law enforcement agencies had neither the manpower nor the money to submit their fingerprint records to the FBI. The system, while still identifying many criminals, was inefficient.

During the 1960s and 1970s, the FBI began researching methods for using computers to perform fingerprint identification. These early efforts led to the Integrated Automated Fingerprint Identification System (IAFIS). Thanks to the IAFIS, instead of using ink and card stock, new fingerprints are now digitally scanned directly into a computer. In addition, existing fingerprint cards can be scanned and captured in the database. The IAFIS made fingerprint searching and identification much quicker through the use of its computer searching algorithms.

In the late 1990s, the FBI released a software package called Remote Fingerprint Editing Software (RFES). The RFES package (shown in Figure 11.18) allows local law enforcement agencies to capture fingerprints electronically and perform searches in the IAFIS database.

As more law enforcement agencies scan old fingerprint cards and post them to the IAFIS, the odds of finding a fingerprint match increases. In fact, criminal investigators are submitting fingerprints from old, unsolved cases to the IAFIS in attempts to find a match to fingerprints collected in cases where no suspects were found by conventional methods. In Pennsylvania in 2001, this approach resulted in the arrest and conviction of a suspect for a murder that had been unsolved for 14 years!

Developing an efficient, nationwide system of fingerprint identification would not be possible without the use of computerized databases. Databases allow the information to be stored, sorted, and retrieved quickly. Next time you're watching a police drama on television, don't forget that databases are working behind the scenes to identify the criminals.

FIGURE 11.18

Law enforcement agencies can use the FBI's Remote Fingerprint Editing Software (RFES) to search the IAFIS database, capture fingerprints from new suspects, or scan in information from existing fingerprint cards.

OUTPUTTING DATA

How do I get data out of a database? The most common form of output for any database is a printed report. Businesses routinely summarize the data within their databases and compile **summary data reports**. For instance, your school prints a grade report for you at the end of a semester that shows the classes you took and the grades you received.

Database systems can also be used to **export** data to other applications. Exporting data involves putting it into an electronic file in a format that another application can understand. For example, the query shown in the wizard in Figure 11.16 may be used to generate a list of mailing labels. In that case, the query output would be directed to a file that could be easily **imported** into Microsoft Word so that the data could be used in the Mail Merge tool to generate labels.

In the next section, we'll look at the operation of relational databases and explore how relationships are established among tables in these databases.

Relational Database Operations

As explained earlier, relational databases operate by organizing data into various tables based on logical groupings. For example, all student address and contact information (phone numbers, e-mail addresses, and so on) would be grouped into one table. Since all of the data in a relational database will not be stored in the same table, a methodology must be implemented to *link* data between tables.

In relational databases, the links between tables that define how the data is related are referred to as **relationships**. To establish a relationship between two tables, both tables must have a common field (or column). Common fields contain the same data (such as social security numbers), as shown in Figure 11.19.

(a) Student Information Table : Table

SS#	First Name	Last Name	Home Address	City	State	Zip Code	Telephone
234-56-7891	Li	Chan	123 Main Street	Tuba City	NV	49874-7643	(736) 555-8421
456-78-9123	Diane	Coyle	745 Station Drive	Springfield	MA	18755-5555	(402) 555-3982
123-45-6789	Jennifer	Evans	123 Oak Street	Gotham City	PA	19999-8888	(215) 555-1345
567-89-1234	Donald	Lopez	3421 Lincoln Court	Spalding	ND	87564-2546	(612) 555-9312
678-91-2345	Harold	Schwartz	756 Fulton Blvd.	West Lake	MA	18745-4433	(402) 555-3294
345-67-8912	William	Wallace	654 Front Street	Locust Glen	MI	67744-3584	(413) 555-4021

Common field in each table

(b) Roster Master Table : Table

Registraton Code #	Class Code	Student SS#
15	LAN 330	234567891
16	REL 216	234567891
17	ENG 102	234567891
18	CIS 111	123456789
19	ENG 101	123456789
20	HIS 103	123456789
21	CHE 140	123456789
22	PSY 110	123456789
23	HIS 401	567891234
24	SOC 310	567891234
25	PSY 110	567891234
26	ENG 102	678912345
27	PSY 110	678912345
28	HIS 204	345678912
29	PEH 125	345678912
30	SOC 220	345678912

FIGURE 11.19

(a) The Student Information Table and the (b) Roster Master Table share the common field of student social security number. This allows a relationship to be established between the two tables.

RELATIONAL DATABASE OPERATIONS

	1	Class Registration - Fall 2004								
	2										# of
	3	SS #	Last Name	First Name	Home Address	City	State	Zip Code	Class Code	Class Name	Credits
	4	234567891	Chan	Li	123 Main Street	Tuba City	NV	49874-7643	LAN 330	Japanese 1	3
	5	234567891	Chan	Li	123 Main Street	Tuba City	NV	49874-7643	REL 216	Early Buddhism	3
	6	234567891	Chan	Li	123 Main Street	Tuba City	NV	49874-7643	ENG 102	English Comp 2	3
	7	456789123	Coyle	Diane	745 Station Drive	Springfield	MA	18755-5555			
	8	123456789	Evans	Jennifer	123 Oak Street	Gotham City	PA	19999-8888	CIS 111	Programming	3
	9	123456789	Evans	Jennifer	123 Oak Street	Gotham City	PA	19999-8888	ENG 101	English Comp 1	3
	10	123456789	Evans	Jennifer	123 Oak Street	Gotham City	PA	19999-8888	HIS 103	Western Civ	3
	11	123456789	Evans	Jennifer	123 Oak Street	Gotham City	PA	19999-8888	CHE 140	Chemistry	4
	12	123456789	Evans	Jennifer	124 Oak Street	Gotham City	PA	19999-8888	PSY 110	Intro to Psychology	3
	13	567891234	Lopez	Donald	3421 Lincoln Court	Spalding	ND	87564-2546	HIS 401	16th Century Europe	3
	14	567891234	Lopez	Donald	3421 Lincoln Court	Spalding	ND	87564-2546	SOC 310	Interpersonal Relationships	3
	15	567891234	Lopez	Donald	3421 Lincoln Court	Spalding	ND	87564-2546	PSY 110	Intro to Psych	3
	16	678912345	Schwartz	Harold	756 Fulton Blvd.	West Lake	MA	18745-4433	ENG 102	English Comp 2	3
	17	678912345	Schwartz	Harold	756 Fulton Blvd.	West Lake	MA	18745-4433	PSY 110	Intro to Psych	3
	18	345678912	Wallace	William	654 Front Street	Locust Glen	MI	67744-3584	HIS 204	Scottish History	3
	19	345678912	Wallace	William	654 Front Street	Locust Glen	MI	67744-3584	PEH 125	Fencing 1	3
	20	345678912	Wallace	William	654 Front Street	Locust Glen	MI	67744-3584	SOC 220	Negotiation	2

FIGURE 11.20

Data from unrelated topics is located in the Class Registration List. The column headings in blue are related to student contact data while the column headings in red relate to course enrollment information. To construct a database, these topics should be contained in separate tables.

NORMALIZATION OF DATA

How do I decide what tables I need and what data to put in them? You create database tables (or files) for two reasons: to hold unique data about a person or thing and to describe unique events or transactions. In databases, the goal is to reduce data redundancy by recording data only once. This process is called **normalization** of the data. Yet the tables must still work well enough together to enable you to retrieve the data when you need it. Tables should be grouped using logical data that can be identified uniquely.

Let's look at an example. In Figure 11.20, the Class Registration List contains a great deal of data related to individual students and their course registration. However, each table in a relational database should contain a *related group* of data on a *single topic*. There are two distinct topics in this list: student contact information and student registration information. Therefore, this list needs to be divided into two tables so the distinct data (contact data and registration data) can be categorized appropriately.

The Student Information Table shown in Figure 11.21 organizes all of the student contact information found in Figure 11.20 into a separate table.

Student Information Table : Table

SS#	First Name	Last Name	Home Address	City	State	Zip Code
234-56-7891	Li	Chan	123 Main Street	Tuba City	NV	49874-7643
456-78-9123	Diane	Coyle	745 Station Drive	Springfield	MA	18755-5555
123-45-6789	Jennifer	Evans	123 Oak Street	Gotham City	PA	19999-8888
567-89-1234	Donald	Lopez	3421 Lincoln Court	Spalding	ND	87564-2546
678-91-2345	Harold	Schwartz	756 Fulton Blvd.	West Lake	MA	18745-4433
345-67-8912	William	Wallace	654 Front Street	Locust Glen	MI	67744-3584

FIGURE 11.21

Student contact data is grouped in the Student Information Table and needs to be entered only once for each student. The primary key for each record is a unique social security number.

Class Registration List Fall 2004 : Table

SS #	Class Code	Class Name	# Of Credits
123456789	PSY 110	Intro to Psychology	3
123456789	CIS 111	Programming	3
123456789	ENG 101	English Comp 1	3
123456789	HIS 103	Western Civ	3
123456789	CHE 140	Chemistry	4
234567891	LAN 330	Japanese 1	3
234567891	ENG 102	English Comp 2	3
234567891	REL 216	Early Buddhism	3
345678912	SOC 220	The Art of Negotiation	2
345678912	HIS 204	Scottish History	3
345678912	PEH 125	Fencing 1	3
567891234	PSY 110	Intro to Psychology	3
567891234	SOC 310	Interpersonal Relationships	3
567891234	HIS 401	16th Century Europe	3
678912345	ENG 102	English Comp 2	3
678912345	PSY 110	Intro to Psychology	3

FIGURE 11.22

Although it contains related data (registration information), this table still contains a great deal of duplicate data and no usable primary key.

Names and course numbers must be duplicated for each student taking the course

Also, there is no unique field that can be used as a primary key. SS# cannot be used as the primary key since the same SS# will be entered on multiple records when a student enrolls in more than one course.

Notice that the information for each student needs to be shown only once instead of multiple times as in the list in Figure 11.20. The unique primary key for this table is the student's social security number. Student information might be needed in a variety of instances and by a variety of departments but it needs to reside only in this one database table, which many departments of the school can share.

Next, we could put the registration data found in Figure 11.20 for each student in a separate table, as shown in Figure 11.22. There is no need to repeat student name and address data in this table. Instead, each student can be identified by his or her social security number. However, there are problems with this table. Each class name and class code has to be repeated for every student taking the course, and there is no unique field that can be used as a primary key for this table.

What can be done to fix the table in Figure 11.22? In Figure 11.22, we have identified more data that should be grouped logically into another separate table: class code and class name. Therefore, we should create another table for just this information. This will allow us to avoid repeating class names and codes. Figure 11.23 shows the Course Master Table. Note that the class code is unique for every course and acts as a primary key in this table.

To solve the other problems with the table in Figure 11.22, you need a way to uniquely identify each student registration for a specific course. This can be solved by creating a course registra-

Course Master Table : Table

Class Code	Class Name	Credits
CHE 140	Chemistry	4
CIS 111	Programming	3
ENG 101	English Comp 1	3
ENG 102	English Comp 2	3
HIS 103	Western Civ	3
HIS 204	Scottish History	3
HIS 401	16th Century Europe	3
LAN 330	Japanese 1	3
PEH 125	Fencing 1	3
PSY 110	Intro to Psychology	3
REL 216	Early Buddhism	3
SOC 220	The Art of Negotiation	2
SOC 310	Interpersonal Relations	3

FIGURE 11.23

Related information about courses (class code, class name, and the credits for the class) is grouped logically in one table. The unique class code is the primary key.

RELATIONAL DATABASE OPERATIONS   451

Roster Master Table : Table

Registration Code #	Class Code	Student SS#
15	LAN 330	234567891
16	REL 216	234567891
17	ENG 102	234567891
18	CIS 111	123456789
19	ENG 101	123456789
20	HIS 103	123456789
21	CHE 140	123456789
22	PSY 110	123456789
23	HIS 401	567891234
24	SOC 310	567891234
25	PSY 110	567891234
26	ENG 102	678912345
27	PSY 110	678912345
28	HIS 204	345678912
29	PEH 125	345678912
30	SOC 220	345678912

FIGURE 11.24

The Roster Master Table shows only pertinent data related to a student's registration. Only three fields are needed: the registration code number (which is the unique primary key), the class code, and the student's social security number. This approach greatly minimizes duplicate data.

tion number that will be unique and assigned by the database as records are entered. Figure 11.24 shows the resulting Roster Master Table.

How do I get the data in the tables to work together now that it is split up? The entire premise behind relational databases is that *relationships* are established among the tables to allow the data to be shared. As noted earlier, to establish a relationship between two tables, the tables must have a common field (column). This usually involves the primary keys of a table.

For instance, to track registrations by student in the Roster Master Table in Figure 11.24, the social security number of the student is the logical piece of data to use. The social security number is the primary key in the Student Information Table in Figure 11.21; however, in the Roster Master Table, the social security number is called a **foreign key**—the primary key of another table that is included for purposes of establishing relationships with that other table. Figure 11.25, on the next page, shows the relationships that exist among the Course Master, Roster Master,

and Student Information Tables. Relationships among tables can be established whenever you need them.

Since relationships are vital to the operation of the database, it is important to ensure that there are no inconsistencies in the data entered in the common fields of two tables. Each foreign key (Student SS# in the Roster Master Table in Figure 11.24) entered into a table must be a valid primary key from the related table (SS# from the Student Information Table in Figure 11.21).

For instance, if 392-13-5684 is not a valid social security number for any student listed in the Student Information Table, then it should not be entered into the Roster Master Table. Each entry in the Roster Master Table has to correspond to a student (linked by his or her SS#) in the Student Information Table. If this requirement is not applied to foreign keys, a relationship cannot be established between tables.

How do I ensure that a foreign key field contains a valid primary key from the related table? To apply this restraint, when defining a relationship in a database, you have the option of enforcing referential integrity for that relationship. **Referential integrity** means that for each value in the foreign key of one table, there is a corresponding value in the primary key of the related table.

For instance, if you attempt to enter a record in the Roster Master Table with a SS# of 156-78-4522 and referential integrity is being enforced, the database will check to ensure that a record with SS# 156-78-4522 exists in the Student Information Table. If the corresponding record does not exist, an error message will be displayed. Establishing referential integrity between two tables helps prevent inconsistent data from being entered.

All of the data that is collected in databases needs to be stored and managed. In the next section, we'll explore the types of systems where databases are typically used today.

SOUND BYTE
IMPROVING AN ACCESS DATABASE

In this Sound Byte, you'll learn how to create input forms, queries, and reports to simplify maintenance of your CD database. You'll follow along step by step using Microsoft Access wizards to create and modify queries to suit your needs.

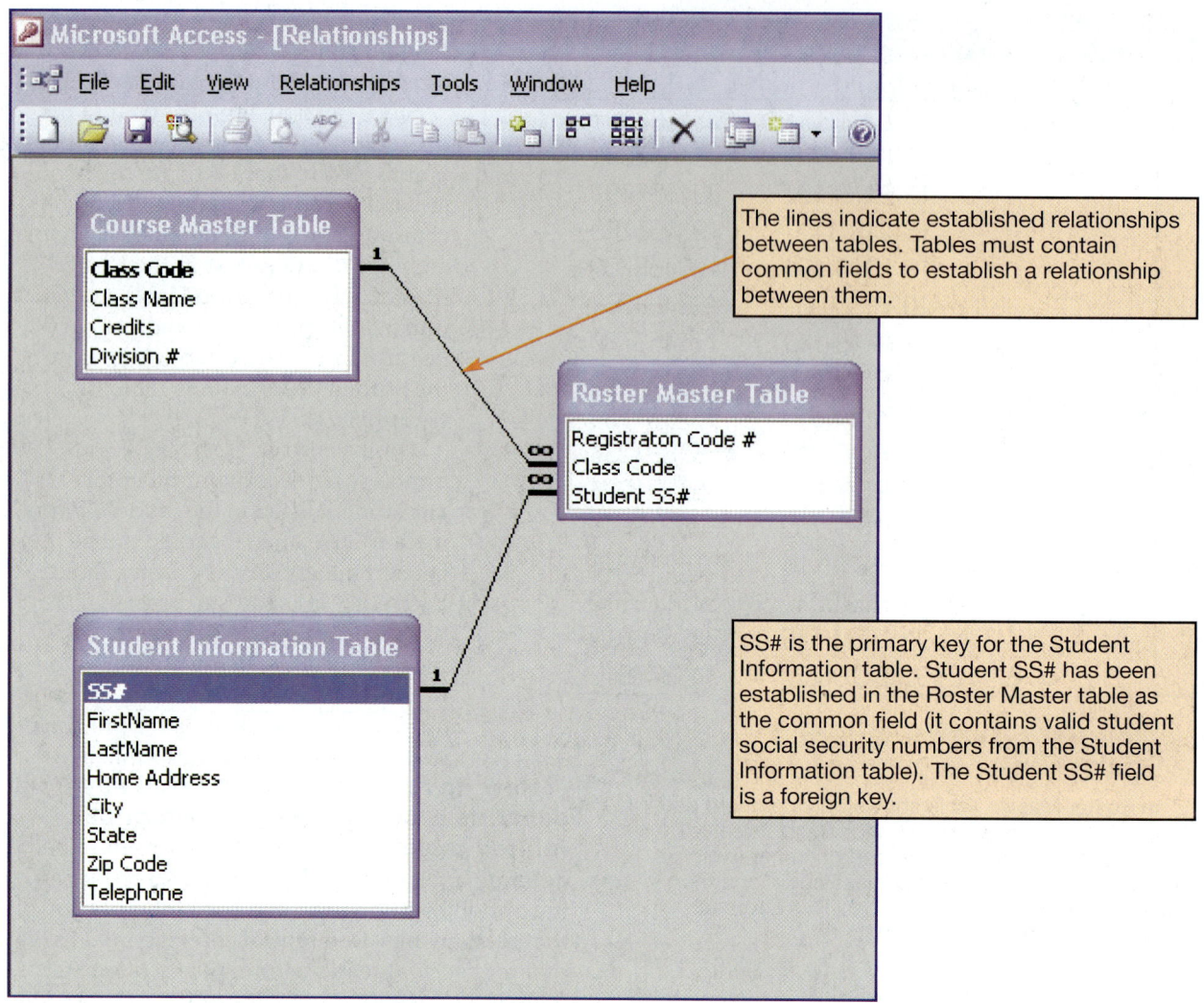

FIGURE 11.25

The Relationships screen in Microsoft Access displays a visual representation of the relations established between tables. Notice that foreign keys in related tables do not have to have the same field names as the primary keys in the other table. They merely have to contain the same type of data.

Data Storage

At the simplest level, data is stored in a single database on a database server and you retrieve the data as needed. This works fine for small databases and when all of the data you are interested in is in a single database. But problems arise when the data you need is not all in one convenient spot. Large storage repositories called data warehouses and data marts help solve this problem.

DATA WAREHOUSES

What is a data warehouse? A data warehouse is a large-scale electronic repository of data that contains and organizes all the data related to an organization in one place. Individual databases contain a wealth of information, but each database's information usually pertains to one topic.

For instance, the order database at Amazon.com contains information about book orders, such as name, address, payment information, and book name. However, the order database does not contain information on inventory levels of books, nor does it list suppliers where out-of-stock books can be obtained. Data warehouses, therefore, consolidate information from disparate sources to present an enterprisewide view of business operations.

Is data in a data warehouse organized the same way as in a normal database? Data in the data warehouse is organized by

TRENDS IN IT

COMPUTERS IN SOCIETY: Need Cash? Use Databases to Find Your Property

Did you ever lose a check before cashing it and forget to get it replaced (or figure you couldn't)? Did you ever relocate quickly and forget to get your security deposit back on your apartment? Did your grandmother open a bank account in your name and forget to tell you about it? These are all examples of unclaimed property. And thanks to the Internet and databases, it has never been easier to look for this "found money."

In most states, unclaimed property (bank accounts, security deposits on apartments, uncashed checks, and so on) is required by law to be turned over to the state treasury for safekeeping after a certain period of time. The state treasury is required to hold the property for a period of time (sometimes forever) to see if the rightful owner can make a claim. Until a few years ago, searching for your unclaimed property meant sifting through mountains of paper records at the state capitol.

But today, many states have unclaimed property databases deployed on the Web. Go to the official state Web site for the state where you think you may have unclaimed property and search on the terms "unclaimed property" or "abandoned property." California's site, located at www.sco.ca.gov/col/ucp, features a simple search box where you can enter your name to search for your pot of gold (see Figure 11.26). If you find an item that might be yours, follow the procedures set down by that state to make a claim and prove ownership.

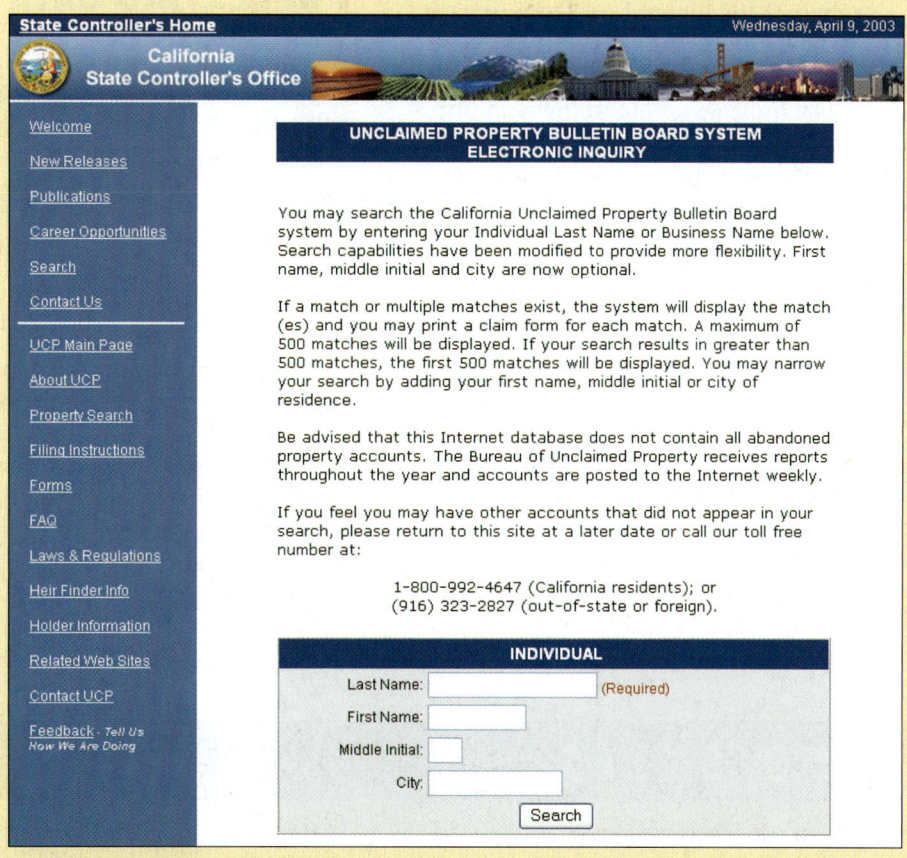

FIGURE 11.26

The state of California's unclaimed property Web site, at www.sco.ca.gov/col/ucp, might be just the place to find your missing funds.

subject. Most databases focus on one specific operational aspect of business operations. For example, insurance companies sell many types of insurance, such as life, automobile, and homeowners' insurance. Different divisions of the insurance company are responsible for each type of insurance and track the policies they sell in different databases (one for automobile insurance policy sales and one for life insurance policy sales, for example), as shown in Figure 11.27.

These databases capture specific information about each type of policy. The Automobile Policy Sales database captures information about driving accident history, car model, and the age and gender of the drivers since this is pertinent to the pricing of car insurance policies. The Life Insurance Policy Sales database captures information about the age and gender of the policyholder and whether the insured smokes, but does not include details about cars or driving records.

However, total policies sold (and the resulting revenue generated) is critical to the management of the insurance company no matter what type of policy is involved. Therefore, an insurance company's data warehouse would have a subject called Policy Sales Subject (as shown in Figure 11.27) that would contain information about *all* policies sold throughout the company. The Policy Sales Subject is a database that contains information from the other databases the company maintains. However, all data in the Policy Sales Subject database is specifically related to policy sales.

From the Policy Sales Subject database, it is easy for managers to produce comprehensive reports such as the Total Policy Sales Report, as shown in Figure 11.27, which can contain information pertaining to all policy sales.

Are data warehouses much larger than conventional databases? Data warehouses, like conventional warehouses, are vast repositories of information. The data contained within them is not operational in nature, but rather archival. Data warehouse data is **time-variant data**, meaning it doesn't all pertain to one period in time.

The warehouse contains current values, such as amounts due from customers, as well as **historical data**. If you want to examine the buying habits of a certain type of customer, you need data about both current and prior purchases. Having time-variant data in the warehouse allows you to analyze the past, examine the present in light of historical data, and make projections about the future.

POPULATING DATA WAREHOUSES

How are data warehouses populated with data? Source data for data warehouses can come from three places:

1. Internal sources (such as company databases)
2. External sources (suppliers, vendors, and so on)
3. Customers or visitors to the company Web site

Internal data sources are obvious. Sales, billing, inventory, and customer databases all provide a wealth of information. However, internal information is not contained exclusively in databases. Spreadsheets and other ad hoc analysis tools may contain data that needs to be loaded into the warehouse.

External data sources include vendors and suppliers that often provide data regarding prod-

BITS AND BYTES

Measuring Large Databases

Some databases take up only a few megabytes of space. However, businesses generate very large databases. It is widely accepted that the average business doubles the amount of data it accumulates every year. So how big are a large corporation's databases? Many large databases are now measured in terabytes. A terabyte is 2^{40} bytes or over 1 trillion bytes of data. If the bytes were typed pages of paper, a terabyte of pages would form a stack 51 miles high!

But with data doubling every year, soon we'll be measuring databases in petabytes (that's 1,024 terabytes or 2^{50} bytes). Now the stack of paper is about 52,000 miles high. And computer scientists are already prepared for even more data: an exabyte is a staggering 1,024 petabytes or 2^{60} bytes. Large government scientific research projects already have databases spanning hundreds of terabytes. If databases double in size every year, it won't be long before there is a petabyte of data stored somewhere near you.

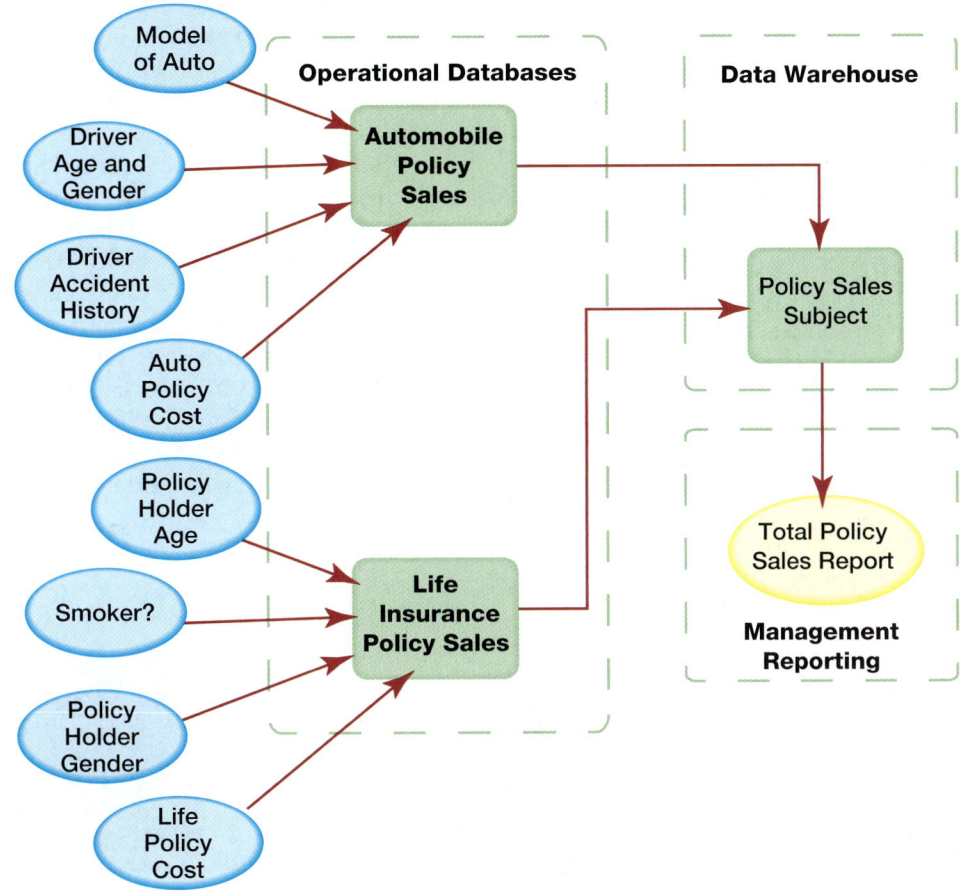

FIGURE 11.27

Data from individual databases is drawn together under appropriate subject headings in a data warehouse. Managers can then produce comprehensive reports that would be impossible to create from the individual databases.

uct specifications, shipment methods and dates, electronic billing information, and so on. In addition, a virtual wealth of customer (or potential customer) information is available by monitoring the clickstream of the company Web site.

What is a clickstream and why is it important? Companies can use software on their Web sites to capture information about each click that users make as they navigate through the site. This information is referred to as **clickstream data**. Monitoring the clickstream helps managers assess the effectiveness of a Web site. Using clickstream data-capture tools, a company can determine what pages users visit most often, how long users stay on each page, which sites directed users to the company site, and the user demographics. This data can provide valuable clues to what a company needs to improve on its site to stimulate sales.

DATA STAGING

Does all source data fit into the warehouse? No two source databases are the same. Therefore, although two databases might contain similar information (such as customer names and addresses), the format of the data is most likely different in each database. Therefore, source data must be "staged" before entering the data warehouse. **Data staging** consists of three steps:

1. Extraction of the data from source databases
2. Transformation (reformatting) of the data
3. Storage of the data in the warehouse

456   CHAPTER 11   BEHIND THE SCENES: DATABASES AND INFORMATION SYSTEMS

Many different software programs and procedures may have to be created to extract the data from varied sources and to reformat it for storage in the data warehouse. The nature and complexity of the source data determines the complexity of the data staging process; it is different for every data warehouse.

Once the data is stored in the data warehouse, how can it be extracted and used? Managers can query the data warehouse in much the same way you query an Access database. However, since there is more data in the warehouse, significantly more flexible tools are needed to perform such queries. Online analytical processing (OLAP) software provides standardized tools for viewing and manipulating data in a data warehouse. The key feature of OLAP tools is that they allow for flexible views of the data, which the software user can easily change.

DATA MARTS

Is finding the right data in a huge data warehouse difficult? Looking for the data you need in a data warehouse can be daunting when there are terabytes of data. Therefore, small slices of the data warehouse, called **data marts**, are often created. Whereas data warehouses have an enterprisewide depth, the information in data marts pertains to a single department.

For instance, if you work in the sales department, you need accurate sales-related information at your fingertips—and you would not want to wade through customer service data, accounts payable data, and product shipping data to get it. Therefore, a data mart that contains information relevant only to the sales department can be created to make the task of finding this data easier. The entire data warehousing process is illustrated in Figure 11.28.

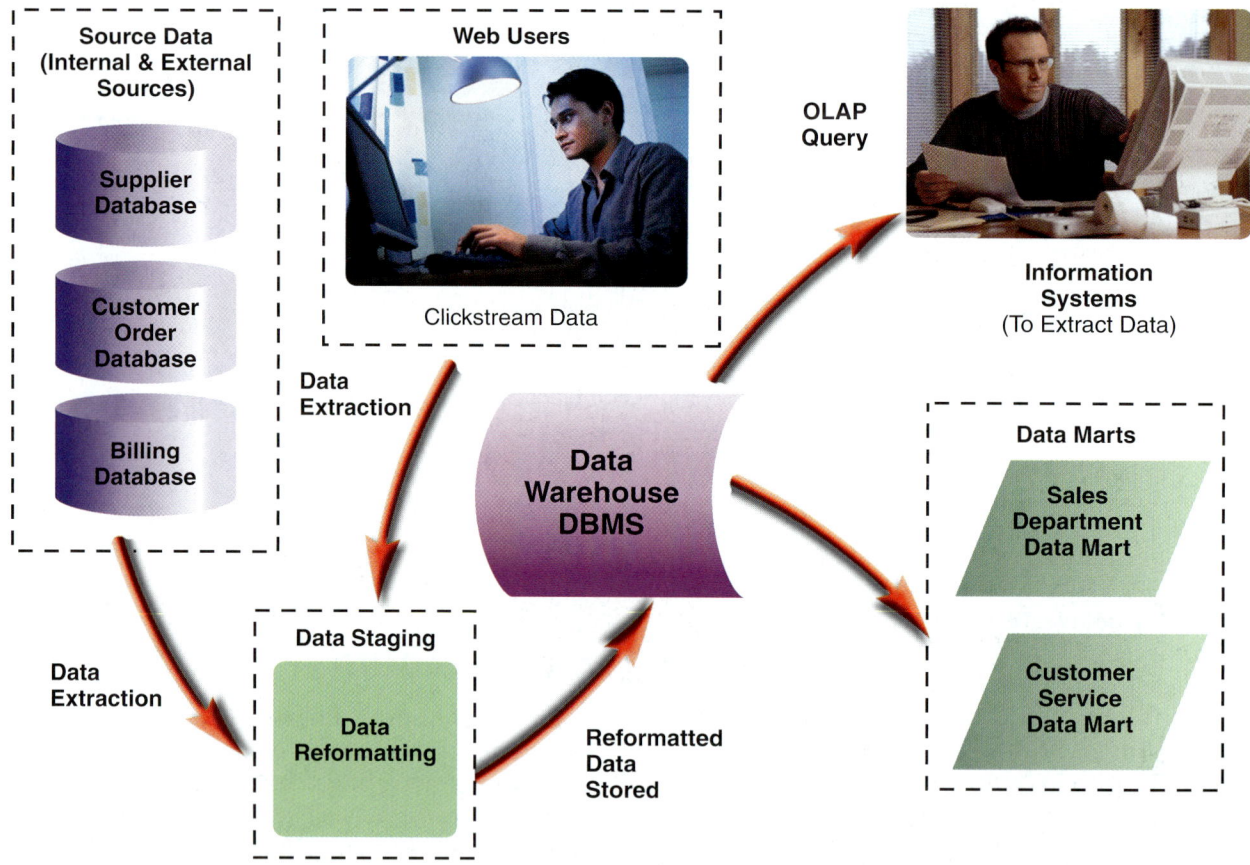

FIGURE 11.28

Shown here is an overview of the data warehouse process. Data staging is vital as different data must be extracted and then reformatted to fit the data structure defined in the data warehouse DBMS. Data can be extracted using powerful OLAP query tools or it can be stored in specialized data marts for use by specific employee groups.

Managing Data: Information Systems

Making intelligent decisions about developing new products, creating marketing strategies, and buying raw materials requires timely, accurate information. **Information systems** are software-based solutions used to gather and analyze information. A system that delivers up-to-the-minute sales data on books to the computer of Amazon.com's president is one example of an information system. Databases, data marts, and data warehouses are integral parts of information systems since they store the information that makes information systems functional.

All information systems perform similar functions, including acquiring data, processing that data into information, storing the data, and providing the user with a number of output options with which to make the information meaningful and useful, as shown in Figure 11.29. Most information systems fall into one of four categories: office support systems, transaction processing systems, management information systems, and decision support systems. Each type of system almost always involves the use of one or more databases.

OFFICE SUPPORT SYSTEMS

What does an office support system accomplish? An **office support system (OSS)** is designed to assist employees in accomplishing their day-to-day tasks and to improve communications. Microsoft Office is an example of an OSS as it assists employees with routine tasks such as maintaining an employee phone list in Excel.

Modern OSSs include software tools with which you are probably familiar, including

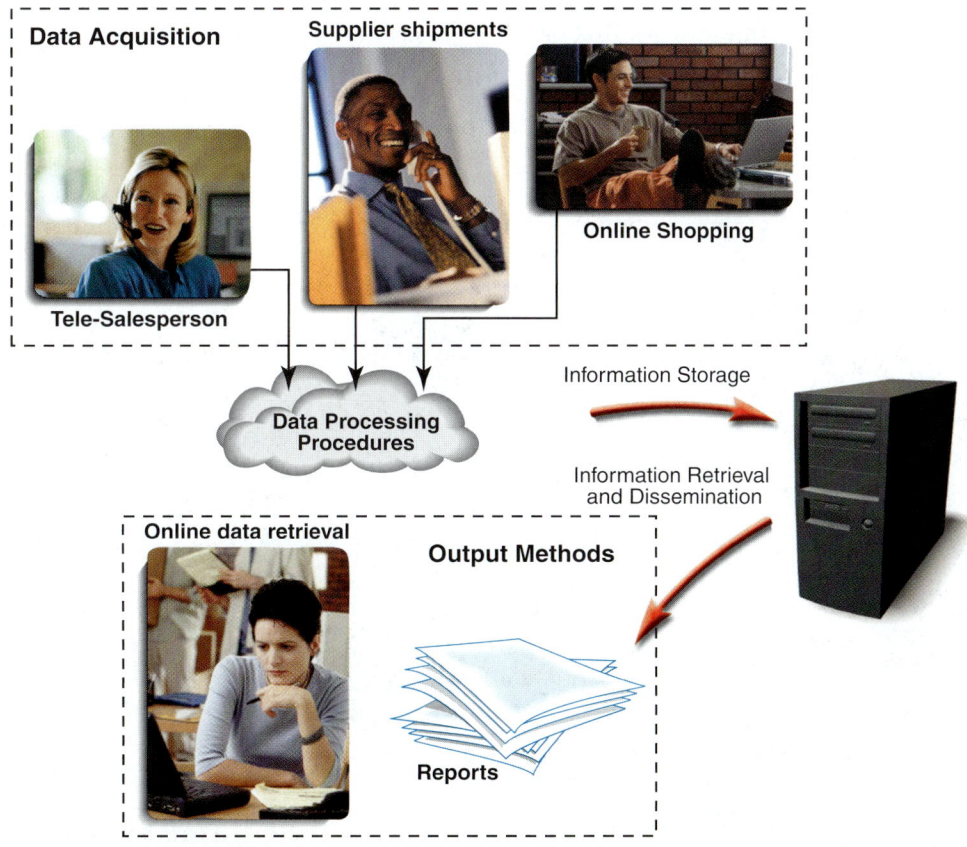

FIGURE 11.29

All information systems perform similar functions, including acquiring data, processing that data into information, storing the data, and providing the user with a number of output options with which to make the information meaningful and useful.

e-mail, word processing, spreadsheet, database, and presentation programs. OSSs had their roots in manual, paper-based systems that were developed before computers. After all, maintaining a company phone listing was necessary long before computers were invented. A paper listing of employee phone extensions typed by an administrative assistant is an example of an early OSS. A modern OSS system might publish this directory on the company's intranet (its internal network).

TRANSACTION PROCESSING SYSTEMS

What is a transaction processing system? A **transaction processing system (TPS)** is used to keep track of everyday business activities. For example, at your college, transactions that occur frequently include registering students for classes, accepting tuition payments, mailing course listings, and printing course catalogs. Your college has TPSs in place to track these types of activities.

When computers were introduced to the business world, they were often put to work first as TPSs. Computers were much faster at processing large chunks of data than previous manual systems. Imagine having clerks typing up tuition invoices for each student at a 10,000-student university. Obviously, a computer can print invoices much quicker from a database.

How do transactions enter a TPS? Transactions can be entered manually or electronically. When you call a company and order a sweater, for example, the call taker enters your data into a TPS. When you purchase gasoline at a pay-at-the-pump terminal, the pump captures your credit card data and transmits it to a TPS, which automatically records a sale (gallons of gasoline and dollar value). Transactions are either processed in batches or in real time. Various departments in an organization then access the TPSs to extract the information they need to process additional transactions, as shown in Figure 11.30.

What is batch processing? **Batch processing** means that transaction data is accumulated until a certain point is reached, then a number of transactions are processed all at once. Batch processing is appropriate for activities that are not time sensitive, such as developing a mailing list to mail out the new course catalogs that students have requested. A mailing label could be printed for each person as he or she requests a catalog, but it is more efficient to batch the requests and process them all at once when the catalogs are ready to be addressed.

How does real-time processing work? For most activities, processing and recording transactions in a TPS occur in real time. **Real-time processing** means that the database is queried and updated while the transaction is taking place. For instance, when you register for classes, the registration clerk checks to make sure seats are still available for the classes you want and records your registration in the class immediately. This **online transaction processing (OLTP)** ensures that the data in the TPS is as up to date as possible.

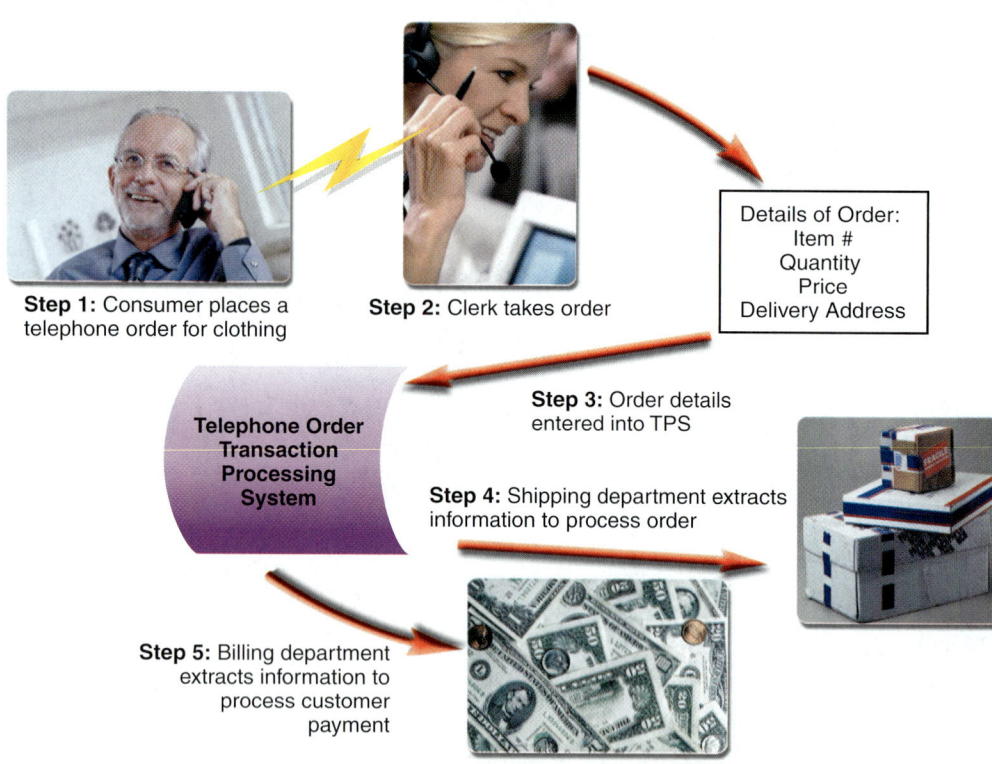

FIGURE 11.30

Transaction processing systems (TPSs) help capture and track critical business information needed to successfully complete business transactions, such as the selling of merchandise via the telephone.

MANAGEMENT INFORMATION SYSTEMS

What is a management information system? A **management information system (MIS)** provides timely and accurate information that enables managers to make critical business decisions. MISs were a direct outgrowth of TPSs. Managers quickly realized that the data contained in TPSs was an extremely powerful tool only if the information could be organized and output in a useful form. Today's MISs are therefore often built in as a feature of TPSs.

What does an MIS provide that a TPS does not? The original TPSs were usually designed to output detail reports. A **detail report** provides a list of the transactions that occurred during a certain time period. For example, during registration periods at your school, the registrar might receive a detail report that lists the students who registered for classes each day. Figure 11.31a shows an example of a detail report on daily enrollment.

Going beyond the detail reports provided by TPSs, MISs provide summary reports and exception reports. **Summary reports** provide a consolidated picture of detailed data. These reports usually include some calculation (totals) or visual displays of information (such as charts and graphs). Figure 11.31b shows an example of a summary report displaying total daily enrollment.

Exception reports show conditions that are unusual or that need attention by users of the system. The registrar at your college may get an exception report when all sections of a course are full, indicating that it may be time to schedule additional sections. Figure 11.31c shows an example of such an exception report.

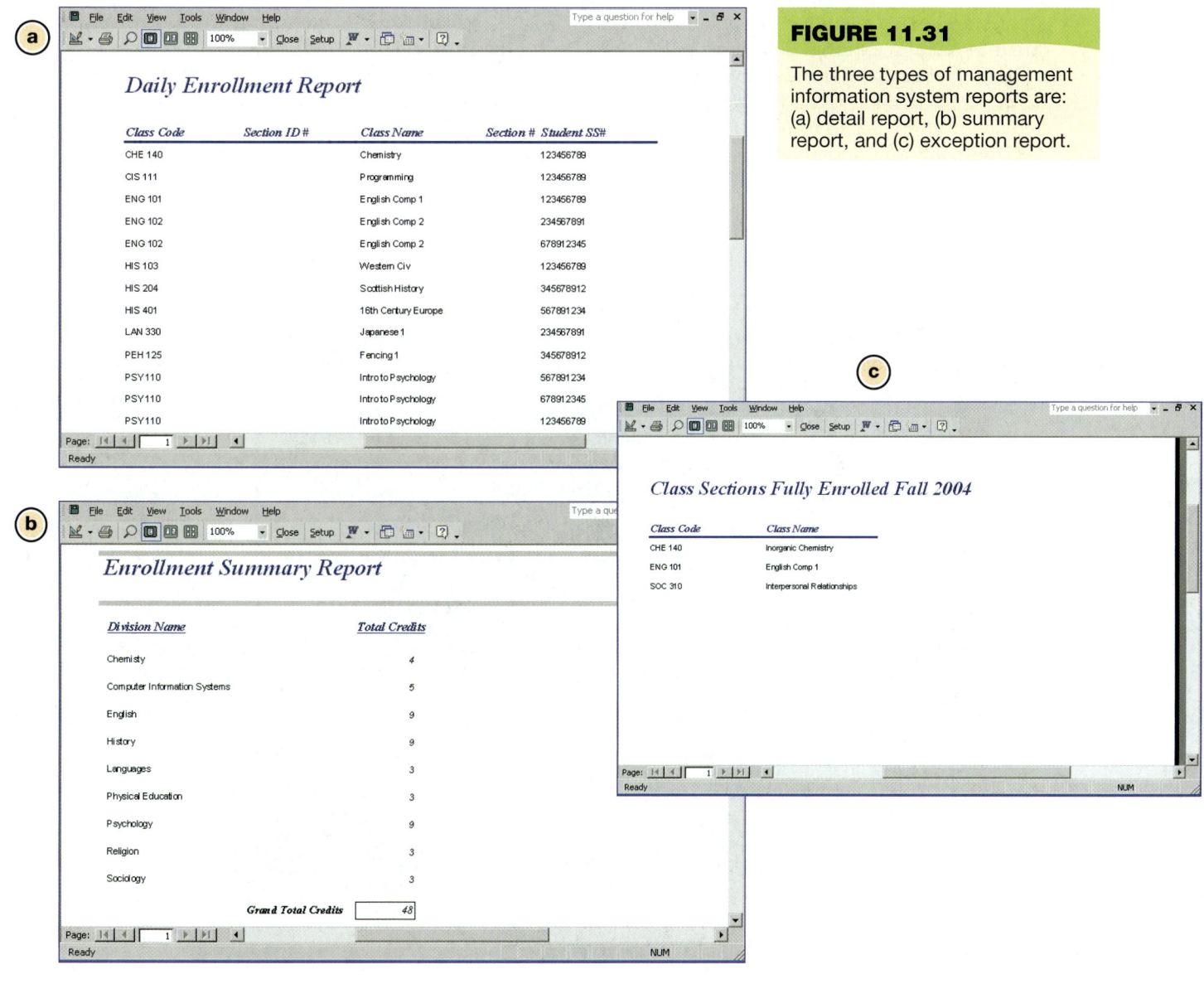

FIGURE 11.31

The three types of management information system reports are: (a) detail report, (b) summary report, and (c) exception report.

DECISION SUPPORT SYSTEMS

What is a decision support system? A **decision support system (DSS)** is designed to help managers develop solutions for specific problems. A DSS for a marketing department might provide statistical information on customer attributes (such as income levels, buying patterns, and so on) that would assist managers in making decisions regarding advertising strategy. Not only does a DSS use data from databases and data warehouses, it also enables users to add their own insights and experiences and apply them to the solution.

What does a decision support system look like? Database management systems, while playing an integral part of a DSS, are supplemented by additional software systems in a DSS. In a DSS, the user interface provides the means of interaction between the user and the system. An effective user interface must be easy to learn. The other major components of a DSS are internal and external data sources, model management systems, and knowledge-based systems. As shown in Figure 11.32, all of these systems work together to provide the user of the DSS with a broad base of information upon which to base decisions.

INTERNAL AND EXTERNAL DATA SOURCES

What are internal and external data sources for DSSs? Data can be fed into the DSS from a variety of sources. Internal data sources are maintained by the same company that operates the DSS. For example, internal TPSs can provide a wealth of statistical data about customers, ordering patterns, inventory levels, and so on. External data sources include any source not owned by the company that owns the DSS, such as customer demographic data purchased from third parties, mailing lists, or statistics compiled by the federal government. Internal and external data sources provide a stream of data that is integrated into the DSS for analysis.

MODEL MANAGEMENT SYSTEMS

What function does a model management system perform? A **model management system** is software that assists in building management models in DSSs. A management model is an analysis tool that provides a view of a particular business situation (through the use of internal and external data) for the purposes of

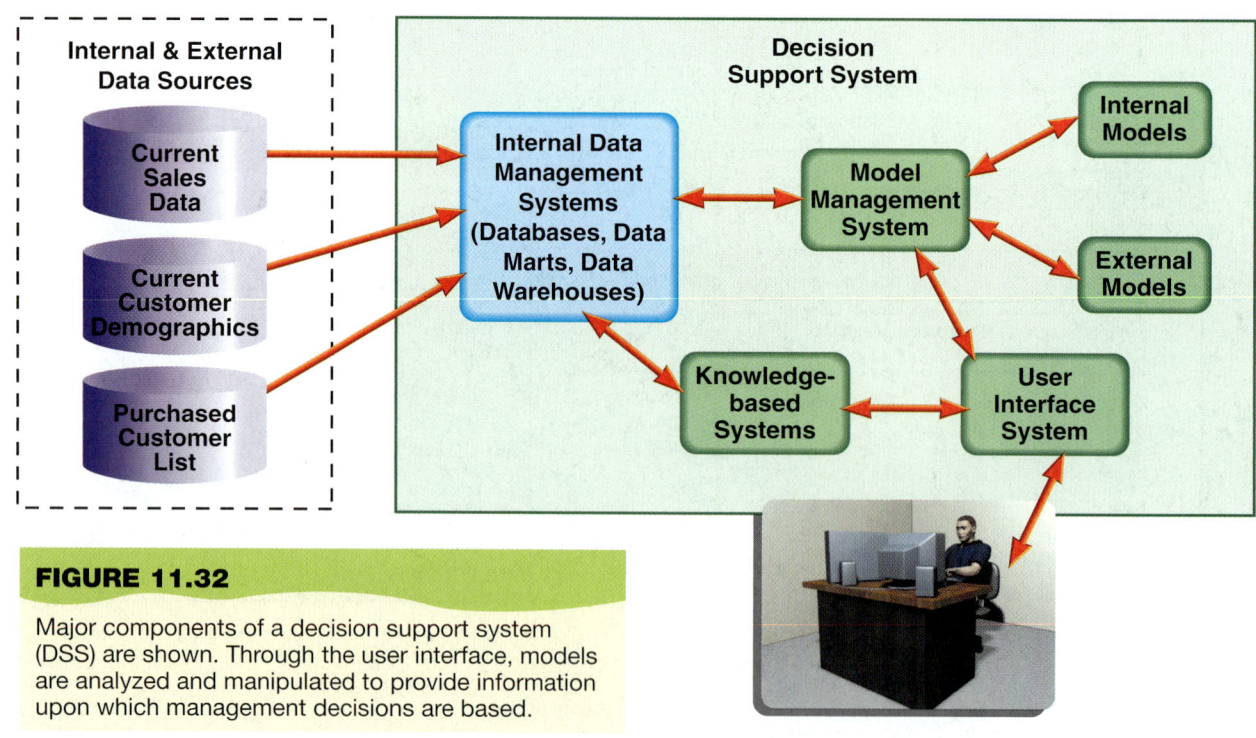

FIGURE 11.32

Major components of a decision support system (DSS) are shown. Through the user interface, models are analyzed and manipulated to provide information upon which management decisions are based.

decision making. Models can be built to describe any business situation, such as the classroom space requirements for next semester or a listing of alternative satellite campus locations.

Internal models are developed inside the organization (such as a spreadsheet that shows current classroom utilization on a college campus). External models are purchased from third parties (such as statistics about student populations for two-year college students in the United States). Model management systems typically contain financial and statistical analysis tools used to analyze the data provided by models or to create additional models.

KNOWLEDGE-BASED SYSTEMS

What is a knowledge-based system and how is it used in DSSs? A **knowledge-based system** provides additional intelligence that supplements the user's own intellect and makes the DSS more effective. It can be an **expert system**, which tries to replicate the decision-making processes of human experts to solve specific problems. For example, an expert system might be designed to take the place of a physician in remote locations such as a scientific base on Antarctica. A physician expert system would ask the patient about symptoms just as a live physician would, and the system would make a diagnosis based on the algorithms programmed into it.

Another type of knowledge-based system is a **natural language processing (NLP) system**. NLP systems enable users to communicate with computer systems using a natural spoken or written language as opposed to using computer programming languages. Handicapped individuals who cannot use a keyboard benefit greatly from NLP systems since they can just speak to the computer and have it understand what they are saying without using specific computer commands. Using an NLP system can simplify the user interface, making it much more efficient and user friendly. The speech-recognition feature of Microsoft Office 2003 is a type of NLP system.

All knowledge-based systems fall under the science of artificial intelligence. **Artificial intelligence (AI)** is the branch of computer science that deals with attempting to create computers that think like humans. To date, no computers have been constructed that can replicate the thinking patterns of a human brain. The failure to achieve artificial intelligence is primarily due to the fact that scientists still do not know how humans store and integrate knowledge and experiences to form human intelligence. But there is much research going on in the field of artificial and human intelligence that could lead to the development of truly intelligent machines within your lifetime.

How does a knowledge-based system help in the decision-making process? Databases and the models provided by model management systems tend to be very analytical and mathematical in nature. If we solely relied on databases and models to make decisions, the answers would be derived with a "yes or no" mentality, allowing no room for human thought. Fortunately, human users are involved in these types of systems, providing an opportunity to inject human judgment and experience into the decision-making process.

The knowledge-based system also provides an opportunity to introduce experience into the mix. Knowledge-based systems support the concept of fuzzy logic. Normal logic is very rigid: if "x" happens then "y" will happen. **Fuzzy logic** allows the interjection of experiential learning into the equation by considering probabilities. Whereas an algorithm in a database has to be specific, an algorithm in a knowledge-based system could state: if "x" happens, 70 percent of the time "y" will happen.

For instance, managers at Amazon.com may find it extremely helpful if their DSSs informed them that 40 percent of customers who bought a certain book also bought the sequel. This could suggest that designing a discount program for sequels bought with the original book might spur sales. Fuzzy logic allows a system to be more flexible and to consider a wider range of possibilities than with conventional algorithmic thinking.

Data Mining

Just because you captured data in an organized fashion and stored it in a certain format that seems to make sense doesn't mean that an analysis of the data will automatically reveal everything you need to know. Trends can sometimes be hard to spot if the data is not organized or analyzed in a unique way. To make data work harder, companies employ data mining techniques.

Data mining is the process by which great amounts of data are analyzed and investigated. The objective is to spot significant patterns or trends within the data that would otherwise not be obvious. For instance, through mining student enrollment data, a school may discover that 40 percent of new nursing degree students are Hispanic.

Why do businesses mine their data? The main reason businesses mine data is to understand their customers better. If a company can better understand the types of customers who buy their products and what motivates them to do so, they can market effectively by concentrating their efforts on the populations that are most likely to buy.

How do businesses mine their data? Data mining enables managers to sift through data in a number of ways. Each method produces different information that managers can then base their decisions on. The following are five things managers do to make their data meaningful:

1. **Classification**. To analyze data, managers need to classify it. Therefore, before mining, managers define data classes that they think will be helpful in spotting trends. They then apply these class definitions to all unclassified data in order to prepare it for analysis. For example, "good credit risk" and "bad credit risk" are two data classes managers could establish to determine whether to grant mortgages to applicants. Managers would then identify factors (such as credit history and yearly income) they could use to classify applicants as "good" or "bad."

2. **Estimation**. When managers classify data, the record either fits the classification criteria or it doesn't. Estimation enables managers to assign a value, based on some criteria, to data. For example, assume a bank wants to send out credit card offers to people who are likely to be granted a credit card. The bank may run the customers' data through a program that assigns them a score based on where they live, their household income, and their average bank balance. This provides managers with an estimate of the most likely credit card prospects so that they can include them in the mailing.

3. **Affinity Grouping or Association Rules.** When mining data, managers can also determine which data goes together. In other words, they can apply affinity grouping or association rules to the data. For example, suppose analysis of a sales database indicates that two items are bought together 70 percent of the time. Based on this data, managers might decide that these items should be pictured on the same page in the next mail order catalog they send out.

4. **Clustering.** Clustering involves organizing data into similar subgroups, or clusters. It is different from classification in that there are no predefined classes. The data mining software makes the decision about what to group together and it is up to managers to determine whether the clusters are meaningful. For example, the data mining software may identify clusters of customers with similar buying patterns. Further analysis of the clusters may reveal that certain socioeconomic groups have similar buying patterns.

5. **Description and Visualization.** Often, the purpose of data mining is merely to describe data so managers can visualize it. Sometimes having a clear picture of what is going on with the data helps people to interpret it in new and different ways. For example, if based on large amounts of data we found out that left-handed males who live in suburban environments never take automotive technology courses, it would most likely spark a heated discussion about the reasons why. It would certainly provide plenty of opportunities for additional study on the part of psychologists, sociologists, and college administrators!

You may have noticed that products are frequently moved around in supermarkets. This is usually the result of data mining. With electronic scanning of bar codes, each customer's purchase is recorded in a database. By classifying the data and using cluster analysis, supermarket managers can determine which products people usually purchase with other products. The store then places these products close to each other so that shoppers can find them easily. For instance, if analysis shows that people often buy coffee with breakfast cereal, it would make sense to place these items in the same aisle.

Summary

1. **What is a database and why is it beneficial to use databases?** Databases are electronic collections of related data that organize data so that it can be more easily accessed and manipulated. Properly designed databases reduce data redundancy by ensuring relevant data is only recorded in one place. This also helps eliminate data inconsistency. Databases also enable multiple users to share and access information at the same time.

2. **What components make up a database?** A category of information in a database is stored in a *field*. Each field is identified by a field name. Fields are assigned a data type that indicates what type of data can be stored in the field. Common data types include text, numeric, computational, date, memo, object, and hyperlink. A group of related fields is called a *record*. A group of related records is called a *table* or *file*.

3. **What types of databases are there?** The three major types of databases currently in use are relational, object-oriented, and object-relational. With relational databases, tables of data are linked by a common field. Object-oriented databases contain not only data but also instructions about how that data is to be manipulated or processed. Object-relational databases contain characteristics of both relational and object-oriented databases.

4. **What do database management systems do?** Database management systems are application software (such as Oracle or Microsoft Access) that interact with the user, other applications, and the database itself to capture and analyze data. The main operations of a DBMS are: creating databases, entering data, browsing data, sorting data, querying data, and outputting data. To extract records from a database, you use a query language. Almost all relational databases today use SQL. However, most DBMSs include wizards that allow you to query the database without learning a query language.

5. **How do relational databases organize and manipulate data?** Relational databases operate by organizing data into tables based on logical groupings. Since all of the data in a relational database is not stored in the same table, a methodology must be implemented to *link* data between tables. In relational databases, these links between tables are referred to as *relationships*. To establish a relationship between two tables, both tables must have a common field (or column).

6. **What are data warehouses and data marts and how are they used?** A data warehouse is a large electronic repository of data that attempts to contain and organize all data related to an organization in one place. Data warehouses often contain information from multiple databases. Because it can be difficult to find information in a data warehouse, small slices of the data warehouse, called data marts, are often created.

7. **What is an information system and what types of information systems are used in business?** Information systems, software-based solutions used to gather and analyze information, fall into four categories: An office support system assists employees in accomplishing tasks and to improve communications. A transaction processing system is used to keep track of everyday business activities. A management information system provides information that enables managers to make decisions. A decision support system is designed to help managers solve problems.

8. **What is data mining and how does it work?** Data mining is the process by which large amounts of data are analyzed to spot hidden trends. Through processes such as classification, estimation, clustering, affinity grouping, and description, data is organized so that managers can use it to identify trends.

Key Terms

alphabetic check	p. 441	join queries	p. 445
artificial intelligence	p. 461	knowledge-based systems	p. 461
batch processing	p. 458	management information system (MIS)	p. 459
browsing	p. 442		
clickstream data	p. 455	memo fields	p. 435
completeness check	p. 441	metadata	p. 439
computed field (computational field)	p. 435	model management system	p. 460
		natural language processing (NLP) system	p. 461
consistency check	p. 441		
data dictionary (database schema)	p. 437	normalization	p. 449
		numeric check	p. 441
data inconsistency	p. 430	numeric fields	p. 434
data integrity	p. 433	object fields	p. 435
data marts	p. 456	object-oriented database	p. 436
data mining	p. 461	object-relational databases	p. 436
data redundancy	p. 430	office support system (OSS)	p. 457
data staging	p. 455	online transaction processing (OLTP)	p. 458
data type (field type)	p. 434		
data warehouse	p. 452	primary key (key field)	p. 435
database administrator (database designer)	p. 433	query language	p. 443
		range checks	p. 440
database management system (DBMS)	p. 437	real-time processing	p. 458
		record	p. 435
database query	p. 443	referential integrity	p. 451
databases	p. 430	relational algebra	p. 444
date fields	p. 435	relational database	p. 436
decision support system (DSS)	p. 460	relations	p. 436
default values	p. 438	relationships	p. 448
detail report	p. 459	select query	p. 444
exception reports	p. 459	sort (index)	p. 442
expert systems	p. 461	structured (analytical) data	p. 436
export	p. 448	structured query language (SQL)	p. 443
field	p. 434		
field constraints	p. 440	summary data reports	p. 448
field name	p. 434	summary reports	p. 459
field size	p. 435	table (file)	p. 435
foreign key	p. 451	text fields	p. 434
fuzzy logic	p. 461	time-variant data	p. 454
historical data	p. 454	transaction processing system (TPS)	p. 458
hyperlink fields	p. 435		
imported	p. 448	unstructured data	p. 436
information systems	p. 457	validation	p. 439
input form	p. 439	validation rules	p. 440

Buzz Words

Word Bank

- database
- decision support system
- data consistency
- transaction processing system
- SQL
- memo field
- completeness check
- field name
- relational algebra
- primary key
- select query
- text field
- table
- join query
- object field
- record
- data mining
- data mart
- metadata
- field
- field size
- numeric field
- list
- data dictionary
- data type
- data warehouse

Instructions: Fill in the blanks using the words from the Word Bank above.

When constructing a database (1)_____, it is important to ensure each record is identified uniquely. A(n) (2)_____ should be established as a unique field to be included with each record. In a database, digits such as 1234 are normally stored in a(n) (3)_____ but could also be stored in a(n) (4)_____ if calculations do not need to be performed on the number. Extremely lengthy textual data is stored in a(n) (5)_____ whereas video files are appropriately stored in a(n) (6)_____ The (7)_____ fully describes each field in the database and its attributes. Data used to describe other data in this manner is referred to as (8)_____.

Queries are used to prepare data for viewing or printing. A(n) (9)_____ displays requested information from one table. For displaying information from multiple tables, a(n) (10)_____ must be used. The most popular query language in use today is (11)_____. Queries generated by this language make use of English-language statements driven by the mathematical principles of (12)_____.

When individual databases are not sufficient to maintain all the data that needs to be tracked, a(n) (13)_____ should be installed. Databases are often key components of (14)_____s, which record routine business activities. A(n) (15)_____ utilizes databases and other related systems to assist management with building business models and making critical decisions.

Organizing Key Terms

Instructions: This chapter introduced many new terms and concepts. In the illustration below, fill in each of the blanks with words from the Word Bank and the chapter in order to show how categories of ideas fit together.

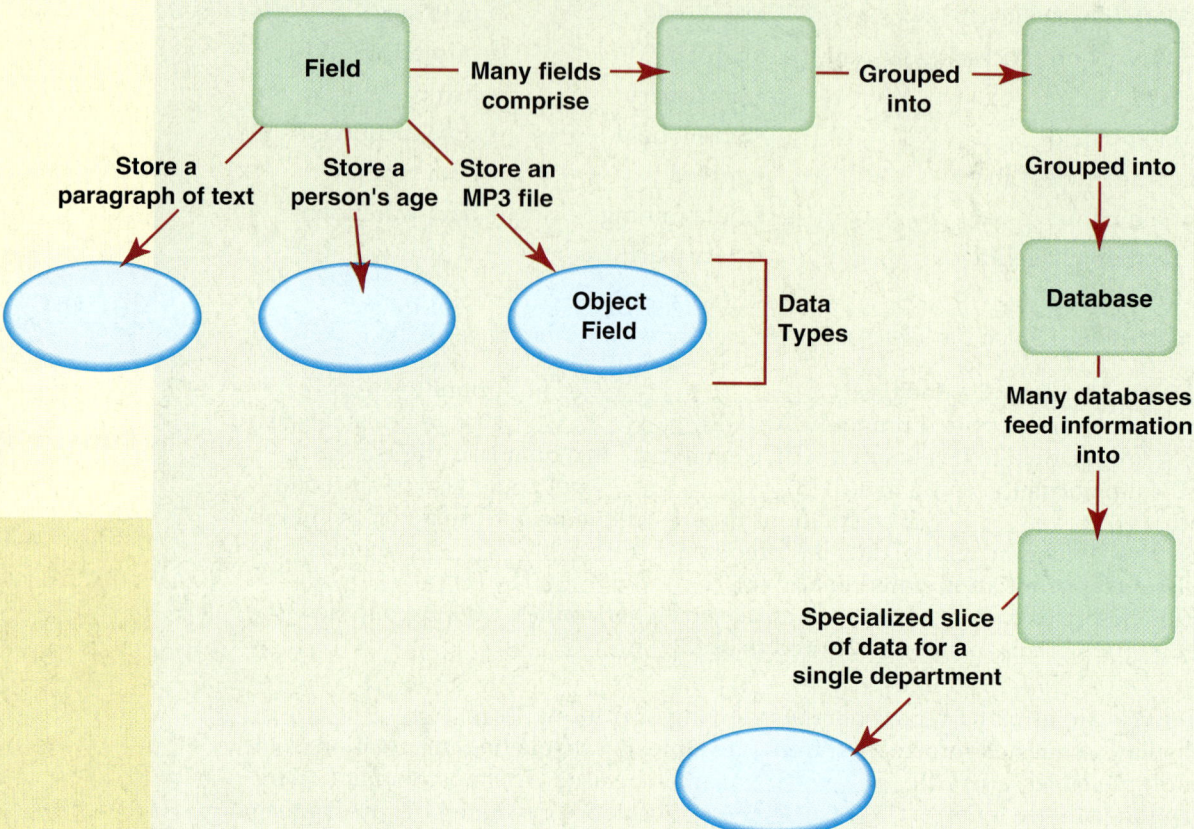

Making the Transition to . . . Next Semester

1. Many schools try to maintain contact with former students to keep them informed of new programs and courses being offered at the institution. Visit your school's alumni office and determine the following:

   a. Is a separate database maintained for the purposes of communicating with alumni?
   b. What type of database software is used?
   c. What data is captured in this database?
   d. How often is the data contained in the database verified with the alumni?

2. If the alumni office doesn't maintain a separate database for the purposes of communicating with graduates, use the previous questions as a guideline for developing an alumni database design for your school.

3. Imagine that you own 700 CDs and want to track them in a database. Determine the following:

   a. What fields do you need in your database for categorizing and tracking your CDs?
   b. What fields do you need for capturing information about your friends to whom you have loaned CDs?

Making the Transition to . . . The Workplace

1. You are a summer intern in the information services group of an office supply manufacturer. At the weekly staff meeting, the chief information officer (CIO) indicates that the president is requesting information about sales of all lines of goods for the past 10 years. Unfortunately, data more than three years old is not maintained in the current sales database and has been archived to tape. A volunteer is needed (and everyone looks to you) to extract the data from the tapes and prepare the needed reports.

   As you are spending your 16th evening extracting the data, the CIO mentions that manufacturing, shipping, and accounts receivable may need their own reports that include different data but span the same time period as your current assignment. The CIO assures you that most of the information is available on tape, but if not, it can be located in the mountains of paper stored in the old warehouse. You know that this tedious work could be avoided by the introduction of a data warehousing system!

   a. Prepare a data warehousing plan for the CIO. Describe briefly the benefits of data warehousing and provide an overview of the process.
   b. For sales, manufacturing, shipping, and accounts receivable data, suggest the types of information that should be captured in the data warehouse.

2. You have a part-time job in the IT department at your school. The school's fund-raising foundation has been manually maintaining detailed records of donors and fund-raising activities. The school's president has recently asked the CIO to computerize the foundation's records. The CIO has asked for your help in designing the tables for the donor database. The following information is captured in the manual records:

   Donor name, address, and phone number
   Amount of donation
   Name of fund-raising campaign (Student Scholarship Fund, New Technology Center Building Fund, Bell Tower Fund, etc.)

   a. What specific fields should be set up for the electronic database? Make sure to suggest fields that should be included but are currently not in the manual system.
   b. Assume three tables are set up: Donor Contact Information, Campaign (which includes the names of all the specific fund-raising campaigns), and Donation (which tracks individual donations by donor). What is an appropriate primary key to use for the Donor Contact Information table? (Note that social security numbers are not currently captured.)
   d. Do you think it would be feasible to collect social security numbers of current donors to use as primary keys?
   e. For the Donation table, what should be used as a primary key? How will two donations by the same donor be differentiated?

Critical Thinking Questions

Instructions: Albert Einstein used "Gedanken experiments," or critical thinking questions, to develop his theory of relativity. Some ideas are best understood by experimenting with them in our own minds. The following critical thinking questions are designed to demand your full attention but require only a comfortable chair—no technology.

1. Internet databases abound with personal information about you. You probably provided some of this information, but it may have been sold to other companies. Other information about you may have been obtained without your knowledge while you surfed on Web sites. Consider the following:

    a. Is it ethical for a company to sell personal information (such as household income) that you voluntarily gave to it?
    b. Is gathering information about people's surfing and buying habits by tracking their clicks through a Web site an invasion of privacy?
    c. Should Web sites be legally required to inform users that they are tracking surfing habits? Why or why not?

2. The Internal Revenue Service (IRS) maintains large databases about taxpayers that include a wealth of information on personal income that can be easily sorted by geographic location. This information would be of great value to marketing professionals for targeting marketing programs to consumers. Currently, the IRS is prohibited from selling this information to third parties. However, the IRS (and other government agencies) are under increasing pressure to find ways to increase revenue or decrease expenses.

    a. Do you favor a change in the laws that would permit the IRS to sell names and addresses with household income information to third parties? Why or why not?
    b. Would it be acceptable for the IRS to sell income information to marketing firms if it did not include personal information (such as names and addresses) but only included income statistics for certain geographic areas? How is this better (or worse) than selling personal information?
    c. How would you feel about the IRS marketing financial products (such as tax software) directly to consumers? Would this be a conflict of interest with the IRS's main mission (the collection of tax revenue and enforcement of tax compliance)?

3. After the terrorist attacks on September 11, 2001, some U.S. citizens began demanding more scrutiny of foreign nationals and people wishing to immigrate to the United States. In response to these concerns the Department of Defense launched the Total Information Awareness (TIA) Program. The program was created to develop data mining techniques to probe massive federal databases as well as commercial and private employment, medical, and financial databases. The objective was to spot trends that would identify people who were threats to national security. Initially, the public did not complain, but as more stories surfaced about the program, citizens began to resent this invasion of privacy.

    a. Research the TIA program and determine its current status. Is the program going ahead or has it been terminated? If the program has been discontinued, has another replaced it?
    b. Do you think the government should institute a program like TIA?
    c. Which is more important to you: safeguarding your privacy or protecting the United States from terrorists? Why?

Team Time

Problem:

Many schools use student data information systems that were installed more than 10 years ago. When preparing to transition to a newer system, additional information is often identified that needs to be captured in the new system that was not recorded in the old system (such as e-mail addresses). Also, in preparation for transferring legacy data to a new system, the data often needs to be "groomed." Grooming data means verifying the data accuracy, ensuring data consistency, and correcting problems.

Task:

Your class has volunteered to assist the IT department with the transition to a new student information system. Customers (students) often provide unique perspectives and should always be consulted (when possible) during the design and implementation of new systems. The school administration feels your input into the design process will enhance the usability of the new system.

Process:

Divide the class into small groups.

STEP 1: Your group should examine your school's current student information system (from a user perspective). Identify the data that is being captured from students (when enrolling or registering for courses). Compile a list of suggestions for data that does not need to be captured (but currently is captured) and for additional data that should be gathered (but is not currently). Determine the extent to which student services can be accessed via the Internet and suggest services that require Internet accessibility but are not currently offered.

STEP 2: Present your group's findings to the class. Compare your suggestions to those of other groups. Be sure to address the needs of all groups of students (residents, commuters, and online students).

STEP 3: Prepare a list of recommendations for improvements to the current student information system for the director of student affairs. Clearly indicate how the proposed changes will benefit both students and the school employees who are interacting with the system.

Conclusion:

Colleges and universities are competing for students more fiercely than ever before. Those institutions that listen to their students and find innovative ways to serve their needs will have a distinct advantage in attracting and retaining students. Listening to customers is a basic principle of business that educational institutions should not overlook.

Becoming Computer Fluent

Everyone loves the jewelry you make and you've decided to start selling it. You have some customers already but plan to put your work in a few art shows and make some deals with area stores to feature it. Keeping track of all this information is going to be important but not your idea of fun, so you decide to hire a database consultant to design a database for this new business. Write a letter to the consultant describing what you expect from her. What kind of database do you wish to create? What kind of information do you need to be able to pull from it? How should that information be formatted? How do you imagine the business, and the database, growing?

Materials on the Web

In addition to the review materials presented here, you'll find extra materials on the book's companion Web site (**www.prenhall.com/techinaction**) that will help reinforce your understanding of the chapter content. These materials include:

Sound Byte Lab Guides

For each Sound Byte mentioned in the chapter, there is a corresponding lab guide located on the book's companion Web site. These guides review the material presented in the Sound Byte and direct you to various Web resources that examine the material. The Sound Byte Lab Guides for this chapter include:

- Creating an Access Database
- Improving an Access Database

True/False and Multiple Choice Quizzes

The book's Web site includes a True/False and a Multiple Choice quiz for this chapter. You can take these quizzes, automatically check the results, and e-mail the results to your instructor.

Web Research Projects

The book's Web site also includes a number of Web research projects for this chapter. These projects ask you to search the Web for information on computer-related careers, milestones in computer history, important people and companies, emerging technologies, and the applications and implications of different technologies.

OBJECTIVES

After reading this chapter, you should be able to answer the following questions:

- What are the advantages of a business network? (p. 474)

- How does a client/server network differ from a peer-to-peer network? (pp. 474–475)

- What are the different classifications of client/server networks? (pp. 475–477)

- What components are needed to construct a client/server network? (p. 478)

- What do the various types of servers do? (pp. 478–481)

- What are the various network topologies (layouts) and why is network topology important in planning a network? (pp. 481–486)

- What types of transmission media are used in client/server networks? (pp. 486–489)

- What software needs to be running on computers attached to a client/server network and how does this software control network communications? (p. 491)

- How do network adapters enable computers to participate in a client/server network? (pp. 491–494)

- What devices assist in moving data around a client/server network? (pp. 494–498)

- What measures are employed to keep large networks secure? (pp. 498–502)

for information from the Internet and all messages being sent through the Internet pass through the communications server. Since Internet traffic is substantial at most organizations, the communications server has a heavy workload.

Often, the communications server is the only device on the network connected to the Internet. E-mail servers, Web servers, and other devices needing to communicate with the Internet usually route all their traffic through the communications server. Providing a single point of contact with the outside world makes it easier to secure the network from hackers.

WEB SERVERS

What function does a Web server perform? A **Web server** is used to host a Web site available through the Internet. Web servers run specialized software, such as Apache (open-source server software) or Microsoft Internet Information Server, that allows them to host Web pages. Not every large network will have a Web server. Many companies use an Internet service provider (ISP) to host their corporate Web sites instead.

Network Topologies

Just as buildings have different floor plans depending on their uses, networks have different blueprints delineating their layout. **Network topology** refers to the physical arrangement of computers, transmission media (cable), and other network components. Since networks have different uses, not all networks will have the same topology.

In this section, we'll explore the most common network topologies (bus, ring, and star) and discuss when each topology is used. As you'll see, the type of network topology used is important because it can affect the network's performance and scalability. Knowing how the basic topologies work and the strengths and weakness of each will help you understand why particular network topologies were chosen on the networks you use.

BUS TOPOLOGY

What does a bus topology look like? In a **bus** (or **linear bus**) **topology,** all computers are connected in sequence on a single cable, as shown in Figure 12.6. This topology is deployed most often in peer-to-peer networks (not client/server networks). Each computer on the bus network can communicate with every other computer on the network directly. However, only one computer can transmit data at a time. This is because all computers on the bus network must share the available bandwidth (or data transfer capacity) of the transmission media (wires or cables connecting the devices).

The effect is the same as having a group of three people (Diane, Michelle, and Lee) sitting in a room having a conversation. For the conversation to be effective, only one person can speak at a time; otherwise, they would not be able to hear and understand each other. Therefore, if Diane is speaking, Michelle and Lee must wait until she finishes before presenting their ideas, and so on.

Since two signals can't be transmitted at the same time on a bus network, an **access method** has to be established to control which computer is allowed to use the transmission media at a certain time. Computers on a bus network behave the same way as a group of people having a conversation. The computers "listen" to the network data traffic on the media. When no other computer is transmitting data (that is, when "conversation" stops), the computer knows it is allowed to transmit data on the media. This means of taking turns "talking" avoids **data collisions,** which happen when two computers send data at the same time and the sets of data collide somewhere in the media. When data collides, it is often lost or irreparably damaged.

BITS AND BYTES

Too Much Data? Here Comes the SAN

Databases can become so large that conventional database servers can't handle the information flowing in and out of them. A storage area network (SAN) is specifically designed to store and disseminate large amounts of data to client computers or servers. SANs are made up of several network attached storage (NAS) devices, which are specialized devices attached to a network whose sole function is to store and disseminate data. Picture a computer with nothing but hard drives in it and you have a pretty good idea of what a NAS device looks like. NAS devices are the filing cabinets of the new millennium and exist merely to store the huge amounts of data network users generate. Although they behave like dedicated servers, NAS devices have their own operating systems and file storage algorithms. Because they don't perform any network services other than storage and retrieval, they can do so quickly and efficiently.

482 CHAPTER 12 BEHIND THE SCENES: NETWORKING AND SECURITY

> **FIGURE 12.6**
> A Linear Bus Topology

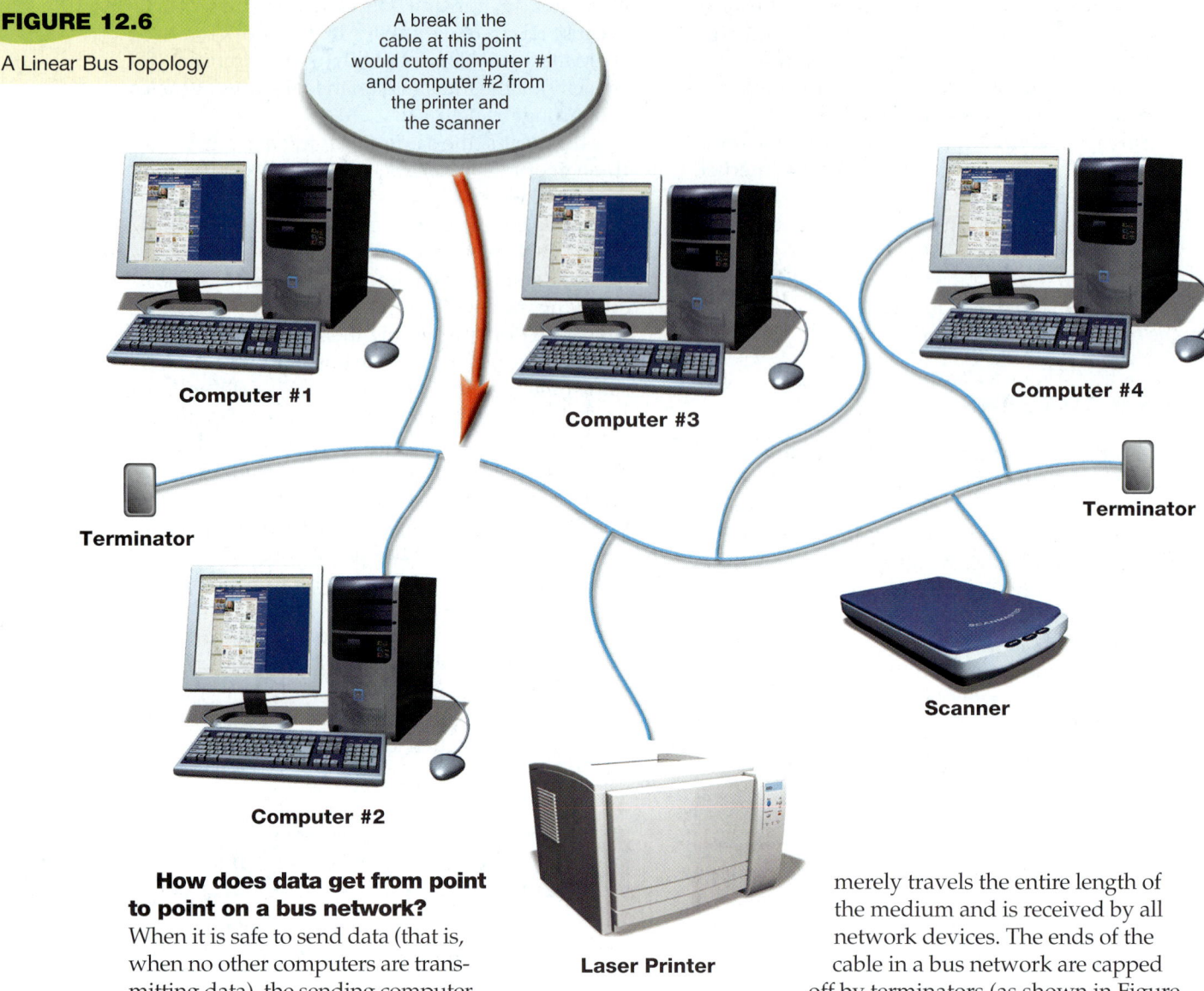

How does data get from point to point on a bus network?
When it is safe to send data (that is, when no other computers are transmitting data), the sending computer broadcasts the data onto the media. The data is broadcast throughout the network to *all* devices connected to the network. The data is broken into small segments called **packets**. Each packet contains the address of the computer or peripheral device to which it is being sent. Each computer or device connected to the network listens for data that contains its address. When it "hears" data addressed to it, it takes the data off the media and processes it.

For example, say your computer needs to print something on the printer attached to the network. Your computer "listens" to the network to ensure no other nodes are transmitting. It then sends the print job out onto the network. When the printer "hears" a job addressed to it (the print job your computer just sent), it pulls the data off the network and executes the job.

The devices (nodes) attached to a bus network do nothing to move data along the network. This makes a bus network a **passive topology**. The data merely travels the entire length of the medium and is received by all network devices. The ends of the cable in a bus network are capped off by terminators (as shown in Figure 12.6). A **terminator** is a device that absorbs the signal so it is not reflected back onto parts of the network that have already received it.

What are the advantages and disadvantages of bus networks?
The simplicity and low cost of configuring a bus network is the major reason why this topology is deployed most often in P2P networks. The major disadvantage is that if there is a break in the cable, the bus network is effectively disrupted, as some computers are cut off from others on the network.

Also, since transmission signals degrade as the distance of the cable increases, a bus network is difficult to expand to a large number of users. And since only one computer can communicate at a time, adding a large number of nodes to a bus network limits performance and causes delays in sending data. Therefore, you rarely see a bus network deployed except in very small networks that are not expected to grow.

RING TOPOLOGY

What does a ring topology look like? Not surprisingly, given its name, the computers and peripherals in a **ring** (or **loop**) **topology** are laid out in a circle, as shown in Figure 12.7. Data flows around the circle from device to device in one direction only. Because data is passed using a special data packet called a **token**, this type of topology is commonly referred to as a **token-ring topology**.

How is a token used to move data around a ring? A token is passed from computer to computer around the ring until it is grabbed by a computer that needs to transmit data. The computer holds onto the token until it is done transmitting data. Only one computer on the ring can "hold" the token at a time, and only one token exists on each ring.

If a computer (or node) has data to send, it waits for the token to be passed to it. It then takes the token out of circulation and sends data to its destination. When the receiving node receives a complete transmission of the data, it sends an acknowledgment to the sending node. The sending node then generates a new token and starts it going around the ring again. This **token method** is the access method that ring networks use to avoid data collisions.

A ring topology is an **active topology** since each node on the network is responsible for retransmitting the token or the data to the next node on the ring. Large ring networks have the capability to use multiple tokens, which help move more data faster.

Is a ring topology better than a bus topology? A ring topology provides a fairer allocation of network resources than does a bus topology. By using a token, a ring network allows all nodes on the network to have an equal chance to send data. One "chatty" node cannot as easily monopolize the network bandwidth since after sending a batch of data, it must pass the token on.

In addition, the ring topology's performance remains acceptable even with large numbers of users. However, if one computer fails on a ring network it will bring the entire network to a halt

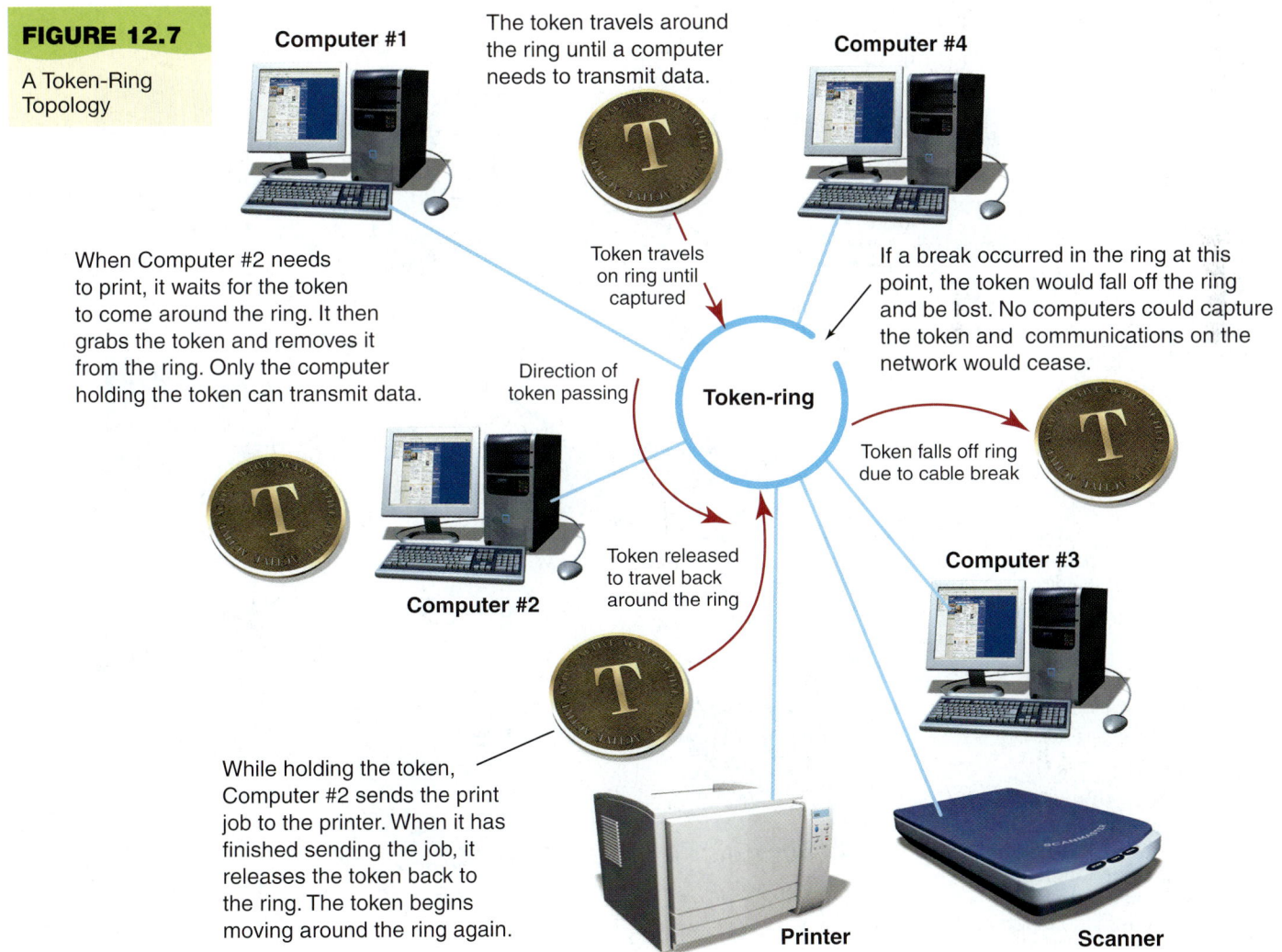

FIGURE 12.7 A Token-Ring Topology

since that computer is unavailable to retransmit tokens and data. Problems in the ring can also be hard for network administrators to find. It's easier to expand a ring topology than a bus topology, but adding a node to a ring does cause the ring to cease to function while the node is installed.

STAR TOPOLOGY

What is the layout for a star topology? A **star topology** is the most widely deployed client/server network layout in businesses today since it offers the most flexibility. In a star topology, the nodes connect to a central communications device called a *hub,* thus forming a star, as shown in Figure 12.8. The hub receives a signal from the sending node and retransmits it to all other nodes on the network. The network nodes examine data and only pick up the transmissions addressed to them. Because the hub retransmits data signals, a star topology is an active topology. (We'll discuss hubs in more detail later in this chapter.)

Many star networks use the Ethernet protocol. Networks using the Ethernet protocol are by far the most common type of network in use today. Although many students think that Ethernet is a type of network topology, it is actually a communications protocol. A topology is a physical design of a network whereas a **protocol** is a set of rules for exchanging communication. Therefore, an Ethernet network can be set up using a bus, ring or a star topology.

For example, assume that your class has to send a message to the class next door. You decide to arrange your class in a straight line from your classroom to the other classroom. Each student will whisper the message to the next student in the line until the message is eventually passed to a student in the other classroom. The arrangement of the students in a straight line is your topology. The passing of the message from student to student using the English language is your protocol.

How do computers on a star network avoid data collisions? Because most star networks are Ethernet networks, they use the method

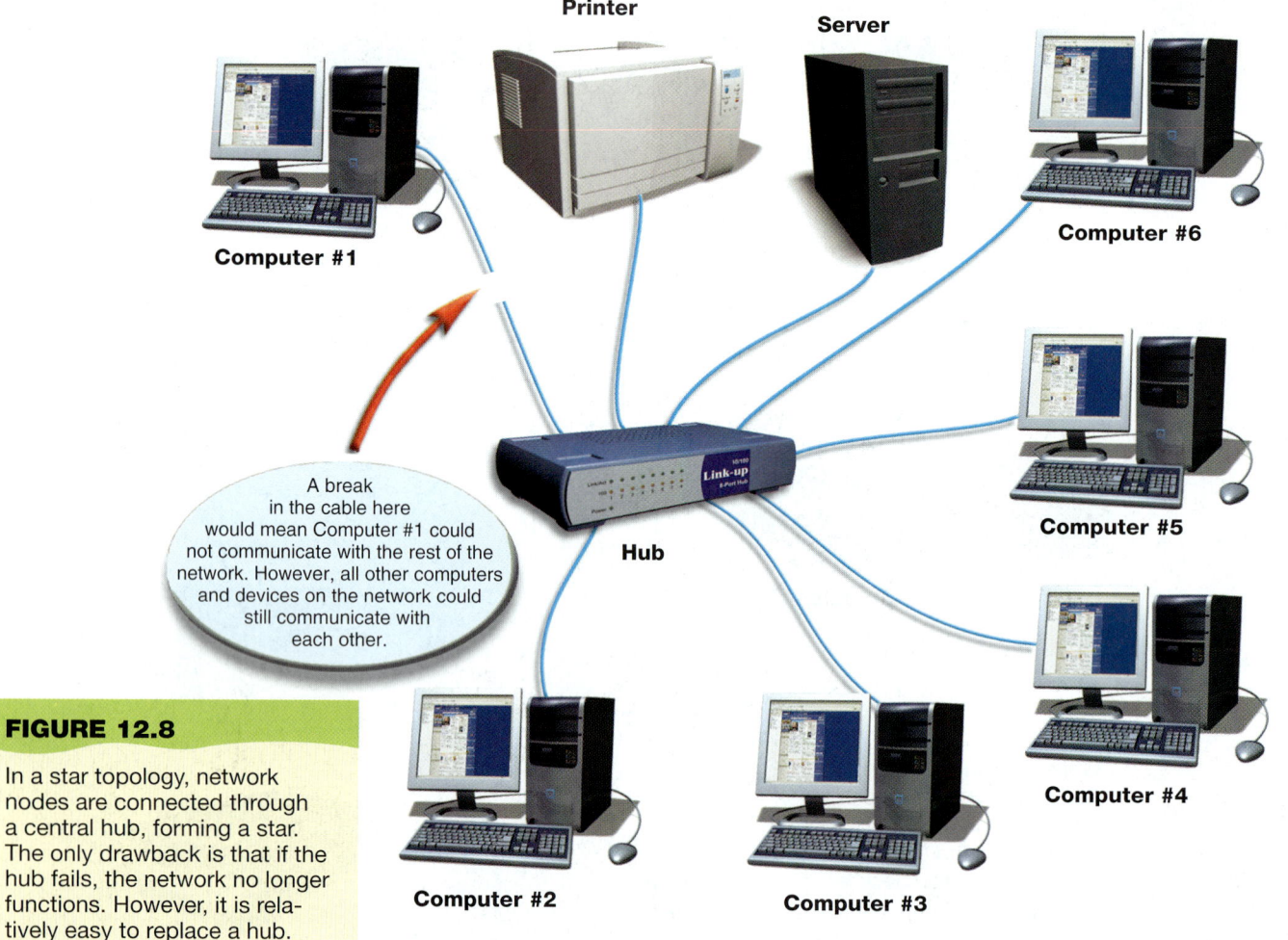

FIGURE 12.8

In a star topology, network nodes are connected through a central hub, forming a star. The only drawback is that if the hub fails, the network no longer functions. However, it is relatively easy to replace a hub.

used on all Ethernet networks to avoid data collisions: **CSMA/CD** (short for Carrier Sense, Multiple Access with Collision Detection). With CSMA/CD, a node connected to the network listens (that is, has carrier sense) to determine that no other nodes are currently transmitting data signals. If the node doesn't hear any other signals, it assumes it is safe to transmit data. All devices on the network have the same right (that is, they have multiple access) to transmit data when they deem it safe. It is therefore possible for two devices to begin transmitting data signals at the same time. If this happens, the two signals collide.

What happens when the signals collide? As shown in Figure 12.9, when signals collide, a node on the network detects the collision. It then sends a special signal called a **jam signal** to all network nodes, alerting them that a collision has occurred. The nodes then stop transmitting and wait a random amount of time before retransmitting their data signals. The wait time needs to be random, otherwise both nodes would start transmitting at the same time and another collision would occur.

What are the advantages and disadvantages of a star topology? Due to the complexity of the layout of star networks, they require more cable and are often more expensive than bus or ring networks. However, star topologies are generally considered to be superior to a ring topology because if one computer fails it doesn't affect the rest of the network. This is extremely important in a large network, where one disabled computer affecting the operations of several hundred other computers would be totally unacceptable.

It is also easy to add nodes to star networks, and performance remains acceptable even with large numbers of users. In addition, the centralization of communications (through a hub) makes troubleshooting and repairs on star networks easier for network technicians. As opposed to searching for a particular length of cable that broke in a ring network, technicians can usually pinpoint a communications problem just by examining the hub.

FIGURE 12.9

Avoiding Data Collisions on a Star (Ethernet) Network

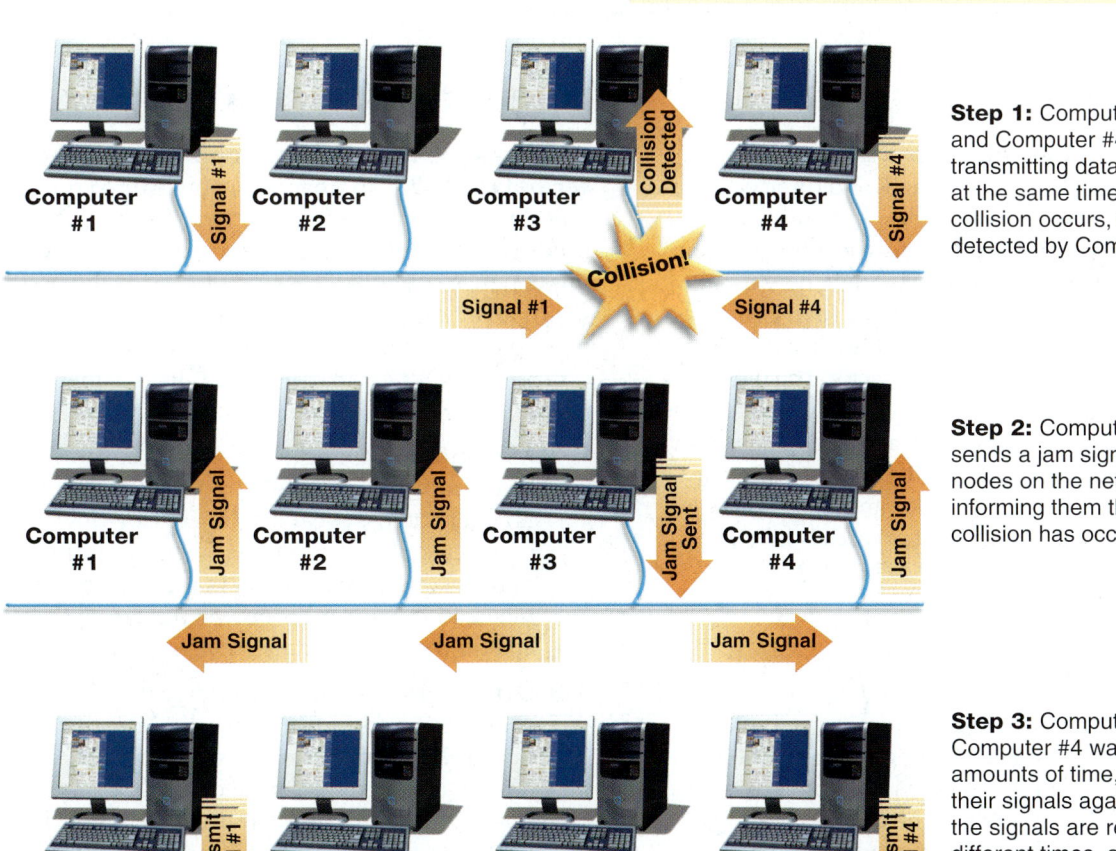

FIGURE 12.10 Advantages and Disadvantages of Bus, Ring, and Star Topologies

TOPOLOGY	ADVANTAGES	DISADVANTAGES
BUS	Uses a minimal amount of cabling Easy, reliable, and inexpensive to install	Breaks in the cable can disable the network Large numbers of users will greatly decrease performance due to high volumes of data traffic
RING	Allocates access to the network fairly Performance remains acceptable even with large numbers of users	Adding or removing nodes disables the network Failure of one computer can bring down the entire network Problems in data transmission can sometimes be difficult to find
STAR	Failure of one computer does not affect other computers on the network Centralized design simplifies troubleshooting and repairs Easy to add additional computers or network segments as needed (high scalability) Performance remains acceptable even with large numbers of users	Requires more cable and is often more expensive than a bus or ring topology The hub is a central point of failure. If it fails, all computers connected to that hub are affected.

COMPARING TOPOLOGIES

So which topology is the best one? Figure 12.10 lists the advantages and disadvantages of bus, ring, and star topologies. In all but the smallest networks, star topologies are the most common. Since networks are constantly adding new users, the ability to add new users simply (that is, by installing a new hub) without affecting users already on the network is the deciding factor. The networks you'll encounter at school and in the workplace will almost certainly be laid out in a star topology. However, bus topologies are still the most common layout for simple home networks, and ring topologies are popular when fair allocation of network access is a major requirement of the network.

Transmission Media

When constructing a house, a variety of building materials are available, depending on the needs of the builder. Similarly, when building a network, there are different types of media network engineers can use. **Transmission media**, whether it is cable or wireless communications technology, comprise the routes data takes to flow between devices on the network. Without transmission media, network devices would be unable to communicate.

WIRED TRANSMISSION MEDIA

What types of cable are commonly used for networks? In Chapter 7, you learned that most home networks use either twisted pair cable (phone wire or Ethernet) or electrical wires as transmission media. For business networks, the three main cable types that are used today are twisted pair, coaxial, and fiber optic. Although each type is different, they share many common factors that need to be considered when choosing a cable type:

- **Maximum run length.** Each type of cable has a maximum run length over which signals sent across it can be "heard" by devices connected to it. Therefore, when designing a network, network engineers must accurately measure the distances between devices to ensure that appropriate cable is selected.

- **Bandwidth.** As you learned in earlier chapters, **bandwidth** (also called **throughput** or **data transfer rate**) is the amount of data that can be transmitted across a transmission medium in a certain amount of time. Each cable is different and is rated by the maximum bandwidth it can support. Bandwidth is measured in bits per second, which represents how many bits of data can be transmitted along the cable each second.

- **Bend radius (flexibility).** When installing cable, it is often necessary to bend the cable around corners, surfaces, and so on. The bend radius of the cable defines how many degrees a cable can be bent in a 1-foot segment before it is damaged. If many corners need to be navigated when installing a network, network engineers use cabling with a high bend radius.

- **Cable cost.** The cost per foot of different types and grades of cable varies widely. Cable selection may have to be made on the basis of cost if adequate funds are not available for the optimal type of cabling.

- **Installation costs.** Certain cable (such as twisted pair, which is used in home networks) is easy and inexpensive to install. Fiber-optic cable requires special training and equipment to install, which increases the installation costs.

- **Susceptibility to interference.** Signals traveling down a cable are subject to two types of interference. Electromagnetic interference (EMI), caused by the cable being exposed to strong electromagnetic fields, can distort or degrade signals on the cable. Fluorescent lights and machinery with motors or transformers are the most common sources of EMI emissions. Cable signals can also be disrupted by radio frequency interference (RFI), which is usually caused by broadcast sources (television and radio signals) being located near the network. Cable types are rated as to how well they resist interference.

- **Signal transmission methods.** Coaxial cable and twisted pair cable both send electrical impulses down conductive material to transmit data signals. Fiber-optic cable transmits data signals as particles of light.

In the sections that follow, we'll discuss the characteristics of each of the three major types of cable. We will also discuss the use of wireless media as an alternative to cable.

Twisted Pair Cable
What does twisted pair cable look like? **Twisted pair cable** should be familiar to you since the telephone cable (or wire) in your home is one type of twisted pair cable. Twisted pair cable consists of pairs of copper wires twisted around each other and covered by a protective jacket (or sheath). The twists are important since they cause the magnetic fields that form around the copper wires to intermingle, which makes them less susceptible to outside interference. It also reduces the amount of crosstalk interference, or the tendency of signals on one wire interfering with signals on a wire next to it.

If the twisted pair cable contains a layer of foil shielding to reduce interference, it is known as **shielded twisted pair (STP) cable**. If it does not contain a layer of foil shielding, it is known as **unshielded twisted pair (UTP) cable**, which is more susceptible to interference. Figure 12.11 shows illustrations of both types of twisted pair cable. Due to its lower price, UTP is more widely used, unless significant sources of interference must be overcome (such as in a production environment where machines create magnetic fields). However, there are different standard categories of UTP cable from which to choose.

What types of UTP cable are available? The two most common types of UTP cable in use today are Category 5E (CAT 5E) and Category 6 (CAT 6). CAT 6 cable can handle a bandwidth of 1 gigabit per second (Gbps), whereas CAT 5E can handle a bandwidth of just 200 megabits per second (Mbps).

Unless severe budget constraints are in place, network engineers usually install the highest-bandwidth cable possible since reinstalling cable later (and the subsequent tearing up of walls and ceilings) can be very expensive. Therefore, since the fall of 2002, when the standard for CAT 6 cable was approved, new cable runs in businesses have

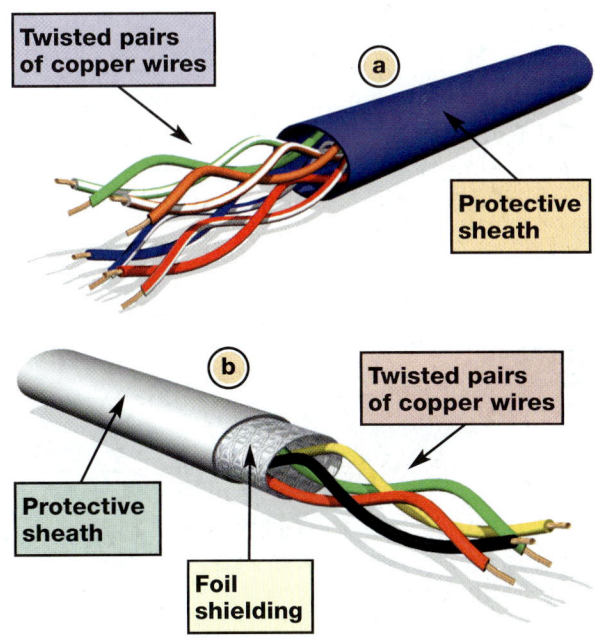

FIGURE 12.11

Anatomy of (a) unshielded twisted pair (UTP) cable and (b) shielded twisted pair (STP) cable.

been made with CAT 6 cable. Home networks that use twisted pair cable generally use CAT 5E cable since it's less expensive.

Coaxial Cable

What does coaxial cable look like? **Coaxial cable** should be familiar to you if you have cable television, as most cable television installers use coaxial cable. Coaxial cable (as shown in Figure 12.12) consists of four main components:

1. The core (usually copper) is in the very center and is used for transmitting the signal.
2. A solid layer of nonconductive insulating material (usually a hard, thick plastic) surrounds the core.
3. A layer of braided metal comes next to reduce EMI and RFI interference with signals traveling in the core.
4. Finally, an external jacket of lightweight plastic covers the internal cable components to protect them from damage.

Although coaxial cable used to be the most widely used cable in business networks, advances in twisted pair cable shielding and transmission speeds, as well as twisted pair's lower cost, have reduced the popularity of coaxial cable.

Are there different types of coaxial cable? The two main coaxial cable types are Thinnet and ThickNet. Thinnet is the cable used by the cable TV company to wire your home and is usually covered by a black plastic jacket. ThickNet, usually distinguished by a yellow jacket, is similar to Thinnet only it is more rigid and better shielded to protect against interference. ThickNet, because it is better shielded, is used in industrial settings where there is a lot of electrical interference. Thinnet is used in homes because it is cheaper and because most houses do not have significant sources of interference (such as industrial machinery).

Fiber-Optic Cable

What does fiber-optic cable look like? As shown in Figure 12.13, **fiber-optic cable** is composed of a glass fiber (or a bundle of fibers) that comprises the core of the cable (where the data will be transmitted). Cladding, a protective layer made of glass or plastic, is wrapped around the core to protect it. Finally, for additional protection, an outer jacket (sheath) is added, often made of durable materials such as Kevlar (the substance used to make bulletproof vests). Data transmissions can only pass through fiber-optic cable in one direction. Therefore, there are usually at least two cores located in each fiber-optic cable to allow for transmission of data in both directions.

How does fiber-optic cable differ from twisted pair and coaxial cable? As noted earlier, the main difference between fiber-optic cable and other types of cable is the method of signal transmission. Twisted pair and coaxial cable use copper wire to conduct electrical impulses. In a fiber-optic cable, electrical data signals from network devices (client computers, peripherals, and so on) are converted to light pulses before they are transmitted. Since EMI and RFI do not affect light waves, fiber-optic cable is virtually immune to interference.

WIRELESS MEDIA OPTIONS

What wireless media options are there? Although the word *wireless* implies no wires, in businesses, **wireless media** are usually add-ons to extend or improve access to a wired network.

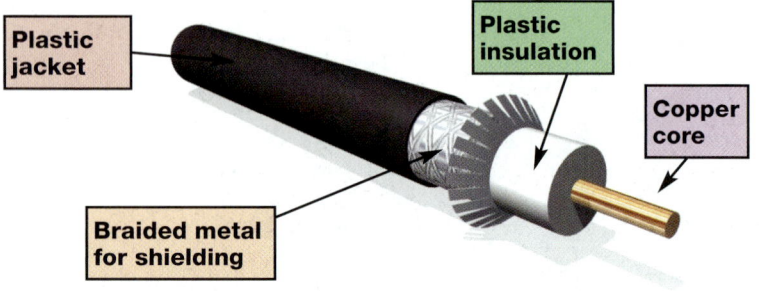

FIGURE 12.12

Coaxial cable consists of four main components: the core, an insulated covering, a braided metal shielding, and a plastic jacket.

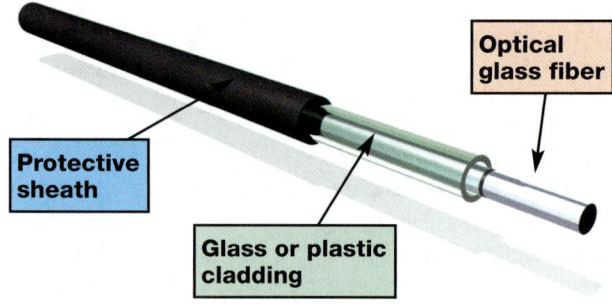

FIGURE 12.13

Fiber-optic cable is made up of a glass fiber (or a bundle of fibers), a glass or plastic cladding, and a protective sheath.

In the corporate environment, wireless access is often provided to give employees a wider range to their working area. For instance, if conference rooms offer wireless access, employees can bring their laptops to meetings and gain access to the network during the meeting. However, when they go back to their offices, they may connect to the regular wired network through a wired connection. So today's corporate networks are often a combination of wired and wireless media.

As you learned in earlier chapters, wireless devices must use the same communications standard to communicate with each other. Wireless networks in the United States are currently based on the **802.11 standard**, also known as **Wi-Fi** (short for Wireless Fidelity), established by the Institute of Electrical and Electronics Engineers (IEEE). Wireless devices attached to networks using the 802.11 standard communicate with each other using radio waves.

The 802.11 standard is actually divided into a number of separate standards. The 802.11b standard is in widespread use and supports a maximum throughput of 11 Mbps. Since manufacturers using the 802.11b standard produced the first cost-effective wireless devices, these caught on quickly in the late 1990s for home and corporate network use. Windows XP also supports 802.11b.

With a maximum throughput of 54 Mbps, the 802.11a and 802.11g standards are fast gaining on 802.11b. Affordable 802.11g devices are even widely available for home use now. For networks in the United States, deploying 802.11a- and g-compliant devices is not usually a problem, and most corporate networks being installed in the United States today are using one of these standards.

Meanwhile, in Europe, HiperLAN2 has been developed as a competing standard for 802.11a and g. Sanctioned by the European Telecommunications Standards Institute, HiperLAN2 supports a throughput of 54 Mbps. Communications giants Ericsson and Nokia are backers of the HiperLAN2 standard.

COMPARING TRANSMISSION MEDIA

So which medium is best for client/server networks? Network engineers specialize in the design and deployment of networks and are responsible for selecting network topology and media types. Their decision as to which transmission medium the network will use is based on the topology selected, the length of the cable runs needed, the amount of interference present, and the need for wireless connectivity.

Figure 12.14 compares the attributes of the major cable types. Most large networks have a mix of media. For example, coaxial cable may be appropriate for the portion of the network that traverses the factory floor where interference from magnetic fields is significant. However, unshielded twisted pair cable may work fine in the general office area. And wireless media may be required in conference rooms and other areas where employees are likely to connect their laptops or where it is impractical or expensive to run cable.

FIGURE 12.14 Comparison of Characteristics of Major Cable Types

CABLE CHARACTERISTICS	TWISTED PAIR (CAT 5E)	COAXIAL (THINNET)	COAXIAL (THICKNET)	FIBER OPTIC
Maximum Run Length	328 feet (100 m)	607 feet (185 m)	1640 feet (500 m)	Up to 62 miles (100 km)
Bandwidth	200 Mbps	10 Mbps	10 Mbps	100 Mbps to 2 Gbps
Bend Radius (Flexibility)	No limit	360 degrees/feet	30 degrees/feet	30 degrees/feet
Cable Cost	Very low	Low	Moderate	High
Installation Cost	Very low	Low	Slightly higher than Thinnet	Most expensive due to installation training required
Susceptibility to Interference	High	Low	Very low	None (not susceptible to EMI and RFI)

TRENDS IN IT

EMERGING TECHNOLOGIES:
Virtual Network Computing Is Here and It's Free

Have you ever been in someone else's office or dorm room on another floor of your building and needed to access something on your computer? Virtual Network Computing (VNC) allows you to access the information on your computer when you are away from it!

The AT&T Laboratories, located at Cambridge University, have developed a free software program that allows you to access your computer from another computer. VNC consists of two software parts: a "server" and a "viewer." To use VNC, you run the software "server" program on your desktop computer before you leave it. This makes your computer accessible from other computers running the appropriate viewer software.

You then install and run the "viewer" software (which fits on a floppy disk) on the computer from which you wish to control your desktop computer. The good news is that the software works on virtually any platform. This means that your desktop computer can be running from Windows, for instance, but you can access it remotely from a Mac or UNIX machine. The only requirement is that a TCP/IP connection (the normal Internet connection used on most computers) must exist between the two machines.

Once you take control of your desktop computer from the remote computer, everything you do (opening and saving files, and so on) is reflected on your desktop computer. When you return to your desk, your computer will reflect all the changes you made to it at the remote computer. This can be a tremendous time saver if you are frequently away from your desk and need to access your computer. Figure 12.15 shows a UNIX desktop being accessed remotely from a Mac.

For more information on Virtual Network Computing, visit **http://www.uk.research.att.com/vnc/start.html**.

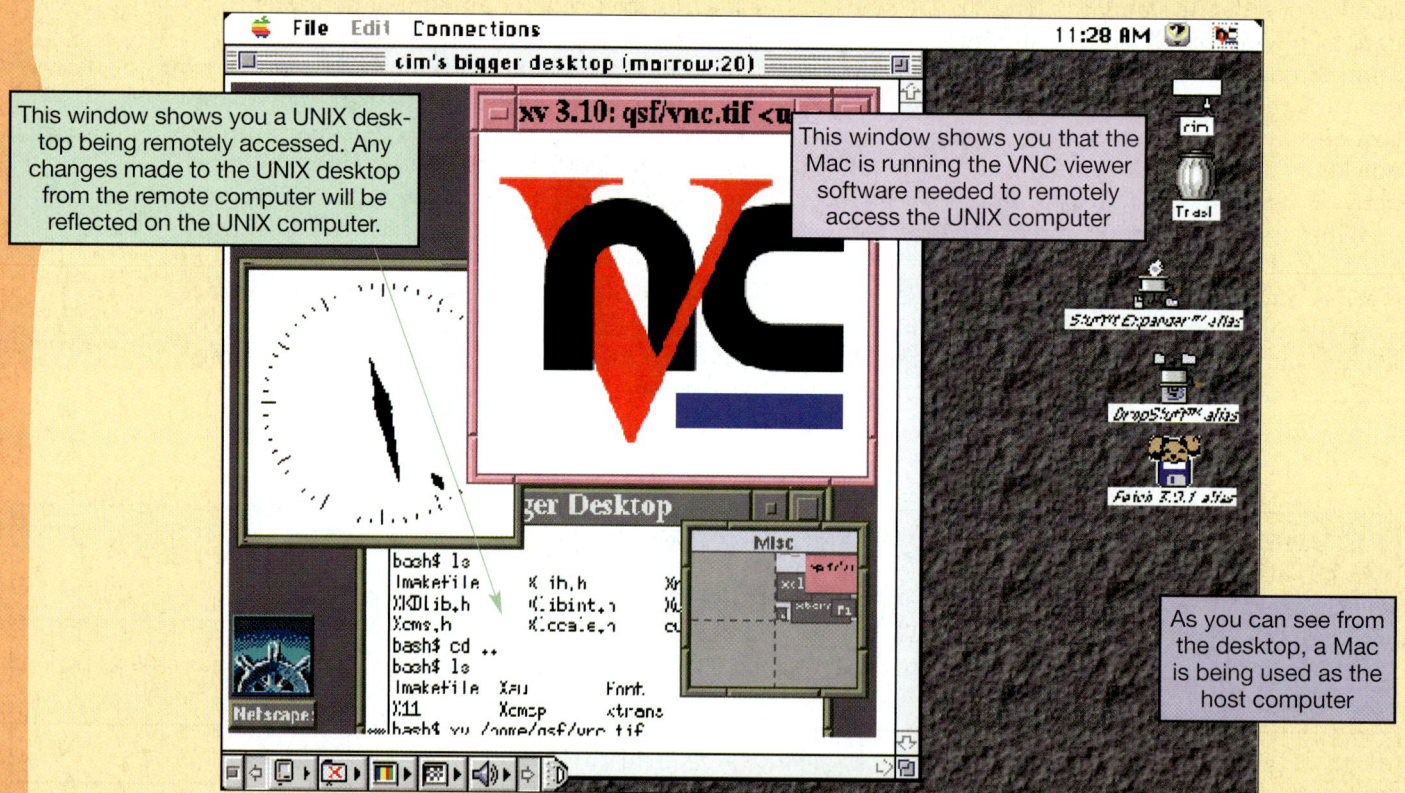

FIGURE 12.15
Using VNC, a UNIX desktop is being remotely accessed from a Macintosh (host).

Network Operating Systems

Merely connecting computers and peripherals with media does not create a client/server network. Special software, known as a **network operating system (NOS)**, needs to be installed on each computer and server connected to the network in order to provide the services necessary for them to communicate. The NOS provides a set of common rules (a protocol) that controls communication between devices on the network. The major NOSs on the market today include Windows Server 2003, UNIX, and Novell NetWare.

Do peer-to-peer networks need special NOS software? The software that P2P networks require is built into the Windows and Macintosh operating systems. Therefore, there is no need to purchase specialized NOS software.

How does NOS software differ from operating system software? Operating system (OS) software is designed to facilitate communication between the software and hardware components of your computer. NOS software is specifically designed to provide server services, network communications, management of network peripherals, and storage. To provide network communications, the client computer must run a small part of the NOS in addition to the OS. Windows XP is an OS and is installed on home computers. Because it also has some NOS functionality, client computers (in a client/server network) that have Windows XP installed as the OS do not need an additional NOS. Windows Server 2003 is an NOS that is deployed on servers in a client/server network.

How does the NOS control network communications? Each NOS has its own proprietary communications language, file management structure, and device management structure. The NOS also sets and controls the protocols (rules) for all devices wishing to communicate on the network. Many different proprietary networking protocols exist, such as Novell's Internetwork Packet Exchange (IPX), Microsoft's NetBIOS Extended User Interface (NetBEUI), and the Apple File Protocol (AFP). These protocols were developed for a specific vendor's operating system. For example, IPX was developed for networks running the Novell NOS. Proprietary protocols such as these do not work with another vendor's NOS.

However, since the Internet uses an open protocol (called TCP/IP) for communications, many corporate networks use TCP/IP as their standard networking protocol regardless of the manufacturer of their NOS. All modern NOSs support TCP/IP. (We'll discuss TCP/IP in more detail in Chapter 13.)

Can a network use two different NOSs? Many large corporate networks use several different NOSs at the same time. This is because different NOSs provide different features, some of which are more useful in certain situations than others. For instance, although the employees of a corporation may be using a Microsoft Windows environment for their desktops and e-mail, the file servers and print servers may be running a Novell NOS.

Since NOSs use different internal software languages to communicate, they can't communicate directly with each other. However, if both NOSs are using the same protocol (such as TCP/IP), they can pass information between the networks and it can be interpreted by the other network.

Network Adapters

As we noted in Chapter 7, client computers and peripherals need an interface to connect with and communicate on the network. **Network adapters** are devices that perform specific tasks to allow computers to communicate on a network. Certain network adapters are installed *inside* computers and peripherals as expansion cards. These adapters are referred to as *network interface cards (NICs)*.

Although you could use network adapters that plug into USB ports on a client/server network, most network adapters will be NICs. That's because external devices are more susceptible to damage.

What do network adapters do? Network adapters perform three critical functions:

1. **They generate high-powered signals to enable network transmissions.** Digital signals generated inside the computer are fairly low powered and would not travel well on network media (cable or wireless technology) without network adapters. Network adapters convert the signals from inside the computer to higher-powered signals that have no trouble traversing the network media.

2. **They are responsible for breaking the data down into packets and preparing them for transmission across the network.** They are also responsible for receiving incoming data packets and, in accordance with networking protocols (rules), reconstructing them, as shown in Figure 12.17 on page 494.

DIG DEEPER

THE OSI MODEL: DEFINING PROTOCOL STANDARDS

The Institute of Electrical and Electronics Engineers (IEEE) has taken the lead in establishing recognized worldwide networking protocols, including a standard of communications called the Open Systems Interconnect (OSI) Reference Model. The OSI model, which was quickly adopted as a standard throughout the computing world, provides the protocol guidelines for all modern networks. All modern NOS protocols are designed to interact in accordance with the standards set out by the OSI model.

The OSI model divides communications tasks into seven distinct processes called *layers*. Each layer of an OSI network has a specific function. Figure 12.16 shows the layers of the OSI model and their functions. Each layer knows how to communicate with the layer above and below it.

This layering approach makes communications more efficient since specialized pieces of the NOS perform specific tasks. It is akin to assembly line manufacturing. Producing thousands of cars per day would be difficult if one person had to build a car on his or her own. By splitting up the work of assembling a car into specialized tasks (such as installing the engine, bolting on the bumpers, and so on), people who are exceptionally good at certain tasks can be assigned to do them and achieve higher efficiency. This is how the OSI layers work. By handling specialized tasks and communicating only with the layers above and below them, the layering approach makes communications more efficient.

Let's look at how each OSI layer functions by following an e-mail you create and send to your friend:

Application Layer

- The *application layer* handles all interaction between the application software and the network. It translates the data from the application into a format that the presentation layer can understand. For example, when you send an e-mail, the application layer takes the e-mail message you created in Microsoft Outlook, translates it into a format your network can understand, and passes it to the presentation layer.

Presentation Layer

- The *presentation layer* reformats the data so that the session layer can understand it. It also handles data encryption (changing the data into a format that makes it harder to intercept and read the message) and compression if required. In our e-mail

FIGURE 12.16 The Layers of the OSI Model and Their Functions

LAYER NAME	LAYER FUNCTION
APPLICATION LAYER	Handles all interfaces between the application software and the network Translates user information into a format the presentation layer can understand
PRESENTATION LAYER	Reformats data so that the session layer can understand it Compresses and encrypts data
SESSION LAYER	Sets up a virtual (not physical) connection between the sending and receiving devices Manages communications sessions
TRANSPORT LAYER	Creates packets Handles packet acknowledgment
NETWORK LAYER	Determines where to send the packets on the network
DATA LINK LAYER	Assembles the data into frames, addresses them, and sends them to the physical layer for delivery
PHYSICAL LAYER	Transmits (delivers) data onto the network so it can reach its intended address

Materials on the Web

In addition to the review materials presented here, you'll find extra materials on the book's companion Web site (**www.prenhall.com/techinaction**) that will help reinforce your understanding of the chapter content. These materials include:

Sound Byte Lab Guides

For each Sound Byte mentioned in the chapter, there is a corresponding lab guide located on the book's companion Web site. These guides review the material presented in the Sound Byte and direct you to various Web resources that examine the material. The Sound Byte Lab Guides for this chapter include:

- What's My IP Address? (and Other Interesting Facts about Networks)
- A Day in the Life of a Network Technician

True/False and Multiple Choice Quizzes

The book's Web site includes a True/False and a Multiple Choice quiz for this chapter. You can take these quizzes, automatically check the results, and e-mail the results to your instructor.

Web Research Projects

The book's Web site also includes a number of Web research projects for this chapter. These projects ask you to search the Web for information on computer-related careers, milestones in computer history, important people and companies, emerging technologies, and the applications and implications of different technologies.

OBJECTIVES

After reading this chapter, you should be able to answer the following questions:

- Who manages and pays for the Internet? (p. 516)

- How do the Internet's networking components interact? (pp. 517–521)

- What data transmissions and protocols does the Internet use? (pp. 522–524)

- Why are IP addresses and domain names important for Internet communications? (pp. 524–529)

- What are FTP and Telnet and how do I use them? (pp. 529–530)

- What are HTML and XML used for? (pp. 530–535)

- How does e-mail and instant messaging work and how are messages kept secure? (pp. 536–540)

- What will the Internet of the future look like? (pp. 541–543)

CHAPTER 13

Behind the Scenes:
The Internet: How It Works

**TECHNOLOGY IN ACTION:
KNOWING HOW THE INTERNET WORKS**

At this point in your studies, you know what the Internet is, and you've certainly used some of its features, such as the Web and e-mail. So why do you need to know how the Internet works? Most people can drive a car without knowing how an internal combustion engine is designed and built. However, a more thorough understanding of auto mechanics is useful when you're making decisions about buying a car or when you need to fix it when it breaks down. Similarly, understanding the mechanics behind the Internet can assist you in a number of ways.

In the business world, even if you aren't making a career out of information technology, you'll be interacting with coworkers in IT departments on a regular basis. Communicating with IT employees is much easier if you speak a common language. For example, when the IT manager tells you that the sales system you just proposed won't be feasible because the "router" can't handle the volume of outgoing requests, it would be helpful for you to understand what a "router" is and does.

Gaining a thorough understanding of the Internet and its capabilities can also help you determine whether you want to pursue additional coursework in Web development and networking. Creating and maintaining Web sites and the equipment that allows Internet connectivity takes many people with different talents. If you understand the jobs of the various individuals who support the infrastructure and services provided by the Internet, you'll be better able to assess whether an Internet-related career is right for you.

This chapter will build on what you learned in Chapter 3 and take you behind the scenes of the Internet. We'll look at who manages the Internet and discuss in detail how the Internet works and the various standards it follows. Along the way, we'll go behind the scenes of some Internet communication features, such as e-mail and instant messaging services, and discuss just how safe these features are and what you can do to make your communications even more secure.

FIGURE 13.1 Major Organizations that Play a Role in Internet Governance and Development

ORGANIZATION NAME	ORGANIZATION AND PURPOSE	WEB ADDRESS
Internet Society (ISOC)	Professional membership society comprising more than 11,000 organizations and individuals. Provides leadership for the orderly growth and development of the Internet.	www.isoc.org
Internet Engineering Task Force (IETF)	A subgroup of ISOC made up of individuals and organizations that research new technologies for the Internet to improve capabilities or to keep the infrastructure functioning smoothly.	www.ietf.org
Internet Architecture Board (IAB)	Technical advisory group to IETF. Provides direction for the maintenance and development of the protocols that are used on the Internet.	www.iab.org
Internet Corporation for Assigned Names and Numbers (ICANN)	Organization responsible for management of the Internet's domain name system and the allocation of IP addresses.	www.icann.org
World Wide Web Consortium (W3C)	Consortium whose 450 member organizations set standards and develop protocols for the Web.	www.w3.org

The Management of the Internet

You learned about the history of the Internet in Chapter 3. But to keep the Internet functioning at peak efficiency, it must be governed and regulated.

Who owns the Internet? Although the U.S. government funded the development of the technologies that spawned the Internet, no one really owns it. The individual local networks that constitute the Internet are all owned by different individuals, universities, government agencies, and private companies. Government entities, such as the National Science Foundation (NSF) and NASA, and many large, privately held companies all own pieces of the communications infrastructure (the high-speed data lines that transport data between networks) that makes the Internet work.

Does anyone manage the Internet? While no single entity owns all of the individual networks that participate in the Internet, the Internet would grind to a halt without some sort of organization. Therefore, a number of nonprofit organizations and user groups, each with a specialized purpose, are responsible for management. Figure 13.1 shows the major organizations that play a role in the governance and development of the Internet.

Many of the functions handled by these nonprofit groups were previously handled by U.S. government contractors since the Internet developed out of a U.S. government military project. However, because the Internet now serves the global community, not just the United States, passing off responsibilities to organizations with global memberships is helping to speed along the internationalization of the Internet. Through close collaboration of the organizations listed in Figure 13.1 (and a few others), the Internet's vast collection of users and networks is managed.

Who pays for the Internet? You do! The U.S. government pays for a large portion of the Internet infrastructure as well as funds research and development for new technologies. The primary source of these funds is your tax dollars. Originally, U.S. taxpayers footed the entire bill for the Internet, but as the Internet grew and organizations were formed to manage it, businesses, universities, and other countries began paying for Internet infrastructure and development.

Internet Networking

Despite being the largest network on earth, the Internet's response to our requests for information seems almost magical at times. By just entering a URL into your browser address box, you can summon up information that is stored on servers half a world away. But there is no "magic" involved, just a series of communication transactions that allow the Internet to function as a global

INTERNET NETWORKING 517

network. In this section, we'll explore the various networks that make up the Internet and examine the workings of Internet data communications.

Is the Internet officially considered a network? Although the Internet can connect individual computers to each other, the Internet is really a network of networks. The word *internet* (with a lowercase "i," not to be confused with the Internet with an uppercase "I") was originally used to describe a connection between two or more networks. The word eventually became capitalized (Internet) and associated with the worldwide network of computer systems that grew explosively in the 1990s. The Internet does behave like a network in that it follows a set of communications protocols and is used for transferring data between computers. A **protocol** is simply a set of rules for communicating. All computers connected to the Internet need to use common protocols so they can understand one another.

CONNECTING ISPS

How are computers connected to the Internet? As you learned in Chapter 3, to connect individual computers or networks to the Internet, home users and businesses use Internet service providers (ISPs). As shown in Figure 13.2, ISPs are classified in a hierarchy that consists of three tiers: Tier 1, Tier 2, and Tier 3.

At the heart of the Internet are **Tier 1 ISPs,** located in the green zone in Figure 13.2. Tier 1 ISPs route a large percentage of the traffic on

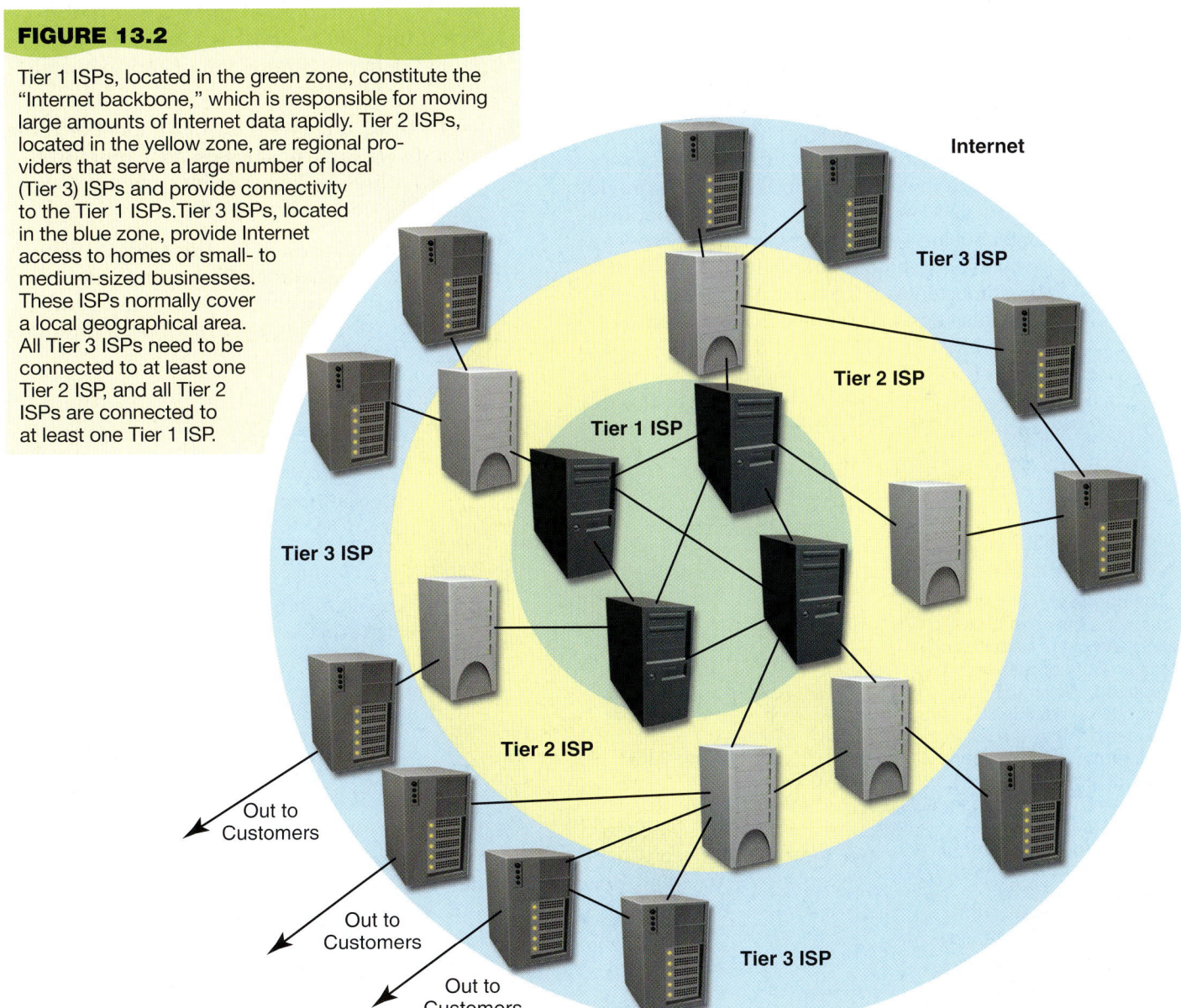

FIGURE 13.2

Tier 1 ISPs, located in the green zone, constitute the "Internet backbone," which is responsible for moving large amounts of Internet data rapidly. Tier 2 ISPs, located in the yellow zone, are regional providers that serve a large number of local (Tier 3) ISPs and provide connectivity to the Tier 1 ISPs. Tier 3 ISPs, located in the blue zone, provide Internet access to homes or small- to medium-sized businesses. These ISPs normally cover a local geographical area. All Tier 3 ISPs need to be connected to at least one Tier 2 ISP, and all Tier 2 ISPs are connected to at least one Tier 1 ISP.

the Internet and have extremely high-speed connections with other ISPs, sometimes in the 2.5 to 10 gigabits per second (Gbps) range. The high-speed communications lines Tier 1 ISPs use are referred to as the **Internet backbone.** There are dozens of Tier 1 ISPs, each of which is required to be directly connected to *all other* Tier 1 ISPs. For example, AT&T and Sprint have subsidiaries that are Tier 1 ISPs. Tier 1 ISPs are also normally connected to a large number of Tier 2 ISPs and span international borders.

Tier 2 ISPs, located in the yellow zone in Figure 13.2, usually have a regional or national focus. Therefore, to enable their customers to reach any possible point on the *global* Internet, Tier 2 ISPs must route at least a portion of their traffic through the global Tier 1 ISPs. Information flow between the green zone (Tier 1 ISPs) and the yellow zone (Tier 2 ISPs) occurs using high-speed data lines but usually with less bandwidth than the internal connections within the green zone's Tier 1 ISPs. Large companies and universities often connect directly to a Tier 2 ISP.

The thousands of **Tier 3 ISPs,** which are located in the blue zone in Figure 13.2, provide Internet access to homes or small- to medium-sized businesses. These ISPs normally cover a local geographical area. All Tier 3 ISPs need to be connected to at least one Tier 2 ISP. The ISP you're using at your home is most likely a Tier 3 ISP.

T LINES

Are the data lines connecting ISPs faster than DSL or cable connections? Most high-speed communications between ISPs are achieved using *T lines*. **T lines** are high-speed fiber-optic communications lines that are designed to provide much higher throughput than conventional voice (telephone) and data (DSL) lines. T-lines come in a variety of speeds:

- A **T-1 line** can support 24 simultaneous voice or data channels and achieve a maximum throughput of 1.544 megabits per second (Mbps). Businesses or Tier 3 ISPs often use T-1 lines to connect to the Internet due to the large volume of Internet traffic they experience. If a business's bandwidth requirements grow, it can upgrade to higher-capacity T lines that are merely bundles of T-1 lines.

- **T-2 lines** are composed of four T-1 lines and deliver a throughput of approximately 6.3 Mbps.

- **T-3 lines**, often used by Tier 1 and Tier 2 ISPs and very large businesses, are a bundle of 28 T-1 lines. T-3 lines deliver a whopping 44.736 Mbps of bandwidth.

Still, this isn't enough for Tier 1 ISPs to communicate with each other. They usually need to use **T-4 lines**, which contain 168 T-1 lines and provide an astounding 274.176 Mbps of through-

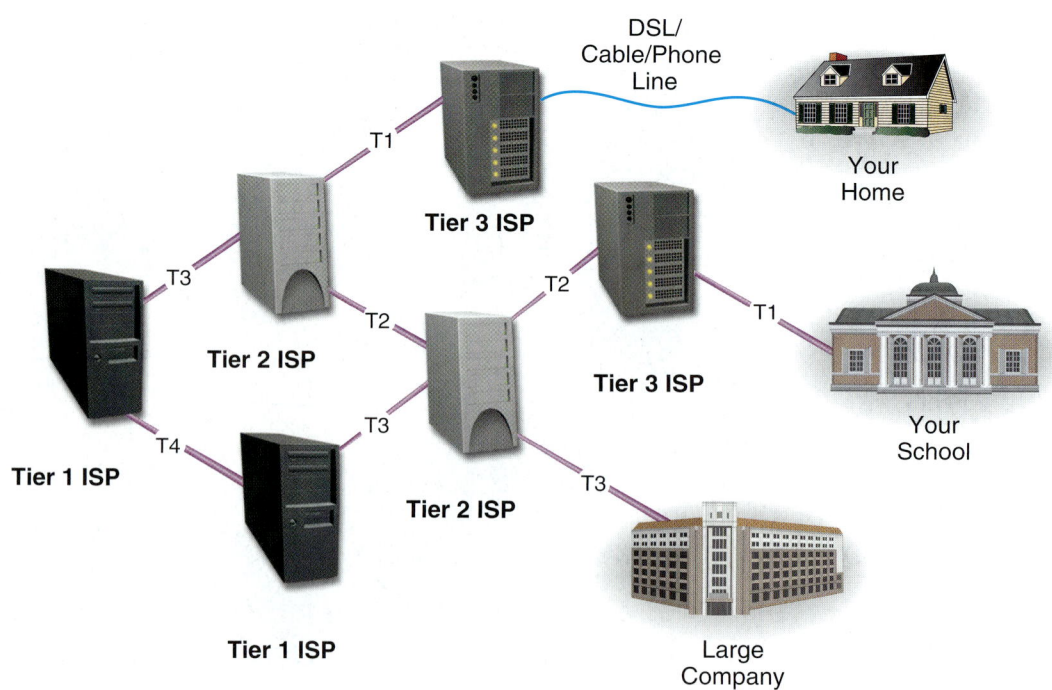

FIGURE 13.3

The bandwidth of the connections between ISPs and end users depends on the amount of data traffic required. While traffic at your home wouldn't require a T line, the volume of Internet traffic at your school probably requires at least a T-1 line to move data to the school's ISP. Large companies with thousands of employees usually need T-2 or T-3 lines running to their ISPs. Since Tier 1 ISPs are moving a great deal of data, they need the robust bandwidth of T-4 lines.

put. Figure 13.3 illustrates how individual users, organizations, and ISPs use the various T lines to send and receive data.

NETWORK ACCESS POINTS

How are the ISPs connected to each other? The points of connection between ISPs are known as **network access points (NAPs)**. NAPs contain groups of routers specifically designed to move large amounts of data quickly between networks. As you'll recall from earlier chapters, **routers** are devices that send data packets between networks. Since the Internet is really a large collection of connected networks, routers are needed to move data throughout the Internet. Large backbone providers or third-party telecommunications companies maintain these groups of routers, or NAPs.

As you can see in Figure 13.4, a Tier 1 ISP can connect to many NAPs at a number of diverse locations throughout the world (in this figure, a Tier 1 ISP connects to a NAP in Denver, London, and Miami). This provides many points where Tier 2 ISPs can connect to a Tier 1 ISP. Many ISPs can connect through the same NAP. Tier 2 ISPs pay for third-party high-speed leased communications lines (such as T-1s and T-3s) to connect their networks to the Tier 1's NAPs.

In addition to the fees the third parties charge for leased lines, Tier 2 ISPs are charged for access to the Tier 1 ISPs based on the amount of bandwidth connecting the two. Bandwidth charges are like tolls on the highway. The higher the volume of vehicles, the more tolls are paid. To reduce bandwidth charges from the high-speed ISPs, Tier 2 ISPs often connect directly to each other if there is a high volume of traffic passing between them (see Figure 13.5 on page 520). For example, say it costs $1 for each message sent by a Tier 2 ISP through a Tier 1 ISP. Assume two Tier 2 ISPs send 100 messages to each other a month. If they can establish a direct connection between them (say for $50 a month), they don't have to send their messages through the Tier 1 ISP and they save $100 a month (less the $50 for the direct connection).

POINTS OF PRESENCE

How do individuals connect to an ISP? Whether dialing up through a conventional modem or connecting through high-speed access (such as cable or DSL), individual Internet users enter an ISP through a **point of presence (POP)**, which is basically a bank of modems (shown in Figure 13.6 on page 521) through which many users can connect to an ISP simultaneously. ISPs maintain multiple POPs throughout the geographic area they serve.

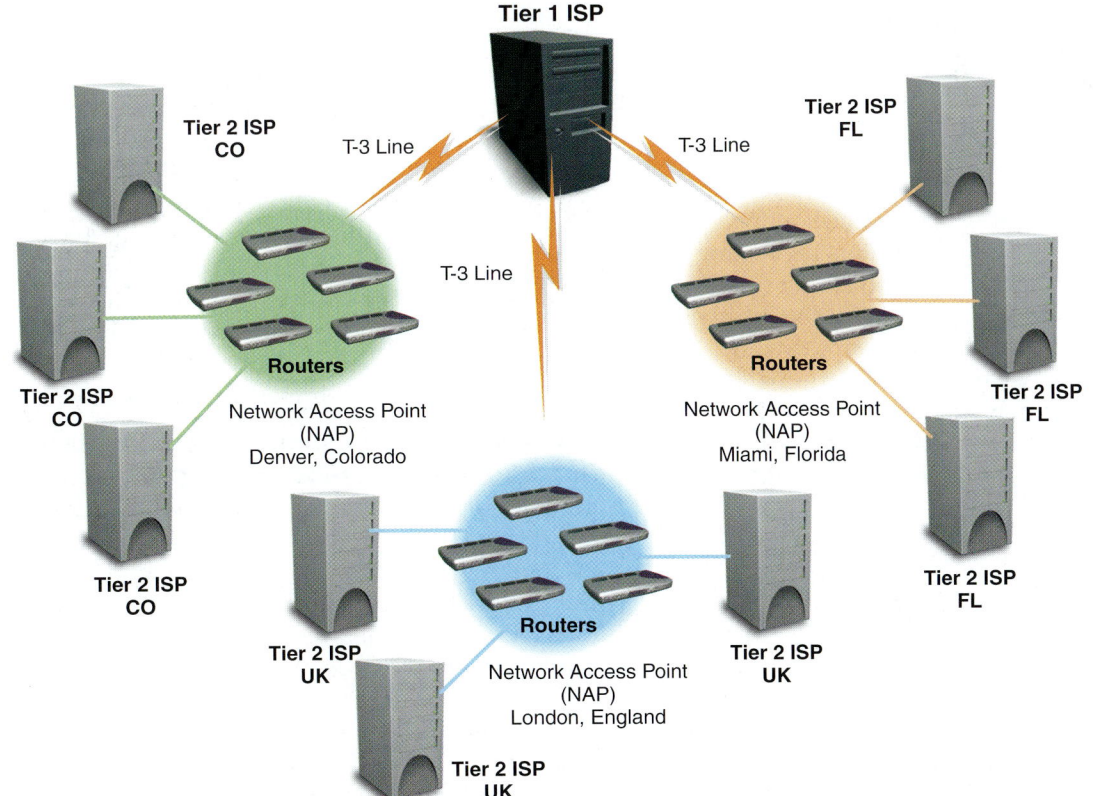

FIGURE 13.4

NAPs, essentially large collections of routing equipment, provide a central point of connection where many Tier 2 ISPs can connect efficiently to a Tier 1 ISP.

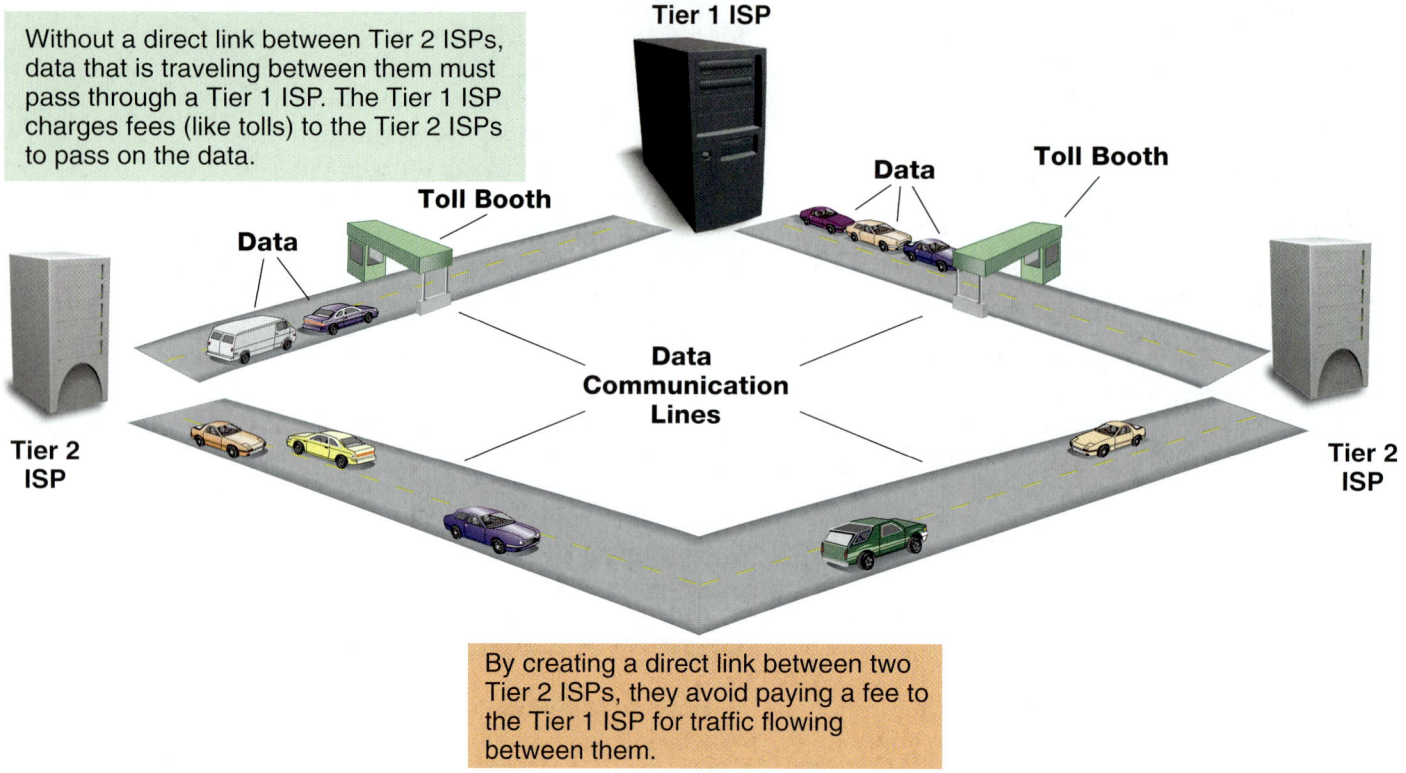

FIGURE 13.5

Just like cars traveling a toll road, sending data between Tier 2 ISPs and Tier 1 ISPs costs money. To reduce costs, some Tier 2 ISPs build their own data communications "highways" to reduce data charges.

THE NETWORK MODEL OF THE INTERNET

What type of network model does the Internet use? The majority of Internet communications follows the **client/server model** of network communications, which we defined in earlier chapters as client computers requesting services and servers providing (serving up) those services to the clients. In the case of the Internet, the *clients* are devices such as computers and PDAs using browsers (or other interfaces) that request services (Web pages and so on). There are various types of *servers* deployed on the networks that make up the Internet from which clients can request services:

- **Web servers** are computers running specialized operating systems that allow them to host Web pages (and other information) and provide requested Web pages to the clients.
- **Commerce servers** host software that allows you to purchase goods and services over the Web. These servers generally use special security protocols to protect sensitive information (such as credit card numbers) from being intercepted.
- **File servers** are deployed to provide remote storage space or to act as a repository for files that users can download. For example, Yahoo! provides a file storage option for its users called Yahoo! Briefcase. Storing files in your personal "briefcase" (file folder) allows you to access these files from anywhere you can access the Web with a browser. When you use the briefcase feature of Yahoo!, you are storing files remotely on a file server.

Do all Internet connections take place in a client/server mode? Certain services on the Internet operate in a peer-to-peer mode, as depicted in Figure 13.7. For example, Kazaa is a popular file-sharing service through which Internet users can exchange files. Kazaa and other file-sharing services require the user's computer to act as *both* a client *and* a server. When requesting files from another user, the computer behaves like a client. It switches to server mode when it in turn provides a file stored on its hard drive to another computer.

INTERNET NETWORKING   521

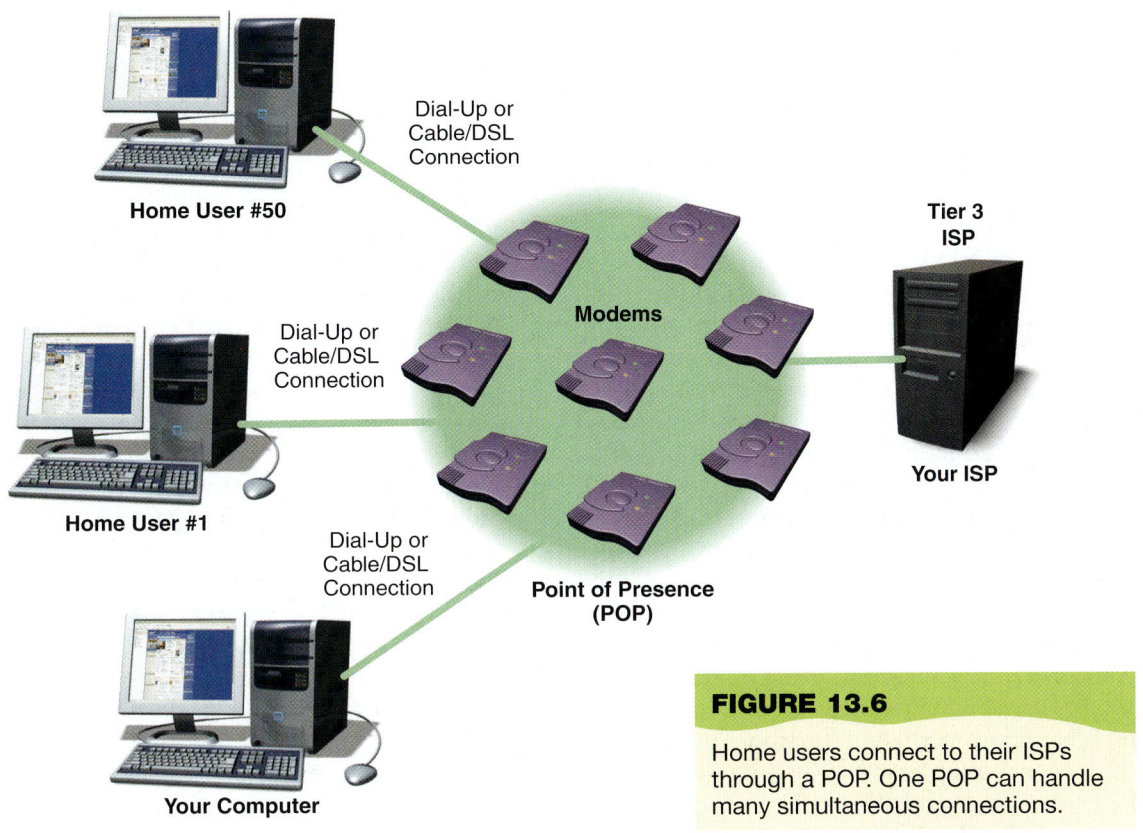

FIGURE 13.6
Home users connect to their ISPs through a POP. One POP can handle many simultaneous connections.

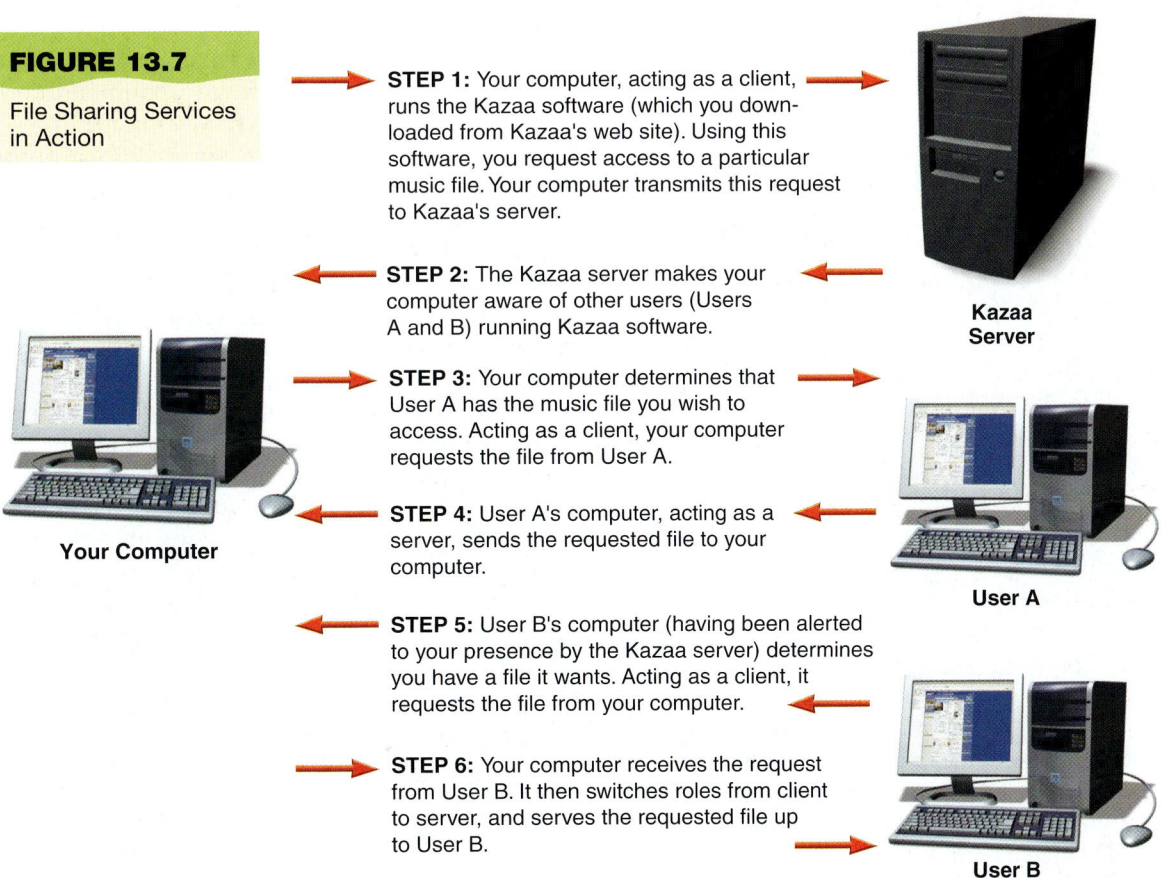

FIGURE 13.7
File Sharing Services in Action

STEP 1: Your computer, acting as a client, runs the Kazaa software (which you downloaded from Kazaa's web site). Using this software, you request access to a particular music file. Your computer transmits this request to Kazaa's server.

STEP 2: The Kazaa server makes your computer aware of other users (Users A and B) running Kazaa software.

STEP 3: Your computer determines that User A has the music file you wish to access. Acting as a client, your computer requests the file from User A.

STEP 4: User A's computer, acting as a server, sends the requested file to your computer.

STEP 5: User B's computer (having been alerted to your presence by the Kazaa server) determines you have a file it wants. Acting as a client, it requests the file from your computer.

STEP 6: Your computer receives the request from User B. It then switches roles from client to server, and serves the requested file up to User B.

Data Transmission and Protocols

Just like any other network, the Internet follows standard protocols to send information between computers. A **computer protocol** is a set of rules for accomplishing electronic information exchange. If the Internet is the information superhighway, then protocols are the driving rules.

To accomplish the early goals of the Internet, protocols needed to be written and agreed upon by users. The protocols needed to be **open systems**, meaning their designs would be made public for access by any interested party. This was in direct opposition to the **proprietary** (or private) **systems** that were the norm at the time.

As we mentioned in earlier chapters, when common communications protocols (rules) are followed, networks can communicate even if they have different topologies, transmission media, or operating systems. The idea of an open-system protocol is that anyone can use it on their computer system and be able to communicate with any other computer using the same protocol. The three biggest Internet tasks (e-mail, Web surfing, and file transfer) are all being done the same way on any system that is following accepted Internet protocols.

Were there problems developing an open-systems Internet protocol? Agreeing on common standards was relatively easy. However, the tough part was developing a new method of communication since the current technology, *circuit switching*, was not efficient for computer communication. Circuit switching has been used since the early days of the telephone for establishing communication. In **circuit switching**, a dedicated connection is formed between two points (two people on telephones) and the connection remains active for the duration of the transmission. This method of communication is extremely important when communications must be received in the order in which they were sent (such as in telephone conversations).

When applied to computers, however, circuit switching is inefficient. Computer processing and communication take place in bursts. As a computer processor performs the operations necessary to complete a task, it transmits data in a group (or burst). The processor then begins working on its next task and ceases to communicate with output devices or other networks until it is ready to transmit data in the next burst. Circuit switching is inefficient and wasteful for computers since the circuit either would have to remain open (and therefore unavailable to any other system) with long periods of inactivity, or it would have to be reestablished for each burst.

PACKET SWITCHING

If they can't use circuit switching, what do computers use to communicate? **Packet switching** is the communications methodology that makes computer communication efficient today. Packet switching does not require a dedicated communications circuit to be maintained. With packet switching, data is broken into smaller chunks (called **packets**) and sent over various routes at the same time. When the packets reach their destination, they are reassembled by the receiving computer. This technology resulted from one of the original goals of creating the Internet: if Internet nodes are disabled or destroyed (such as through an act of warfare or terrorism), the data can travel an alternative route to reach its destination.

What information does a packet contain? Packet contents vary depending on the protocol being followed. At a minimum, all packets must contain (1) an address to where the packet is being sent, (2) reassembling instructions if the original data was split between packets, and (3) the data that is being transmitted.

Sending a packet is sort of like sending a letter. Assume you are sending a large amount of information in written format from your home in Philadelphia to your aunt in San Diego. The information is too large to fit in one small envelope, so you mail three different envelopes to your aunt. Each envelope includes your aunt's address, a return address (your address), and the information being sent inside it. The pages of the letters sent in each envelope are consecutively numbered so your aunt will know in which order to read them.

Each envelope may not find its way to San Diego by the same route (the letters may be routed through different post offices), but they will all eventually arrive in your aunt's mailbox. Your aunt will then reassemble the message (put the pages of the letters in numerical order) and read it. The process of sending a message through the Internet works in much the same way. This is illustrated in Figure 13.8, which traces an e-mail message sent from a computer in Philadelphia to a computer in San Diego.

Why do packets take different routes and how do they decide which route to use? The routers that connect ISPs with each other monitor traffic and decide on the most effi-

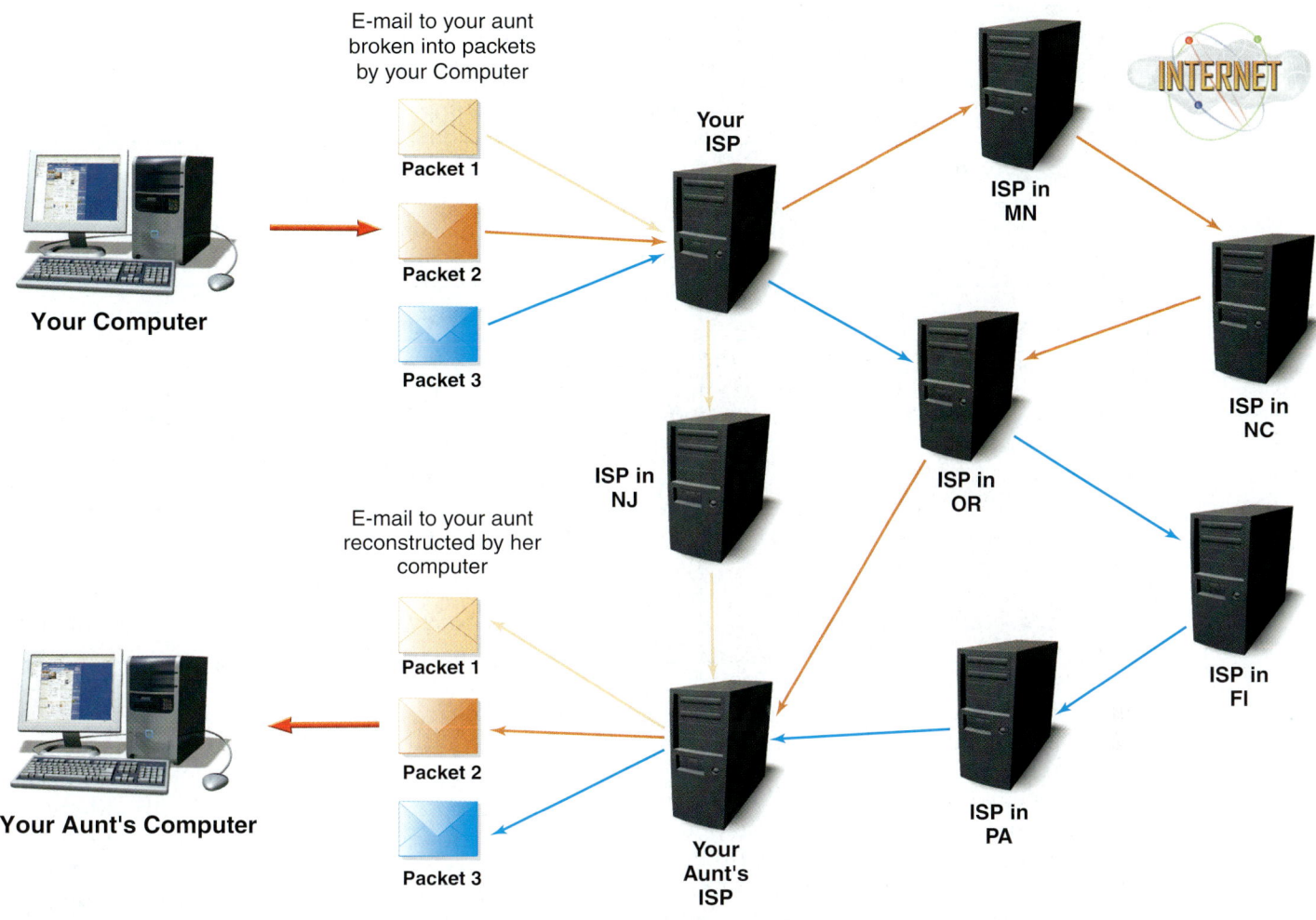

FIGURE 13.8

Packets can each follow their own route to their final destination. Sequential numbering of packets ensures they are reassembled in the correct order at their destination.

cient route for packets to take to their destination. The router works the same as a police officer during a traffic jam. When routes are clogged with traffic, police officers are deployed in areas of congestion to direct you to an alternate route to get to your destination.

TCP/IP

What protocol does the Internet use for transmitting data? Although many protocols are available on the Internet, the main suite of protocols used is **TCP/IP**. The suite is named after the original two protocols that were developed for the Internet: the **Transmission Control Protocol (TCP)** and **Internet Protocol (IP)**. Although most people think that the TCP/IP suite consists of only two protocols, it actually comprises many interrelated protocols, the most important of which are listed in Figure 13.9 on the following page.

Which particular protocol actually sends the information? The Internet Protocol (IP) is responsible for sending the information from one computer to another. The IP is like a postal worker who takes a letter (a packet of information) that was mailed (created by the sending computer) and sends it on to another post office, which in turn routes it to the addressee (receiving computer). The postal worker never knows whether the recipient actually receives the letter, nor does he or she care. The only thing the postal worker knows is that the letter was handed off to an appropriate post office that will assist in completing the delivery of the letter.

FIGURE 13.9 The Main Protocols Contained in the TCP/IP Protocol Suite

PROTOCOL NAME	MAIN PROTOCOL FUNCTION
Internet Protocol (IP)	Sends data between computers on the Internet
Transmission Control Protocol (TCP)	Prepares data for transmission and provides for error checking and resending lost data
User Datagram Protocol (UDP)	Prepares data for transmission—no resending capabilities
File Transfer Protocol (FTP)	Enables files to be downloaded to a computer or uploaded to other computers
Telnet	Allows logging in to a remote computer and working on it as if sitting in front of it
Hypertext Transfer Protocol (HTTP) and Secure HTTP (S-HTTP)	Transfers HTML data from servers to browsers
Simple Mail Transfer Protocol (SMTP)	Used for transmission of e-mail messages across the Internet

IP Addresses and Domain Names

As you learned in Chapter 3, each computer, server, or device (router, etc.) connected to the Internet is required to have a unique number identifying it, called an **IP address**. IP addresses fulfill the same function as street addresses. For example, to get directions to John Doe's house in Walla Walla, Washington, you have to know his address. John might live at 123 Main Street, which is not a unique address (many towns have a Main Street). But 123 Main Street, Walla Walla, WA 99362 is a unique address.

What helps make it unique? The numeric zip code provides unique identification for a specific geographic area. The Internet Corporation for Assigned Names and Numbers (ICANN) is responsible for allocating IP addresses to network administrators, just like the U.S. Postal Service is responsible for assigning zip codes to geographic areas.

What does an IP address look like? A typical IP address is expressed as follows:

    197.24.72.157

Expressing an IP address like this is called a **dotted decimal number**. However, computers work with binary numbers. The same IP address in binary form would be as follows:

    11000101.00011000.01001000.10011101

The four numbers in the dotted decimal notation are each referred to as an **octet**. This name derives from the fact that each number would have eight positions when shown in binary form. Since there are 32 positions available for IP address values (four octets with eight positions each), IP addresses are considered 32-bit numbers. A position is either filled by a 1 or a 0, resulting in 256 (2^8) possible values for each octet. Values start at 0 (not 1); therefore, each octet can have a value from 0 to 255. The entire 32-bit address can represent 4,294,967,296 values (or 2^{32}), which is quite a few Internet addresses!

Will we ever run out of IP addresses? When the original IP addressing scheme, **Internet Protocol version 4 (IPv4)**, was created, no one foresaw the explosive growth of the Internet in the 1990s. (Of course, the Internet wasn't yet the exciting visual medium that it is today.) Therefore, four billion values for an address field seemed like enough to last forever. However, as the Internet grew rapidly, it quickly became apparent that we were going to run out of IP addresses before the new millennium.

What is being done to make sure we have enough IP addresses in the future? **Internet Protocol version 6 (IPv6)** is a proposed IP addressing scheme that makes IP addresses longer, thereby proving more available IP addresses. It uses eight groups of 16-bit numbers, referred to as *hexadecimal notation* (or *hex* for short), which you learned about in Chapter 9. An IPv6 address would have the following format:

    0000:0000:0000:0000:0000:0000:0000:0000

IP ADDRESSES AND DOMAIN NAMES

Hex addressing provides a much larger field size that will enable a much larger number of IP addresses (approximately 340 followed by 36 zeros). This should be a virtually unlimited supply.

Is IPv6 being used today? IPv6 has been slow to catch on. The complexities involved with adopting IPv6 include designing a new TCP/IP protocol. Not many vendors have rolled out IPv6 protocols yet. The good news is that IPv6 will be backward compatible with IPv4, so there will be no need to change current IPv4 systems.

How does my computer get an IP address? You learned in Chapter 7 that IP addresses are either assigned *statically* or *dynamically*. **Static addressing** means that the IP address for a computer never changes and is most likely assigned manually by a network administrator. **Dynamic addressing**, in which your computer is assigned an address from an available pool of IP addresses, is more common today. A connection to an ISP could use either method. If your ISP uses static addressing, you were assigned an IP address when you applied for your service and had to manually configure your computer to use that address. More often, though, your ISP assigns your computer a temporary (dynamic) IP address.

How exactly are dynamic addresses assigned? Dynamic addressing is normally handled by the **Dynamic Host Configuration Protocol** (**DHCP**), which belongs to the TCP/IP protocol suite. DHCP takes a pool of IP addresses and shares them with hosts on the network on an as-needed basis. ISPs don't need to maintain a pool of IP addresses for *all* of their subscribers since not everyone is logged onto the Internet at one time. Thus, when a user logs on to an ISP's server, the DHCP server assigns that user an IP address for the duration of the session. Similarly, when you log on to your computer at work in the morning, the DHCP protocol assigns your computer an IP address. These temporary IP addresses may or may not be the same from session to session.

What are the benefits of dynamic addressing? Although having a static address would seem to be convenient, dynamic addressing provides for better security measures to keep hackers out of computer systems. Imagine how hard it would be for burglars to find your home if you changed your address every day!

DOMAIN NAMES

I've been on the Internet, so why have I never seen IP addresses? Computers are fantastic at relating to IP addresses and other numbers. However, humans remember names better than they remember strings of numbers. (Would you rather call your friend 1236231 or Maria?) As the Web was being formed, a naming system needed to be developed to allow people to work with names instead of numbers. Hence *domain names* were born. As you learned in Chapter 3, a **domain name** is simply a name that takes the place of an IP address, making it easier for people to remember it. You've most likely visited www.yahoo.com. Yahoo.com is a domain name. The server where Yahoo!'s main Web site is deployed has an IP address (such as 123.45.67.89), but it's much easier for you to remember to tell your browser to go to Yahoo.com than it is to type in the nine-digit IP address.

How are domains organized? Domains are organized by level. As you'll recall from

BITS AND BYTES

What's Your IP Address?

Curious as to what your IP address is? In Windows XP, start the command prompt (located under Accessories). At the c:>, type ipconfig and hit Enter. A screen similar to Figure 13.10 will be displayed that shows, among other things, the IP address your PC is currently using.

```
C:\>ipconfig

Windows IP Configuration

Ethernet adapter Network Bridge (Network Bridge):

        Connection-specific DNS Suffix  . : walngs01.pa.comcast.net
        IP Address. . . . . . . . . . . . : 192.168.0.3
        Subnet Mask . . . . . . . . . . . : 255.255.255.0
        Default Gateway . . . . . . . . . : 192.168.0.1

C:\>
```

FIGURE 13.10

Running ipconfig at the command prompt reveals your computer's IP address.

DIG DEEPER

MAKING THE CONNECTION: CONNECTION-ORIENTED VS. CONNECTIONLESS PROTOCOLS

The Internet Protocol is responsible only for *sending* the packets on their way. The packets are created by either the TCP or the **UDP (User Datagram Protocol)** protocols. You don't decide whether to use TCP or UDP. The choice of protocol was made for you by the authors of the computer programs you are using or by the other protocols (such as those listed in Figure 13.9) that will interact with your data packet.

As we explained earlier, data transmission between computers is very efficient if connections do not need to be established (as in circuit switching). However, there are benefits to maintaining a connection, such as less data loss. The difference between TCP and UDP is that TCP is a *connection-oriented protocol*, while UDP is a *connectionless protocol*.

A **connection-oriented protocol** requires two computers to exchange control packets, which set up the parameters of the data exchange session, prior to sending packets that contain data. This process is referred to as **handshaking**. TCP uses a process called a **three-way handshake** to establish a connection, as shown in Figure 13.11a. Perhaps you need to report sales figures to your home office. You phone the sales manager and tell him or her that you are ready to report your figures. The

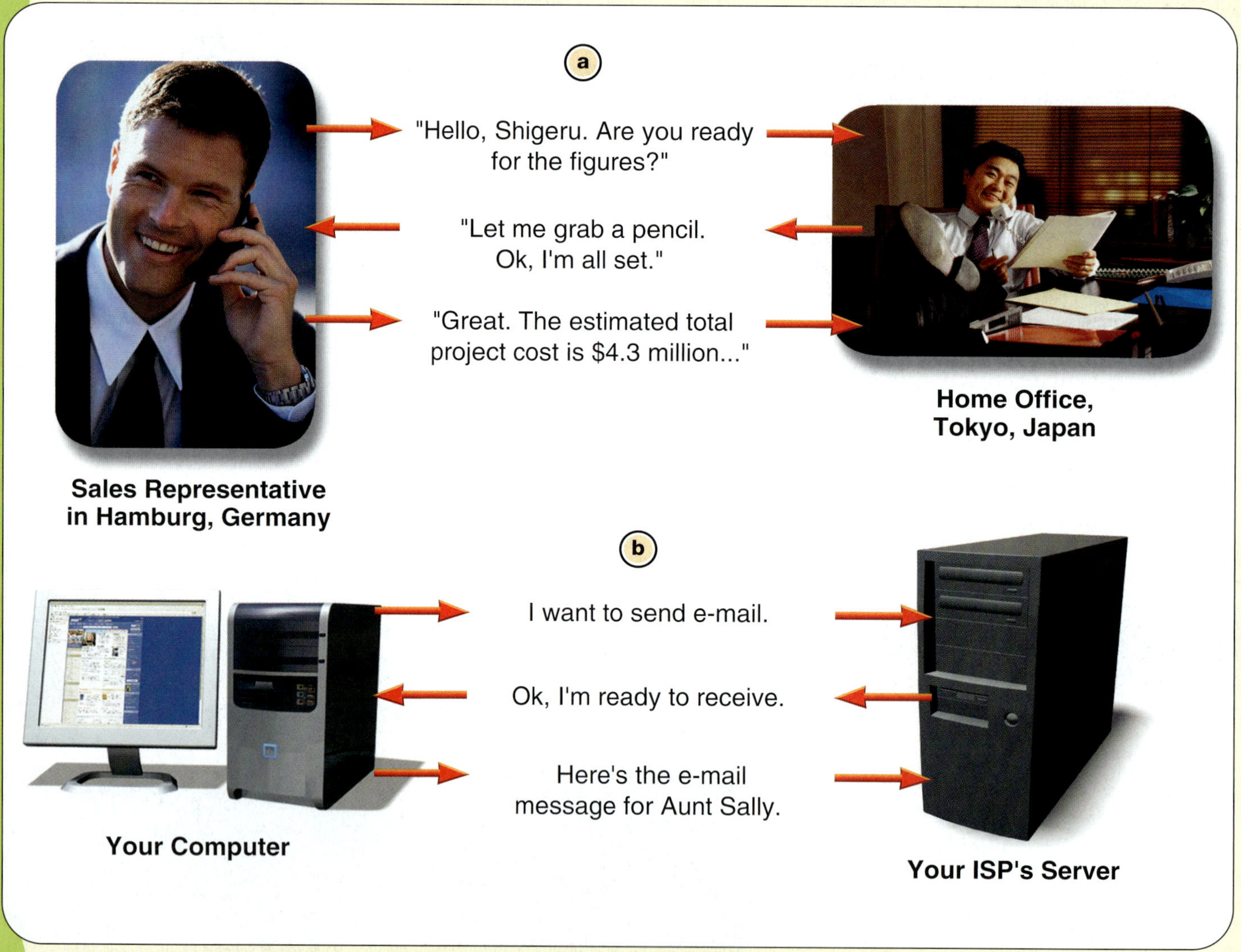

(a)

"Hello, Shigeru. Are you ready for the figures?"

"Let me grab a pencil. Ok, I'm all set."

"Great. The estimated total project cost is $4.3 million..."

Sales Representative in Hamburg, Germany

Home Office, Tokyo, Japan

(b)

I want to send e-mail.

Ok, I'm ready to receive.

Here's the e-mail message for Aunt Sally.

Your Computer

Your ISP's Server

FIGURE 13.11

(a) Colleagues in Hamburg and Tokyo are establishing communication via a three-way handshake. (b) Here, two computers are establishing communication in the same way.

sales manager then prepares to receive the information by getting a pencil and a piece of paper. By confirming he or she is ready and by your beginning to report the figures, a three-way (three-step) handshaking process is completed.

Your computer does the same thing when it wishes to send an e-mail through your ISP, as shown in Figure 13.11b. It establishes a connection to the ISP and announces it has e-mail to send. The ISP server responds that it is ready to receive. Your computer then acknowledges the ready state of the server and begins to transmit the e-mail.

A **connectionless protocol** does not require any type of connection to be established or maintained between two computers that are exchanging information. Just like mailing a letter, the data packets are sent without notifying the receiving computer or receiving any acknowledgment that the data was received. UDP is the Internet's connectionless protocol.

Besides establishing a connection, TCP provides for reliable data transfer. Reliable data transfer means that the application that uses TCP can rely on this protocol to deliver all the data packets to the receiver free from errors and in the correct order. TCP achieves reliable data transfer by using acknowledgments and providing for the retransmission of data, as shown in Figure 13.12.

Assuming two systems, X and Y, have established a connection, when Y receives a packet from X, it sends a **positive acknowledgment (ack)** when it receives a data packet that it can read. If X does not receive an ack in an appropriate period of time, it resends the packet. If the packet is unreadable (damaged in transit), Y sends a **negative acknowledgment (nack)** to X, indicating the packet was not received in understandable form. X will then retransmit that packet. Acknowledgments assure that the receiver has gotten a complete set of data packets. If a packet is unable to get through after being resent several times, the user is generally presented with an error message indicating the communications were unsuccessful.

You may wonder why you wouldn't always want to use a protocol that provides for reliable data transfer. On the Internet, speed is often more important than accuracy. For certain applications (such as e-mail), it's very important that your message be delivered completely and accurately. For streaming multimedia, it's not always important to have every frame delivered accurately because most streaming media formats provide for error correcting due to data loss. It is, however, extremely important for streaming media to be delivered at a high rate of speed; otherwise, playback quality can be affected. Therefore, a protocol like TCP using handshakes and acknowledgments would probably not be appropriate when viewing a movie trailer over the Internet.

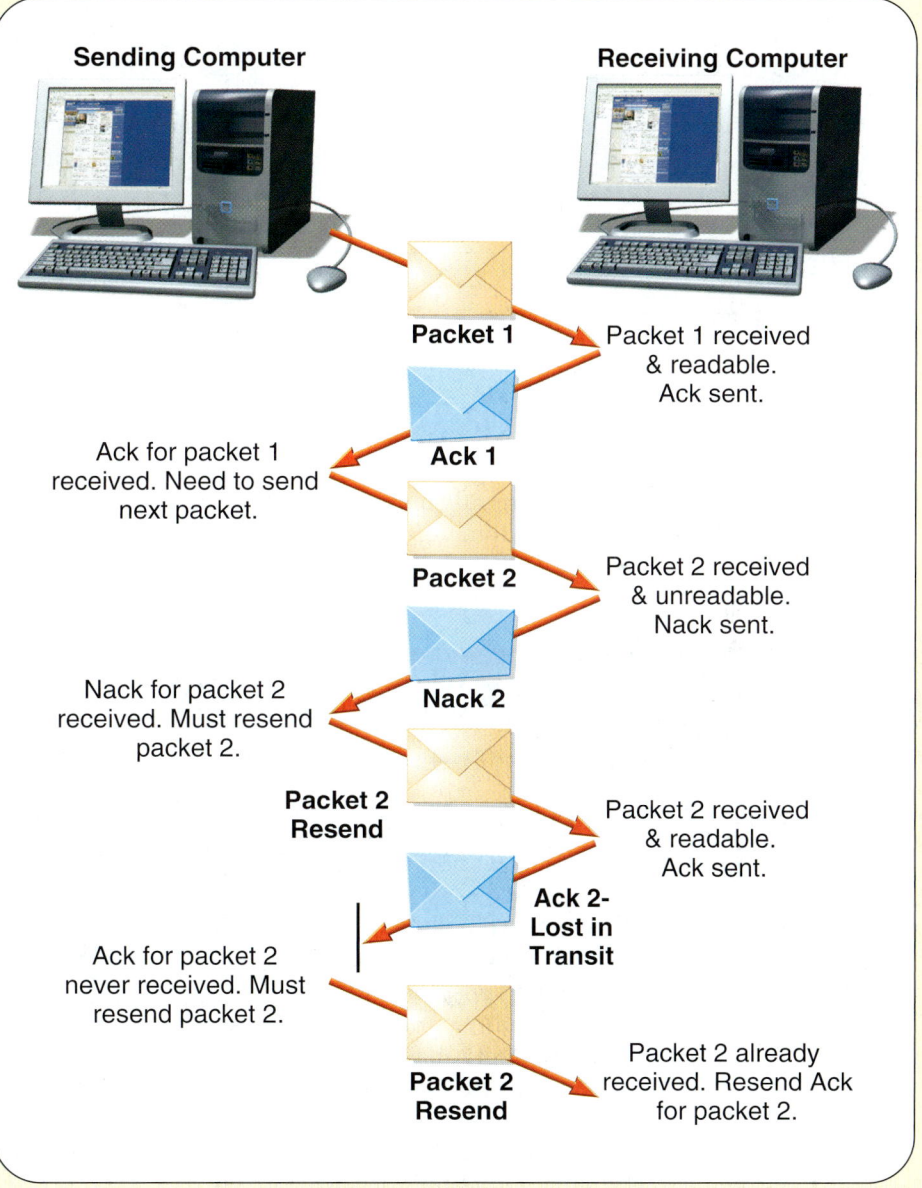

FIGURE 13.12

Packet Acknowledgment in Action

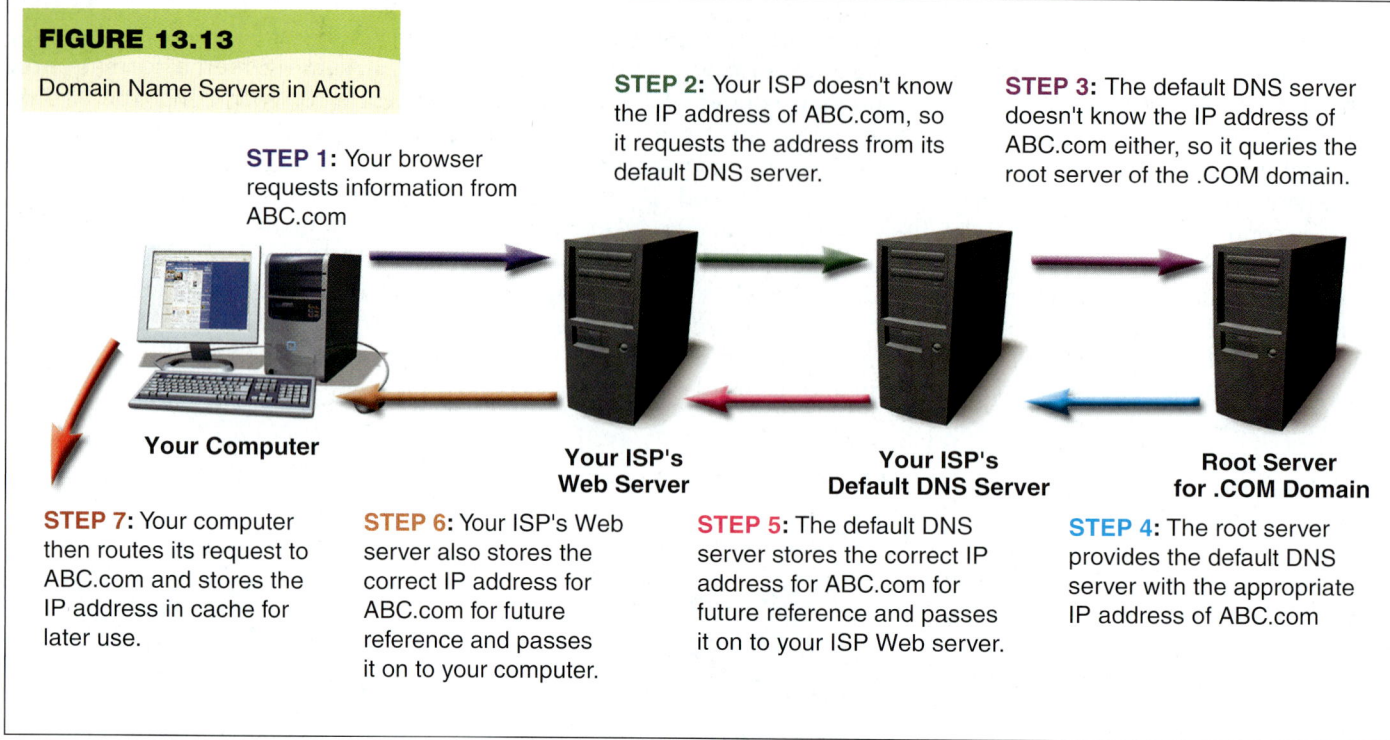

FIGURE 13.13 Domain Name Servers in Action

STEP 1: Your browser requests information from ABC.com

STEP 2: Your ISP doesn't know the IP address of ABC.com, so it requests the address from its default DNS server.

STEP 3: The default DNS server doesn't know the IP address of ABC.com either, so it queries the root server of the .COM domain.

STEP 7: Your computer then routes its request to ABC.com and stores the IP address in cache for later use.

STEP 6: Your ISP's Web server also stores the correct IP address for ABC.com for future reference and passes it on to your computer.

STEP 5: The default DNS server stores the correct IP address for ABC.com for future reference and passes it on to your ISP Web server.

STEP 4: The root server provides the default DNS server with the appropriate IP address of ABC.com

Chapter 3, the portion of the domain name farthest to the right (after the dot) is the top-level domain (TLD). The TLDs are standardized pools established by ICANN (such as COM and ORG). (Refer back to Figure 3.10 in Chapter 3 for a list of the TLDs that are currently approved and in use.) Within the top-level domains are many **second-level domains**. In the .com domain, there are popular sites such as Amazon.com, Yahoo.com, and Microsoft.com. Each of the second-level domains needs to be unique within that particular domain, but not necessarily unique to all top-level domains. For example, Mycoolsite.com and Mycoolsite.org could be registered as separate domain names.

Who controls domain name registration? ICANN assigns companies or groups to manage domain name registration. Since names can't be duplicated within a top-level domain, one company is assigned to oversee each TLD and maintain a listing of all registered domains. For instance, Network Solutions oversees the .com domain and provides a database that lists all the registered domains and their contact information. You can look up any .com domain at **www.networksolutions.com/cgi-bin/whois/whois** to see if it is registered and who owns it. Country-specific domains are controlled by groups in those countries. You can find a complete list on the Internet Assigned Numbers Authority Web site at **www.iana.org**.

DOMAIN NAME SERVERS

How does my computer know the IP address of another computer? Say you want to get to Yahoo.com. To do so, you type the

BITS AND BYTES
What Is an Internet Cache?

Your Internet cache is a section of your hard drive that stores information that you may need again for surfing (such as IP addresses, frequently accessed Web pages, and so on). Caching of domain name addresses in domain name servers helps speed up Internet access time since the domain name server doesn't have to constantly query master domain name servers for TLDs. However, caches do have limited storage space, so entries are only held in the cache for a fixed period of time and then are deleted. The time component associated with cache retention is known as the Time To Live (TTL). Without caches, surfing the Internet would take a lot longer.

URL www.yahoo.com into your browser. However, the URL is not important to your computer; only the IP address of the computer hosting the Yahoo! site is. When you enter the URL in your browser, your computer must convert the URL to an IP address. To do this, your computer consults a database maintained on a **domain name server** (**DNS**), which functions like a phone book for the Internet.

Your ISP's Web server has a default domain name server (one that is convenient to contact) that it goes to when it needs to translate a URL to an IP address (illustrated in Figure 13.13). Your ISP or network administrator defines the default DNS. If the default DNS does not have an entry for the domain name you requested, it queries another DNS (perhaps maintained by a Tier 2 or Tier 1 ISP).

If all else fails, it contacts one of the many *root domain name servers* maintained throughout the Internet. The **root domain name servers** know the location of all the domain name servers that contain the master listings for an entire TLD. Your default DNS receives the information from the master DNS (say for the .com domain), then stores that information in its cache for future use and communicates the appropriate IP address to your computer.

Other Protocols: FTP and Telnet

The TCP/IP protocol suite contains numerous protocols, although some of them are not used very often. Other commonly used protocols on the Internet are the *File Transfer Protocol (FTP)* and *Telnet*.

How does FTP work? The **File Transfer Protocol** (**FTP**) allows users to share files that reside on local computers with remote computers. If you're attempting to download files via FTP from your local computer, the FTP client program (most likely a Web browser) first establishes a TCP session with the remote computer. FTP provides for authentication and password protection so you may be required to log in to an FTP site with a user name and password.

Can I upload files with FTP? Most FTP sites allow you to upload files. To do so, you either need a browser that handles FTP transfer (current versions of Internet Explorer and Netscape do) or you need to obtain an FTP client application. Many FTP client programs are available as freeware or shareware. Search-

BITS AND BYTES

Anonymous FTP Archives

Many FTP sites are set up as *anonymous* FTP archives. These sites contain files you can retrieve for free without identifying yourself. When connecting to an anonymous FTP site, your browser automatically sends "anonymous" as your username and provides your e-mail address as the password. Once you are connected, you can download any files (using Windows XP) on the anonymous site just by right-clicking on their link and selecting the Copy to Folder option. Note, however, that most anonymous sites don't permit the uploading of files.

ing on the term "FTP" on **www.download.com** will produce a list of programs to choose from. FTP Voyager is a popular FTP shareware program you can try for free and later pay for if you want to continue using it.

What is Telnet? Telnet is both a protocol for connecting to a remote computer and a TCP/IP service that runs on a remote computer to make it accessible to other computers. At colleges, students often use Telnet to connect to mainframe computers via their personal computers. The Telnet client application (which runs on your personal computer) connects to the Telnet server application (running on a remote computer). Telnet allows you to take control of a remote computer (the server) with your computer (the client) and manipulate files and data on the server as if you were sitting in front of it.

How do I use Telnet? To establish a Telnet session, you need to know the domain name or IP address of the computer to be connected to via Telnet. In addition, logon information (ID and password) is generally required. You can start Telnet in Windows by clicking the Start button in the task bar, selecting the Run command, entering "telnet" in the dialog box that opens, and clicking the OK button. This will then display the window shown in Figure 13.14 on page 530. Typing ?/ at the command prompt displays the available Telnet commands. To connect to a remote computer, type "open" and the host name (or IP address) of the remote computer and follow the logon instructions (which vary from system to system).

FIGURE 13.14

This Telnet command window shows the commands it has available.

HTTP, HTML, and Beyond

Although most people think that the Internet and the Web are the same thing, the World Wide Web (WWW or the Web) is a grouping of protocols and software that *resides* on the Internet (which is a collection of linked networks). The Web provides an engaging interface for exchanging graphics, video, animations, and other multimedia on the Internet.

Did the same people who invented the Internet invent the Web? The Web was invented many years after the original Internet. In 1989, Tim Berners-Lee, a physicist at the European Organization for Nuclear Research (CERN), wanted a method for linking his research documents together so that other researchers could access them. In conjunction with Robert Cailliau, Berners-Lee developed the basic architecture of the Web and created the first Web browser. The original browser could only handle text and was only usable on computers running the NeXT operating systems (a commercially unsuccessful OS), which limited its usage. So Berners-Lee put out a call to the Internet community to assist with development of browsers for other platforms.

In 1993, the National Center for Supercomputing Applications (NCSA) released the Mosaic

BITS AND BYTES

Why Should You Run the Latest Version of Browser Software?

As new file formats are developed for the Web, browsers need new plug-ins to properly display them. Constantly downloading and installing plug-ins can be a tedious process. Although many Web sites provide links to sites that enable you to download plug-ins, not all do, resulting in frustration when you can't display the content you want.

When new versions of browsers are released, they normally include the latest versions of popular plug-ins. Corrections of security breaches are typically included in these versions of browser software as well. Therefore, upgrading to the latest version of your browser software provides for safer, more convenient Web surfing.

Fortunately, most upgrades are free. If you're using Netscape, go to **www.netscape.com** and click on the download link. If you're using IE, go to **www.microsoft.com** and search for the term "Internet Explorer Support." You'll then see a link that will enable you to download the current version of IE.

browser for use on the Macintosh and Windows operating systems. Mosaic could display graphics as well as text. As the popularity of this browser grew, Marc Andreessen, the leader of the Mosaic development team, formed a company called Mosaic Communications (later renamed Netscape Communications) with Jim Clark. Within 6 months, many of the developers from the original Mosaic project at the NCSA were working for this new company. The Netscape browser (version 1.0) was released by the company in December 1994. This new browser featured improvements in usability over Mosaic and quickly became the dominant Web browser. The launch of Netscape heralded the beginning of the Web's monumental growth.

HTTP AND SSL

What Internet protocol does a browser use to send requests? The **Hypertext Transfer Protocol (HTTP)** was created especially for the transfer of hypertext documents across the Internet. **Hypertext** documents are documents in which text is linked to other documents or media (such as video clips, pictures, and so on). Clicking on a specific piece of text (called a *hyperlink*) that has been linked elsewhere will take you to the linked file.

When the browser sends a request, does it do anything to make the information secure? Commerce servers use security protocols to protect sensitive information from being intercepted by hackers. One common protocol is the **Secure Sockets Layer (SSL)**, which provides for the encryption of data transmitted via TCP/IP protocols such as HTTP. All major Web browsers support SSL.

There are at least two indications you are using SSL. When on a site using SSL, your browser displays a padlock icon in the status bar at the bottom of the browser. Also, when a URL begins with https:// instead of http://, that Web site is requiring information to be sent using SSL encryption. On a Web site that is using SSL, you can be secure in the knowledge that the information you are sending (such as your credit card number) is encrypted and would be extremely difficult (if not impossible) to decode.

HTML

How are Web pages formatted? A Web page is merely a text document that is formatted using the **Hypertext Markup Language (HTML)**.

HTML is not a programming language; rather, it is a set of rules for marking up blocks of text so that a browser knows how to display them. Blocks of text in HTML documents are surrounded by a pair of **tags** (such as <b> and </b> to indicate bolding). These tags and the text between them are referred to as **elements**. The elements are interpreted by the browser and appropriate effects are applied to the text. The following is an element from an HTML document:

```
<i> This should be italicized.</i>
```

The browser would display this element as:

This should be italicized.

The first tag <i> tells the browser that the text following it should be italicized. The end </i> tag indicates that the browser should cease applying italics to the text. Note that tags can be combined in a single element, such as:

```
<b><i>This should be bolded and italicized.</b></i>
```

The browser would display this element as:

This should be bolded and italicized.

Obviously, the <b> tag indicates bolding.

Tags for creating hyperlinks appear as follows:

```
<a href = http://www.prenhall.com/phit/> Prentice Hall Information Technology Site</a>
```

(<a href = http://www.prenhall.com/phit/>) defines the link's destination. The </a> tag indicates the end of the hyperlink element. The text in between the two tags (Prentice Hall Information Technology Site) is the link label. The link label is the text (or image) that will be displayed on the Web page as clickable text for the hyperlink.

Can I see the HTML coding of a Web page? HTML documents are merely text documents with HTML tags applied to them. If you want to look at the HTML coding behind your favorite Web page, just right-click anywhere on the page and a dialog box will appear. Select the View Source option and the HTML code for that page will be displayed, as shown in Figure 13.15 on the following page.

SOUND BYTE
CONSTRUCTING A SIMPLE WEB PAGE

Creating simple Web pages using Microsoft Word is relatively easy. In this Sound Byte, you'll learn the basics of Web page creation by setting up a Web site featuring a student resume.

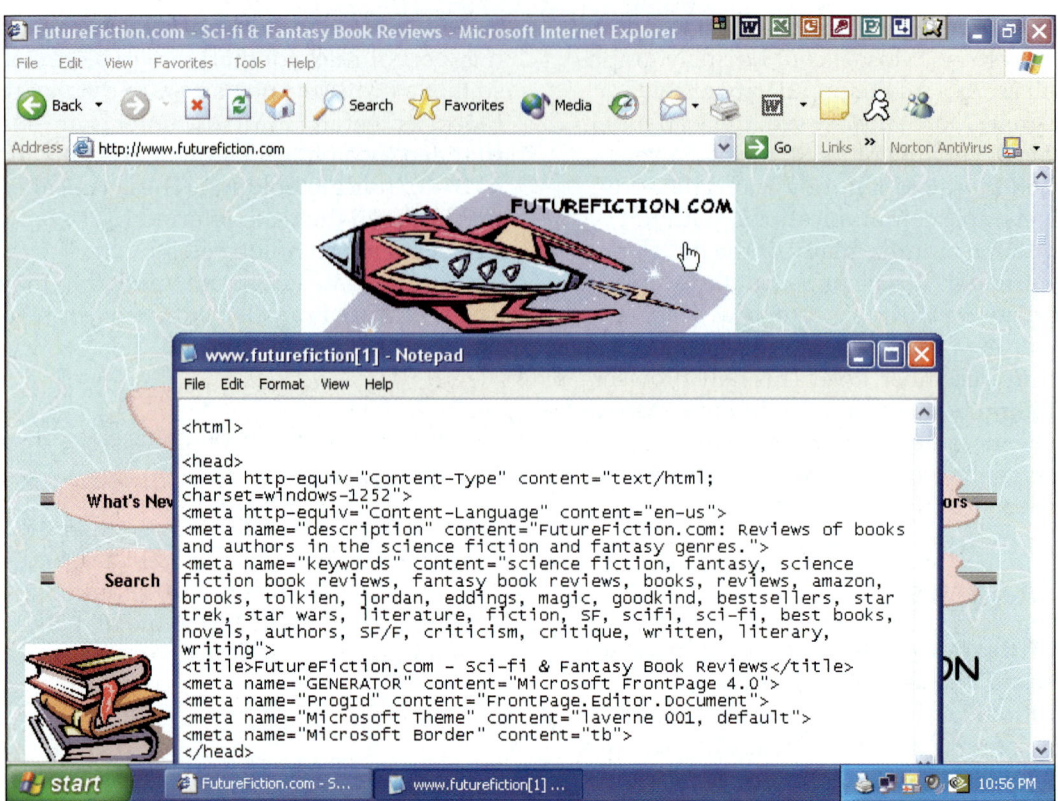

FIGURE 13.15

Displaying the source code in a browser opens up a Notepad window containing the HTML text document.

THE COMMON GATEWAY INTERFACE

Can you use HTML to do everything you need to do on a Web page? Since HTML was originally designed to link text documents, HTML by itself can't do all the amazing things we expect modern Web pages to do. As we mentioned above, HTML is not a programming language; rather, it's a set of tags for determining how text is displayed and where elements are placed. Fortunately, the limitations of HTML were recognized early and the *Common Gateway Interface (CGI)* was developed.

Most browser requests merely result in a file being displayed in your browser (such as the eBay.com main Web page). Displaying a file is fine if you're just going to be reading text. However, to make a Web site interactive, you may need to execute a software program to perform a certain action (such as gathering a name and address and adding it to a database). The **Common Gateway Interface (CGI)** provides a methodology by which your browser can request that a program file be executed (or run) instead of just being delivered to the browser. This allows functionality beyond the simple displaying of information.

CGI files can be created in almost any programming language, and the programs created are often referred to as **CGI scripts**. Common languages that are used to create CGI scripts are PERL, C, and C++. Since programming languages are very powerful, almost any task can be accomplished by writing a CGI script. You have probably encountered CGI scripts on Web pages without realizing it. Have you ever left an entry in a guest book on a Web page? Have you used a search engine to create a customized results page based on keywords you entered? Have you filled out a form to be added to a mailing list? All of these tasks are commonly done using CGI scripts.

How do CGI programs get executed? On most Web servers, a directory called **cgi-bin** is created by the network administrator who configures the Web server. All CGI scripts are placed into this directory. The Web server knows that all files in this directory are not to be merely read and sent, but also need to be executed. Since these programs are run on the Web server as opposed to running inside your browser, they are referred to as **server-side** programs.

For instance, a button on a Web site may say "Click Here to Join Mailing List" (see Step 1 in Figure 13.16). Clicking this button may call a script file (perhaps called mailinglist.pl) from the cgi-bin directory on the Web server hosting the site (Step 2). This file generates a form that is sent to your browser, which includes fields for a name and e-mail address, and a button that says "Submit" (Step 3). After filling in the fields and pushing the Submit button, the mailinglist.pl program sends the information back to the server. The server then records the information in a database (Step 4).

CLIENT-SIDE APPLICATIONS

Aside from CGI scripts, are there other ways to make a Web site interactive? Sometimes running programs on the server is not optimal. Server-side program execution can require many communication sessions between the client and the server to achieve the goal. Often it is more efficient to run programs on your computer (the client). Therefore, *client-side applications* were created A **client-side application** is a computer program that runs on the client and requires no interaction with a Web server. Client-side applications are fast and efficient because they run at your desktop and don't depend on sending signals back and forth to the Web server.

Two main types of client-side methods exist. The first involves embedding programming language code directly within the HTML code of a Web page using an **HTML embedded scripting language**. The most popular embedded language is **JavaScript**, which was developed through the joint efforts of Netscape and Sun Microsystems. It is often confused with the Java programming language because of the similarity in the name. Although they share some common elements, the two languages function very differently.

Pure HTML documents don't respond to user input. However, through the use of JavaScript, HTML documents can be made responsive to mouse clicks and typing. When JavaScript code is embedded in an HTML document, it is downloaded with the HTML page to the browser. All actions dictated by the embedded JavaScript commands are executed on the client computer (the one with the browser). Without JavaScript, Web pages would be pretty lifeless.

The second type of client-side application is an **applet**, a small program that resides on a server. When requested, a compiled version of the program is downloaded to the client computer and executed there. The Java language is the most common language used to create applets for use in browsers. The applets can be requested from the server when a Web page is loaded and they will be run once they're downloaded to the client computer.

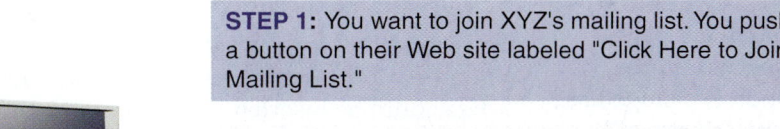

STEP 1: You want to join XYZ's mailing list. You push a button on their Web site labeled "Click Here to Join Mailing List."

STEP 2: Pushing the button tells the XYZ.com server to execute a CGI program (called mailinglist.pl) located in the cgi-bin directory on the server.

Your Computer

STEP 3: The server executes the CGI program mailing list.pl, which generates a form that is sent to your browser.

XYZ.com Server

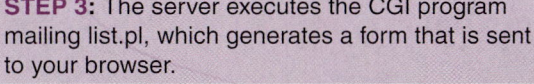

STEP 4: You enter your mailing list information and push the "Submit" button on the mailing list form. This sends your information back to the server. The server saves your information into its database.

FIGURE 13.16

Information Flow When a CGI Program Is Executed

FIGURE 13.17

Deployment of a Java Applet to a Computer

Although there can be some delay in functionality while waiting for the Java applet to download to the client, once the applet arrives it can execute all its functions with-out further communication with the server. Games are often sent to your browser as applets. In Figure 13.17, your browser makes contact with the game site (Chess.com) and makes your request to play a game of chess (Step 1). The Web server returns the Java applet (Step 2) that contains all the code to run the game on your computer. Your computer executes the applet code and the game runs on your computer.

XML

Can I create my own HTML tags to fit my special needs? For HTML, you're required to use the standard predefined tags that constitute the HTML standard. This works fairly well for displaying information on Web pages, but is less optimal if two Web pages need to exchange information. Information exchange has become much more common with the rise of business-to-business (B2B) electronic commerce. B2B transactions involve two businesses selling products and services to each other without a retail customer being involved. Since HTML was not designed for information exchange, *Extensible Markup Language (XML)* was created.

How is XML different from HTML? **Extensible Markup Language (XML)** is a set of tools you can use to create your own markup language. In a sense, it is a more flexible version of HTML. Instead of being locked into standard tags and formats for data, users can build their own markup languages to accommodate particular data formats and needs.

For example, three pieces of typical information that need to be captured for an e-commerce transaction are credit card number, price, and zip code. In HTML, the paragraphs tags (<p> and </p>) are used to define text and numeric elements. Almost anything can fall between these tags and be treated as a paragraph. So, in our example, the HTML code would appear as follows:

```
<p>1234567890123456</p> (credit card number)
<p>12.95</p> (price)
<p>19422</p> (zip code)
```

The browser will interpret this data as separate paragraphs. But the paragraph tags tell us nothing about the data contained within them. Without the labels added (not part of the HTML code), we may not realize what data was contained within. Also, <p> tags don't provide any methodology for data validation. Credit card numbers are 16 numbers long. But any length of data could be inserted between <p> and </p> tags. How would we know if the credit card number was a valid length? The answer lies in creating tags that are specific to the task at hand, and that actually describe the data contained within them. Here's how our data might look in XML:

```
<credit card number>
1234567890123456</credit card number>
<price>12.95</price>
<zip code>19422</zip code>
```

We have created the tags we need for data capture. Our XML specification provides a tag called "credit card number" that is used exclusively for credit card data. As well as defining

TRENDS IN IT

CAREERS: If You Build It, Will They Come? Web Development and Design Careers

Although Web development careers span a vast array of job functions, many students seeking to pursue these opportunities insist on seeking information on becoming a Webmaster. However, in the modern Internet economy, Webmasters are quickly becoming a thing of the past.

As the term implies, *Webmaster* originally meant the person who was solely responsible for the Web site at a company. In 1994, you might have been a Webmaster because you were the only person in the company who knew HTML code. At that time, Web sites were static and changed little from day to day. In the 21st century, Web sites are large, dynamic, and key to many companies' business strategies. Users are demanding and expect fresh content and new services to be offered frequently. Therefore, most modern businesses' Web sites cannot be supported by just one person.

Teams that support today's Web sites are usually organized into four areas:

1. *Web server administrators* are specialized versions of network administrators. Their primary responsibility is to install and maintain the Web servers that host a company's Web pages. Web server administrators have specialized training in Apache and Microsoft IIS Web servers as well as general networking training.

2. *Content creators* generate the words and images on the Web. Journalists, writers, and editors prepare an enormous amount of Web content, while video producers, graphic artists, and animators create Web-based multimedia. These individuals have a thorough understanding of their own fields as well as XHTML and JavaScript. They also need to be familiar with capabilities and limitations of modern Web development tools so they know what the Web publishers can accomplish.

3. *Web publishers* build Web pages to deploy the material content creators develop. They wield the Web tools (such as Macromedia's Dreamweaver and Microsoft's FrontPage) that develop the Web pages and create links to databases (using products such as Oracle and SQL Server) to keep information flowing between users and the Web page. They must possess a solid understanding of client- and server-side Web languages (XHTML, XML, Java, JavaScript, ASP, and PERL) and development environments such as .NET.

4. The *customer interaction team* provides feedback to a Web site's customers. Answering e-mail, sending requested information, funneling questions to appropriate personnel (technical support, sales, and so on), and providing suggestions to Web publishers for site improvements are major job responsibilities. Extensive customer service training is essential to effectively work in this area.

As you can see, there are many opportunities to be part of a modern Web development team, and not all of these jobs require extensive programming capabilities. They do all require hard work, imagination, and attention to detail. The Web will only be as useful as you make it! We'll talk more about Web-related IT careers in the Technology in Focus "Careers in IT."

---

how the browser will display information tagged as <credit card number>, you can also require that the tagged data be numeric and 16 numbers long. With XML, you can achieve the data validation necessary for ensuring data is exchanged accurately between applications.

Does XML work with HTML? The World Wide Web Consortium (WC3) has recently released a new language that combines elements from both XML and HTML, called the **Extensible Hypertext Markup Language (XHTML)**. XHTML, borrowing heavily from XML, has much more stringent rules than HTML regarding tagging (for instance, all elements require an end tag). XHTML is the development environment of choice for Web developers today, although many Web sites will probably never be converted from HTML documents unless absolutely necessary.

SOUND BYTE
THE BEST UTILITIES FOR YOUR COMPUTER

In this Sound Byte, you'll explore various utilities that you can use to avoid Internet annoyances. You'll learn how to install and use specific utilities and find out where to download some of them for free.

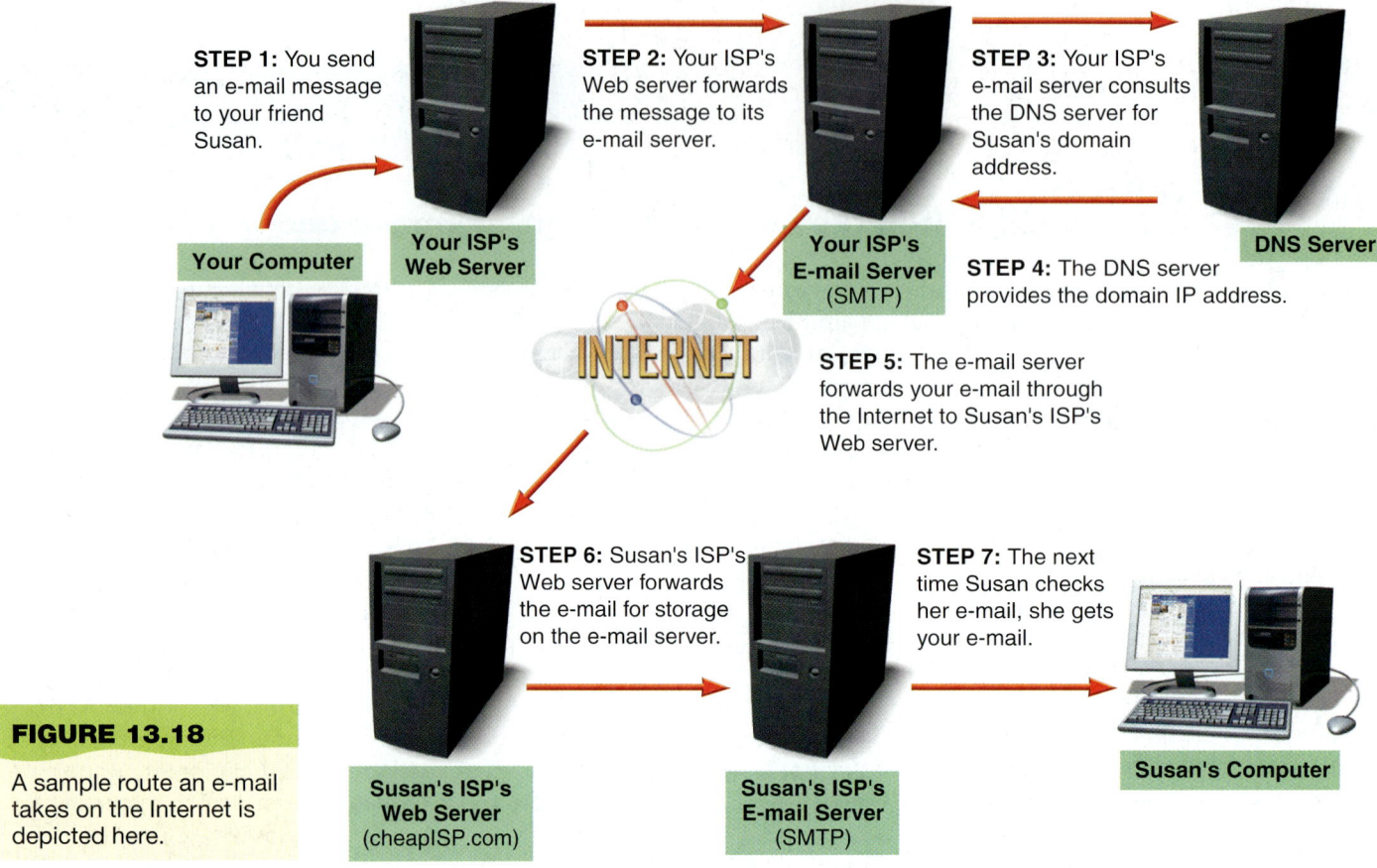

FIGURE 13.18
A sample route an e-mail takes on the Internet is depicted here.

Communications via the Internet

A new communications revolution was started when Internet use began to explode in the mid-1990s. The volume of Internet e-mail is growing exponentially every month, instant messaging is a major method of communication, and the popularity of Internet telephony is also on the rise. In the following sections, we'll explore all of these communications media in more detail and show you how to keep your information exchanges efficient and secure.

E-MAIL

Who invented e-mail? In 1971, Ray Tomlinson, a computer engineer who worked on the development of the ARPANET (the precursor to the Internet) for the U.S. government, created e-mail. E-mail grew from a simple program that Tomlinson wrote to enable computer users to leave text messages for each other on a single machine. The logical extension to this was sending text messages between machines on the Internet. Tomlinson created the convention of using the @ sign to distinguish between the mailbox name and the destination computer. E-mail became the most popular application on ARPANET, and by 1973, it accounted for 75 percent of all data traffic.

How does e-mail travel the Internet? Just like other kinds of data that flows along the Internet, e-mail has its own protocol. The **Simple Mail Transfer Protocol (SMTP)** is responsible for sending e-mail along the Internet to its destination. As in most other Internet applications, e-mail is a client/server application. Two primary kinds of e-mail client software are in use today. Although client-based e-mail software, such as Microsoft Outlook, AOL, and Eudora, continues to be very popular, Web-based e-mail software, in use on sites such as Yahoo! and Hotmail, has also grown in popularity.

Client-based software is installed on your computer and all functions are supported and run from your computer. Web-based software is launched from a Web site; the programs and features are stored on the Web and are accessible anywhere you have access to an Internet connection. No matter which type of client software you use, your mail will pass through **e-mail servers**—specialized servers whose sole function is to store, process, and send e-mail—on the way to its destination.

Where are e-mail servers located? If your ISP provides you with an e-mail account, it

COMMUNICATIONS VIA THE INTERNET 537

runs an e-mail server that uses SMTP. For example, as shown in Figure 13.18, say you are sending an e-mail message to your friend Susan. Susan uses Cheapisp.com as her ISP. Therefore, your e-mail to her is addressed to **Susan@cheapisp.com**. When you send the e-mail message, your ISP's Web server receives it and passes it to your ISP's e-mail server.

The e-mail server reads the domain name (cheapisp.com) and communicates with a domain name server (DNS) to determine the location of Cheapisp.com. Once the address is located, the e-mail message is forwarded to Cheapisp.com through the Internet and arrives at a mail server maintained by Susan's ISP. The e-mail is then stored on Susan's ISP's e-mail server. The next time Susan logs on to her ISP and checks her mail, she will receive your message.

If e-mail was designed for text messages, why are we able to send files as attachments? SMTP was designed to handle text messages. When the need arose to send files via e-mail (in the early 1970s), a program had to be created to convert binary files to text. The text that represented the file was appended to the end of the e-mail message. When the e-mail arrived at its destination, the recipient had to run another program to translate the text back into a binary file. Uuencode and uudecode were the two most popular programs used for encoding and decoding binary files.

This was fine in the early days of the Internet when users were mostly computer scientists. However, when the Internet started to become popular in the early 1990s, it became apparent that a simpler methodology was needed for sending and receiving files. The **Multipurpose Internet Mail Extensions (MIME)** specification was introduced in 1991 to simplify attachments to e-mail messages. All e-mail client software now uses this protocol for attaching files.

E-mail is still being sent in text, but the e-mail client using the MIME protocol now handles the encoding and decoding for the users. For instance, in Yahoo! mail, on the Attach Files screen you merely browse to the file you want to attach (located somewhere on your hard drive), select the file, and press the "attach files" button. Unbeknownst to you, the Yahoo! e-mail client encodes the file for transmission.

E-MAIL SECURITY: ENCRYPTION AND SPECIALIZED SOFTWARE

If e-mail is sent in regular text, can other people read my mail? E-mail is very susceptible to being read by unintended parties since it's sent in plain text. Also, copies of your e-mail message may exist (temporarily or permanently) on numerous servers as it makes its way through the Internet. Two options exist for protecting your sensitive e-mail messages: *encryption* and *secure data transmission software*.

How do I encrypt my e-mail? **Encryption** refers to the process of coding your e-mail so that only the person with the key to the code (the intended recipient) can decode (or decipher) and read the message. Secret codes for messages can be traced almost to the dawn of written language. The military and government espionage agents are big users of codes and ciphers. The trick is making the coding system easy enough to use so that everyone who needs to communicate with you can.

There are two basic types of encryption: *private-key* and *public-key*. In **private-key encryption**, only the two parties involved in sending the message have the code. This could be a simple shift code where letters of the alphabet are shifted to a new position (see Figure 13.19). For example, in a two-position right-shift code, the letter a becomes c, b becomes d, and so on. Or it could be a more complex substitution code (a = h, b = r, c = g, etc.). The main problem with private-key encryption is key security. If someone steals a copy of the code, the code is broken.

In **public-key encryption,** two keys, known as a **key pair**, are created. You use one key for coding and the other for decoding. The key for coding is generally distributed as a **public key**. You can place this key on your Web site, for instance. Anyone wishing to send you a message codes it using your public key.

FIGURE 13.19

Sample Code Using a Two-Position Right Shift

A = C	N = P
B = D	O = Q
C = E	P = R
D = F	Q = S
E = G	R = T
F = H	S = U
G = I	T = V
H = J	U = W
I = K	V = X
J = L	W = Y
K = M	X = Z
L = N	Y = A
M = O	Z = B

The word C O M P U T E R using the two-position code at the left now becomes:

E Q O R W V G T

This is difficult to interpret without the code key at the left.

When you receive the message, you use your **private key** to decode it. You are the only one who ever possesses the private key and therefore it is very secure. The keys are generated in such a way that they can only work with each other. The private key is generated first and then the public key is generated by running it through a complex mathematical formula. The computations are so complex that they are considered unbreakable. Both keys are necessary to decode a message; if one key is lost, the other key cannot be used by itself.

What type of encryption is used on the Internet? Public-key encryption is the most commonly used encryption on the Internet. Tried and true public-key packages, such as **Pretty Good Privacy (PGP)**, are available for download at sites such as www.download.com, and you can usually use them free of charge (although there are now commercial versions of PGP). After obtaining the PGP software, you can generate key pairs to provide a private key for you and a public key for the rest of the world.

What does a key look like? A key is a binary number (a string of 1s and 0s). Keys vary in length depending upon how secure they need to be. A 10-bit key has ten positions and might look like this:

1001101011

Longer keys are more secure because they have more possible values. A 10-bit key provides 1,024 different possible keys while a 40-bit key allows for 1,099,511,627,776 possible values. The key and the message are run through a complex algorithm in the encryption program (such as PGP) that converts the message into unrecognizable code. Each key will turn the message into a different code.

Is my private key really secure? Because of the complexity of the algorithms used to generate key pairs, it is impossible to deduce the private key from the public key. However, that doesn't mean your coded message can't be cracked. A brute force attack, as you learned in Chapter 12, occurs when hackers try every possible key combination to decode a message. This type of attack can allow hackers to deduce the key and decode the message.

What is considered a safe key then? In the early 1990s, 40-bit keys were thought to be totally resistant to brute force attacks and were the norm for encryption. But in 1995, a French programmer used a unique algorithm of his own and 120 workstations simultaneously to attempt to break a 40-bit key. He succeeded in just eight days. Since then, 128-bit keys have become the standard. Even using supercomputers, no one has yet to crack a 128-bit key. It is believed that even with the most powerful computers in use today it would take hundreds of billions of years to crack a 128-bit key.

How else can I protect my e-mail? Using encryption doesn't always solve the other problems associated with e-mail. Messages leave a trail as they travel through the Internet, and copies of messages can exist on servers for long periods of time. In addition, immediate reading of sensitive documents is often essential, but encryption software doesn't provide a means for confirming your messages have been delivered. To combat these issues, companies like Securus Systems Ltd. (www.safemessage.com) have developed secure data transmission software (called SafeMessage) that works outside of the conventional SMTP mail servers.

How is SafeMessage software used? Both parties wishing to send secure messages install the SafeMessage software. When messages are to be sent, a secure point-to-point connection is established between the sender and the recipient's e-mail boxes. Proprietary protocols with encryption, not SMTP, are used to send

BITS AND BYTES

Using the Internet for Phone Calls

Many conventional telephone service providers use **Internet telephony**, also called **Voice over IP (VoIP)**, to transmit conventional long-distance phone calls. VoIP consists of transmitting phone calls over the same data lines and networks that make up the Internet. In fact, you've probably used such a connection without even knowing it when making a long-distance call.

So why do phone companies use VoIP? As mentioned earlier, conventional phone calls use a process called circuit switching in which a connection between two phones is established and maintained during the entire call. The bandwidth is dedicated to that call even if there are gaps in the conversation. With VoIP, calls use the same packet switching technology as other messages traveling the Internet. Therefore, bandwidth is not dedicated to a single phone call, so more messages can be sent over the same bandwidth. Using less bandwidth per call means the telephone provider can send more calls over the same network and thereby increase billings.

the messages. Additional options are provided such as delivery confirmation, message shredding (destruction of messages on command), and the ability to have messages erase themselves after a set period of time. Although not free, this type of software is catching on in the business community where fear of industrial espionage is high.

INSTANT MESSAGING

What do I need to run instant messaging? Instant messaging (IM) is another client/server application that can run on a PC or Mac. AOL Instant Messenger (AIM), ICQ, Yahoo! Messenger, and Windows Messenger are the top four instant messaging applications in use today. No matter which one you chose, you need to run the appropriate client software on your computer.

How does instant messaging work? The client software running on your computer makes a connection with the chat server via your Internet connection, as shown in Figure 13.20. Once contact is established, you can log into the server with your name and password (the first time you can sign up for a free account). The client software provides the server with connection information (such as the IP address) for your computer. The server then consults the list of contacts ("Buddies" or friends) that you have previously established in your account, and checks to see if any of your contacts are online. If any are, the server sends a message back to your client providing the necessary connection information (the IP addresses) for your friends. You can now click on your friends' names to establish a chat session with them.

Since both your computer and your friend's computer have the connection information (the IP addresses) for each other, the server isn't involved in the chat session. Chatting takes place directly between the two computers via the Internet.

Can I communicate with my friend who uses ICQ if I use AOL Instant Messenger? Each instant messaging service uses its own proprietary software and file format. Therefore, instant messaging services aren't compatible with each other. Fortunately, since the services are free, you can belong to several at the same time, allowing you to communicate with all of your friends even if they don't use your favorite IM program.

Your Computer Running Instant Messenger (Client Software) for Chatting

STEP 1: Using its Instant Messenger client software, your computer queries the chat server to determine which of your Buddies are online.

STEP 2: The chat server provides your computer with the IP addresses of your Buddies who are online.

Chat Server

STEP 3: Your computer and your Buddy's computer can communicate directly once the chat server has provided your computer with the IP address of your Buddy's computer.

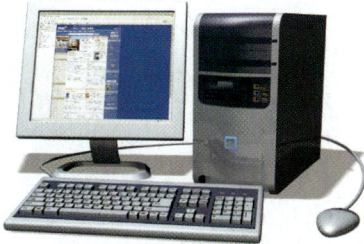

Your Buddy's Computer Running the Same Instant Messenger (Client Software) for Chatting

FIGURE 13.20

How an Instant Messaging Program Works

TRENDS IN IT

COMPUTERS IN SOCIETY:
Have You Ever Used an Extranet or a Virtual Private Network?

As mentioned in Chapter 12, *intranets* are private corporate networks that are used exclusively by employees of the company to facilitate information sharing, database access, group scheduling, videoconferencing, or other employee collaboration. Sometimes, though, restricting access only to employees doesn't meet the company's needs. In these cases, an *extranet* is employed.

Extranets are pieces of intranets that only certain corporations or individuals can access. The owner of the extranet decides who will be permitted to access it. Customers and suppliers are typical entities that would benefit from accessing information on an extranet. Extranets are useful for allowing Electronic Data Interchange (EDI). EDI provides for the exchanging of large amounts of business data (such as orders for merchandise) in a standardized electronic format. Other uses of extranets include providing access to catalogs and inventory databases, and sharing information of use to partners or industry trade groups.

Due to security concerns, intranets and extranets often utilize *virtual private networks* to keep information secure. A **virtual private network (VPN)** utilizes the public Internet communications infrastructure to build a secure, private network between various locations. Although wide area networks (WANs) can be set up using private leased communications lines, these lines are expensive and tend to increase in price as the distance between points increases. VPNs use special security technologies and protocols that enhance security, allowing data to traverse the Internet as securely as if it were on a private leased line. Installing and configuring a VPN requires special hardware such as VPN optimized routers and firewalls. In addition, VPN software must be installed on users' PCs.

The main technology for achieving a VPN is called *tunneling*. Data packets are placed inside new data packets. The format of the new data packets is encrypted and only understood by the sending and receiving hardware, known as *tunnel interfaces*. The hardware is optimized to seek efficient routes of transmission through the Internet. This provides a high level of security and makes information much more difficult to intercept and decrypt.

Imagine you have to deliver a message to a branch office. You could have one of your employees drive to the other office and deliver the message. But suppose he has to go through a bad neighborhood or has never been to the office before? The messenger could be waylaid by a carjacker or could become hopelessly lost. Using a VPN (as shown in Figure 13.21) is the equivalent of hiring a limousine and an armed guard to drive your employee through a private tunnel directly to the destination. Of course, a VPN avoids the enormous cost associated with this method!

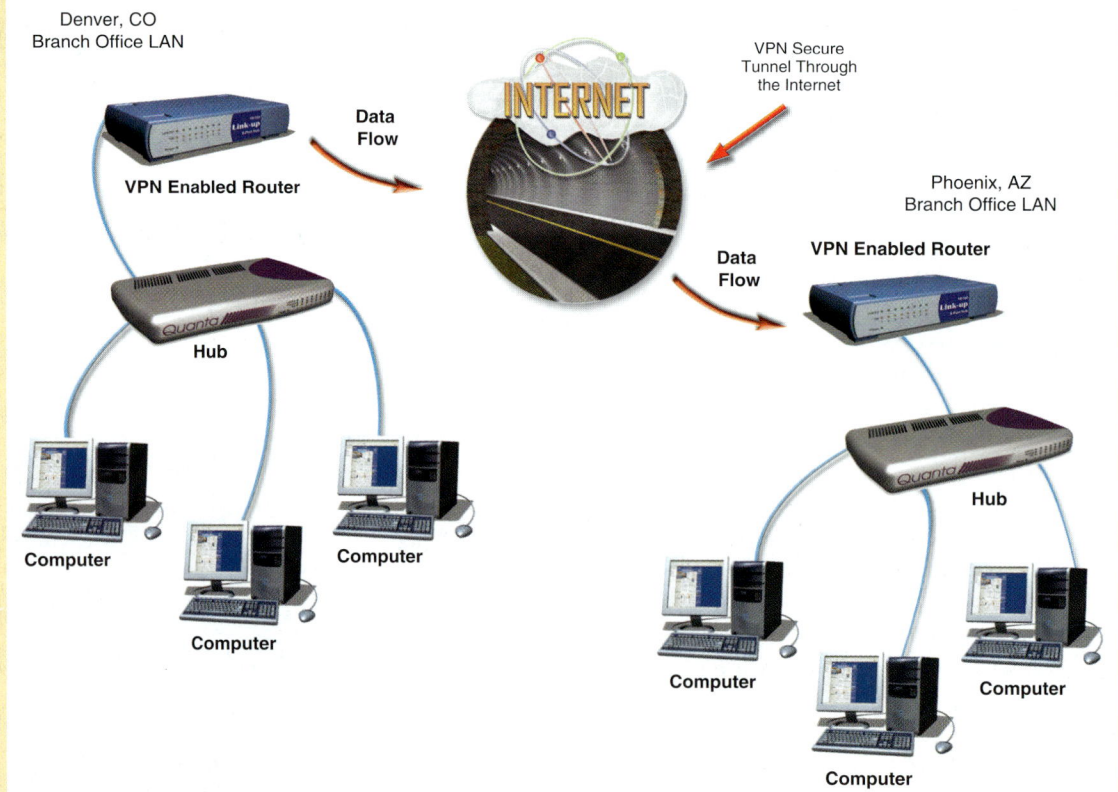

FIGURE 13.21

Local area networks (LANs) in different cities can communicate securely over the Internet using VPN technology.

Is sending an instant message secure? Most instant messaging services do not use a high level of encryption for their messages, if they bother to use encryption at all. Therefore, it is not a good idea to send sensitive information via instant messaging as it is susceptible to interception and possible misuse by hackers.

The Future of the Internet

What does the future hold in store for the Internet? For a certainty, the Internet of the future will have higher bandwidth, offer increased services, and reach more of the world's population than it does today. Two major projects currently under way in the United States to develop advanced technologies for the Internet are the *Large Scale Networking (LSN)* program and the *Internet2*. In addition, plans are underway to wire the solar system—the next logical expansion of the Internet once the planet Earth is fully connected. And all the while, new Internet experiences are being developed for us all to enjoy.

THE LARGE SCALE NETWORKING PROGRAM AND INTERNET2

What are the objectives of the Large Scale Networking and Internet2 programs? Out of a project entitled the Next Generation Internet (which ended in 2002), the U.S. government created the **Large Scale Networking (LSN)** program (**http://www.itrd.gov/iwg/lsn.html**). The objective of LSN is to fund the research and development of cutting-edge networking technologies. Major goals of the program are the development of enhanced wireless technologies and increased network throughput.

The **Internet2** is an ongoing project sponsored by more than 200 universities (supported by government and industry partners) to develop new Internet technologies and disseminate them as rapidly as possible to the rest of the Internet community. The Internet2 backbone supports extremely high-speed communications (up to 9.6 Gbps), which provides an excellent test bed for new data transmission technologies. It is hoped that the Internet2 will solve the major problem plaguing the current Internet—lack of bandwidth. Once the Internet2 is fully integrated with the current Internet, greater volumes of information should flow more smoothly.

AN INTERPLANETARY INTERNET?

How might the Internet expand into outer space in the future? Many scientists think that a manned mission to Mars could be a reality within the next 20 years. Before the end of the century, a colony on Mars is not out of the question. To accomplish these lofty goals, many unmanned missions to Mars will need to be undertaken. Reliable communications between Mars and Earth need to be established to ensure efficient communications can take place between spacecraft on route to Mars and ground stations on Earth. Because the Internet allows virtually instantaneous communication between any two points on the Earth, researchers are hoping to take advantage of this technology and create an **interplanetary Internet**, a network that spans the planets.

How would an interplanetary Internet be constructed? Unfortunately, there are a few physical obstacles to overcome. Communications, even via light rays, require line of site to be maintained. The orbital dynamics of planets need to be taken into consideration to ensure that line of site between Earth and Mars is not obstructed for any great length of time. However, this problem has already been addressed on Earth by the installation of the *Deep Space Network*. The **Deep Space Network** comprises three antenna installations located in California, Spain, and Australia. These installations provide for almost continuous transmission of data to outer space regardless of the orbital position of the Earth. A series of seven satellites has been proposed to orbit Mars to provide similar coverage on that end (see Figure 13.22 on page 542).

Will the interplanetary Internet provide the same communications speed as the current Internet? Interplanetary distances preclude the instantaneous communications we're accustomed to on Earth. Light travels at 186,000 miles per second, making data transmission from point to point on the Earth possible in a fraction of a second. However, depending on orbital factors, a point-to-point transmission between Earth and Mars may have to traverse as many as 248 million miles, which does not provide for instantaneous transmission. Because of this delay, current Internet protocols will not work.

Therefore, Vinton Cerf (one of the fathers of the Internet) and a team of other scientists and computer professionals is working on a new transmission protocol known as the *Parcel Transfer Protocol (PTP)*. The **Parcel Transfer Protocol (PTP)** must be designed to keep running even if

For more information on the future of high-speed Internet access, see *Transfer DVDs in Seconds*, a TechTV clip found at www.prenhall.com/techinaction

542 CHAPTER 13 BEHIND THE SCENES: THE INTERNET: HOW IT WORKS

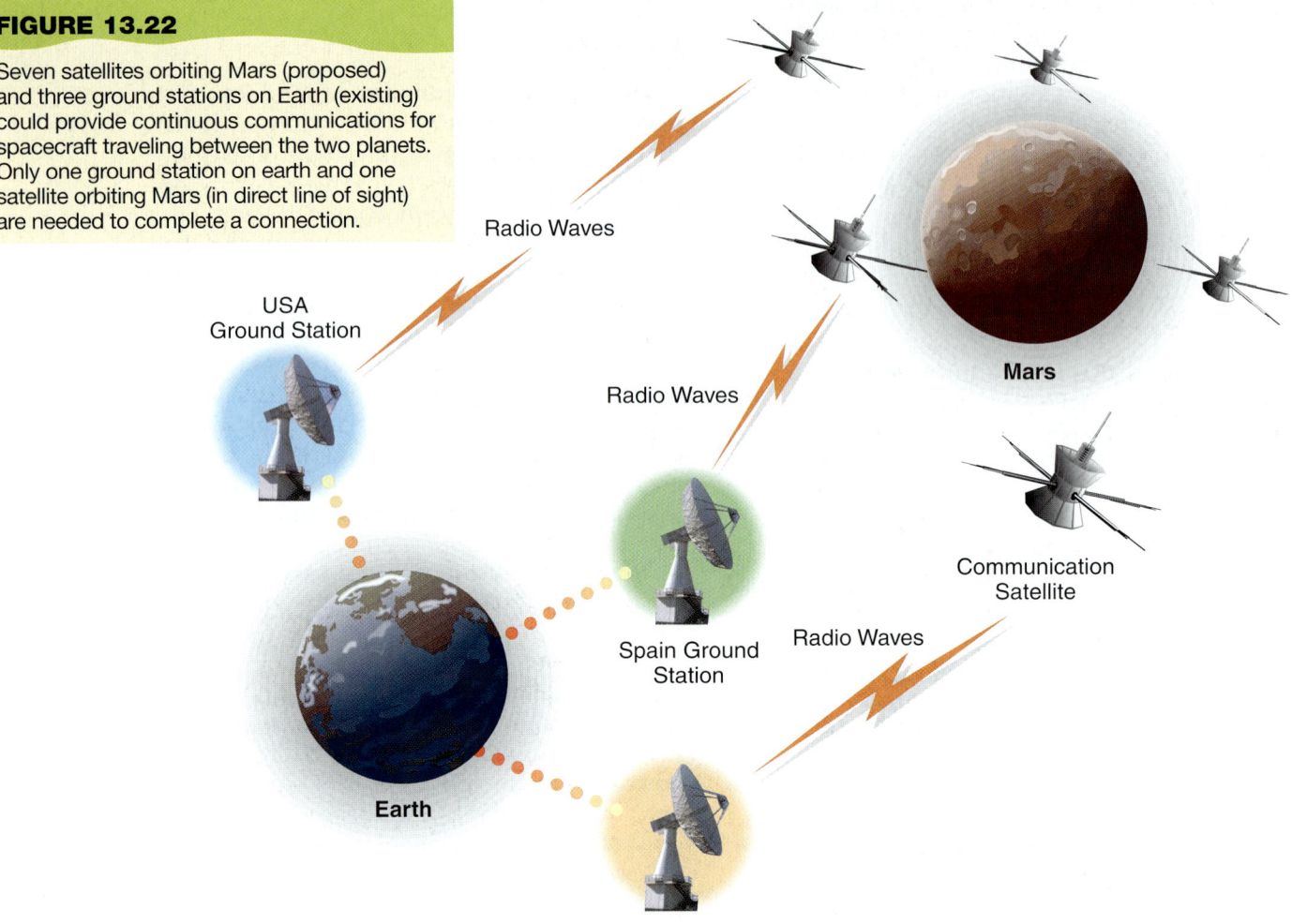

FIGURE 13.22
Seven satellites orbiting Mars (proposed) and three ground stations on Earth (existing) could provide continuous communications for spacecraft traveling between the two planets. Only one ground station on earth and one satellite orbiting Mars (in direct line of sight) are needed to complete a connection.

packets are lost in transmission and to block out noise that can be picked up while data is traversing millions of miles. PTP stores data at the receiver until all data is received and accounted for, then transmits it to the proper destination. In addition, the interplanetary Internet must be made safe from hackers through the use of secure protocols. And due to the distances involved, components must be of much higher quality than those used on Earth since repairs will be difficult and costly.

What are the benefits of installing an interplanetary Internet? Aside from enhanced communication between spacecraft and Earth home bases, Webcams could be installed on other planets, which would allow people on Earth to take virtual tours of distant celestial bodies. The interplanetary Internet would provide the same functionality to astronauts in the International Space Station or to colonists on Mars that the Internet does to regular users here on Earth. You wouldn't want to be on a three-year mission to Mars and not be able to check your e-mail, would you?

THE EXPANDING FEATURES OF THE INTERNET

What new experiences will the Internet provide in the next five years? The Internet today does a good job of allowing us to communicate via e-mail and instant messaging, view multimedia, and shop for goods and services. The Internet of tomorrow will stimulate more of our senses and become an even more pervasive force in our lives.

What other senses will the Internet be able to reach? The goal of virtual reality is to completely mimic the "real world" electronically. When using the Internet today, we use our senses of sight and hearing quite frequently to view various multimedia deployed on the Web.

But what about our other senses such as smell and taste? In 2002, TriSenx (**www.trisenx.com**) launched the Sensory Enhanced Net Experience machine, or the SENX device. This device reads embedded Web page commands and uses specialized software to generate fragrances and aromas. The desktop device plugs into a port on your computer and is priced at under $300. With smell technology conquered, can the sense of taste be far behind? Certainly not! Printer-like devices could generate printed flavor cards in the near future. Several companies are working on bringing these devices to market.

But why would I want to smell and taste the Internet? Imagine watching a *Star Wars* DVD on your computer and actually being able to smell the ozone when a laser cannon evaporates an Imperial TIE fighter. In addition, advertisers could increase sales by inducing consumers to buy or consume products after they've had a "whiff" of them. For example, Internet banner ads could be embedded with the sweet aroma of cookies, which might make you want to dash off to the store to buy some (or order some online). If the smell doesn't do it, you can just download and print an ice-cream flavor card from the Internet and be able to taste it before you buy any. Combined with home delivery of grocery products, you may never need to leave your house again!

How else will the Internet become a more integral part of our lives? In the future, you can expect to use the Internet to assist you with many day-to-day tasks that you do manually today. For example, already popular in upscale housing developments, Internet-enabled appliances and household systems allow your home to virtually run itself. Internet-enabled appliances can communicate over the Internet with each other and with third-party computer systems.

For example, refrigerators can monitor their contents and go online to order more diet soda when they detect that the supply is getting low (see Figure 13.23). Appliances such as ovens can be preloaded with ingredients and be set to have a meal ready at a certain time by following cooking directions and a time set on the Internet. Running late at work? Contact your Internet-aware appliance through the Internet and change your meal delivery time. Meanwhile, Internet heating and cooling systems can monitor weather forecasts and order fuel deliveries when supplies run low or bad weather is expected. These appliances will all become more widespread as the price of equipment drops.

What can the Internet do to save me time at college? Time is a precious commodity for college students. At MIT, a student has created a Web site that monitors the washers and dryers in dorms. Students can log onto the Web site and see which washers and dryers are running and how long it will be until they're available—saving students the time and hassle of running up and down the stairs to check on the status of their laundry. Similar systems are also in place in many college libraries related to real-time notification of reference materials availability—no more wasted trips across campus only to find the last copy of the book you need was just taken off the shelf. Timesaving Internet applications such as these are the wave of the future.

Will I be able to use the Internet in my car? The Global Positioning System (GPS), a satellite navigation system developed by the U.S. military, has taken the automobile industry by storm. Already, many new automobiles are fitted with GPS receivers and interfaces so they can receive data from satellites and help drivers navigate to their destination. Many of these systems sport wireless Internet connectivity for fast access to direction-finding Web sites like MapQuest. In a few years, heads-up displays (which will be visible on the windshield) will enable you to get directions and imaging information without taking your eyes off the road.

The uses for the Internet are only limited by our imaginations and the current constraints of technology. But as you enter the work force, perhaps you will invent the next killer consumer application (the next eBay?) or contribute to the development of key technologies (nanotech circuits?) that will drive the speed of the Internet to new heights. Think about what you will want to use the Internet for tomorrow—then make it a reality!

FIGURE 13.23

What could be more convenient than ordering groceries on the Internet via a screen embedded in your refrigerator (as shown here)?

Summary

1. **Who manages and pays for the Internet?** Management of the Internet is carried out by a number of nonprofit organizations and user groups such as the Internet Corporation for Assigned Names and Numbers (ICANN) and the World Wide Web Consortium (W3C). Currently, the U.S. government funds a majority of the Internet costs.

2. **How do the Internet's networking components interact?** Individual computers or networks connect to the Internet using Internet service providers (ISPs). These providers are ranked in tiers according to the volume of traffic they carry and the speed at which they transfer data. The connections between Tier 1 ISPs are extremely high speed and are known collectively as the Internet backbone. Tier 1 ISPs are connected to Tier 2 ISPs, which are in turn connected to Tier 3 ISPs. Individuals or corporations connect to Tier 2 or Tier 3 ISPs, which have regional or local focus, respectively.

3. **What data transmissions and protocols does the Internet use?** Data is transmitted along the Internet via packet switching. Although many protocols are available on the Internet, the main suite of protocols used to move information along the Internet is TCP/IP. The suite is named after the Transmission Control Protocol (TCP) and the Internet Protocol (IP). While the TCP is responsible for preparing data for transmission, the IP sends data between computers on the Internet.

4. **Why are IP addresses and domain names important for Internet communications?** An IP address is a unique number assigned to all computers connected to the Internet. The IP address is necessary so that packets of data can be sent to a particular location (computer) on the Internet. A domain name is merely a name that stands for a certain IP address and makes it easier for people to remember it.

5. **What are FTP and Telnet and how do I use them?** File Transfer Protocol (FTP) allows users to share files with remote computers. Telnet is both a protocol for connecting to a remote computer and a TCP/IP service that runs on a remote computer to make it accessible to other computers.

6. **What are HTML and XML used for?** The Hypertext Markup Language (HTML) is a set of rules for marking up blocks of text so that a browser knows how to display them. Since HTML was not designed for information exchange, Extensible Markup Language (XML) was created. XML allows users to create their own markup languages to accommodate particular data formats and needs. XML is used in e-commerce for exchanging data between corporations.

7. **How does e-mail and instant messaging work and how are messages kept secure?** Simple Mail Transfer Protocol (SMTP) is the protocol responsible for sending e-mail along the Internet. E-mail passes through e-mail servers whose function is to store, process, and send e-mail to its ultimate destination. ISPs and portals such as Yahoo! maintain e-mail servers to provide e-mail functionality to their customers. Your ISP's e-mail server uses domain name servers to locate the IP addresses for the recipients of the e-mail you send. Encryption software is used to code messages so authorized recipients can decode them.

8. **What does the Internet of the future look like?** The Internet of the future will have higher bandwidth and provide additional services due to projects such as the Large Scale Networking (LSN) program and Internet2. The Internet of the future will also engage more of our senses and be more ingrained into our daily lives with Internet-enabled appliances.

Key Terms

applet	p. 533
CGI scripts	p. 532
cgi-bin	p. 532
circuit switching	p. 522
client/server model	p. 520
client-side application	p. 533
commerce servers	p. 520
Common Gateway Interface (CGI)	p. 532
computer protocol	p. 522
connectionless protocol	p. 527
connection-oriented protocol	p. 526
Deep Space Network	p. 541
domain name	p. 525
domain name server (DNS)	p. 529
dotted decimal number	p. 524
dynamic addressing	p. 525
Dynamic Host Configuration Protocol (DHCP)	p. 525
elements	p. 531
e-mail servers	p. 536
encryption	p. 537
Extensible Hypertext Markup Language (XHTML)	p. 535
Extensible Markup Language (XML)	p. 534
extranets	p. 540
file servers	p. 520
File Transfer Protocol (FTP)	p. 529
handshaking	p. 526
HTML embedded scripting language	p. 533
hypertext	p. 531
Hypertext Markup Language (HTML)	p. 531
Hypertext Transfer Protocol (HTTP)	p. 531
Internet backbone	p. 518
Internet Protocol (IP)	p. 523
Internet Protocol version 4 (IPv4)	p. 524
Internet Protocol version 6 (IPv6)	p. 524
Internet telephony	p. 538
Internet2	p. 541
interplanetary Internet	p. 541
IP address	p. 524
JavaScript	p. 533
key pair	p. 537
Large Scale Networking (LSN)	p. 541
Multipurpose Internet Mail Extensions (MIME)	p. 537
negative acknowledgment (nack)	p. 527
network access points (NAPs)	p. 519
octet	p. 524
open systems	p. 522
packet switching	p. 522
packets	p. 522
Parcel Transfer Protocol (PTP)	p. 541
point of presence (POP)	p. 519
positive acknowledgment (ack)	p. 527
Pretty Good Privacy (PGP)	p. 538
private key	p. 538
private-key encryption	p. 537
proprietary (or private) systems	p. 522
protocol	p. 517
public key	p. 537
public-key encryption	p. 537
root domain name servers	p. 529
routers	p. 519
second-level domains	p. 528
Secure Sockets Layer (SSL)	p. 531
server-side	p. 533
Simple Mail Transfer Protocol (SMTP)	p. 536
static addressing	p. 525
T lines	p. 518
T-1 line	p. 518
T-2 line	p. 518
T-3 line	p. 518
T-4 line	p. 518
tags	p. 531
TCP/IP	p. 523
Telnet	p. 529
three-way handshake	p. 526
Tier 1 ISPs	p. 517
Tier 2 ISPs	p. 518
Tier 3 ISPs	p. 518
Transmission Control Protocol (TCP)	p. 523
UDP (User Datagram Protocol)	p. 526
virtual private network (VPN)	p. 540
Voice over IP (VoIP)	p. 538
Web servers	p. 520

Buzz Words

Word Bank

- URL
- TCP/IP
- T-1
- IP address
- ICANN
- SSL
- circuit switching
- Web server
- packet switching
- HTML
- public-key encryption
- DNS
- ack
- point of presence (POP)
- XML
- SMTP
- applet
- PGP
- HTTP
- Tier 3
- domain names
- protocol
- dynamic addressing
- static addressing
- FTP
- backbone

Instructions: **Fill in the blanks using the words from the Word Bank above.**

As a network administrator, Patricia knows that she can count on the organization (1)_____ to ensure that she has an appropriate range of IP addresses for her work site. Her high-speed connection to her company's (2)_____ ISP was vital to providing the connectivity her employees need to get their jobs done. Recently, the company moved up from a DSL connection to a (3)_____ line due to the high volume of Internet traffic they were generating. Hopefully, Patricia thinks, the government will continue to fund projects such as the Internet2 to continue research to improve the Internet (4)_____, the main highway to the Internet, and other vital technologies.

But Patricia had indulged in enough daydreaming. It was time to ensure that the Internet connection to the bank of modems, or (5)_____ provided by the Tier 3 ISP her company is using, was fully functional before the majority of the employees arrived for work. Since the company sends a tremendous amount of e-mail, old-fashioned (6)_____ technology would never have sufficed for sending messages. Fortunately, the Internet employs (7)_____ to allow messages to be sent over widely varying routes. Of course she knows that the main suite of protocols that controls Internet data traffic is called (8)_____.

After ensuring all was functional, Patricia begins to assist the Web development team with Web page creation. To provide robust interaction with company databases, (9)_____ is being used to code Web pages for the corporate Web site instead of HTML. Unsure of her instructions, she e-mails the director of Web development for clarification, knowing that the (10)_____ protocol will ensure the e-mail is delivered to the director at their U.K. office. Requiring secure communications, she encrypts the e-mail using a (11)_____ algorithm, knowing that the director can retrieve Patricia's key from her personal Web site.

After reading the response to her e-mail, she quickly writes a Java (12)_____ to produce an interactive form to collect customer names and addresses. Using the (13)_____ protocol, Patricia posts her completed Web page to the corporate site. Of course users will view the completed Web page using the (14)_____ protocol. And since the Web page contains potentially sensitive information, Patricia makes sure the (15)_____ protocol is in use to provide added security for the data. Just another exciting day in the life of a network administrator!

Organizing Key Terms

Instructions: *This chapter introduced many new terms and concepts. In the illustration below, fill in each of the blanks with key terms from the chapter in order to show how categories of ideas fit together.*

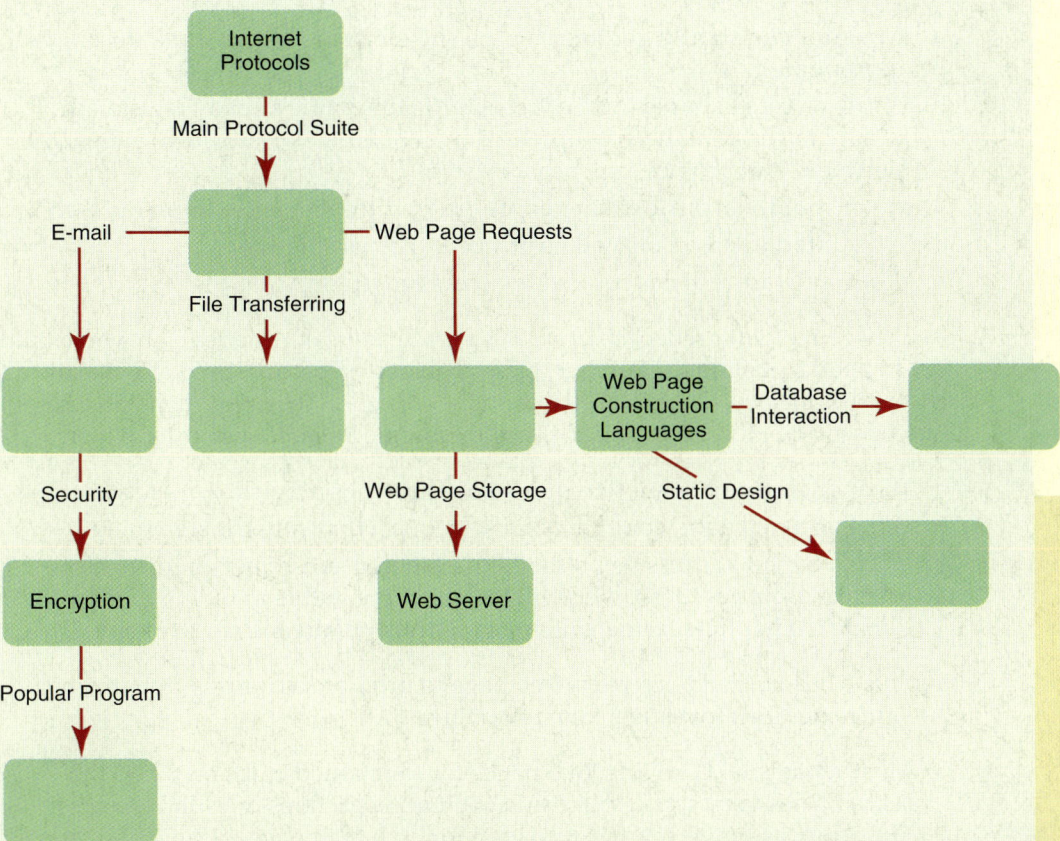

Making the Transition to . . . Next Semester

1. Your American History instructor has asked your group to design a Web site on the Battle of Gettysburg. The site will include textual and graphical information on the battle as well as an interactive quiz. Users will be encouraged to post college term papers on the site as a resource for other students. The following issues need to be addressed for your portion of the project:

    a. What domain should you register the site in? Why do you think this is appropriate?
    b. What is a catchy name for the Web site? Is this name already registered? Be sure to check in the appropriate domain and make sure it isn't already registered and provide proof that is isn't.
    c. How would you ensure that people can find the site once it's on the Web?
    d. What Internet protocols will users need to access the site?
    e. What software will be required to access the site?

2. As president of your school's Future Business Leaders of America club, you would like to build a mailing list for your quarterly newsletter. You would also like to ensure that the newsletter is available on the college's Web site. When you visit the college Web developer, she asks you to provide the following information:

    a. What data do you want to require all online subscribers to provide?
    b. What optional data would you like to request that subscribers provide?
    c. Will an e-mail address be provided to potential subscribers if they wish to make inquiries? If so, who will be reviewing and responding to these e-mails? What time frame will you guarantee for an answer to e-mails?

3. You have been asked to create a Web page to promote a campus club of which you are a member. Investigate the following:

    a. Most schools will provide Web pages for sanctioned clubs on their Web server. Does your school place any content restrictions on club Web sites? How much storage space is provided for club Web pages? Can club Web pages gather data from prospective members (such as contact information) for storage in a database?
    b. Assume your club is unsanctioned and your school will not provide space for a Web page. Using Web sites such as www.ispfinder.com and www.findanisp.com, find three ISPs where you could potentially host the Web site and prepare a chart comparing the following for each ISP:

    - Cost per month for hosting
    - Amount of disk storage space provided
    - Number and type of e-mail accounts provided
    - Data throughput (transfer) allowed per month
    - Level of technical support provided

    c. Which ISP will you recommend to your club and why?

4. With tuition costs rising, you feel it would be appropriate to find a part-time job next semester. Visit some popular employment sites such as www.monster.com and www.careerbuilder.com and:

    a. Search for a part-time job or internship opportunity in your field of study. Provide a listing of opportunities you think would be appropriate for you.
    b. Search for full-time jobs that you might apply for when you graduate. What are the average starting salaries? Do the starting salaries vary by geographic location or are they reasonably similar?

Making the Transition to . . . The Workplace

1. Your employer, a distributor of specialty stereo equipment, offers high-speed Internet access at the office where you work. Thirty-five workstations are connected to the Internet, which employees use to send e-mail and conduct research. Recently, company trade secrets (traced to an employee e-mail) were printed in the local press. The company president has enlisted your help in determining preventative measures to avoid such security breaches in the future. Draft a memo that includes the following:

   a. An employee e-mail policy that requires the use of encryption technologies.
   b. Other suggested uses for the Internet access that would benefit the company employees.

2. The director of your college placement office has suggested that you prepare a resume that can be posted on the new college graduate employment site that the university is developing. It must be in HTML format and include a picture of yourself. Consider the following:

   a. What software would you use to develop the resume?
   b. Would straight HTML be sufficient or would you need to use JavaScript to display the picture?
   c. Aside from your college's Web site, find at least three other Web sites where you could post your resume so employers could see it.

3. Your supervisor just saw a news documentary about Internet addiction on television last night. He is concerned about whether the company's employees might be affected. Research various types of Internet addiction and prepare a report for your boss suggesting ways to minimize the impact of addiction on the employees.

4. At your company, someone was just fired because sensitive information related to a company product was associated with their name on the Internet. Discretion being the better part of valor, you decide to do a search for your name on the Web (in a search engine such as www.google.com) just to see what is out there. Prepare a report on what you found by exploring the following:

   a. Did you find any accurate information about yourself (such as your home page, resume, etc.)? Any erroneous information that you need to correct?
   b. Did you find Web sites or information about other people with the same name as you? Could any of that information be damaging to your reputation if someone thought the other person was you? Provide examples.
   c. Is there information that you found about yourself or others that you think should never be available on the Internet? What types of information should not be available online about individuals?

Critical Thinking Questions

Instructions: Albert Einstein used "Gedanken experiments," or critical thinking questions, to develop his theory of relativity. Some ideas are best understood by experimenting with them in our own minds. The following critical thinking questions are designed to demand your full attention but require only a comfortable chair—no technology.

1. Domain names often spark fierce controversy between competing companies. Legal wrangling over the rights to attractive names such as Buynow.com and Lowprices.com can generate large fees for attorneys. Meanwhile, some famous individuals have had to fight for the right to own domains based on their own names.

    a. Should everyone be entitled to a Web site in a certain domain (say .com) that contains their own name? How would you handle disputes by people who have the exact same name (say two people named John Smith)?
    b. Is it ethical to register a domain name (say Coke.net) just for the purpose of selling it to the organization that may benefit from it the most (such as the Coca-Cola Company)?
    c. Aside from the domains currently approved (such as .com and .org), what domain names do you think would have commercial appeal? What types of Web sites would these domains be used for?

2. Ensuring that computer-based information is secure is a key objective of many companies today.

    a. How would you prepare for a job as a network security specialist?
    b. Aside from computer-based security measures, what physical precautions should be taken to enhance computer data security?

3. Encryption programs built on 128-bit encryption algorithms are currently considered unbreakable.

    a. Since strong encryption programs using 128-bit encryption or better are considered "unbreakable," the U.S. government places restrictions on exports of these encryption products. The government is also considering a requirement that all encryption products should have a "back door" code which would allow government agencies (such as the CIA and FBI) to read encrypted messages? Do you think this should be implemented? Why or why not?
    b. Assuming you figure out how to break 128-bit encryption, should you post that information on the Internet for anyone to use?

Team Time

Problem:

As Web usage increases exponentially, demands are constantly placed on the Internet's infrastructure. Development groups, such as the Internet2 consortium, are designed to foster the innovation of new technologies to improve delivery of Internet services. In this Team Time, you'll research up-and-coming technologies and consider their impact on the Internet.

Task:

Your group has just received an invitation to speak at a technology conference being held at your school. Your topic is cutting-edge Internet technologies. You will be given 20 minutes at the conference to deliver a presentation to a group of approximately 250 students and educators, all interested in (but not necessarily familiar with) technology.

Process:

Break the class into small groups of three or four students. Each group should prepare a report as follows:

STEP 1: Explore sites such as **www.howstuffworks.com**, **www.internet2.edu**, and **www.wired.com** to investigate new technologies. Prepare a list of current innovations that are being developed.

STEP 2: Briefly present your group's findings to the class for debate and discussion. Are any ideas a replacement for the Internet as opposed to an extension of its capabilities? Your objective is to determine which topic would be of most interest to your class for a presentation.

STEP 3: Develop a PowerPoint presentation on the most popular topic. If the technology you chose seems very futuristic or unbelievable, make sure you convey to the class why this will be achievable in the next 50 years.

Conclusion:

Don't be afraid to envision communications media that go beyond the current confines of the Internet. Great inventors, such as Edison, Gates, and Cerf, forced themselves to think "outside the box" instead of being trapped by current technology and engineering limitations. Encourage your friends to "daydream" about the future of data communications . . . the next communications revolution may start with you!

Becoming Computer Fluent

While attending college, you are working at MultiPharm, Inc., which is a small manufacturer of specialty steel products. The new CEO has charged your supervisor with bringing the company into the 21st century by connecting it to the Internet and developing a company Web site. Your supervisor has asked you to help draft a memo to the CEO, laying out the benefits of having an Internet presence.

Instructions: Draft a memo for your boss detailing the benefits of connecting the company to the Internet. Make sure to suggest what types of Internet connections would be appropriate and what type of ISP would be needed. While trying to use as many of the key words from the chapter as you can, ensure that the report can be presented to computer-illiterate managers.

Materials on the Web

In addition to the review materials presented here, you'll find extra materials on the book's companion Web site (**www.prenhall.com/techinaction**) that will help reinforce your understanding of the chapter content. These materials include:

Sound Byte Lab Guides

For each Sound Byte mentioned in the chapter, there is a corresponding lab guide located on the book's companion Web site. These guides review the material presented in the Sound Byte and direct you to various Web resources that examine the material. The Sound Byte Lab Guides for this chapter include:

- Constructing a Simple Web Page
- The Best Utilities for Your Computer

True/False and Multiple Choice Quizzes

The book's Web site includes a True/False and a Multiple Choice quiz for this chapter. You can take these quizzes, automatically check the results, and e-mail the results to your instructor.

Web Research Projects

The book's Web site also includes a number of Web research projects for this chapter. These projects ask you to search the Web for information on computer-related careers, milestones in computer history, important people and companies, emerging technologies, and the applications and implications of different technologies.

TECHNOLOGY IN FOCUS

Careers in IT

It's hard to imagine an occupation in which computers are not used in some fashion. Even such previously low-tech industries as junkyards and fast food use computers extensively for inventory management and commodity ordering. Since computers are a ubiquitous part of our lives, students often consider pursuing a career involving computing. In this Technology in Focus, we'll explore various IT career paths open to you.

WHAT TO CONSIDER FIRST: JOB OUTLOOK

If you want to investigate a career with computers, the first question you probably have is: "Will I be able to get a job?" With all the media hoopla surrounding the demise of dot-com companies in the late 1990s and 2000–2001, many people think the boom in computer-related jobs is over. Imagine this scenario playing out in households across the nation:

> **PARENT:** You need to start applying to colleges soon. Have you thought of what you want to study?
>
> **HIGH SCHOOL STUDENT:** Well, I love computers. Maybe I'll go into programming?
>
> **PARENT:** Hmm. I'm still hearing on the news about all the dot-com companies that have gone out of business. Just today I heard there are 44 percent fewer jobs in IT than there were in the year 2000. Maybe you should consider something else.
>
> **HIGH SCHOOL STUDENT:** Yeah, maybe you're right. How about if I major in communications with a minor in English Lit instead?
>
> **PARENT:** That sounds like a much better plan. Just don't forget your parents when you win your first Pulitzer.

TECHNOLOGY IN FOCUS

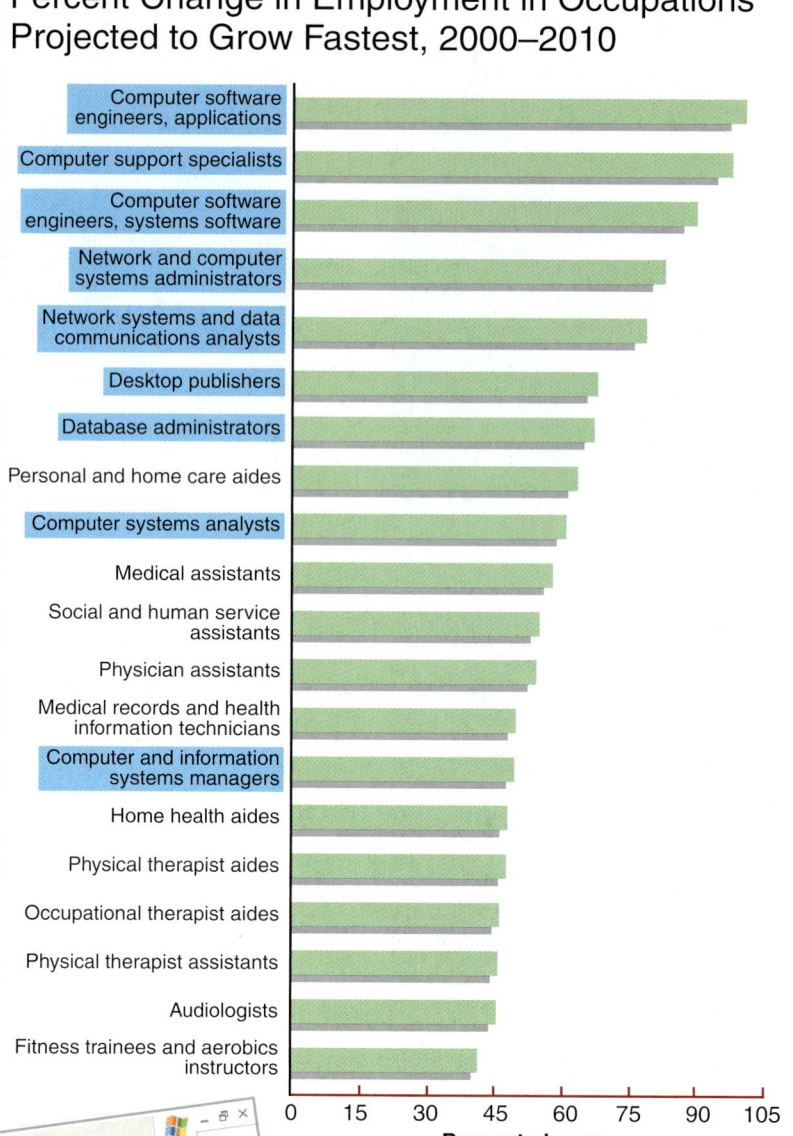

FIGURE 1 The Bureau of Labor Statistics is projecting huge growth in nine different computer occupations, as highlighted in this chart.

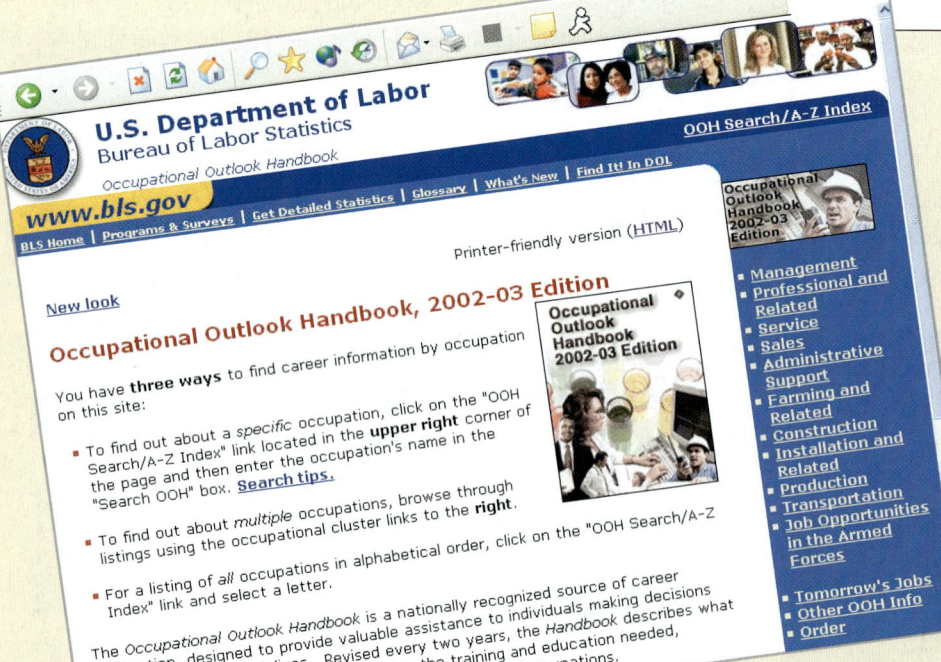

FIGURE 2 The *Occupational Outlook Handbook* is published by the Bureau of Labor Statistics and revised every two years. The handbook describes typical tasks that workers perform in their jobs and provides guidance on the training and education needed to land a particular job. The handbook also projects earnings and job prospects.

SETTING THE RECORD STRAIGHT: COMMON MYTHS ABOUT IT CAREERS

Regardless of whether you choose to pursue a career in IT, you should visit the Bureau of Labor Statistics's Web site at www.bls.gov. It contains a wealth of information about careers, with projections going out as far as eight years. One of the site's most useful features is the *Occupational Outlook Handbook*, as shown in Figure 2. This is a good starting point for researching careers. Aside from projecting job growth in a career field, it describes typical tasks that workers perform and the amount of training and education needed to perform a job. The handbook also provides salary estimates. Your tax dollars are already paying for development of the handbook, so you might as well take advantage of the information it contains.

Well, maybe it is not going to be exactly like that, but many students are concerned about the job outlook. However, current projections of the U.S. Department of Labor's Bureau of Labor Statistics show that 9 out of the top 20 fastest-growing occupations (through 2010) are still in a computer field, as shown in Figure 1. As you can see, there is quite a bit of demand for IT workers.

SETTING THE RECORD STRAIGHT: COMMON MYTHS ABOUT IT CAREERS

Many people have misconceptions about pursuing a career in IT. These misconceptions scare them away from considering a career in computing or convince them to pursue a computing career for the wrong reasons. Do you have any of the following misconceptions?

COMMON MYTHS ABOUT IT CAREERS

MYTH #1: **A degree in computer science means you're going to be rich.** Some students think they're going to walk out of school and announce to prospective employers, "Show me the money!" Computer-related careers often offer high salaries, but choosing a computer career isn't a guarantee you'll get a high-paying job. Just as in other professions, you'll need years of training and on-the-job experience to earn a high salary.

MYTH #2: **Computer professionals are in demand, so you'll never be fired.** Just because employees are scarce in a profession doesn't mean that you'll stay employed if you're doing a poor job. You'll still be fired if you're not proficient at your work.

MYTH #3: **You've got three professional certifications... you're ready to work.** Many freshly minted technical school graduates sporting IT certifications feel ready to jump into a job. But just because you're certified doesn't mean that employers will hire you. Employers routinely cite experience as being more desirable than certification. Experience earned through an internship or a part-time job will make you much more marketable when your certification program is complete.

MYTH #4: **Women are at a disadvantage in an IT career.** Currently, women make up less than 25 percent of the IT workforce. This presents an opportunity for women with IT training since many IT departments are actively seeking to diversify their workforce. Also, although a salary gender gap (difference between what men and women earn for the same job) exists in IT careers, it's less severe than in many other professions.

MYTH #5: **People skills don't matter in IT jobs.** Despite what many people think, IT professionals are not locked in lightless cubicles, basking in the glow of their monitors. Most IT jobs require constant interaction with others, often in team settings. People skills are important, even when you work with computers.

MYTH #6: **Mathematically impaired people need not apply.** It is true that a career in programming involves a fair bit of math, but even if you're not mathematically inclined, you can explore other IT careers. IT positions also value such attributes as teamwork, creativity, leadership ability, and artistic style.

Resolving myths is an important step toward considering a job in IT. But you need to consider other issues related to IT careers before you decide to pursue a particular path.

How Much Will I Make?

You probably want to know actual salary ranges for IT careers. However, IT salaries vary widely depending on experience level and the geographic location of the job. To obtain the most accurate information, research salaries yourself. Job posting sites such as Monster.com can provide guidance, but a better site is Salary.com. As shown in Figure 3, you can use the salary wizard on this site to determine what IT professionals in your town are making compared with national averages. Hundreds of IT job titles are listed so you can fine-tune your search to the specific job you're interested in researching.

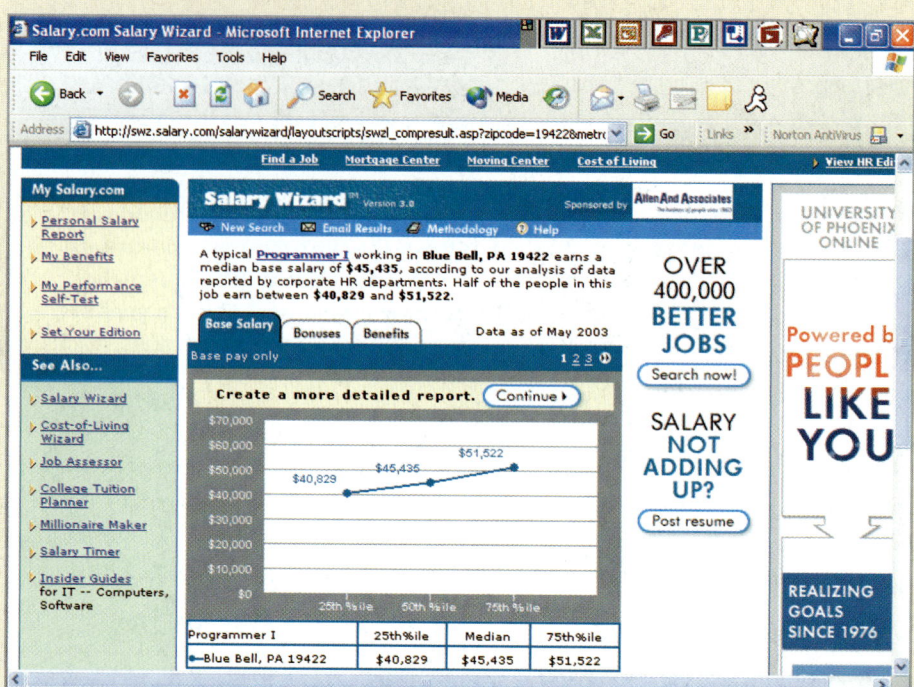

FIGURE 3 The salary wizard at Salary.com shows that for an entry-level programming position (programmer I) in Blue Bell, Pa., you could expect to earn a median salary of $45,435. The wizard is easy to tailor to your location and job preferences.

IS AN IT CAREER RIGHT FOR YOU?

Choosing a career in IT can be a difficult path. Before preparing yourself for an IT career, consider the following:

1. **Salary range.** What affects your salary in an IT position? Your skill set and experience level are obvious answers. However, there are less obvious factors, such as the size of your employer and geographic location. Large companies tend to pay more, so if you're pursuing a high salary, set your sights on a large corporation. But remember that making a lot of money isn't everything. There are other quality-of-life issues to consider, such as job satisfaction.

2. **Possibility of gender bias.** We mentioned earlier that there are still few women in IT careers. Why aren't there more women in IT? Many women view IT departments as Dilbert-like microcosms of antisocial geeks and don't feel they would fit in. Unfortunately, a number of mostly male IT departments suffer from varying degrees of gender bias against women. While some women may thrive on the challenge of enlightening these "male enclaves" and bringing them forward to the 21st century, others find such environments difficult to work in.

3. **Location.** Location in this case refers to the setting in which you work. IT jobs can be office-based, field-based, project-based, or home-based. Not every situation is perfect for every individual. You need to decide what is right

FIGURE 4 **Where Do You Want to Work?**

TYPE OF JOB	LOCATION/HOURS	SPECIAL CONSIDERATIONS
Office-based	Report for work to the same location each day and interact with the same people on a regular basis. Requires regular core hours of attendance (such as 9 to 5).	May require working beyond "normal" working hours. Some positions require workers to be on call 24/7.
Field-based	Travel from place to place, as needed, and perform short-term jobs at each location.	Involves a great deal of travel and working independently.
Project-based	Work at client sites on specific projects for extended periods of time (weeks or months). Contractors and consultants fall into this area.	Can be very attractive to individuals who like workplace situations that vary on a regular basis.
Home-based (Telecommuting)	Work from home.	Involves very little day-to-day supervision and requires an individual who is self-disciplined.

for you based on your personal preferences. Figure 4 summarizes the major jobs types and their locations.

4. **A changing workplace.** In IT, the playing field is always changing. New software and hardware are constantly being developed. It's almost a full-time job to keep your skills up-to-date. You can expect to spend a lot of time in training and self-study trying to learn new systems and techniques to keep your skills current.

5. **Stress.** Longer hours are becoming a norm in the United States, but it is more acute in the IT arena. Where the average American works 42 hours a week, a survey by *InformationWeek* shows that the average IT staff person works 45 hours a week and is on call for another 24 hours a week. On-call time (hours an employee must be available to work in event of a problem) has been increasing in recent years due to more IT systems (such as e-commerce systems) requiring 24/7 availability. As a result, many IT employees feel like they are mice on exercise wheels—except they aren't permitted to stop when they get tired (see Figure 5 on page 560).

The good news is that despite the stress and changing nature of the IT environment, most computing skills are portable from industry to industry. Pharmaceutical companies, banks, steel mills, and retail stores all have computer networks. A networking job in the clothing manufacturing industry uses the same primary skill set as a networking job for a supermarket chain. So if something disastrous happens to the industry you're in now, you should be able to transition to another industry with little trouble.

TECHNOLOGY IN FOCUS

FIGURE 5 Stress comes from multiple directions in IT jobs.

WHAT REALM OF IT SHOULD YOU WORK IN?

Figure 6 on page 561 provides an organizational chart for a modern IT department that should help you understand the careers currently available. The chief information officer (CIO) has overall responsibility for the development, implementation, and maintenance of information systems and infrastructure. Usually the CIO reports to the chief operating officer (COO).

The responsibilities below the CIO are usually grouped into two units: development and integration and technical services. The development and integration unit is responsible for the development of systems and Web sites. The technical services unit is responsible for the day-to-day operations of the company's information systems, including all hardware and software deployed.

In large organizations, responsibilities are distinct and jobs are more narrowly defined. In medium-size organizations, there can be a lot of overlap between position responsibilities. At a small shop, you might end up being the network administrator, database administrator, computer support technician, and help-desk analyst all at the same time. Let's look at each department in more depth and explore typical entry-level jobs for which you can prepare.

Working in Development and Integration

Two distinct paths exist in this division: Web development and systems development. Since everything involves the Web today, there is often a great deal of overlap between these departments.

Web Development
When most people think of Web development careers they usually equate them to being a **Web master**.

However, today's Web masters are usually supervisors with responsibility for certain aspects of Web development. At smaller companies, they may be responsible for tasks that the other folks in a Web development group usually do:

- **Web programmers** create the programming for Web pages and link them to other systems such as databases.

- **Graphic designers** create art and multimedia elements for Web pages and often do page layout.

- **Interface designers** work with graphic designers to create a look and feel for the site and make it easy for you to navigate a site.

- **Web content creators** create the content for Web pages or develop interesting animations, games, or interactive information pieces.

- **Customer interaction technicians** interact with customers on an as-needed basis, dealing with customer inquiries regarding information needed or a product ordered.

The education required varies widely for these jobs. Web programming jobs often expect a four-year college

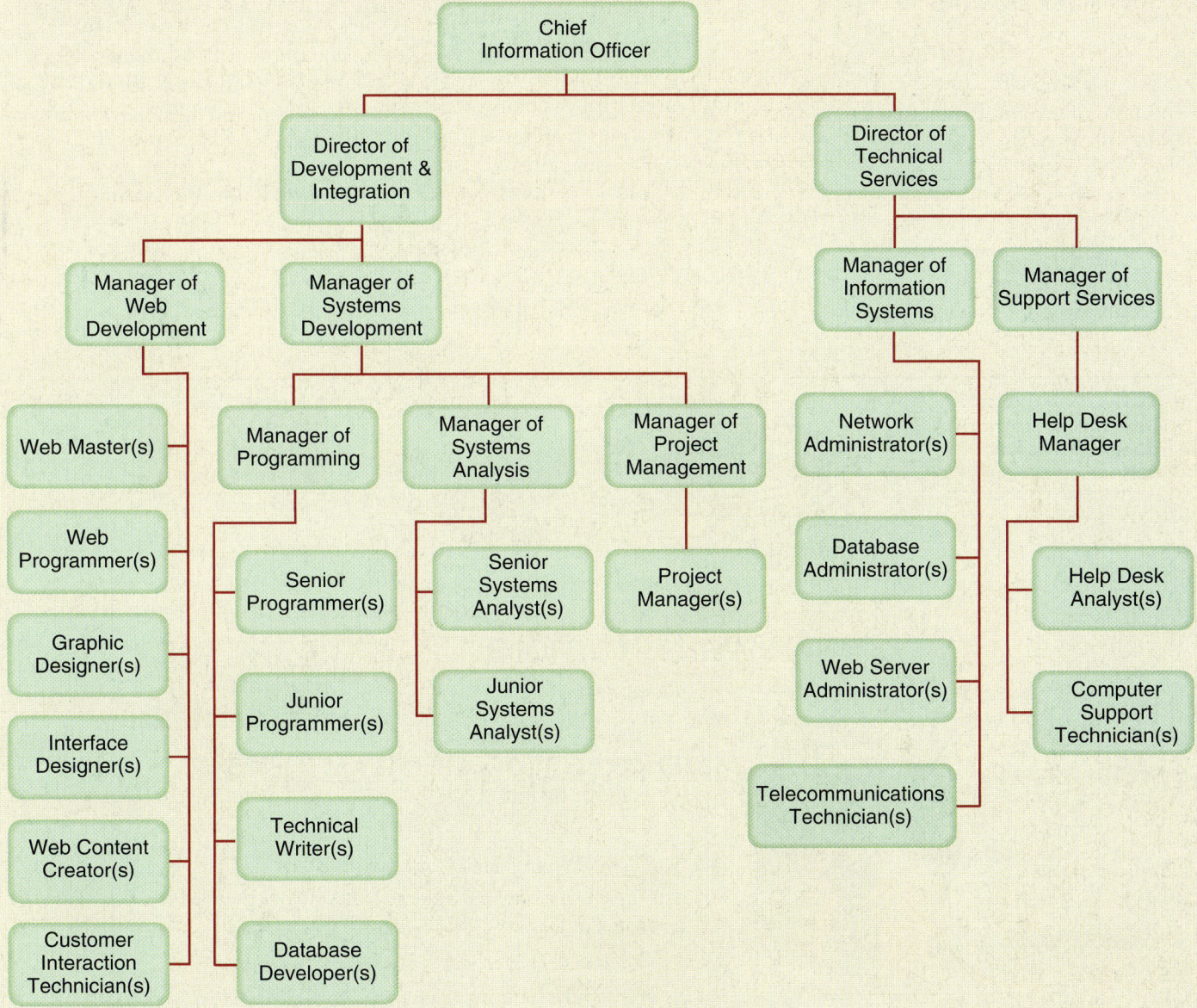

FIGURE 6 This is a typical structure for an IT department at a large corporation.

degree in computer science whereas graphic designers are often hired with two-year art degrees.

Systems Development

Ask most people what systems developers do and they will say "programming," but this is only one aspect of systems development. A variety of jobs are involved:

- **Computer programmers** write programs and interact with other team members and end users to determine the specifications for the programs they create.

- **Systems analysts** work with end users to determine the nature of the problem to be solved and to develop a solution. They then work with programmers to get the system built and tested.

- **Project managers** prepare project budgets, assemble the team, supervise the work, and generate reports to keep upper management informed of a project's progress.

- **Technical writers** generate systems documentation for end users and for programmers who may make modifications to the system in the future.

- **Database developers** design and build databases to support the software systems being developed.

Large development projects may have all of these team members on the project. Smaller projects may require an overlap of positions (such as a programmer also acting as a systems analyst). The majority of these jobs require four-year college degrees in computer science or Management Information Systems (MIS). As shown in Figure 8, team members work together to build a system.

Working in Technical Services

Technical services jobs are vital to keeping IT systems running. These people install and maintain the infrastructure behind the IT systems and work with end users to make sure they can effectively interact with the systems. The two major categories of technical services careers are information systems and support services.

Information Systems

The information systems department keeps the networks and telecommunications up and running at all times. Within the department, you'll find a variety of positions:

- **Network administrators** install and configure servers, design and plan networks, and test new networking equipment.

FIGURE 7 As you can see, it takes a "village" to create and maintain a Web site.

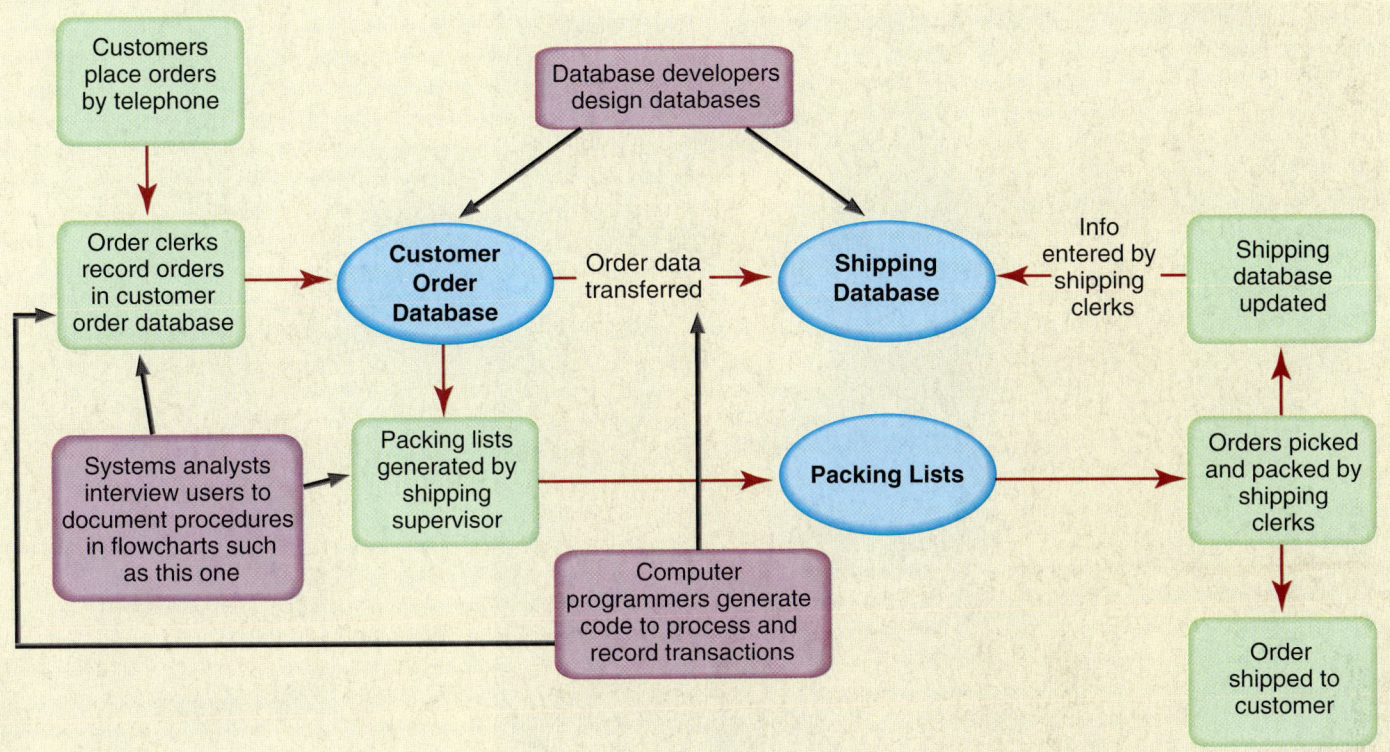

FIGURE 8 Here you see a flowchart of an order-processing system. Each member of the systems development team performs functions critical to the development process (as shown in the purple boxes).

- **Database administrators (DBAs)** install and configure database servers and ensure that the servers provide an adequate level of access to all users.
- **Web server administrators** install, configure, and maintain Web servers and ensure that the company maintains Internet connectivity at all times.
- **Telecommunications technicians** oversee the communications infrastructure, including training employees to use telecommunications equipment. They are often on call 24 hours a day.

Support Services

As a member of the support services team, you interface with users (external customers or employees) and troubleshoot their computer problems. These positions include:

- **Help-desk analysts** staff the phone (or e-mail) and solve problems for customers or employees, either remotely or in person. Often, help-desk personnel are called on to train users on the latest software and hardware.
- **Computer support technicians** go to a user's physical location and fix software and hardware problems. They also often have to chase down faults in the network infrastructure and repair them.

As important as these people are, they often receive a great deal of abuse by angry users whose computers are not working. When working in support services, you need to be patient and have a "thick skin."

Technical services jobs often require two-year college degrees or training at trade schools or technical institutes. At smaller companies, job duties tend to overlap between the two areas.

HOW SHOULD YOU PREPARE FOR A JOB IN IT?

A job in IT requires a robust skill set and formal training and preparation. Most employers today have an entry-level requirement of a college degree, a technical institute diploma, appropriate professional certifications, and/or experience in the field. How can you prepare for a job in IT?

FIGURE 9 At smaller companies, you may be fixing a user's computer in the morning, installing and configuring a new network operating system in the afternoon, and troubleshooting a router problem in the evening (as shown here).

1. **Get educated.** Two- and four-year colleges and universities normally offer three degrees to prepare students for IT careers: computer science, MIS, and computer engineering (although titles will vary). Alternatives to colleges and universities are privately licensed technical (or trade) schools. Generally, these programs focus on building skill sets rapidly and obtaining a job in a specific field, such as Web development or network administration. The main advantage of trade schools is that their programs usually take less time to complete than college degrees. However, to have a realistic chance of employment in other IT fields (such as programming), you should attend a degree-granting college or university.

2. **Investigate professional certifications.** Certifications attempt to provide a consistent method of measuring a base level of skill in a particular area of IT. There are hundreds of IT certifications out there, most of which you get by passing a written exam. Software and hardware vendors (such as Microsoft and Cisco) and professional organizations (such as the Computing Technology Industry Association) often establish certification standards. (Visit www.microsoft.com, www.cisco.com, www.comptia.org, and www.sun.com for more information on types of IT certifications.)

Employees with certifications generally earn more than employees who aren't certified. However, most employers don't view a certification as a substitute for a college degree or a trade school program. You should think of certifications as an extra edge beyond your formal education that will make you more attractive to employers. To ensure you're pursuing the right certification, ask employers which certifications they respect. Or explore online job sites for job postings in your field and see which certifications are listed as desirable or required.

3. **Get experience.** Aside from education, employers want you to have experience, even for entry-level jobs. As you're completing your education, consider getting an internship (or part-time employment) in your field of study. Many colleges will help you find internships and allow you to earn credit toward your degree through internship programs.

HOW DO YOU FIND A JOB IN IT?

Training for a career is not useful unless you can find a job at the end of your training. Here are some tips on getting a job:

1. **Visit your school's placement office.** Many employers recruit directly at schools, and most schools maintain a placement office to assist students in finding jobs. Early in your educational program, visit the placement office. Employees there can help you with resume preparation and job interviewing skills, and provide you with leads for internships and full-time jobs.

2. **Visit online employment sites.** Most IT jobs are advertised online at sites such as Monster.com and CareerBuilder.com (see Figures 10 and 11). Begin searching for jobs on these sites early in your

FIGURE 10 **Online IT Career Resources**

www.computerjobs.com	www.jobcircle.com
www.computeruser.com	www.justtechjobs.com
www.computerwork.com	www.tech-engine.com
www.dice.com	www.techiegold.com

TECHNOLOGY IN FOCUS

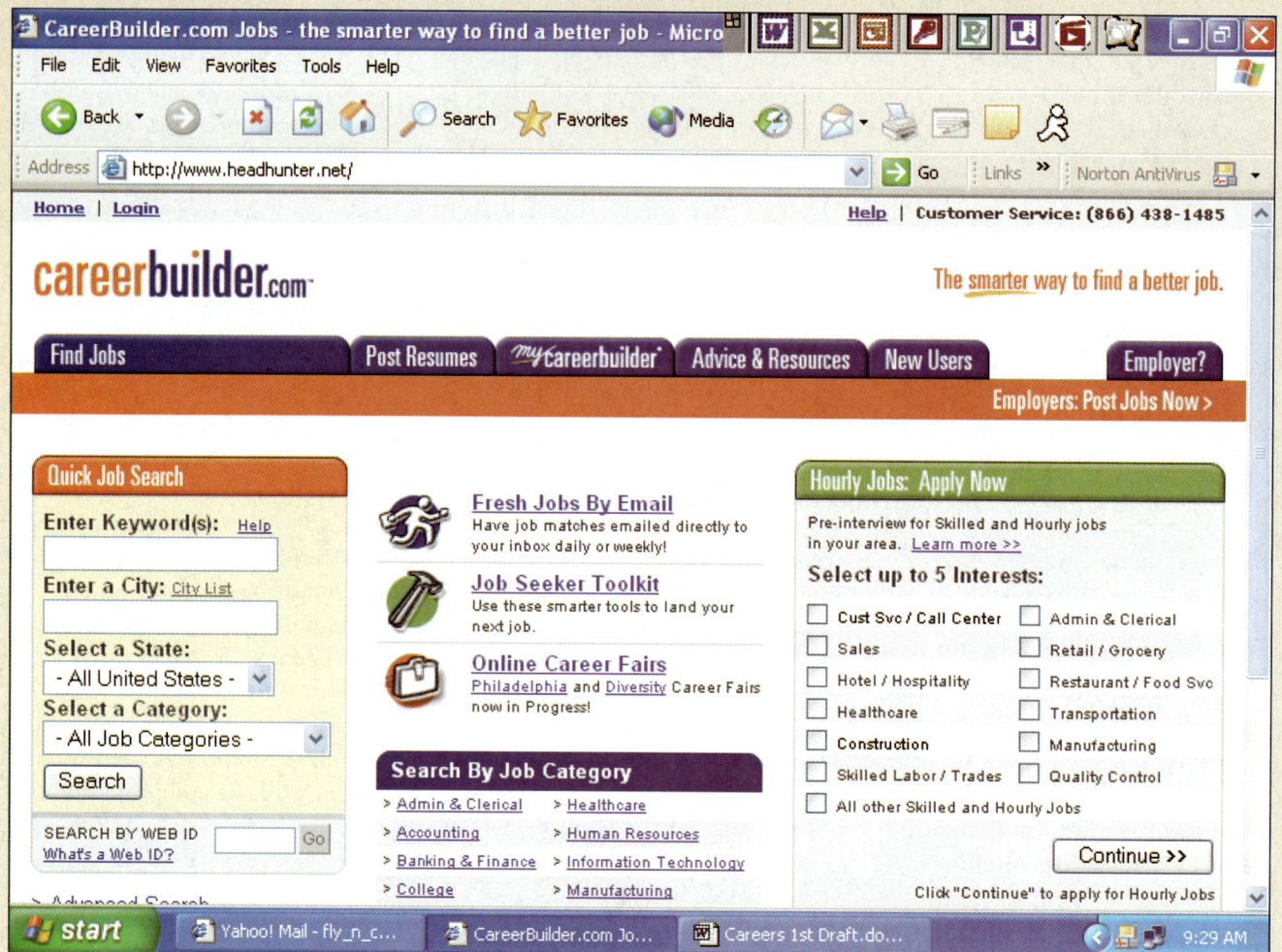

FIGURE 11 Employment sites, such as CareerBuilder.com, allow you to search for specific jobs within a defined geographic area. These sites also allow you to store your resume at the site and apply for positions with one click.

FIGURE 12 *Professional Organizations*

ORGANIZATION NAME	PURPOSE	WEB SITE
Association for Computing Machinery (ACM)	Oldest scientific computing society. Maintains a strong focus on programming and systems development.	www.acm.org
Institute of Electrical and Electronics Engineers (IEEE)	Provides leadership and sets engineering standards for all types of network computing devices and protocols.	www.ieee.org
Association of Information Technology Professionals (AITP)	This association focuses heavily on IT education and development of seminars and learning materials.	www.aitp.org

HOW DO YOU FIND A JOB IN IT?

FIGURE 13 Corporate Web sites often list available jobs. The Merck Pharmaceuticals site shows jobs by area of the world, such as the jobs available at the headquarters in Pennsylvania, shown here. You can often send a resume to apply for a job with one click of the mouse.

education, since the jobs listings detail the skill sets employers require. Focusing on coursework that will provide you with desirable skill sets will make you more marketable when you graduate.

3. **Start networking.** Not computer networking, networking with people! Many jobs are never advertised but are filled by word of mouth. Seek out contacts in your field and discuss job prospects with them. Find out what skills are needed and ask them to recommend others in the industry with whom you can speak. Professional organizations such as the Association for Computing Machinery (ACM) are one way to network. These organizations often have chapters on college campuses and offer reduced membership rates for students. The contacts you make there could lead to your next job. Figure 12 lists major professional organizations you should consider investigating.

4. **Check corporate web sites for jobs.** Many corporate Web sites list current job opportunities. For example, pharmaceutical company Merck features a searchable site (see Figure 13) you can tailor to a specific geographic area. Check company sites you're interested in working for.

The outlook for IT jobs should continue to be positive in the future. We wish you luck with your education and job search.

GLOSSARY

3D sound card Enables a computer to produce a sound that is omnidirectional, or three-dimensional.

802.11 standard A wireless standard established in 1997 by the Institute of Electrical and Electronics Engineers; also known as Wi-Fi (short for Wireless Fidelity), it enables wireless network devies to work seamlessly with other networks and devices.

A

academic fair use A provision that gives teachers and students special consideration regarding copyright violations. As long as the material is being used for educational purposes, limited copying and distribution is allowed.

Accelerated Graphics Port (AGP) bus The AGP bus design was specialized to help move three-dimensional graphics data quickly. It establishes a direct pathway between the graphics card and main memory so that data does not have to travel on the PCI bus, which already handles moving a great deal of system data being moved through the computer.

access method Established to control which computer is allowed to use the transmission media at a certain time.

access time The time it takes a storage device to locate its stored data.

accounting software An application program that helps small business owners manage their finances more efficiently by providing tools for tracking accounting transactions such as sales, accounts receivable payroll, inventory purchases and accounts payable.

active-matrix displays With a monitor using active-matrix technology, each pixel is charged individually, as needed. The result is that an active-matrix display produces a clearer, brighter image than a passive-matrix display.

Active Server Pages (ASP) A scripting environment in which users combine *HTML*, scripts, and reusable Microsoft ActiveX server components to create dynamic Web pages.

active topology A network topology in which each node on the network is responsible for retransmitting the token or the data to other nodes.

algorithm A set of specific, sequential steps that describe in natural language exactly what the computer program must do to complete its task.

alphabetic check Confirms that only textual characters are entered in a database field.

alphanumeric pagers Devices that display numbers and text messages on their screens. Like numeric pagers, alphanumeric pagers do not allow the user to send messages.

Alt key One of the function keys used with other keys for shortcuts and special tasks.

analog-to-digital converter chip Converts analog signals into digital signals.

antivirus software Software that is specifically designed to detect viruses and protect a computer and files from harm.

applet A small program designed to be executed from within another application. Java applets are often run on your computer by your browser through the Java Virtual Machine (an application built into current browsers).

application program interfaces (APIs) Blocks of code in the operating system itself that software applications need in order to interact with it.

application server Acts as a repository for application software.

application software The set of programs on a computer that help a user carry out tasks tasks such as word processing, sending e-mail, balancing a budget, creating presentations, editing photos, taking an online course, and playing games.

arithmetic logic unit (ALU) Part of the CPU that designed to perform mathematical operations like addition, subtraction, multiplication, and division and to perform comparison operations such as greater than, less than, or equal to.

artificial intelligence (AI) The science that attempts to produce computers that display the same type of reasoning and intelligence that humans do.

ASCII (American Standard Code for Information Interchange) code A format for representing each letter or character as an 8-bit (or 1-byte) binary code.

Assembly languages Languages that allow programmers to write their programs using a set of short, English-like commands that speak directly to the CPU and give the programmer very direct control of hardware resources.

Asymmetrical Digital Subscriber Line (ADSL) A typical DSL transmission that downloads (or receives) data from the Internet faster than it uploads (or sends) data.

authentication The process of a user typing in his or her login name and password to gain access to a computer or network.

authentication servers Servers that keep track of who is logging on to the network and which services on the network are available to each user.

B

Back button A button on a Web browser used to return to a Web page viewed previously.

backdoor program A program that allows a hacker to take complete control of a computer without the user's knowledge or permission.

backup utility A utility that creates a duplicate copy of selected data on the hard disk and copies it to another storage device.

bandwidth (also called **throughput** or **data transfer rate**) The amount of data that can be transmitted across a transmission medium in a certain amount of time.

base class The original object class.

base 2 number system A number system that uses two digits, 0 and 1, to represent any value. Also called the binary number system.

base 10 number system (or **decimal notation**) A number system that uses 10 digits, 0 through 9, to represent any value.

base transceiver station A large communications tower with antennas, amplifiers, and receivers/transmitters.

basic input/output system (**BIOS**) A program that manages the data between the operating system and all the input and output devices attached to the system. BIOS is also responsible for loading the OS from its permanent location on the hard drive to RAM.

bastion host A heavily secured server located on a special perimeter network between the company's secure internal network and the firewall.

batch processing Accumulating transaction data until a certain point is reached, then processing those transactions all at once.

569

GLOSSARY

benchmarking A process used to measure performance in which two devices or systems run the same task and the times are compared.

beta versions Early versions of software programs that are still under development. Beta versions are usually provided free of charge in return for user feedback.

binary decisions Decision points that can only be answered in one of two ways: yes (true) or no (false).

binary digit (or **bit**) A digit that corresponds to the on and off states of a computer's switches. A bit contains either a value of 0 or 1.

binary language The language computers use to process information, consisting of only the values 0 and 1.

binary number system The number system used by computers to represent all data. Because it only includes two digits (0 and 1), the binary number system is also referred to as the base 2 number system.

biometric access devices Devices that use some unique characteristic of human biology to identify authorized users.

bistable display A display that has the ability to retain its image even when the power is turned off.

bit depth The number of bits the video card uses to store data about each pixel on the monitor.

black-hat hackers Hackers who use their knowledge to destroy information or for illegal gain.

BlueBoard An application developed by Colligo Networks that enables a PDA display to be used as a drawing board and instantly connected with up to four other PDA users.

Bluetooth Technology that uses radio waves to transmit data over short distances.

Bookmark A feature in the Netscape browser that places a marker of a Web site's URL in an easily retrievable list in the browser's toolbar (called Favorites in Internet Explorer).

Boolean operators Words used to refine your searches. These words—AND, NOT, and OR—describe the relationships between keywords in a search.

boot process (or **start-up process**) Process for loading the operating system into RAM when the computer is turned on.

boot-sector viruses Viruses that replicate themselves into the master boot record of a floppy or hard drive.

breadcrumb list A list that shows the hierarchy of Web pages above the Web page that you are currently visiting. Shown at the top of a Web page, it provides an aid to Web site navigation.

bridges Network devices that are used to send data between two different LANs or two segments of the same LAN.

broadband connections High speed Internet connections including include cable, satellite, and various high-speed uses of traditional phone wires, such as DSL and Integrated Services Digital Network (ISDN).

browsing (1) Viewing database records. (2) "Surfing" the Web.

brute force attacks Attacks delivered by specialized hacking software that attempts to try many combinations of letters, numbers, and pieces of a user ID in an attempt to discover a user password.

Buddy List A list of contacts for instant messaging. To communicate with someone on a Buddy List, both people must be online at the same time.

bus A group of electrical pathways inside a computer that provide high speed communications between various parts of a computer and the CPU and main memory.

business-to-business (B2B) E-commerce transactions between businesses.

business-to-consumer (B2C) E-commerce transactions between businesses and consumers.

byte Eight binary digits (or bits).

C

C The predecessor language of C++, it was developed originally for system programmers by Brian Kernighan and Dennis Ritchie of AT&T Bell Laboratories in 1978. It provides higher-level programming language features (like if statements and for loops) but still allows programmers to manipulate the system memory and CPU registers directly.

C++ The successor language to C, devloped by Bjarne Stroustrup. It used all of the same symbols and keywords as C, but extended the language with additional keywords, better security, and more support for the reuse of existing code through object-oriented design.

cable modem Modulates and demodulates the cable signal into digital data and back again. The cable TV signal and Internet data can share the same line.

cache memory Small blocks of memory located directly on and next to the CPU chip itself that act as holding places for recently or frequently used instructions or data that the CPU needs the most. When these instructions or data are stored in cache memory, the CPU can more quickly retrieve them than if it had to access the instructions or data from RAM.

Carrier Sense, Multiple Access with Collision Detection (CSMA/CD). With CSMA/CD, a node connected to the network listens (that is, has *carrier sense*) to determine that no other nodes are currently transmitting data signals. If the node doesn't hear any other signals, it assumes it is safe to transmit data. All devices on the network have the same right (that is, they have *multiple access*) to transmit data when they deem it safe. It is therefore possible for two devices to begin transmitting data signals at the same time. If this happens, the two signals collide. When signals collide, a node on the network alerts the other nodes.

cathode ray tube (CRT) A picture tube device in a computer monitor; very similar to the picture tube in a conventional television set.

CD-R (Compact Disk-Recordable) disc A portable, optical storage device that can be written to once and can be used with either a CD-R drive or a CD-RW drive.

CD-R drive A drive for reading CD-R discs.

CD-ROM A portable, read-only optical storage device.

CD-ROM drive A drive for reading compact discs (CDs).

CD-RW (Compact Disk-Read/Writable) disc A portable, optical storage device that can be written and rewritten to many times.

CD-RW drive A drive that can both read and write data to CDs.

cells Individual boxes formed by the columns and rows in a spreadsheet. Each cell can be uniquely identified according to its column and row position.

cell phones Telephones that operate over a wireless network. Cell phones can also offer offer Internet access, text messaging, personal information management (PIM) features, and more.

central processing unit (CPU) The part of the system unit that is responsible for data processing (or the "brains" of the computer) it is the largest and most important chip in the computer. It controls all the functions performed by the computer's other components and processes all the commands issued to it by software instructions.

centralized A type of network design where users are not responsible for creating their own data backups nor for providing security for their computers; instead those tasks are handled by a centralized server and software.

chart A graphical representation of numerical values.

chat room An area on the Web where many people come together to communicate online. The conversations are in real time and are visible to everyone in the chat room.

GLOSSARY

cgi-bin A directory where CGI scripts are normally placed.

CGI scripts Computer programs that conform to the Common Gateway Interface (CGI) specification which provides a method for sending data between end-users (using browsers) and Web servers.

circuit switching A method of communication in which a dedicated connection is formed between two points (two people on telephones) and the connection remains active for the duration of the transmission.

classes Categories of objects that define the common properties of all objects belonging to the class.

click-and-brick businesses Traditional stores that have an online presence.

clickstream data Information captured about each click that users make as they navigate through a Web site.

client A computer that requests information (such as your computer when you are connected to the Internet).

client-based e-mail E-mail that is dependent on an e-mail account provided by an ISP and a client software program, such as Microsoft's Outlook or Eudora.

client-side application A computer program that runs on the client and requires no interaction with a Web server.

client/server network A network that consists of client and server computers, in which the clients make requests of the server and the server "serves up" the response.

clip art A precreated gallery of images that is often included with software packages.

clock cycle The "ticks" of the system clock. One cycle equals one "tick."

clock speed The steady and constant pace at which a computer goes through machine cycles, measured in hertz (Hz).

coaxial cable A single copper wire surrounded by layers of plastic insulation and sheathing.

code editing The step in which programmers actually type the code into the computer.

coding Translating an algorithm into a programming language.

cold boot A complete power down and restart of a computer.

command-driven interface One in which the user enters commands in order to communicate with the computer system.

commerce servers Computers that host software that allows consumers to purchase goods and services over the Web. These servers generally use special security protocols to protect sensitive information (such as credit card numbers) from being intercepted.

Common Gateway Interface (CGI) Provides a methodology by which a browser can request that a program file be executed (or run) instead of just being delivered to the browser.

communications server Handles all communications between the network and other networks, including managing Internet connectivity.

Compact Flash Cards about the size of a matchbook that can hold between 64 MB and 1 GB of data.

compilation The process by which code is converted into machine language, the language the CPU can understand.

compiler The program that understands both the syntax of the programming language and the exact structure of the CPU and its machine language. It can "read" the source code and translate the source code directly into machine language.

completeness check Ensures that all fields defined as "required" have data entered into them.

computed field (or **computational field**) A numeric field in a database that is filled as the result of a computation.

computer A data processing device that gathers, processes, outputs, and stores data and information.

computer-aided design (CAD) 3-D modeling programs to create automated designs, technical drawings, and model visualizations.

computer fluent A person who understands the capabilities and limitations of computers, and knows how to use them to accomplish tasks.

computer forensics The application of computer systems and techniques to gather potential legal evidence; a law-enforcement specialty used to fight high-tech crime.

computer network Two or more computers that are connected together via software and communications media so they can communicate.

computer protocol A set of rules for accomplishing electronic information exchange. If the Internet is the information superhighway, then protocols are the driving rules.

computer virus A computer program that attaches itself to another computer program (known as the host program) or pretends to be an innocuous program and attempts to spread itself to other computers when files are exchanged.

connection-oriented protocol Requires two computers to exchange control packets, which set up the parameters of the data exchange session, prior to sending packets that contain data. This process is referred to as handshaking.

connectivity port A port that allows the computer (or other device) to be connected to other devices or systems such as networks, modems and the Internet.

consistency check Comparing the value of data in a field against established parameters to see if the value is reasonable.

consumer-to-consumer (C2C) E-commerce transactions between consumers through online sites such as eBay.com.

control (Ctrl) key One of the function keys that is used in combination with other keys to perform shortcuts and special tasks.

control unit Controls the switches inside the CPU.

cookies Small text files that some Web sites automatically store on a computer's hard drive when a user visits the site.

copyright violation When one person uses another person's material for their own personal *economic* benefit, or when someone diminishes the economic benefits of the originator.

course management software Programs that provide traditional classroom tools such as calendars and grade books over the Internet, as well as areas for students to exchange ideas and information in chat rooms, discussion forums, and e-mail.

CPU usage The percentage of time a CPU is working.

cradle Connects a PDA to a computer using either a USB port or a serial port.

cursor The flashing | symbol that indicates where the next character will be inserted.

custom installation Installing only those features of a software program that a user wants on the hard drive, thereby saving space on the hard drive.

customer relationship management (CRM) software A business program for storing sales and client contact information in one central database.

cybercrime Any criminal action perpetrated primarily through the use of a computer.

cybercriminals Individuals who use computers, networks, and the Internet to perpetrate crime.

cyberterrorists Terrorists who use computers to accomplish their goals.

D

data A number, a word, a picture, or sound that represents a fact or idea.

data collisions When two computers send data at the same time and the sets of data collide somewhere in the media.

data dictionary (or the **database schema**) Defines the name, data type, and length of each field in the database.

data inconsistency Differences in data that is caused when data exists in multiples lists, and not all lists are updated when a piece of data changes.

data integrity When data contained in a database is accurate and reliable.

data marts Small slices of a data warehouse.

data mining The process by which great amounts of data are analyzed and investigated in order to spot significant patterns or trends within the data that would otherwise not be obvious.

data redundancy When the same data exists in more than one place in a database.

data staging A three step process: extracting data from source databases, transforming (reformatting) the data, and storing the data in a data warehouse.

data transfer rate (or **throughput**) The speed at which a storage device transfers data to other computer components, expressed in kilobytes per second or megabytes per second.

data type (or **field type**) Indicates what type of data can be stored in the database field or memory location.

data warehouse A large-scale electronic repository of data that contains and organizes all the data related to an organization in one place.

database Electronic collections of related data that is organized and searchable.

database administrator (or **database designer**) An individual trained in the design, construction and maintenance of databases.

database management system (DBMS) Specially designed application software (such as Oracle or Microsoft Access) that interacts with the user, other applications, and the database to capture and analyze data.

database query A inquiry the user poses to the database so that it provides the data the user wishes to view.

database server Provides client computers with access to information stored in a database.

database software An electronic filing system best used for larger and more complicated groups of data that require more than one table, and where it's necessary to group, sort, and retrieve data, and to generate reports.

date fields Fields in a database that hold date data such as birthdays, due dates, and so on.

debugger A tool that helps programmers step through a program as it runs to locate runtime errors.

debugging The process of running the program over and over to find errors and to make sure the program behaves in the way it should.

decentralized A type of network where users are responsible for creating their own data backups and for providing security for their computers.

decision points Points at which a computer program must choose from an array of different actions based on the value of its current inputs.

decision support system (DSS) A system designed to help managers develop solutions for specific problems.

dedicated servers Used to fulfill one specific function (such as handling e-mail).

Deep Space Network Comprises three antenna installations located in California, Spain, and Australia. These installations provide for almost continuous transmission of data to outer space regardless of the orbital position of the Earth.

default values The values a database will use for fields unless the user enters another value.

denial-of-service (DoS) attack Occurs when legitimate users are denied access to a computer system because a hacker is repeatedly making requests of that computer system to tie up its resources and deny legitimate users access.

derived class A class created based on a previously existing class (i.e., base class). Derived classes inherit all of the member variables and methods of the base class from which they are derived.

desktop The first interaction a user has with the operating system and the first image seen on the monitor. As its name implies, your computer's desktop puts at the user's fingertips all of the elements necessary for a productive work session that are typically found on or near the top of a traditional desk, such as files and folders.

desktop box The most common style of system unit for desktop computers, which sit horizontally on top of a desk.

desktop publishing (DTP) software Programs for incorporating and arranging graphics and text to produce creative documents.

detail report A list showing the individual transactions that occurred during a certain time period.

device driver Software that facilitates the communication between a device and the operating system.

dial-up connection A connection to the Internet via a standard telephone line.

digital ink An extension of the text-entry systems used on PDA devices.

digital pen A device for drawing images and entering text in a tablet PC.

digital signal processor A specialized chip that processes digital information and transmits signals very quickly.

Digital Subscriber Line (DSL) A technology that uses telephone lines to connect to the Internet and provide higher throughput. DSL allows phone and data transmission to share the same telephone line.

Digital Subscriber Line (DSL) modem A modem that uses modulation techniques to separate the types of signals into voice and data signals so they can travel in the right "lane" on the twisted pair wiring. Voice data is sent at the lower speed, while digital data is sent at frequencies ranging from 128 Kbps to 1.5 megabits per second (Mbps).

digital video-editing software Programs for editing digital video.

directories Hierarchical structures that include *files*, *folders*, and *drives*, used to create a more organized and efficient computer.

Disk Cleanup A Windows utility that cleans unnecessary files off your hard drive.

disk defragmenter utilities Utilities that regroup related pieces of files together on the hard disk, allowing the faster retrieval of the data.

Disk Operating System (MS-DOS) A single-user, single-task operating system created by Microsoft. DOS was the first widely installed operating system in personal computers.

distributed denial of service (DDoS) attacks Atomated attacks that are launched from more than one zombie computer at the same time.

docking station Hardware for connecting a portable computing device to printers, scanners, full-size monitors, mice, and other peripherals.

documentation A description of the development and technical details of a computer program, including how the code works, and how the user interacts with the program.

domain name Part of the URL that typically follows www. Domain names consist of two parts: The first part indicates who the site's host is; the second part is a three-letter suffix that indicates the type of organization.

domain name server (DNS) Contains location information for domains on the Internet. Functions like a phone book for the Internet.

dot-matrix printer The first type of computer printer, which has tiny hammerlike keys that strike the paper through an inked ribbon.

dot pitch The diagonal distance, measured in millimeters, between pixels of the same color on the screen. A smaller dot pitch means that there is less blank space between pixels, and thus a sharper, clearer image.

GLOSSARY

dotted decimal number The numbers in an IP address.

double data rate synchronous DRAM (DDR SDRAM) Memory chips that are faster than SDRAM but not as fast as RDRAM.

drawing software (or **illustration software**) Programs for creating or editing two-dimensional line-based drawings.

drive bays Special shelves inside computers designed to hold storage devices.

DSL/cable routers Routers that specifically designed to connect to Digital Subscriber Line (DSL) or cable modems.

dual processor A design that has two separate CPU chips installed on the same system.

DVD drive A drive that allows the computer to read digital video discs (DVDs).

DVD-RW drive A drive that allows the computer to read and write to DVDs.

Dvorak keyboard A leading alternative keyboard that puts the most commonly used letters in the English language on "home keys," the keys in the middle row of the keyboard. It is designed to reduce the distance your fingers travel for most keystrokes, increasing typing speed.

dynamic addressing The process of assigning IP addresses when users log on to their Internet service provider (ISP). The computer is assigned an address from an available pool of IP addresses, which is common today.

Dynamic Host Configuration Protocol (DHCP) Handles dynamic addressing. Part of the TCP/IP protocol suite, DHCP takes a pool of IP addresses and shares them with hosts on the network on an as-needed basis.

dynamic RAM (DRAM) The most basic type of RAM; used in older systems or in systems where cost is an important factor. DRAM offers access times on the order of 60 ns.

E

editor A tool that helps programmers as they enter the code, highlighting keywords and alerting them to typos.

educational software Applications that offers some form of instruction or training.

edutainment Software that both educates and entertains the user.

electronic commerce (e-commerce) Conducting business online for purposes ranging from fund-raising to advertising to selling products.

electronic switches Devices inside the computer that can be flipped between these two states: 1 or 0, on or off.

elements In HTML, elements are the tags and the text between them.

e-mail (electronic mail) Internet-based communication in which senders and recipients correspond.

e-mail server Processes and delivers incoming and outgoing e-mail.

e-mail virus A virus tranmitted via e-mail that often uses the address book in the victim's e-mail system to distribute the virus.

encryption The process of encoding data (ciphering) so that only the person with a corresponding decryption key (the intended recipient) can decode (or decipher) and read the message.

entertainment software Programs designed to provide users with entertainment; computer games make up the vast majority of entertainment software.

ergonomics Refers to how a user sets up his or her computer and other equipment to minimize risk of injury or discomfort.

error handling In programming, the instructions that the program executes should do if the input data is incorrect or another error is encountered.

Ethernet networks Networks that use the Ethernet protocol as the means (or standard) by which the nodes on the network communicate.

Ethernet port A port that is slightly larger than a standard phone jack and transfers data up to 100 Mbps. It is used to connect a computer to a cable modem or a network.

event Every keystroke, every mouse click, and each signal to the printer creates an action, or event, in the respective device (keyboard, mouse, or printer) to which the operating system responds.

exception reports Reports that show conditions that are unusual or that need attention by users of a system.

executable program The binary sequence (code) that instructs the CPU to perform certain calculations.

expansion bus Expands the capabilities of a computer by allowing a range of different expansion cards (such as video cards and sound cards) to communicate with the motherboard.

expansion cards (or **adapter cards**) Circuit boards with specific functions that augment the computer's basic functions as well as providing connections to other devices.

expansion hub A device that connects to one port, such as a USB port, to provide additional new ports, similar to a multiplug extension cord for electrical appliances.

expert system Designed to replicate the decision-making processes of human experts to solve specific problems.

export Putting data into an electronic file in a format that another application can understand.

Extended Industry Standard Architecture (EISA) bus An older expansion bus for connecting devices such as the mouse, modem, and sound cards.

Extensible Hypertext Markup Language (XHTML) A new standard recently released by the World Wide Web Consortium (WC3) that combines elements from both XML and HTML. XHTML has much more stringent rules than HTML regarding tagging (for instance, all elements require an end tag).

Extensible Markup Language (XML) Allows designers to define their own tags, making it much easier to transfer data between Web sites and Web servers.

extension (or **file type**) The three letters that follow the filename after the dot (.) ; the extension identifies what kind of family of files the file belongs to or what application should be used to read the file.

extranets Pieces of intranets that only certain corporations or individuals can access. The owner of the extranet decides who will be permitted to access it.

F

Favorites A feature in Microsoft Internet Explorer that places a marker of a Web site's URL in an easily retrievable list in the browser's toolbar.

fiber-optic line (or **cable**) Lines that transmit data at close to the speed of light along glass or plastic fibers.

field Where a category of information in a database is stored. Fields are displayed in columns.

field constraints Properties that must be satisfied for an entry to be accepted into the field.

fifth-generation languages (5GLs) are considered the most "natural" of languages. With 5GLs, instructions closely resemble human speech or are visual in nature so that little programming knowledge is necessary.

file A collection of related pieces of information stored together for easy reference.

File Allocation Table (FAT) An index of all sector numbers that the drive stores in a table in order to keep track of which sectors hold which files.

file compression utility A program that takes out redundancies in a file to reduce the file size.

file management Providing organizational structure to the computer's contents.

GLOSSARY

file path Identifies the exact location of a file, starting with the drive in which the file is located, and including all folders, subfolders (if any), the filename, and extension.

financial planning software A program for managing finances, such as Intuit's Quicken and Microsoft's Money, which include electronic checkbook registers and automatic bill payment tools.

file servers Computers deployed to provide remote storage space or to act as a repository for files that users can access.

File Transfer Protocol (FTP) Used to upload and download files from one computer to another over the Internet.

filename The first part of a file, similar to our first names; it is generally the name a user assigns to the file when saving it.

firewalls Software programs or hardware devices designed to prevent unauthorized access to computers or networks.

FireWire port (previously called the **IEEE 1394 port**) A port based on a standard developed by the Institute of Electrical and Electronics Engineers (IEEE), with a transfer rate of 400 Mbps. Today, it is most commonly used to connect digital video devices such as digital cameras to the computer.

FireWire 800 One of the fastest ports available, moving data at 800 Mbps.

first-generation languages (1GLs) The actual machine languages of a CPU, the sequence of bits—1s and 0s—that the CPU understands.

flash memory Portable, nonvolatile memory.

flash memory card A form of portable storage. This removable memory card is often used in digital cameras, MP3 players, and PDAs.

floppy disk A portable 3.5 inch storage format, with a storage capacity of 1.44 MB.

floppy disk drive A drive bay for a floppy disk.

flowcharts A visual representation of the patterns an alogrithm comprises.

folder A collections of files.

footprint The amount of physical space on the desk a computer takes up.

For and **Next** Keywords in Visual Basic to implement a loop.

foreign key The primary key of another table that is included for purposes of establishing relationships with that other table.

format To design or change the appearance of a document by changing fonts, font styles, and sizes; adding colors to text; adjusting the margins; adding borders to portions of text or whole pages; inserting bulleted and numbered lists; organizing text into columns, etc.

formula An equation a spreadsheet user builds using addition, subtraction, multiplication, and division, as well as values and cell references.

Forward button A button on a Web browser to return to a Web page after going back using the Back button.

fourth-generation languages (4GLs) are nonprocedural: they specify what is to be accomplished without determining how. Many database query languages and report generators are 4GLs.

frames Containers designed to hold multiple data packets.

freeware Any copyrighted software that can be used for free.

frequently asked questions (FAQs) A list of answers to the most common questions.

full installation Installing all the files and programs from the distribution CD to the computer's hard drive.

function A formula that is preprogrammed into spreadsheet software.

function keys Act as shortcut keys to perform special tasks; they are sometimes referred to as the "F" keys because they start with the letter F followed by a number.

fuzzy logic Allows the interjection of experiential learning into the equation by considering probabilities.

G

gigabyte (GB) About a billion bytes.

gigahertz (GHz) One billion hertz.

Global Positioning System is a system of 21 satellites (plus three working spares), built and operated by the U.S. military, that constantly orbit the earth. They provide information to GPS capable devices to pinpoint locations on the Earth.

Graffiti One of the more popular notation systems for entering data into a PDA.

graphical user interface, or **GUI** (pronounced "gooey") Unlike the command- and menu-driven interfaces used earlier, GUIs display graphics and use the point-and-click technology of the mouse and cursor, making them much more user friendly.

graphics and multimedia software Programs to design and create attractive documents, images, illustrations, and Web pages, as well as three-dimensional models and drawings.

grid computing A form of networking that allows linked computers to use idle processors of other networked computers.

groupware Software that helps people who are in different locations work together via e-mail and online scheduling tools.

H

hacker (also called a **cracker**) Anyone who breaks into a computer system (whether an individual computer or a network) unlawfully.

handshaking The process of two computers exchanging control packets that set up the parameters of a data exchange.

hard disk drive (or just **hard drive**) Holds all permanently stored programs and data; is inside the system unit.

hardware Any part of the computer you can physically touch.

head crash A stoppage of the hard disk drive that often results in data loss.

hexadecimal notation A number system that uses 16 digits to represent numbers; also called a base 16 number system.

hibernation When a computer is in a state of deeper sleep. Pushing the power button awakens the computer from hibernation, at which time the computer reloads everything to the desktop exactly as it was before it went into hibernation.

high-level languages Third-, fourth-, and fifth-generation languages.

historical data Data that shows trends over time.

History list A feature on a browser's toolbar that shows all the Web sites and pages visited over a certain period of time.

hits A list of sites (or results) that match an Internet search.

home page The main or opening page of a Web site.

home phoneline network adapter (also called an **HPNA adapter**) Attaches to computers and peripherals on a phoneline network to enable them to communicate via phone lines.

host Organization that maintains the Web server on which a particular Web site is stored.

hot-swappable bays Provide the ability to remove one drive and exchange it with another drive while the laptop computer is running.

HTML embedded scripting language Used to embed programming language code directly within the HTML code of a Web page.

hubs are simple amplification devices that receive data packets and retransmit them to all nodes on the same network (not between different networks).

hyperlink fields Fields in a database that store hyperlinks to Web pages.

hyperlinks Specially coded text that lets a user jump from one location, or Web page, to another within the Web site or to another Web site altogether by clicking on it.

GLOSSARY

hypertext Text that is linked to other documents or media (such as video clips, pictures, and so on).

Hypertext Markup Language (HTML) A set of rules for marking up blocks of text so that a browser knows how to display them. It utilizes a series of tags that define the display of text on a Web page.

Hypertext Transfer Protocol (HTTP) The protocol a browser uses to send requests to a Web server; created especially for the transfer of hypertext documents across the Internet.

I

icons Pictures that represent an object such as a software application or a file or folder.

identity theft Occurs when someone uses personal information about someone else (such as their name, address, and social security number) to assume their identity for the purpose of defrauding others.

if else Keywords in the programming language C++; used for binary decisions.

image-editing software (sometimes called **photo-editing software**) Programs for editing photographs and other images.

impact printers Printers that have tiny hammerlike keys that strike the paper through an inked ribbon, thus making a mark on the paper. The most common impact printer is the dot-matrix printer.

import To bring data back into an application from another source.

indexer A program in a search engine that organizes the information a spider collects on the Internet into a large database.

Industry Standard Architecture (ISA) bus An older expansion bus for connecting devices such as the mouse, modem, and sound cards.

information Data that has been organized or presented in a meaningful fashion.

information system A system that includes data, people, procedures, hardware, and software and is used to gather and analyze information.

inheritance The ability of a new class of object to automatically pick up all of the data and methods of an existing class, and then extend and customize those to fit its own specific needs.

initial value A beginning point in a loop.

inkjet printer A nonimpact printer that sprays tiny drops of ink onto paper.

input device Hardware device used to enter, or input, data (text, images, and sounds) and instructions (user responses and commands) into a computer; input devices include a keyboard, mouse, scanner, microphone, and digital camera.

input form Provides a view of the data fields to be filled, with appropriate labels to assist database users in populating the database.

instant messaging (IM) services Programs that enable users to communicate in real time with friends who are also online.

instructions The steps and tasks the computer needs to process data into usable information.

instruction set The collection of commands a specific CPU can execute.

integrated circuits (or **chips**) Very small regions of semiconductor material, such as silicon, that support a huge number of transistors.

integrated development environment (IDE) A developmental tool that helps programmers write, compile, and test their programs.

Integrated Services Digital Network (ISDN) A high-speed digital technology for transmitting and receiving digital data at speeds up to 128 Kbps.

integrated software application A single software program that incorporates the most commonly used tools of many productivity software programs into one integrated stand-alone program.

Internet A network of networks and the largest network in the world, connecting millions of computers from more than 65 countries.

Internet2 An ongoing project sponsored by more than 200 universities (supported by government and industry partners) to develop new Internet technologies and disseminate them as rapidly as possible to the rest of the Internet community. The Internet2 backbone supports extremely high-speed communications (up to 9.6 Gbps).

Internet backbone The main pathway of high-speed communications lines through which all Internet traffic flows.

Internet Explorer (IE) A popular graphical browser from Microsoft Corporation for displaying different Web sites, or locations, on the Web; it can display pictures (graphics) in addition to text, as well as other forms of multimedia, such as sound and video.

Internet hoaxes E-mails that contain information that is untrue.

Internet Protocol (IP) A protocol for sending data between computers on the Internet.

Internet Protocol version 4 (IPv4) The original IP addressing scheme.

Internet Protocol version 6 (IPv6) A proposed IP addressing scheme that makes IP addresses longer, thereby proving more available IP addresses. It uses eight groups of 16-bit numbers.

Internet service providers (ISPs) National, regional, or local companies that connect individuals, groups, and other companies to the Internet.

interplanetary Internet A future network that would span the planets in our solar system.

interpreter Translates source code into an intermediate form, line by line. Each line is then executed as it is translated.

interrupt A signal that tells the operating system that it is in need of immediate attention.

interrupt handler A special numerical code that prioritizes requests to the operating system.

interrupt table A place in the computer's primary memory (or random access memory, RAM) where interrupt requests are placed.

intranet A private corporate network that is used exclusively by company employees to facilitate information sharing, database access, group scheduling, videoconferencing, or other employee collaboration.

IP (Internet Protocol) address The means by which all computers connected to the Internet identify each other. It consists of a unique set of four numbers separated by dots, such as 123.45.678.91.

IrDA port A port based on a standard developed by the Infrared Data Association for transmitting data. IrDA ports allow the user to transmit data between two devices via infrared light waves, similar to a TV remote control. IrDA ports have a maximum throughput of 4 Mbps and require that a line of sight be maintained between the two ports.

ISDN modem Sends and receives the digital data to and from a computer over a traditional phone line. Because this modem is already working only with digital data, it doesn't need to translate analog data to digital data like a traditional modem. Instead, it is really a "terminal adapter" as it translates information between different devices that connect to the line.

J

jam signal A special signal sent to all network nodes, alerting them that a collision has occurred.

Java A platform-independent programming language that Sun Microsystems introduced in the early 1990s. It quickly became popular because its object-oriented model allows Java programmers to benefit from its set of existing classes.

Java applets Small Java-based programs.

576 GLOSSARY

JSP (Java Server Pages)   An extension of the Java servlet technology with dynamic scripting capability.

JavaScript   A programming language often used to add interactivity to Web pages. JavaScript is not as full-featured as Java, but its syntax, keywords, data types, and operators are a subset of Java's.

join query   A query that links (or joins) two tables using a common field in both tables and extracts the relevant data from each.

K

kernel (or **supervisor program**)   The essential component of the operating system, responsible for managing the processor and all other components of the computer system. Because it stays in RAM the entire time your computer is powered on, the kernel is called *memory resident*.

kernel memory   The memory that the computer's operating system uses.

key pair   A public and a private key used for coding and decoding messages.

keyboard   Used to enter typed data and commands into a computer.

keywords   (1) Specific words a user wishes to query (or look for) in an Internet search. (2) The set of specific words that have predefined meanings for a particular programming language.

kilobyte (KB)   Approximately 1,000 bytes.

knowledge-based system   Provides additional intelligence that supplements the user's own intellect and makes the decision support system (DSS) more effective.

L

label   Descriptive text that identifies the components of the worksheet in a spreadsheet.

laptop computer (or **notebook computer**)   A portable computer that offers offer a large display and all of the computing power of a full desktop system.

Large Scale Networking (LSN)   A program created by the U.S. government, the objective of which is to fund the research and development of cutting-edge networking technologies. Major goals of the program are the development of enhanced wireless technologies and increased network throughput.

laser printer   A nonimpact printer known for quick and quiet production and high-quality printouts; it uses laser beams to make marks on paper.

latency (or **rotational delay**)   Occurs after the read/write head locates the correct track, then waits for the correct sector to spin to the read/write head.

LCD (liquid crystal display)   Technology used in flat-panel computer monitors.

Level 1 cache   A block of memory that is built onto the CPU chip for the storage of data or commands that have just been used.

Level 2 cache   Located on the CPU chip itself but slightly farther away from the CPU; or it's on a separate chip next to the CPU and therefore takes somewhat longer to access. Level 2 cache contains more storage area than Level 1 cache.

Level 3 cache   On computers with Level 3 cache, the CPU checks this area for instructions and data after it looks in Level 1 and Level 2 cache, but before it makes the longer trip to RAM. The Level 3 cache holds between 2 MB and 4 MB of data

Linux   An open-source operating system based on UNIX. Because of the stable nature of this operating system, it is often used on Web servers.

local area networks (LANs)   Networks in which the nodes are located within a small geographic area.

local buses   Located on the motherboard, these buses run between the CPU and the main system memory.

logic bombs   Computer viruses that execute when a certain set of conditions is met, such as specific dates keyed off of the computer's internal clock.

logical ports   Virtual communications gateways or paths that allow a computer to organize requests for information (such as Web page downloads, e-mail routing, and so on) from other networks or computers.

logical port blocking   When a firewall is configured to ignore all incoming packets that request access to port 25 (the port designated for FTP traffic), so no FTP requests will get through to the computer.

loop   An algorithm that asks a question is asked and if the answer is yes, a set of actions is performed. Once the set of actions has been performed, the question is asked again (creating a loop). As long as the answer to the question is yes, the algorithm will continue to loop around and repeat the set of actions. When the answer to the question is no, the algorithm breaks free of the looping and moves on to the next step.

M

MAC (media access control) address   A physical address similar to a serial number on an appliance that is assigned to each network adapter; it is made up of six two-digit numbers such as 01:40:87:44:79:A5.

Mac OS   Apple Computer's operating system. In 1984, Mac OS became the first operating system to incorporate the user-friendly "point-and-click" technology in a commercially affordable computer. The most recent version of the Mac operating system, Mac OS X, is based on the UNIX operating system. Previous Mac operating systems had been based on their own proprietary program.

machine cycle (or **proccessing cycle**)   The steps a CPU follows to perform its tasks.

machine language   Long strings of binary code used by the control unit to set up the hardware in the CPU for the rest of the operations it needs to perform.

Macromedia Flash   A software product from Macromedia for developing Web-based multimedia.

macros   Small programs that group a series of commands to run as a single command.

magnetic card readers   Devices that read information from a magnetic strip on the back of a credit card-like access card (such as a student ID card). The card reader, which can control the lock on a door, is programmed to admit only authorized personnel to the area.

magnetic media   Portable storage devices that use a magnetized film to store data, such as floppies and Zip disks.

mainframes   Large, expensive computers that support hundreds or thousands of users simultaneously.

management information system (MIS)   A system that provides timely and accurate information that enables managers to make critical business decisions.

mapping programs   Software that provides street maps and written directions to locations nationwide.

master boot record   A program that executes whenever a computer boots up.

megabyte (MB)   About a million bytes.

megahertz (MHz)   One million hertz; hertz is the unit of measure for processor speed, or "machine cycles per second."

memo fields   Text fields in a database that are used to hold long pieces of text.

memory   A component inside the system unit that helps process data into information; memory chips hold (or store) the instructions or data that the CPU processes.

memory bound   A system that is limited in how fast it can send data to the CPU because there's not enough RAM installed.

memory card reader   An external device for reading flash memory cards.

memory effect   A battery must be completely used up before it is recharged or it won't hold as much charge as it originally did.

GLOSSARY

Memory Stick Sony's brand of flash memory; the cards measure just 2 inches x 1 inch and weighing a fraction of an ounce.

menus Lists of commands that appear on the screen.

menu-driven interface One in which the user chooses a command from menus displayed on the screen.

meta search engine A search engine that searches other search engines rather than individual Web sites.

metadata Data that describes other data.

methods (or **behaviors**) Actions associated with a class of objects.

metropolitan area networks (MANs) WANs that link users in a specific geographic area (such as within a city or county).

microphone A device for capturing sound waves (such as voice) and transferring them to digital format on a computer.

MIDI port A port for connecting electronic musical instruments (such as synthesizers) to a computer.

microbrowser Software that makes it possible to access the Internet from a cell phone or PDA.

microprocessors Chips that contain a CPU.

Microsoft Visual Basic (VB) A powerful programming language used to build a wide range of Windows applications. VB's strengths include a simple, quick interface that is easy for a programmer to learn and use. It has grown from its roots in the language BASIC to become a sophisticated and full-featured object-oriented language.

Microsoft Windows The most popular operating system for desktop computers.

mobile switching center A central location that receives cell phone requests for service from a base station.

model management system Software that assists in building management models in DSSs.

modem A device that converts (*mo*dulates) the digital signals the computer understands to the analog signals that can travel over phone lines.

modem card Provides the computer with a connection to the Internet.

modem port A port that uses a traditional telephone signal to connect two computers.

monitor (or **display**) A common output device that displays text, graphics, and video as "soft copies" (copies that can only be seen onscreen).

mobile computing devices Portable electronic tools such as cell phones, personal digital assistants (PDAs), and laptops.

Moore's Law A mathematical rule, named after Gordon Moore, the cofounder of the CPU chip manufacturer Intel, which predicts that the number of transistors inside a CPU will increase so fast that CPU capacity will double every 18 months.

motherboard A special circuit board in the system unit that contains the CPU, the memory (RAM) chips, and the slots available for expansion cards. where the CPU and memory are located. It is the largest printed circuit board; all of the other boards (video cards, sound cards, and so on) connect to it to receive power and to communicate.

mouse Used to enter user responses and commands into a computer.

MP3 player A small portable device for storing MP3 files (digital music).

MS Transcriber A notation system for PDAs that doesn't require special strokes and can recognize both printed and cursive writing with fairly decent accuracy.

Multimedia Message Service (MMS) An extension of Short Message Service (SMS0 that enables messages that include text, sound, images, and video clips to be sent to other phones or e-mail addresses.

multimedia cards (**MMCs**) Thin, small, rugged cards used as portable memory that can hold up to 128 MB of data.

Multipurpose Internet Mail Extensions (MIME) A specification that was introduced in 1991 to simplify attachments to e-mail messages. All e-mail client software now uses this protocol for attaching files. but the e-mail client using the MIME protocol now handles the encoding and decoding for the users.

multitasking When the operating system allows a user to perform more than one task at a time.

multiuser operating system (also known as a **network operating system**) Enables more than one user to access the computer system at one time by efficiently juggling all the requests from multiple users.

#

nanoscience The study of molecules and nanostructures whose size ranges from 1 to 100 nanometers.

nanotechnology The science revolving around the use of nanostructures to build devices on an extremely small scale.

Napster The most well-known file-exchange site for digital music. Peer-to-peer sharing technology allowed users to exchange files, rather than downloading from a public server. Napster was purchased, shutdown, and then reopened for downloading music that a user purchases, in compliance with copyright provisions.

natural language processing (NLP) system A system that enables users to communicate with computer systems using a natural spoken or written language as opposed to using computer programming languages.

negative acknowledgment (nack) What computer Y sends to computer X if the packet is unreadable, indicating the packet was not received in understandable form.

netiquette General rules of etiquette for Internet chat rooms and other online forums.

Netscape Navigator A popular graphical browser from Netscape Communications for displaying different **Web sites,** or locations, on the Web; it can display pictures (graphics) in addition to text, as well as other forms of multimedia, such as sound and video.

network A group of two or more computers (or nodes) that are configured to share information and resources such as printers, files, and databases.

network access points (NAPs) The points of connection between ISPs.

network adapters Allow the computer (or peripheral) to communicate with the network using a common data communication language, or protocol.

network address translation (NAT) A process firewalls use to assign internal IP addresses on a network.

network administrator Someone who has training in computer and peripheral maintenance and repair, networking design, and the installation of networking software.

network architecture The design of a network.

network interface card (NIC) An expansion (or adapter) card that enables a computer to connect with a network.

network navigation devices Devices on a network such as routers, hubs, and switches that move data signals around the network.

network operating system (NOS) Software handle requests for information, Internet access, and the use of peripherals for the rest of the network nodes.

network topology The layout and structure of the network.

New Technology File System (NTFS) A file system in Windows XP, which differs from FAT. NTFS was developed with the Windows NT version and has been used in Windows 2000 and Windows XP.

newsgroup (or **discussion group**) An online discussion forum in which people "post" messages and read and reply to messages from other members of the newsgroup.

nodes Devices connected to a network, such as a computer, a peripheral (such as a printer), or a communications device (such as a modem).

578 GLOSSARY

nonimpact printers Printers that spray ink or use laser beams to make marks on the paper. The most common nonimpact printers are inkjet and laser printers.

nonvolatile storage Permanent storage, as in ROM.

normalization The process of recording data only once to reduce data redundancy.

number system An organized plan for representing a number.

numeric check Confirms that only numbers are entered in the field.

numeric fields Fields in a database that store numbers.

numeric keypad Section of a keyboard that allows a user to enter numbers quickly.

numeric pagers Devices that display only numbers on their screens, telling the user that he or she have received a page and providing the number to call. Numeric pagers do not allow the user to send a response.

O

object A class. Each object in a specific class shares similar data and methods with the other objects in the class.

object fields Fields in a database that hold objects such as pictures, video clips, or entire documents.

object-oriented analysis The process when programmers first identify all of the categories of inputs that are part of the problem the program is trying to solve.

object-oriented database A database that stores data in objects, not in tables.

object-relational database A hybrid between a relational and an object-oriented database. It is based primarily on the relational database model, but it is better able to store and manipulate unstructured data such as audio and video clips.

octet The four numbers in the dotted decimal notation of an IP address.

office support system (OSS) A system designed to assist employees in accomplishing their day-to-day tasks and to improve communications, such as Microsoft Office.

online service providers (OSPs) Internet access providers such as America Online (AOL) that have their own proprietary online content, and often offer special services and areas that only their subscribers can access.

online transaction processing (OLTP) The immediate processing of user requests or transactions.

open-source program A program that is available for developers to use or modify as they wish; it is typically free of charge.

open systems Systems whose designs are public for access by any interested party.

operating system (OS) System software that controls the way in which a computer system functions, including the management of hardware, peripherals, and software.

operators The coding symbols that represent the fundamental actions of the language.

optical media Portable storage devices that use a laser to read and write data, such as CDs and DVDs.

optical mouse Instead of a rollerball, an optical mouse uses an internal sensor or laser to control the mouse's movement. The sensor sends signals to the computer, telling it where to move the pointer on the screen.

organic light-emitting displays (OLEDs) These displays, currently used in some Kodak cameras, use organic compounds that produce light when exposed to an electric current.

output device A device that sends processed data and information out of a computer in the form of text, pictures (graphics), sounds (audio), and video.

P

packets Small segments of data that is bundled in order to be sent over transmission media. Each packet contains the address of the computer or peripheral device to which it is being sent.

packet filtering A process firewalls perform to filter out packets sent to specific logical ports.

packet screening Involves examining incoming data packets to ensure they were originated by or are authorized by valid users on the internal network.

packet sniffer A program that looks at (or sniffs) each packet as it travels on the Internet.

packet switching A communications methodology in which data is broken into smaller chunks (called packets) and sent over various routes at the same time. When the packets reach their destination, they are reassembled by the receiving computer.

page file The file the operatings system build on the hard drive when it is using virtual memory to allow processing to continue.

paging If the data and/or instructions that have been placed in the swap file are needed later, the operating system swaps them back into active RAM and replaces them in the hard drive's swap file with less active data or instructions.

paging device (or a **pager**) A small wireless device that allows a user to receive and sometimes send numeric (and sometimes text) messages on a small display screen.

painting software Programs for modifying photographs.

Palm OS One of the two main operating systems for PDAs, made by 3Com.

parallel port A port that sends data between devices in groups of bits at speeds of 92 Kbpst. Parallel ports are often used to connect printers to computers.

parallel processing A networke computer environment in which each computer work on a portion of the same problem simultaneously.

Parcel Transfer Protocol (PTP) A new transmission protocol under development by a team of scientists and computer professionals. The protocol must be designed to keep running even if packets are lost in transmission and to block out noise that can be picked up while data is traversing millions of miles. PTP stores data at the receiver until all data is received and accounted for, then transmits it to the proper destination.

Pascal The only modern language that was specifically designed as a teaching language; it is no longer often taught at the college level.

passive-matrix displays With a monitor that uses passive-matrix technology, electrical current passes through the liquid crystal solution and charges groups of pixels, either in a row or a column. This causes the screen to brighten with each pass of electrical current and subsequently fade.

passive topology When data merely travels the entire length of the medium and is received by all network devices.

path (or **subdirectory**) The information following the slash in a URL.

path separators The backslash marks (\) used by Windows and DOS. Mac files use a colon (:) and UNIX and Linux use the forward slash (/) as the path separator.

PC cards (sometimes called **PCMCIA cards**, for **Personal Computer Memory Card International Association**) Provide the ability to add fax modems, network connections, wireless adapters, USB 2.0 and FireWire ports, and other capabilities primarily to a laptop.

peer-to-peer (P2P) network A network in which each node connected to the network can communicate directly with every other node on the network.

peer-to-peer (P2P) sharing The process of users transferring files between computers.

Peripheral Component Interconnect (PCI) buses Expansion buses that connect directly to the CPU and support such devices as network cards and sound cards. They have been the standard bus for much of the past decade and continue to be redesigned to increase their performance.

peripheral devices Devices such as monitors, printers, and keyboards that connect to the system unit through ports.

GLOSSARY

personal area networks (PANs) are used to connect wireless devices (such as Bluetooth-enabled devices) in close proximity to each other.

personal digital assistant (PDA) A small device that allows a user to carry digital information. Often called palm computers or handhelds, PDAs are about the size of a hand and usually weigh less than 5 ounces.

personal firewalls Firewalls specifically designed for home networks.

personal information manager (PIM) software Programs such as Microsoft Outlook or Lotus Organizer. that strive to replace the various management tools found on a traditional desk, such as a calendar, address book, notepad, and to-do lists.

phoneline networks Networks that use conveetional phonelines to connect the nodes in a network.

physical memory The amount of RAM that is actually sitting on memory modules in a computer.

pipelining A technique that allows the CPU to work on more than one instruction (or stage of processing) at a time, thereby boosting CPU performance.

pixels Illuminated, tiny dots that create the images you seen on a computer monitor. Pixels are illuminated by an electron beam that passes back and forth across the back of the screen very quickly—60 to 75 times a second—so that the pixels appear to glow continuously.

plagiarism When someone uses someone else's ideas or words and represent them as their own.

platform The combination of a computer's operating system and processor. The two most common platform types are the PC and the Apple Macintosh.

plotters Large printers that use a computer-controlled pen to produce oversize pictures that require precise continuous lines to be drawn, such as in maps or architectural plans.

plug-and-play Technology that allows the operating system, once the system is booted up, to automatically recognize any new peripherals and configure them to work with the system.

plug-in (or **player**) A small software program that "plugs in" to a Web browser to enable a specific funtion; for example, to view and hear some multimedia files on the Web.

Pocket PC (formerly Windows CE) One of the two main operating systems for PDAs, made by Microsoft.

point of presence (POP) A bank of modems through which many users can connect to an ISP simultaneously.

pointer The I-beam or arrow that appears on the computer screen.

port An interface through which external devices are connected to the computer.

portability The ability to move a completed solution easily from one type of computer to another.

portal A subject directory on the Internet that is part of a larger Web site that focuses on offering its visitors a variety of information, such as the weather, news, sports, and shopping guides.

positive acknowledgment (**ack**) What computer Y sends when it when it receives a data packet that it can read from computer X.

postcardware (or **e-mailware**) While not charging a fee, some developers release free software and request that users mail them a postcard or send them an e-mail message to thank them for their time in developing the software and to give them their opinion of it.

powerline networks Networks that use the electrical wiring in a home to connect the nodes in the network.

powerline network adapter Attached to each computer or peripheral that is part of a powerline network.

power-on self test (POST) The first job BIOS performs, ensuring that essential peripheral devices are attached and operational. This process consists of a test on the video card and video memory, a BIOS identification process (during which the BIOS version, manufacturer, and data are displayed on the monitor), and a memory test to ensure memory chips are working properly.

power supply Used to regulate the wall voltage to the voltages required by computer chips; it is housed inside the system unit.

presentation software An application program for creating dynamic slide shows, such as Microsoft PowerPoint or Corel Presentations.

Pretty Good Privacy (PGP) A public-key package.

primary key (or a **key field**) The unique field that each database record must have.

print server Manages all client-requested printing jobs for all printers on the network.

printer A common output device that creates tangible or hard copies of text and graphics.

private-key encryption A procedure in which only the two parties involved in sending a message have the code. This could be a simple shift code where letters of the alphabet are shifted to a new position.

problem statement A very clear description of what tasks the computer program must accomplish and how the program will execute these tasks and respond to unusual situations. It is the starting point of programming work.

processor speed The number of operations (or cycles) the processor completes each second, measured in hertz (Hz).

productivity software Programs that enable a user to perform various tasks generally required in home, school, and business. This category includes word processing, spreadsheet, presentation, personal information management (PIM), and database programs.

program development life cycle (PDLC) A number of stages, from conception to final deployment, a programming project follows.

programming The process of translating a task into a series of commands a computer will use to perform that task.

programming language A kind of "code" for the set of instructions the CPU knows how to perform.

programs Instruction sets that provide a means for users to interact with and use the computer, all without specialized computer programming skills.

project management software An application program such as Microsoft Project that helps project managers easily create and modify project management scheduling charts.

proprietary software A program that is owned and controlled by the company it is created by or for.

proprietary (or **private**) **systems** Systems whose design is not made available for public access.

protocol (1) A set of rules for exchanging data and communication. (2) The first part of the URL indicating the set of rules used to retrieve the specified document. The protocol is generally followed by a colon, two forward slashes, www (indicating World Wide Web), and then the domain name

prototype A small model of a computer program, often built at the beginning of a large project.

proxy server Acts as a go-between for computers on the internal network and the external network (the Internet).

pseudocode A text-based approach to documenting an algorithm.

public-key encryption A procedure in which the key for coding is generally distributed as a public key that may be placed on a Web site. Anyone wishing to send a message codes it using the public key. The recipient decodes the message with a private key.

query language Language used to retrieve and display records. A query language consists of its own vocabulary and sentence structure, used to frame the requests.

queue If more than one print job is waiting, a line, or queue, is formed so that the printer can process the requests in order.

QWERTY keyboard A keyboard that gets its name from the first six letters on the top-left row of alphabetic keys on the keyboard.

R

Rambus DRAM (RDRAM) The fastest and the most recent entry on the market. RDRAM is found on multimedia machines as its speed is necessary to transfer large multimedia files efficiently.

random access memory (RAM) The computer's temporary storage space or short-term memory. It is located as a set of chips on the system unit's motherboard and its capacity is measured in megabytes, with most modern systems containing around 256 MB to 512 MB of RAM.

rapid application development (RAD) A method of system development in which developers create a prototype first and they generate system documents as they use and remodel the product.

read only memory (ROM) A set of memory chips located on the motherboard that stores data and instructions that cannot be changed or erased; it holds all the instructions the computer needs to start up.

read/write heads The read/write heads move from the outer edge of the spinning platters to the center, up to 50 times per second, to retrieve (read) and record (write) the magnetic data to and from the hard disk.

real-time operating system (RTOS) A program with a specific purpose that must guarantee certain response times for particular computing tasks, or the machine's application is useless. Real-time operating systems are found in many types of robotic equipment.

real-time processing The process of querying a database and updating it while the transaction is taking place.

recalculate The ability of spreadsheet software to automatically refigure all functions and formulas in the spreadsheet when assumptions are changed.

record In database software, a collection of related fields.

reference software A software application that acts as a source for reference materials, such as the standard atlases, dictionaries, and thesauruses.

referential integrity Means that for each value in the foreign key of one table, there is a corresponding value in the primary key of the related table.

refresh rate (sometimes referred to as **vertical refresh rate**) The number of times per second the electron beam scans the monitor and recharges the illumination of each pixel.

registers Special memory storage areas built into the CPU.

registry Contains all the different configurations (settings) used by the OS as well as by other applications.

relational algebra The use of English-like expressions that have variables and operations, much like algebraic equations.

relational database Organizes data in table format by logically grouping similar data into relations (or tables that contain related data).

relations Tables that contain related data.

relationships In relational databases, the links between tables that define how the data is related.

repeaters Devices that are installed on long cable runs to amplify the signal.

resolution The clearness or sharpness of an image, which is controlled by the number of pixels displayed on the screen.

restore point The snapshot of the entire system's settings that Windows XP creates every time the computer is started, or when a new application or driver is installed.

reusability The ability to reuse existing classes of objects from other projects, enabling programmers to produce new code quickly.

ring (or loop) topology Networked computers and peripherals that are laid out in a circle. Data flows around the circle from device to device in one direction only.

root directory The C: drive, which is the top of the filing structure of the computer system.

root domain name servers Servers that know the location of all the name servers that contain the master listings for an entire top-level domain.

routers Devices that route packets of data between two or more networks.

runtime (or **logic**) **errors** The kinds of errors in the problem logic that are only caught when the program executes.

S

Safe Mode A special diagnostic mode designed for troubleshooting errors that occur during the boot process.

sampling rate The number of times per second the music is measured and converted to a digital value. Sampling rates are measured in kilobits per second.

Satellite Internet A way to connect to the Internet via a small satellite dish, which is placed outside the home and connects to a computer with **coaxial cable**. The satellite company then sends the data to a satellite orbiting the earth. The satellite, in turn, sends the data back to the satellite dish and to the computer.

scalable network Allows the easy addition of users without affecting the performance of the other network nodes (computers or peripherals).

ScanDisk A Windows utility that checks for lost files and fragments as well as physical errors on the hard drive.

screensavers Animated images that appear on a computer monitor when no user activity has been sensed for a certain time.

script kiddies Amateur hackers without sophisticated computer skills, typically teenagers, who don't create programs used to hack into computer systems; instead, they use tools created by skilled hackers that enable unskilled novices to wreak the same havoc as professional hackers.

scripts Lists of commands (miniprograms) that are executed without the user's knowledge.

scrollbars On the desktop, bars that appear at the side or bottom of the screen that control which part of the information is displayed on the screen.

search engine A set of programs that searches the Web for specific words (or keywords) you wish to query (or look for) and then returns a list of the Web sites on which those keywords are found.

second-generation languages (2GLs) Also known as assembly languages.

second-level domains Fall within top-level domains of the Internet. Each second-level domain needs to be unique within that particular domain, but not necessarily unique to all top-level domains.

Secure Digital A newer type of memory card that is faster and offers encryption capabilities so data is secure even if the user loses the card. The stamp-sized Secure Digital cards can hold up to 256 MB of data.

Secure Sockets Layer (SSL) A protocol that provides for the encryption of data transmitted via TCP/IP protocols such as HTTP. All major Web browsers support SSL.

seek time The time it takes for the read/write heads to move over the surface of the disk, between tracks, to the correct track.

select query Displays a subset of data from a table based on the criteria the user specifies.

semiconductor Any material that can be controlled to either conduct electricity or act as an insulator (not allow electricity to pass through).

serial port Allows the transfer of data, one bit at a time, over a single wire at speeds of up to 56 Kbps; it is often used to connect modems to the computer.

GLOSSARY 581

server-side Programs that run on the Web server as opposed to running inside a browser.

servers Computers that provide resources to other computers connected in a network.

shareware Software that allows users to "test" software (run it for a limited time free of charge).

shielded twisted pair (STP) cable Twisted pair cable that contains a layer of foil shielding to reduce interference.

Short Message Service (SMS) (often just called **text messaging**) Technology that enables short text messages (up to 160 characters) to be sent over mobile networks.

Simple Mail Transfer Protocol (SMTP) A protocol for sending e-mail along the Internet to its destination.

single-user, multitask operating system Allows only person to work on a computer at a time, but it can perform a variety of tasks simultaneously.

single-user, single-task operating system Allows only one user on a computer to perform just one task at a time.

smart battery A rechargeable lithium ion battery used in a PDA that can report the number of minutes of battery life remaining.

SmartMedia A type of flash memory card that is especially thin and light. SmartMedia cards can hold up to 128 MB of data.

software The set of computer programs or instructions that that tells the computer what to do and enables it to perform different tasks.

software licenses Agreements between the user and the software developer that must be "accepted" prior to installing the software on a computer.

software piracy Violating a software license agreement by copying an application onto more computer than the license agreement permits.

software suite A collection of software programs that have been bundled together as a package.

software updates (or **service packs**) Small downloadable software modules to repair errors identified in commercial program code.

sort (or **index**) Organizing a database into a particular order.

sound card An expansion card that attaches to the motherboard inside the system unit that enables the computer to produce sounds.

source code The instructions programmers write in a higher-level language.

spam Unwanted or "junk" e-mail.

speech-recognition software (or **voice-recognition software**) Translates spoken words into typed text.

speech-recognition system A computer that can be operated through a microphone, with a user telling it to perform specific commands (such as to open a file) or to translate your spoken words into data input.

spider (also known as a **crawler** or **bot**) A program that constantly collects information on the Web, following links in Web sites and reading Web pages. Spiders got their name because they crawl over the Web using multiple "legs" to visit many sites simultaneously

spooler A program that helps coordinate all print jobs currently being sent to the computer printer.

spreadsheet software An application program such as Microsoft Excel or Lotus 1-2-3 that enables a user to do calculations and numerical analyses easily.

standby mode When a computer' more power-hungry components, such as the monitor and hard drive, are put in idle.

star topology The most widely deployed client/server network layout in businesses. In a star topology, the nodes connect to a central communications device called a hub, thus forming a star. The hub receives a signal from the sending node and retransmits it to all other nodes on the network. The network nodes examine data and only pick up the transmissions addressed to them. Because the hub retransmits data signals, a star topology is an active topology.

statements Sentences in a code.

static addressing Means that the IP address for a computer never changes and is most likely assigned manually by a network administrator.

static RAM (SRAM) A faster type of RAM. In SRAM, more transistors are used to store a single bit, but no capacitor is needed.

stealth viruses Viruses that temporarily erase their code from the files where they reside and hide in the active memory of the computer.

storage devices Devices such as hard disk drives, floppy disk drives, and CD drives for storing data and information.

streaming audio Technology that enables audio files to be fed to a browser continuously, rather than downloading an entire file before it can be listened to.

streaming video Technology that continuously feeds the video file to abrowser large files can be watched as they download instead of first having to download the files completely.

structured (analytical) data (such as "Bill" or "345")

Structured Query Language (SQL) The most popular database query language today.

subject directory A structured outline of Web sites organized by topics and subtopics. Yahoo! is a popular subject directory.

subwoofer A special type of speaker designed to more faithfully reproduce low-frequency sounds.

summary data reports Reports that present a consolidated picture of the detailed data within a database.

supercomputers Specially designed computers that can perform complex calculations extremely rapidly. They are used in situations where complex models requiring intensive mathematical calculations are needed (such as weather forecasting or atomic energy research).

swap file (or **page file**) A temporary storage area on the hard drive where the operating system "swaps out" or moves the data or instructions from RAM that have not been recently used; this process takes place when more RAM space is needed.

switch A "smart" hub. It makes decisions, based on the MAC address of the data, as to where the data is to be sent.

Symbian OS A popular operating system for full-featured cell phones.

Symmetrical Digital Subscriber Line (SDSL) A DSL transmisstion that uploads and downloads data at the same speed.

synchronizing The process of updating data so the files on different systems are the same.

synchronous DRAM (SDRAM) Much faster than DRAM, it is the current standard and provides the level of performance most home users require.

syntax An agreed-upon set of rules defining how a programming language must be structured.

syntax errors Violations of the strict, precise set of rules that define a programming language.

system clock A computer's internal clock.

system development life cycle (SDLC) An organized process (or set of steps) for developing a system.

system evaluation The process of looking at a computer's subsystems, what they do, and how they perform to determine whether the computer system has the right hardware components to do what the user ultimately wants it to do.

system files The main files of the operating system.

System Restore A utility in Windows XP that lets you restore your system settings back to a specific date when everything was working properly.

system software The set of programs that enables a computer's hardware devices and application software to work together; it

582 GLOSSARY

includes the operating system and utility programs.

system unit The metal or plastic case that holds all the physical parts of the computer together, including the computer's processor (its brains), its memory, and the many circuit boards that help the computer function.

T

T lines High-speed fiber-optic communications lines that are designed to provide much higher throughput than conventional voice (telephone) and data (DSL) lines.

T-1 lines High-speed fiber-optic communications lines that can support 24 simultaneous voice or data channels and achieve a maximum throughput of 1.544 megabits per second (Mbps).

T-2 lines High-speed fiber-optic communications lines composed of four T-1 lines that deliver a throughput of approximately 6.3 Mbps.

T-3 lines Fiber-optic communications lines often used by Tier 1 and Tier 2 ISPs and very large businesses; they consist of a bundle of 28 T-1 lines. T-3 lines deliver 44.736 Mbps of bandwidth.

T-4 lines Fiber-optic communications lines that contain 168 T-1 lines and provide 274.176 Mbps of throughput.

table In database software, a group of related records.

tablet PC A portable computer that includes two special technologies: advanced handwriting recognition and speech recognition.

tags In HTML, a way to indicate how the text should look, such as and to indicate bolding.

Task Manager A Windows utility that checks on a program if it has stopped working, exits out of the nonresponding program.

Task Scheduler A Windows utility that allows the user to schedule tasks to automatically run at predetermined times, with no interaction necessary on the user's part.

tax-preparation software An application program such as Intuit's TurboTax and H&R Block's TaxCut for preparing your state and federal taxes. Each program offers a complete set of tax forms and instructions as well as expert advice on how to complete each form.

TCP/IP (Transmission Control Protocol/Internet Protocol) The main suite of protocols used on the Internet.

telephony software Software that combined with the Internet, speakers, and a microphone, turns a computer into a high-tech phone and answering service.

Telnet Both a protocol for connecting to a remote computer and a TCP/IP service that runs on a remote computer to make it accessible to other computers.

templates Forms included in many productivity applications that provide the basic structure for a particular kind of document, spreadsheet, or presentation

terminator A device that absorbs the signal so it is not reflected back onto parts of the network that have already received it.

test condition A check to see if a loop is completed.

testing plan In the problem statement, a plan that lists specific input numbers the program would typically expect the user to enter. It then lists the precise output values that a perfect program will return for those input values.

text fields Fields in a database that can hold any combination of alphanumeric data (letters or numbers) and are most often used to hold text.

thermal printer A printer that works by either melting wax-based ink onto ordinary paper (in a process called thermal wax transfer printing) or by burning dots onto specially coated paper (in a process called direct thermal printing).

third-generation languages (3GLs, also called **high-level languages)** 3GLs use symbols and commands to help programmers tell the computer what to do, making 3GL languages easier to read and remember. Programmers are relieved of the burden of having to understand everything about the hardware of the computer to give it directions. In addition, 3GLs allow programmers to name storage locations in memory with their own names so they are more meaningful to them.

three-way handshake A process TCP uses to establish a connection.

Tier 1 ISPs Internet service providers that route a large percentage of the traffic on the Internet and have extremely high-speed connections with other ISPs, sometimes in the 2.5 to 10 gigabits per second (Gbps) range.

Tier 2 ISPs Internet service providers that usually have a regional or national focus. Therefore, to enable their customers to reach any possible point on the global Internet, Tier 2 ISPs must route at least a portion of their traffic through the global Tier 1 ISPs.

Tier 3 ISPs Internet service providers that provide Internet access to homes or small- to medium-sized businesses. These ISPs normally cover a local geographical area. All Tier 3 ISPs need to be connected to at least one Tier 2 ISP.

toggle key A key whose function changes each time it's pressed; it "toggles" between two function.

token A special data packet used to pass data in a token-ring network.

token method The access method that ring networks use to avoid data collisions.

token-ring topology A network layout in which data is passed using a special data packet called a token.

toolbars On the desktop, groups of icons collected together in a small box.

thrashing A condition of excessive paging in which the operating system becomes sluggish.

time-variant data Data that doesn't all pertain to one period in time, such as data in a data warehouse.

top-down design A systematic approach in which a programming problem is broken down into a series of high-level tasks.

top-level domain (TLD) The three-letter suffix in the domain name (such as .com or .edu) that indicates the kind of organization the host is.

touchpad A small, touch-sensitive screen at the base of the keyboard. To use the touchpad, a user simply move your finger across the pad.

tower configuration A style of system unit on a deskotp computer, which typically sits vertically on the floor below a desk.

trackball mouse A mouse with a rollerball on top instead of on the bottom. Since a user moves the ball with his or her fingers, it doesn't require much wrist motion, so it's considered healthier on the wrists than a traditional mouse.

trackpoint A small, joysticklike nub that allows a user to move the cursor with the tip of his or her finger.

transaction processing system (TPS) A system used to keep track of everyday business activities.

transceiver In a wireless network, a device that translates the electronic data that needs to be sent along the network into radio waves and then broadcasts these radio waves to other network nodes.

transmission media The cable or wireless lines that transport data on a network.

transistors Electrical switches that are built out of layers of a special type of material called a semiconductor.

Transmission Control Protocol (TCP) A protocol for preparing data for transmission; it provides for error checking and resending lost data.

Trojan horse A computer program that appears to be something useful or desirable (like a game or a screen saver), but at the same time does something malicious in the background without the user's knowledge.

twisted pair wiring (or **cable**) Telephone lines made up of copper wires that are twisted around each other and surrounded by a plastic jacket.

two-way pagers Devices that support both receiving and sending text messages.

U

UDP (User Datagram Protocol) Prepares data for transmission—no resending capabilities.

Unicode An encoding scheme that uses 16 bits instead of the 8 bits used in ASCII. Unicode can represent more than 65,000 unique character symbols, enabling it to represent the alphabets of all modern languages and all historic languages and notational systems.

Uniform Resource Locator (URL) A Web site's unique address, such as www.microsoft.com.

UNIX An operating system originally conceived in 1969 by Ken Thompson and Dennis Ritchie of AT&T's Bell Labs. In 1974, the UNIX code was rewritten in the standard programming language C. Today there are various commercial versions of UNIX.

unshielded twisted pair (UTP) cable The most popular transmission media option for Ethernet networks. UTP cable is composed of four pairs of wires that are twisted around each other to reduce electrical interference.

unstructured data includes nontraditional data such as audio clips (including MP3 files), video clips, pictures, and extremely large documents. Data of this type is known as a binary large object (BLOB) since it is actually encoded in binary form.

USB (Universal Serial Bus) port A port that can connect a wide variety of peripherals to the computer, including keyboards, printers, Zip drives, and digital cameras. USB 2.0 transfers data at 480 Mbps and is approximately 40 times faster than the original USB port.

user interface Part of the operating system provides that enables a user to interact with the computer.

Utility Manager A utility in the Accessories folder of Windows XP that allows the user to magnify the screen image; the user can also have screen contents read, or display an onscreen keyboard.

utility programs Small programs that perform many of the general housekeeping tasks for the computer, such as system maintenance and file compression.

V

vacuum tubes Used in early computers, vacuum tubes act as computer switches by allowing or blocking the flow of electrical current.

validation The process of ensuring that data entered into the database is correct (or at least reasonable) and complete.

validation rules Rules that are set up in a database to alert the user to clearly wrong entries.

value Numeric data either entered in to a spreadsheet directly or as a result of a calculation.

variable declaration Tells the operating system that the program needs to allocate storage space in RAM.

variables Input and output items that a computer program manipulates.

VBScript A subset of Visual Basic; also used to introduce interactivity to a Web page.

video card (or **video adapter**) An expansion card that is installed inside a system unit to translate binary data (the 1s and 0s your computer uses) into the images viewed on the monitor.

video RAM (VRAM) The random access memory included with a video.

videoconferencing Technology allows a person sitting at a computer and equipped with a personal video camera and a microphone to transmit video and audio across the Internet (or other communications medium). All computers participating in a videoconference need to have a microphone and speakers installed so that participants can hear one another.

virtual memory The process of optimizing RAM storage by borrowing hard drive space.

virtual private network (VPN) Utilizes the public Internet communications infrastructure to build a secure, private network between various locations.

virtual reality programs Software that turns an artificial environment into a realistic experience.

virus signatures Signatures in files that are portions of the virus code that are unique to a particular computer virus.

visual programming A technique for automatically writing code when the programmer says the layout is complete. It helps programmers produce a final application much more quickly.

Voice over IP (VoIP) The transmission of phone calls over the same data lines and networks that make up the Internet. Also called Internet telephony.

voice pager Device that offers all the features of a numeric pager but also allows the user to receive voice messages.

volatile storage Tempory storage, such as in RAM; when the power is off, the data in volatile is cleared out.

W

walkie-talkie A two-way radio device that sends and receives sound messages. Users cannot send and receive messages at the same time.

warm boot The process of restarting the system while it's powered on.

Web browser Software that allows a user to access the Web.

Web-based e-mail Uses the Internet as the "client"; therefore, a user can access a Web-based e-mail account from any computer that has access to the Web—no special client software is needed.

Web clipping Technology for extracting the information from a Web site and formatting it so it is more useful on smaller PDA displays.

Web-enabled The ability for a device, such as a desktop computer, laptop, or mobile device, to access the Internet.

Web page authoring software Programs for designing interactive Web pages, without knowing any HTML code.

Web server Web servers Computers running specialized operating systems that allow them to host Web pages (and other information) and provide requested Web pages to the clients.

Web site A location on the Web.

"what-if" analysis Testing the affects that different options have on a spreadsheet, using the automatic recalculation.

white-hat hackers Hackers who break into systems just for the challenge of it (and who don't wish to steal or wreak havoc on the systems). They tout themselves as experts who are performing a needed service for society by helping companies realize the vulnerabilities that exist in their systems.

wide area network (WAN) A network made up of LANs connected over long distances.

Wi-Fi (short for Wireless Fidelity) The 802.11 standard established by the IEEE.

wildcards Symbols used in an Internet seach when the user is unsure of the keyword's spelling, or if a word can be spelled in different ways or may contain different endings. The asterisk (*) is used to replace a series of letters and the percent sign (%) to replace a single letter in a word.

windows In a graphical user interface, rectangular boxes that contain programs displayed on the screen.

Windows key A function key specific to the Windows operating system. Used alone, it brings up the Start menu; however, it's used most often in combination with other keys as shortcuts.

Windows Explorer The program in Microsoft Windows that helps a user manage files and folders by showing the location and contents of every drive, folder, and file on the computer.

584 GLOSSARY

wireless access point A device similar to a hub in an Ethernet network. It takes the place of a wireless network adapter and helps to relay data between network nodes.

Wireless Application Protocol (WAP) The standard that dictates how handheld devices will access information on the Internet.

wireless DSL/cable routers Enables wireless and wired nodes to be connect to the same network.

wireless Internet service provider Providers such as Verizon or T-Mobile that offer their subscribers wireless access to the Internet.

Wireless Mark-up Language (WML) A format writing content viewed on a cellular phone or PDA that is text-based and contains no graphics.

wireless media Add-ons to extend or improve access to a wired network.

wireless network A network that uses radio waves instead of wires or cable as its transmission media.

wireless network adapter These adapters are required for each each node on a wireless network.

wireless network interface cards (wireless NICs) Cards installed in the system that connect with wireless access points on the network.

wizards Step-by-step guides that walk you through the necessary steps to complete a complicated task.

word processing software Programs to create and edit written documents such as papers, letters, and resumes.

word size The number of bits a computer can work with at a time.

worksheet The basic element in a spreadsheet program; it is a grid consisting of columns and rows.

World Wide Web (WWW or the Web) The part of the Internet uses the most. What distinguishes the Web from the rest of the Internet is (1) its use of a special language (called *HTML*, or *Hypertext Markup Language*) that allows different computers to talk to and understand each other while connected to the Web and (2) its use of special links (called *hyperlinks*) that enable users to jump from one place to another in the Web.

worm A program that attempts to travel between systems through network connections to spread infections. Worms can run independently of host file execution and are active in spreading themselves.

Z

Zip disk A portable storage medium with storage capacities ranging from 100 MB to 750 MB.

Zip disk drive A drive bay for a Zip disk.

Zombies Computers that are controlled by hackers who used a backdoor program to take control.

INDEX

Symbols and Numbers

@ (at) sign in e-mail addresses, 536
&& (double ampersand) operator (C++), 403
<< (double lesser than) operator (C++), 403
// (double slash), in C++ code, 404
1GLs (first-generation programming languages), 400
2GLs (second-generation programming languages), 400
3-D modeling, 128
3D sound cards, 232
3GIO (Third Generation I/O) (Intel), 362
3GLs (third-generation programming languages), 401
4GLs (fourth-generation programming languages), 401
5GLs (fifth-generation programming languages), 401
100 Base-T (Fast Ethernet) ports, 234
802.11 (Wi-Fi) standard, 262, 489

A

abbreviations, text messaging, 316
academic discounts, 134
academic fair use, 89
Academicsuperstore.com, 134
Accelerated Graphics Port (AGP) buses, 362–363
access, Internet
 communications protocols, 73–75
 multiple access points, 251
Access (Microsoft), 114
 Field Properties box, 438
 Field Properties Table, 442
 Relationships screen, 452
 Simple Query Wizard, 446
 SQL View window, 446
access points
 linear bus topologies, 481
 for wireless networks, 494
access restrictions
 access privileges, 499
 authentication, 498–499
 card readers, 500
 biometric devices, 301
 importance of, 274–275
 passwords, 275–278, 299–301
 PDA bomb software, 300
 personal firewalls, 301
access time, 359
accessibility tools and utilities, 119, 199–200
accessories for laptop computers, 335
Accessories folder (Windows XP), 199–200
Accounting (Peachtree), 122
accounts, e-mail, 86
ack (positive acknowledgement), connectionless protocols, 527
ACM (Association for Computing Machinery), 417
Acrobat (Adobe), 95, 135, 190
ActionScript (Flash), 415
Active Server Pages (ASP), 414–415
active topologies, 483
active-matrix displays (LCD monitors), 39
adapters, network, 256
 adapter cards, 48
 for phone line networks, 258
 for power line networks, 257
Add/Remove Programs utility (Control Panel), 136, 193–194
address books, 112
addresses
 and firewall protection, 276–278
 domain names, 525, 528–529
 e-mail, 536–537
 IP (Internet Protocol), 67
 IPv4 versus IPv6, 524–525
 MAC (media access control), 495–496
 for Web sites, 67
Adobe
 filename extensions, 190
 Acrobat, 95, 135, 190
 Encore DVD, 165
 Illustrator, 126
 PageMaker, 124
 Photoshop, 124, 158
 Premiere, 127, 163, 165
ADSL (Asymmetrical Digital Subscriber Line), 70
Advanced Micro Devices (AMD) microprocessors, 354
Advanced Research Projects Agency (ARPANET), 66
Advanced Simulation and Computing Initiative (ASCI), 365
advertising
 adware, 304–305
 spam, 85
.aero domains, 78
affinity grouping (data mining), 462
AGP (Accelerated Graphics Port) buses, 362–363
AI (artificial intelligence), 15, 461
Aiken, Howard, 383
air flow, importance of, 295
Aladdin Systems add/remove utility, 193
alarms, 298
alcohol-based biofuels, 317
algebra, relational (SQL), 444
algebraic operators, 403
algorithm development (PDLC), 393–394
 decision points, 394–396
 flowcharts, 396
 object-oriented analysis, 398
 pseudocode, 396–397
 top-down design, 397–398
Allen, Paul, 376, 379
AlltheWeb search engine, 80
alphabetic checks (databases), 440–441
alphanumeric data, 434–435
 phone numbers in, 434
alphanumeric pagers, 313
Alt key, 34
Alt+F4 keyboard shortcut, 34
Altair 8800 computer, 376
AltaVista search engine, 80
Alto computer (Xerox), 380
ALU (arithmetic logic unit) (CPU), 214, 359
Amazon.com data warehouse, 452, 455

586 AMD (ADVANCED MICRO DEVICES)

AMD (Advanced Micro Devices) microprocessors, 354
America Online (AOL)
  Instant Messenger, 90, 539
  web-based e-mail accounts, 86
America Online (AOL), 75, 90
American National Standards Institute (ANSI):ASCII code, 352
American Standard Code for Information Interchange (ASCII) code, 352
Ammann, Simon, 132
"&&" (ampersand, double) operator (C++), 403
amplified speakers, 231–232
analog data
  converting to digital, 151–153
  analog video, 160–161
analysis phase (SDLC), 389
analytical data, storing, 436
Analytical Engine (Babbage), 382
AND operator, 82
AND statements (SQL), 445
Andreessen, Marc, 531
animations
  forensic, 10
  players for, 95
  in slide shows, presentations, 113
anonymous FTP archives, 529
ANSI (Amercan National Standards Institute)
  ASCII code, 352
anti-pop-up software, 304
anti-shake feature (digital video), 162
antispam software, 87
antivirus software, 281–282, 302
AOL (America Online)
  Instant Messenger, 90, 539
  web-based e-mail accounts, 86
APIs (application program interfaces), 179
APL (A Programmer's Language), 403
Apple Computer Company
  Apple I, II, and III, 377
  founding of, 377
  G4 pipelining, 364
  iMac, 44
  iPod MP3 player, 317–318
  Lisa, 381
  Macintosh, 20–21, 172, 381
  naming of, 377
  microprocessors for, 354
  QuickTime, 95, 164
applets (Java), 412, 533–534

appliances, household
  Internet-enabled, 543
  operating systems for, 168–170
  programs for, 416
application layer (OSI networks), 492
application program interfaces (APIs), 179
application servers (client/server networks), 479
application software, 19, 108, 168
  accounting software, 122
  communications software, 132–133
  database software, 114–115
  education and reference, 129–130
  entertainment software, 131
  general business software, 122–124
  graphics and multimedia, 124–128
  help systems for, 133–134
  history of, 380
  interactive, 533–534
  managing of, by operating system, 179
  personal financial software, 121–122
  PIM (personal information manager) software, 113
  presentation software, 113–114
  productivity tools, 115–116
  programming languages for, 410
    Web applications, 413–415
    Windows applications, 410–413
  purchase options, 116–117
  software suites, 116–117, 120–121
  speech-recognition software, 118–119
  spreadsheet software, 110–112
  word processing software, 108–110
Applied Digital Solution, 14
Aquabot (Aqua Products), 12
  buses, 361–362
  CPUs (central processing units), 365
  networks, 253–267
archival printing, 159
Archos Multimedia Entertainment Center, 318
ArcSoft image-editing software, 126
arithmetic logic unit (ALU), 214, 359
ARPANET (Advanced Research Projects Agency), 66, 536
arrow keys, 34
artificial intelligence (AI), 15, 401, 461
arts, use of computers in, 7

ASCI (Advanced Simulation and Computing Initiative), 365
ASCII (American Standard Code for Information Interchange) code, 352
Ask Jeeves (Teoma) search engine, 80, 82
ASP (Active Server Pages), 414–415
assembly language, 357–358, 400–401
Association for Computing Machinery (ACM), 417
association rules (data mining), 462
Asymmetrical Digital Subscriber Line (ADSL), 70
AT&T, 74, 173
Atansoff-Berry Computer (ABC), 382–384
Athlon XP (AMD) microprocessor, 354
Atlanta Ballet, 7
attachments (e-mail), 279, 537
auctions, online, 91
audio
  input devices, 37–38
  with instant messaging, 88
  MP3 players, 317, 319–321
  output devices, 42
  sending to cell phones, 316
  sound cards, 232–233
  speakers, 231
  storing in databases, 436
  streaming, 94–95
authentication
  biometric devices, 301
  passwords, 299–301
  PDA bomb software, 300
  process for, 182, 498–499
  servers (client/server networks), 478, 500
autocorrect feature, 109
autofocus feature, 156
automobile GPSs (Global Position Systems), 543
AvantGo, 326
AVERAGE function, 111

B

Babbage, Charles, 382
Back button, 78
back panel (system units), 46–47
backbones, Internet (Tier 1), 67, 74, 517–518
backdoor programs, 272
background images (desktops), 194

CABLE SERVICES 587

backing up data
  data file backups, 306
  frequency, 306
  importance of, 305, 502
  online, 307
  program file backups, 306
  utilities for, 198, 307
backward software compatibility, 136
bandwidth, 255
  cable, 486
  Internet telephony, 538
  ISP connections, 518
banking online, 91
base 10 number system, 349–351
base 2 number system, 350–352
base class, 399
base receiver station (cell phones), 315
BASIC (Beginners All-purpose Symbolic Instruction Code) programming language, 379, 401
basic input/output system (BIOS)
  activation, 181
bastion hosts, 501
batch processing (TPS), 458
batteries
  digital video cameras, 161
  laptop computers, 334–335
  PDAs (personal digital assistants), 322, 324
beep codes, 183
behaviors, in object-oriented analysis, 398
benchmarking, 322
Berners-Lee, Tim, 530
beta testing, 135–136, 409
bidding, programs for, 123
bill payment tools, 121–122
binary (base 2) number system, 17, 350–353
binary decision systems, 394, 402
binary language, 17, 348
binary large object (BLOB) data type, 436
binary switches, 348–349
biofuel cells, 317
biomedical chip implants, 14
biometric authentication devices, 301, 500–501
BIOS (basic input/output system)
  activation, 181
bioterrorism, 269
bistable screens, 53
bit depth, 229
bitmap files, 190

bits (binary digits), 17, 351–352, 361
Bits and Bytes feature, 22
bits per second (bps), 486
.biz domains, 78
black-hat hackers, 270
Blackboard course management software, 129
BLOB (binary large object) data type, 436
BLS (Bureau of Labor Statistics) job outlook projections, 556
BlueBoard (Colligo Networks), 326
Bluetooth technology, 235
  Bluetooth-enabled devices, 476
  with laptop computers, 334
  with PDAs (personal digital assistants), 326
.bmp files, 190
bold formatting, 34
bomb software, 300
bookmarks 79
bookseller training tools, 214
Boolean operators
  examples of, 82
  data types for, 403
boot process, 180–181
  cold versus warm boots, 45, 182
  handling errors during, 182–183
  Safe Mode, 183–184
  steps in, 180–181
  versus installation, 182
boot-sector viruses, 279
bootstrap, 181
Borg, 14
bot programs (search engines), 80
boxes, desktop, 44
bps (bits per second), 486
breadcrumb lists, navigating web sites using, 78
breaks, importance of taking, 51
Bricklin, Dan, 380
bridges (client/server networks), 496–498
broadband communications, 68
  cable Internet, 70
  DSL (Digital Subscriber Lines), 69–70
  emerging technologies, 74
  ISPs (Internet service providers), 74
  and P2P network performance, 267
brokerage, online, 91
browsers, Web, 66
  choosing, 75–76, 531
  for database data, 442

features, 76
  history, 381
  microbrowsers, 316
  navigating, 76–79
  updating, 530
brute force attacks, 499
B2B (business-to-business)
  e-commerce transactions, 90, 534
B2C (business-to-consumer)
  e-commerce transactions, 90
Buddy Lists, 90
budgeting software, 121–122
buffer, printer, 177
bugs, computer, 383
Bureau of Labor Statistics (BLS) job outlook projections, 556
burn-in, 193
bus (linear bus) network topology, 481–483, 486
bus width, 361–362
buses (CPU), 361–362
business applications
  CRM (customer relationship management) software, 123
  databases, 429
  data mining, 461–462
  data warehouses, 454
  management and planning tools, 122–123
  mapping programs, 123
  specialized programs, 123
business network topology, 484–485
Business Plan Pro (Palo Alto Software), 122
Business Software Alliance, 139
business-to-business (B2B)
  e-commerce transactions, 90, 534
business-to-consumer (B2C)
  e-commerce transactions, 90
buttons, on mouse devices, 35–36, 78
Buttterfly.net, 269
buying
  computers, 4, 238, 240
  software, 134–136
bytes, 17, 352, 454

C: drive, 187
C programming language, 401, 412
cable Internet, 70–71
cable locks, 298
cable modems, 261
cable services, 69

cable transmission media, cabling, 486
  bandwidth, 255, 486
  bend radius (flexibility), 487
  coaxial cable, 255, 488
  comparisons among, 489
  cost per foot, 487
  fiber-optic cable, 255, 488
  installation costs, 487
  interference, 487
  run length, 486
  signal transmission methods, 487
  twisted pair, 69, 255, 487
  wireless, 488–489
  coaxial, 72, 255
  UTP (unshielded twisted pair), 260
cache, Internet, 528
cache memory, 356–357
CAD (computer-aided design) software, 128
cages, 298
Cailliau, Robert, 530
calculations
  computed/computational fields, 434
  numeric data types for, 434
  operators for, 403
calculators
  Babbage's Analytical Machine, 382
  converting decimal to binary notation, 350
  Hollerith Tabulating Machine, 383
  Pascalene, 382
  Z1, 383
calendars, spreadsheets as, 112
California Integrated Seismic Network (CISN), 503
California unclaimed property Web site, 453
cameras, digital, 154
  camcorders, 160, 164–165
  downloading images from, 156–157
  flash memory cards, 155
  and image-editing software, 126
  image quality, 154–155
  prices, 155
  using, 155–156
  versus traditional cameras, 159
  video, 160–165
  video capture devices, 160
  video-editing software, 127–128
Canon
  inkjet printer, 158
  PowerShot camera, 154

Canvas (Deneba), 127
Caps Lock key, 34
car computers, operating systems for, 169
carbon nanotubes, 336
careers
  and computer fluency, 6–11
  IT (information technology), 555–558
carrier sense (star networks), 485
Carrier Sense, Multiple Access with Collision Detection (CSMA/CD), 485
cascading windows, 185
case sensitivity (filenames), 191
CAT 5E, CAT6 UTP twisted pair cable, 260, 488
Catch-A-Call, 69
cathode ray tube (CRT) monitors, 38–40, 52
CD-R (Compact Disk-Recordable) disks, 226
CD-ROM (Compact Disk-Read Only Memory) disk and drives, 18, 46
  care and maintenance, 226
  copying CDs, ethical issues, 228
  how they work, 225
  speed, 226
  storage capacity, 224
  types and features, 226
  versus DVDs, 226
CD-RW (Compact Disk-Read/Writable) disksk, 226
Celeron (Intel) microprocessor, 354
cell phones
  bistable displays, 53
  ISPs (Internet service providers), 75
  operating systems, 168–170
cells (spreadsheets), 110–111
cellular phones, 312–314
  Internet connectivity, 316
  mechanics and technology, 315
  multimedia messages, 316
  text messaging, 316
central offices (CO), and DSL performance, 70
central processing unit (CPU), 18, 49, 353
  advanced designs, 212, 363
  buses, 361–362
  dual-processor architecture, 365
  evaluating, 214–216
  machine cycle, 355–359
  manufacturers of, 354
  Moore's Law, 212
  multimedia processing, 365

  multiprocessor systems, 178
  parallel-processing, 365
  pipelining, 363–364
  RAM (random access memory), 359–361
  supercomputers, 365
  upgrading, 216–217
centralized networks, 475
Centrino (Intel), 333
Cerf, Vinton (PTP protocol), 66, 541
CERN (European Organization for Nuclear Research), 530
CGI (Common Gateway Interface), 532–533
cgi-bin directories, 532
chairs, adjustable, 50
char data type (C++), 403
characters
  binary notation for, 351–352
  data types for, 403
  forbidden, in filenames, 191
charts, creating from spreadsheets, 112
chat rooms, 87–88, 90
check-writing tools, 121–122
chip implants, biomedical, 14
chips
  analog-to-digital converters, 315
  integrated circuits, 349
Cinepak codec, 165
circuit switching, 522
CISN (California Integrated Seismic Network), 503
cities, domain names for, 78
Clark, Jim, 531
classes, base and derived (OOP), 398–399
classification approach (data mining), 462
cleaning, maintenance
  CDs and DVDs, 226
  floppy and Zip disks, 225
  keyboards, 33
  monitors, 39
  mouse devices, 36
CleanSweep (Norton), 193
cleanup, of disk contents, utilities for, 193, 195–196, 198
Clear Type feature (Windows XP), 193
click-and-brick businesses, 90
clickstream data-capture tools, 455
client-based e-mail accounts, 86, 536
client-side applications, 533–534
client/server networks, 254, 520
  adapters/interface cards, 478, 491, 494

clients, 66–67, 254, 474–475
commerce servers, 520
file servers, 520
firewalls, 501
Internet, 66
LANs (local area networks), 476
operating systems, 256, 478, 491
resource sharing, 476
scalability), 475
servers, 478–481, 520
topologies, 481–486
transmission media, 478, 486–489
versus P2P (peer-to-peer) networks, 475
WANs (wide area networks), 476–477
wireless connections, 476
clip art, 109, 127
clock cycle (system clock), 355
clock speed (system clock)
 bus clock speed, 361–362
 system clock, 215–216, 355
closing current window, 34
clothing, electronic, 320
clustering approach (data mining), 462
CMOS memory, 181
coaxial cable, 72, 255
 signal transmission methods, 487
 Thinnet versus ThickNet, 488
COBAL programming language, 401
Codd, E.F.
 QL development, 444
 relational model, 436
Code Red worm, 280
code, program
 coding tools, 406–408
 compiling, interpreting, 404–406
 debugging, 408
 reusable components, 404, 406
 source code, 174
 writing, 401, 404–405
codebooks (speech-recognition software), 118
codecs (compression/decompression), 165
cold booting, 45, 182
college networks, 498–502
Colligo Networks BlueBoard, 326
collision domains, 497
collisions, data, 484
color
 display quality, and bit depth, 229
 of clip art, editing, 127

 in printer output, 42
color laser printers, 42
column charts, 112
vcolumns (databases), 434
.com domains, 78
.com filename extension, 405
command-driven user interface, 176
vcommands, for Telnet, 529
comments
 adding to presentations, 114
 in program code, 404
commerce, electronic (e-commerce), 90
commerce servers, 520
Commodore PET 2001 computer, 378
Common Gateway Interface (CGI), 532–533
communication speeds. See also data transfer rates
 DSL (Digital Subscriber Lines), 70
 T-lines, 518
communications
 broadband, 74
 Internet
  chat rooms, 87
  data transfer rates, 68
  e-mail, 83–87, 536–539
  how it works, 66
  instant messaging, 88, 539, 541
  netiquette, 88
  newsgroups, 90
  protocols for, 76–77
  telephony, 538
 interplanetary, 541–542
 servers (client/server networks), 480
 software, 132–133
Compact Flash cards, 46
 with MP3 players, 319
 with PDAs (personal digital assistants), 325
Compaq computer, 378
compatibility
 CPU upgrades, 217
 networking devices, 262
 software ugrades,136
compilers, 383, 404–405
Complete Planet subject directory, 80
completeness checks (databases), 440–441
compressing files
 digital video, 165
 filename extensions, 190
 graphics images, 155

 utilities for, 193–195
CompuServe, 75
computer bugs, 383
computer careers
 content creators, 535
 customer interaction teams, 535
 database administrators and designers, 433
 database analysts and administrators, 390
 graphic designers, interface designers, 390
 information technology, 555–556, 558
 IT security professionals, 283
 and knowledge of the Internet, 515
 learning programming, tips for, 417
 network administrators, 499
 network engineers, 390
 programmers, 390
 project managers, 390
 systems analysts, 390
 technical writers, 390
 Web developers, 535
 Web publishers, 535
 Web server administrators, 535
computer crashes, troubleshooting, 237–238
computer crime, 139, 275–278
computer fluency
 and career options, 6–11
 attributes, 4–5
 defined, 3
 ethical issues, 15–16
 home uses, 11–12
computer forensics, 9–10
computer manufacturers, training resources, 214
computer network, defined, 252. See also networking, networks
computer programming, 387
computer protection software, 303
computer protocols. 522–526, 529
Computer Science courses, 417
computer upgrades
 audio subsystem, 231–233
 CPU subsystem, 214–217
 determining ideal system, 213
 identifying computer uses, 213–214
 memory (RAM) subsystem, 217–220
 performance and reliability issues, 237–239
 port connectivity, 233–237

storage subsystem, nonvolatile storage, 220–227
system evaluation, 214
and technological advances, 212
training needs, 214
versus buying new system, 238, 240
video subsystem, 228–231
computer viruses, 278–279, 302
 boot-sector viruses, 279
 logic bombs, 280
 macro viruses, 280
 multipartite, 281
 polymorphic, 280
 protecting against, 281–282
 script viruses, 280
 stealth, 281
 symptoms of infection, 280
 Trojan horses, 280
 worms, 280
computer-aided design (CAD) software, 128
computerized databases, 429
ComputerJobs Web site, 417
computers. *See also* hardware; input devices; output devices; software
 addresses, 67
 binary language, 17
 boot process, 180–182
 buying, 4, 238–239
 ergonomics, 50–51
 functions, 16
 history of, 348–349, 376–381
 instructions, 353
 laptops, 36–37, 312, 332–335
 mainframes, 21
 PCs (personal computers), 168, 354
 naming during network setup, 268
 protecting, 299–307
 servers, 21
 setting up, 31
 supercomputers, 21
 system units, 44–50
 tablet computers, 329–331
computing, mobile, 311
Computrace computer tracing software, 299
configuration settings (boot process), 182
connectionless protocols, 527
connections, Internet
 broadband, 68–69
 cable Internet, 70–71
 cellular phones, 316
 comparison table, 73

connection speed, 68
 dial-up, 68–69
 DSL (Digital Subscriber Lines), 69–70
 emerging technologies, 74
 ISDN (Integrated Services Digital Network), 71
 ISP (Internet service providers). 518–521
 for PDAs (personal digital assistants), 326
 satellite Internet, 72
connectivity, enhancing, 474
connectivity ports, 46–47
connectors, for networks, 255
consistency checks (databases), 440–441
consumer-to-consumer (C2C) e-commerce transactions, 90
content creators (Web sites), 535
Control key, 34
Control Panel
 Add/Remove Programs feature, 136
 customizing desktop display, 192–194
 installing/uninstalling programs from, 194
Control Program for Microcomputers (CP/M), 379
control unit (CPU), 356
control, of networks, 253
 client/server networks, 254
 P2P (peer-to-peer) networks, 253
controls, for games, 37
cookies, 92, 94
.coop domains, 78
copying
 CDs and DVDs, ethical issues, 228
 copyright violation, 89
 files, 191
 text, 34
Corel Corporation
 CoralDRAW, 127
 Paradox, 114
 Presentations, 113
 Word Perfect, Word Perfect Office, 108
 filename extensions, 190
 programs bundled with, 120
 speech-recognition software, 118
costs
 digital video, 160
 of Internet connections, 73
 mobile devices, 311–312
 and network type, 267

paging devices, 313
 of printers, 42
 Web-based e-mail storage, 86
countries, domain names for, 78
course management software, 129
C++ programming language, 401, 412
 code for household appliances, 416
 code sample, 404–405
 data types, 403
 if else keywords, 402
 operators, 403
 security keywords, 402
 writing comments, 404
CP/M (Control Program for Microcomputers), 379
CPU (central processing unit), 18, 49, 353
 advanced designs, 363
 buses, 361–362
 in cellular phones, 313
 dual-processor architecture, 365
 evaluating, 214–216
 growing chip capacity, 212
 for laptop computers, 332
 machine cycle, 355–359
 manufacturers, 354
 Moore's Law, 212
 multimedia processing, 365
 for multiprocessor systems, 178
 parallel-processing, 365
 PDAs (personal digital assistants, 322
 pipelining, 363–364
 RAM (random access memory), 359–361
 supercomputers, 365
 tablet PCs, 331
 upgrading, 216–217
 viewing usage of, 216
crackers, hackers
 computer access approaches, 274–275
 denial of service (DoS) attacks, 273–274
 identity theft, 270–272
 Trojan horses, 272–273
cradles, 325
crashes, computer, troubleshooting, 237–238
crawler programs (search engines), 80
Creative Labs NOMAD MuVo MP3 player, 318
crime, computer, 269
 computer viruses, 278–281

denial of service (DoS) attacks, 273–274
hacker access approaches, 274–275
hackers, crackers, 270, 272
identity theft, 270–272
protecting against, firewalls, 275–278
software piracy, 139
Trojan horses, 272–273
criminal justice system, role of computers in, 447
CRM (customer relationship management) programs, 123
CRT (cathode ray tube) monitors, 38–39
upgrading to LCD monitors, 230
versus LCD (liquid crystal display) monitors, 40, 52
Crusoe 5800 (Transmeta) processor, 331
CSMA/CD (Carrier Sense, Multiple Access with Collision Detection), 485
C2C (consumer-to-consumer) e-commerce transactions, 90
Ctrl+Alt+Del keyboard shortcut, 45, 197
Ctrl+B keyboard shortcut, 34
Ctrl+C keyboard shortcut, 34
Ctrl+Esc keyboard shortcut, 34
Ctrl+I keyboard shortcut, 34
Ctrl+N keyboard shortcut, 34
Ctrl+O keyboard shortcut, 34
Ctrl+P keyboard shortcut, 34
Ctrl+S keyboard shortcut, 34
Ctrl+V keyboard shortcut, 34
Ctrl+X keyboard shortcut, 34
cursors, 34
custom software installation, 137
customer interaction teams (the Web), 535
customer relationship management (CRM) programs, 123
Cute FTP file transfer software, 76
cutting text, 34
cybercrime, cybercriminals, 269
computer access approaches, 274–275
computer viruses, 278–281
cyberterrorism, 272
denial of service (DoS) attacks, 273–274
hackers, crackers, 270, 272
identity theft, 270–272
Trojan horses, 272–273

D

DAC (digital-to-analog converter) chips, 152
damage, computer, sources of
physical hazards, 295
power outages, 297
power surges, 295–296
troubleshooting utilities, 193
dancer movement, computer analysis of, 7
data
analog versus digital, 151–152
analog-to-digital conversions, 151–153
backing up, 198, 305–307
binary, 17
in data warehouses, 454–456
in databases
accessing, 433
querying, 443–446
sorting and indexing, 442
viewing, 442
defined, 17
digital, 153
metadata, 439
mining, 6, 461–462
normalizing, 449–451
numeric, 111
in object-oriented analysis, 398
packets, 256
sources for
DSS (decision support systems), 460
TPS (transaction processing systems), 458
storing, 452–456
structured versus unstructured, 436
time-variant versus historical, 454
validating, 534
versus information, 16, 430
data access speed, 226
data collisions
bus networks, 481
preventing using bridges, 497
ring networks, 483
data dictionaries, 437–439
data input (databases). See also input devices
data validation, 439–441
entry errors, inconsistencies, 430
input forms, 439
maintaining integrity, 433
data line surge suppressors, 296
data link layer (OSI networks), 492

data management information systems, 457–461
data marts, 456
data output. See also output devices
data processing, 18
data redundancy, 430–431
data retrieval, 217
data staging (data warehouses), 455
data transfer, data transfer rates, 68, 255
AGP (Accelerated Graphics Port) buses, 362
cable media, 71, 486
Bluetooth technology, 234–235
connectivity ports, 46–47
DSL (Digital Subscriber Line), 70
Ethernet ports, 233, 259–260
expansion buses, 361
FireWire (IEEE 1394) ports, 233
hard disk drives, 221–222
home networks, comparisons among, 267
Internet connections, 73–74
IrDA ports, 233–234
ISDN (Integrated Services Digital Networks), 71
MIDI ports, 235
packets, 256
parallel ports, 46, 233–234
PDAs (personal digital assistants), 325
phone line networks, 258
power line networks, 256–257
P2P networks, comparisons among, 267
satellite Internet, 73
serial ports, 233–234
tablet PCs, 330
USB (universal serial bus) ports, 46, 233–234
UTP cables, 260
wireless devices, 75, 262
wired media, 486
data types, 434
BLOB (binary alrge object), 436
computed/computational fields, 434
date fields, 434–435
field size, 435
hyperlink fields, 434–435
in relational databases, 436
memo fields, 434–435
numeric fields, 434
object fields, 434–435
in programming languages, 403
text fields, 434

database administrators, 433
database analysts and administrators, 390
database designers, 433
database queries (SQL)
  query languages, 401
  types and structure, 443–445
  wizards for, 446
database schema, 437
database servers (client/server networks), 480
databases
  advantages of, 429
  creating, 437
  data dictionaries, 438
  data input, 439
  data integrity, 433
  data marts, 456
  data mining, 461–462
  data output, 448
  data validation, 439–441
  data warehouses, 452, 454–456
  DBMS (database management systems), 437
    creating, 437–439
    data input, 439
    data validation, 439–441
    exporting data, 448
    querying data, 443–446
    viewing, sorting data, 442
  defined, 429, 439
  design and management, 433, 437
  fields, 433–435
  flexibility of, 433
  functions of, 429–432, 447, 453
  information sharing, 432
  NAS (network attached storage) devices, 481
  normalizing data, 449–451
  object-oriented databases, 436
  object-relational databases, 436–437
  records, 435
  referential integrity, 451
  relational databases, 436, 448, 451–452
  reports, 448
  SANs (storage area networks), 481
  software for, 114–115
  storage space requirements, 454
  versus lists, 430–431
  viewing and sorting data, 442
data-flow diagrams, 389
date fields, 434–435
DBMS. *See also* databases
DC-RW drives, 46

DDoS (distributed denial of service) attacks, 273
debugging programs, 408
decentralized networks, 475
decimal (Base 10) numbers
  representing in binary systems, 351
  converting to binary numbers, 350
  negative numbers, 351
  notation for, 349
decision points (algorithms)
  binary decisions, 394
  loops, 395–396
decision support systems (DSS), 460–461
decode stage (CPU machine cycle), 355, 357–359
decompressing
  digital signals, 315
  files, 195
dedicated servers (client/server networks), 478–480
Deep Space Network, 541–542
default values (database fields), 438
defragmenter utilities, 193, 195–197, 237
Delete key, 34
deleting
  files, 191–192
  software, versus uninstalling, 137
Dell Computer training resources, 214
demodulating digital signals, 68
Deneba Canvas, 127
denial of service (DoS) attacks, 273–274
derived class, 399
description approach (data mining), 462
design phase
  CPU design, 363
  database design, 433
  during systems development, 389
desktop computers
  microprocessors for, 354
  operating systems, for, 168–173
  setting up, 50–51
  system units, 44–50
  versus laptop computers, 335
desktop publishing (DTP) programs, 124–125
desktops
  customizing displays, 192–194
  features and elements, 184–185
  icons on, 138

  opening multiple windows, 185–186
  screen resolution, 231
Details view (My Documents), 188–189
development phase (SDLC), 389
devices, device drivers
  comparison table, 312
  defined, 257
  for laptop computers, 312, 332–335
  locating, 179
  managing of, by operating system, 179
  mobile, 310–316
  for MP3 players, 312–313, 317, 319–321
  paging devices, 312–313
  for PDAs (personal digital assistants), 312, 321–326
  for tablet PCs, 312, 329–331
  testing during boot process, 181
  troubleshooting boot problems, 183
  versus PnP (Plug and Play), 179
devices, storage
  CD and DVD disk drives, 225–226
  flash memory cards, 224
  floppy and Zip disks, 224–225
  hard disk drives, 221–223
  nonvolatile, 220–221
  upgrading, 226–227
DHCP (Dynamic Host Configuration Protocol), 525
diagnostic tools, Safe Mode, 183–184
diagrams, technical, software for, 126–127
dial-up connections, 68–69
dictionary, data (databases), 437–439
Difference Engine (Babbage), 382
Dig Deeper feature, 22
digital cameras, 154
  downloading images from, 156–157
  flash memory cards, 155
  image-editing software, 126
  image quality, 154–155
  in PDAs (personal digital assistants), 322
  operating systems for, 168–170
  ports for, 48
  prices, 155
  using, 155–156
digital data, 151–153. *See also* data2
digital divide, 366
digital entertainment, 151–159
digital fingerprint readers, 301

digital pens, 329–330
Digital Print Order Form (DPOF), 160
Digital Research, 379
digital signal processors, 315
digital signals
　modulating/demodulating, 68
　processors for, 315
Digital Subscriber Lines (DSL), 69–70, 261–262
digital video
　editing, 127–128, 163, 165
　file formats, 164–165
　prices, 160
　saving to DVD, 165
　versus analog video (camcorders), 164–165
　video cameras, 160–162
digital-to-analog converter (DAC) chips, 152
digitizing
　photographs, digitizing process, 151–153
　video images, 161
DIMMs (Dual Inline Memory Modules), 218
directories
　file, 187–188
　subject (Internet), 79, 80, 82
discounts for software, 134
discussion groups, newsgroups, 90
disk check-and-repair utilities, 197
Disk Cleanup (Windows) utility, 193, 195–198, 237
disk defragmenter utilities, 193, 195–197, 237
Disk Doctor (Norton), 197
Disk First Aid utility, 197
Disk II floppy disk drive, 379
Disk Operating System (DOS; MS-DOS), 170, 379
Disk Properties menu (Windows), 197
disk speeds, hard drives, 223
Display Properties dialog box (desktop), 231
display screens
　smart displays, 331
displays
　color quality, 229
　desktop, customizing, 192–194, 231
　magnifying, 199
　and monitor size, 230
　for PDAs (personal digital assistants), 322

smart displays, 331
　for tablet PCs, 329
distributed denial of service (DDoS) attacks, 273
DNS (domain name servers), 528–529
.doc files, 190
docking stations, 330
documentation
　defined, 409
　during systems development, 389
documents
　creating, 108
　digitizing, 156
　formatting, 109
　printing, 177
　saving, 110
　Web (HTML), 531–532
Dogpile search engine, 80, 82
Dolby Digital 5.1, 232
domain names
　domain name servers (DNS), 528–529
　in e-mail communications, 537
　registry for, 528
　in URLs, 77–78
DoS (denial of service) attacks, 273–274
DOS (Disk Operating System), 170, 379
dot pitch (monitors), 231
dot-matrix printers, 40–41
dots per inch (dpi), 42
dotted decimal notation (IP addresses), 524–525
Download.com, 135
downloading
　audio files, ethical issues, 319–321
　FTP (File Transfer Protocol) for, 529
　images from digital cameras, 156–157
　software, 135–137
dpi (dots per inch), 42
DPOF (Digital Print Order Form), 160
Dragon Naturally Speaking (ScanSoft), 118
DragonBall (Motorola) PDA processor, 322
DRAM (dynamic RAM), 212, 218, 360
drawing software, 126–127
Dreamweaver (Macromedia), 128, 414
Driver Zone Web site, 179

drivers, device, 179
driving and using cell phones, 314
drop-on-demand technology, 43
DSL (Digital Subscriber Lines), 69–70, 261–262
DSS (decision support systems), 460–461
DTP (desktop publishing) programs, 124–125
Dual Inline Memory Modules (DIMMs), 218
dual processor CPU design, 365
Dummies Daily productivity tips, 116
DVD disk drives, 46
　care and maintenance, 226
　copying DVDs, ethical issues, 228
　how they work, 225
　speed, 226
　storage capacity, 224
　versus CDs, 226
DVD players, digital video, 160, 165
DVD Workshop (Ulead), 165
DVD-R (DVD-Recordable) disks, 226
Dvorak keyboards, 32–33
dye-sublimation printing, 159
dynamic addressing, 276, 525
Dynamic Host Configuration Protocol (DHCP), 525
vdynamic RAM (DRAM), 218, 360

E

EarthLink ISP, 74, 86
earthquake prediction data, 503
Easy Install (Aladdin Systems), 193
eBay, 91
EBCDIC (Extended Binary Coded Decimal Interchange Code), 352
Eckert, J. Presper, 383
e-commerce (electronic commerce), 90–92
EDI (Electronic Data Interchange), on extranets, 540
editing
　code, 406
　digital images, photographs, 157–158
　digital video, 162–163, 165
Edtech-cps.com, 134
.edu domains, 78
EducateU (Dell) training site, 214
education
　advanced Internet applications for, 543

computer use in, 10–11
educational software, 129–130
edutainment, 130
EISA (Extended Industry Standard Architecture) buses, 361, 363
electrical appliances and power line networks, 258
electrical hazards, power surges, 295–296
electrical wires, data transmission using, 486
electromagnetic interference (EMI), 487
electronic commerce (e-commerce), 90–92
Electronic Data Interchange (EDI), 540
electronic encyclopedias, 130
velectronic mail. *See also* e-mail
Electronic Numerical Integrator and Computer (ENIAC), 348, 383–384
electronic switches, 348–349
electrotextiles, 320
elements (HTML), 531
e-mail
    accounts for, 86
    forwarding services, 87
    hoaxes, 87
    managing, organizing, 84
    popularity and advantages, 66, 83
    serbvers (client/server networks), 480
    software for, 84
    spam, 85–87, 302–303
    virus infections, 279
embedded scripts, JavaScript, 533
emergency services, software tools, 123
EMI (electromagnetic interference), 487
Empty the Recycle Bin option, 192
Encarta, 130
Encore DVD (Adobe), 165
encryption, 531, 537–538
encyclopedias, electronic, 130
engines, search, 80–83
ENIAC (Electronic Numerical Integrator and Computer), 348, 383–384
entertainment software, 131
Entertainment Software Rating Board (ESRB), 131
entertainment, digital, 151–159
environmental hazards, computer damage from, 295–297
equations in spreadsheets, 111

ergonomics, 50–51
error messages, codes
    during boot process, troubleshooting, 182–184
    hexadecimal notation in, 353
errors in software code
    debugging process, 408
    runtime (logic)errors, 408
    syntax errors, 407–408
    correcting, 392–393, 409
ESRB (Entertainment Software Rating Board), 131
estimation approach (data mining), 462
Ethernet networks
    adapters, 494
    Fast Ethernet, 234
    hubs, 261
    NICs (network interface cards), 259
    RJ-45 ports, 233
    routers, 261–262
    UTP cable, 260
ethical issues
    computer fluency, 15–16
    copying CDs and DVDs, 228
    digital divide, 366
    Internet-related, 89
    MP3 players, 319–321
    software piracy, 139
Eudora, 86
European Organization for Nuclear Research (CERN), 530
events, operating system coordination of, 176–177
evaluation, during systems development, 391
exabytes, 454
Excel (Microsoft), 110, 380
    creating, managing lists in, 430
    filename extensions, 190
    linking spreadsheets to Internet, 113
exception reports (MIS), 459
Excite search engine, 80
executable programs (.exe files), producing, 405
execute stage (CPU machine cycle), 355, 359
Executive (Osborne), 378
expansion buses, 361–362
expansion cards, 48
expansion hubs, 236
expert systems, 461
exporting. *See also* output, output devices

data from databases, 448
digital video, 164–165
Extended Binary Coded Decimal Interchange Code (EBCDIC), 352
Extended Industry Standard Architecture (EISA) buses, 361, 363
Extensible Hypertext Language (XHTML), 535
Extensible Markup Language (XML), 415, 534–535
external data sources
    data warehouses, 454
    DSS (decision support systems), 460
external hard drives, 46, 227
external management models, 461
extracting data, 456
extranets, 540

F

fabrics, electronic, 320
fair use, academic, 89
falls, computer damage from, 295
FAQs (frequently asked questions), 133
FarStone RestoreIT!, 193
Fast Ethernet (100 Base-T), 234
FAT (File Allocation Table), 196
Favorites feature, 79
FBI (Federal Bureau of Investigation) fingerprint identification databases, 447
fees for ISP network access, 519
Fetch file transfer software, 76
fetch stage (CPU machine cycle), 355–357
fiber optic technology, 70, 255, 487–488
fields (databases), 115, 433
    adding, 439
    computed/computational fields, 434
    constraint settings, 440
    data input forms, 439
    date fields, 434–435
    default values, 438
    Field Properties box (Access), 438
    Field Properties Table (Access), 442
    field size, 435
    hyperlink fields, 434–435
    identifying during database creation, 437

in data dictionaries, examples of, 438
memo fields, 434–435
naming, 434
numeric fields, 434
object fields, 434–435
primary key field, 435
size, 435
sorting data in, 442
text fields, 434
validating data in, 439–441
viewing data in, 442
fifth-generation programming languages (5GLs), 401
File Allocation Table (FAT), 196
file compression
digital images, 155
utilities for, 193–195
file exchanges, ethical issues, 320
file formats
choosing during save process, 190
executable files, 405
for digital video, 128, 164–165
namign conventions, 190
file management
hierarchical directory structure, 186–188
identifying storage location, 190–191
keyboard shortcuts, 34
My Documents folder, 188
naming conventions, 190
systems for, history, 380
file paths, 190–191
file servers (client/server networks), 76, 478, 520
file sharing services (Internet), 520–521
file systems, 196
file transfer process, 76, 252
File Transfer Protocol (FTP), 66, 77
anonymous FTP archives, 529
functions, 524
how it works, 529
file types, listing of, 190
filename extensions, 128, 164–165, 190–191
filenames, 190
files. *See also* file management
audio, 317, 319–321
backing up, 198, 306–307
database, 435
defined, 187
managing, 191–192
naming conventions, 190
organizing, 187

storing, 190–191
viewing lists of, 188
filters, for e-mail, spam, 84, 302–303
finances
management software, 121–122
online activities, 91
record-keeping spreadsheets, 112
Find in File (Search) tool, 34
Finder (Macintosh), 187
fingerprint identification databases, 447
fingerprint readers, 301
firewalls, 275–278, 301, 501
FireWire (IEEE 1394) ports, 47, 161–162, 233
First Internet Software House, 134
first-generation computers, 384
first-generation programming languages (1GLs), 400
Flash animations, 415
flash card readers, 156–157
flash memory cards, 46, 155, 224
digital video cameras, 161
MP3 players, 317, 319
PDAs (personal digital assistants), 324
pros and cons, 227
storage capacity, 224
Flash Player (Macromedia), 95
flat-panel monitors, 31, 38–39, 331
flexible organic light-emitting displays (FOLEDs), 52
float data type (C++), 403
floating-point notation, 351
floppy disks, drives, 46, 224
and boot errors, 182
caring for, 225
disk drives for, 18
history, 379, 381
how they work, 224
storage capacity, 224
flowcharts, 389, 396
fluency, computer
and career options, 6–11
attributes, 4–5
ethical issues, 15–16
home uses, 11–12
folders, 187–188
FOLEDs (flexible organic light-emitting displays), 52
For keyword (Visual Basic), 402
forbidden characters in filenames, 191
foreign keys (databases), 451–452
forensics, computer, 9–10

formatting
documents, 109
HTML tags for, 413, 531
keyboard shortcuts, 34
spreadsheets, 112
forms
for data input, 439
templates, 115
formulas
computed/computation fields, 434
in spreadsheets, 111
FORTRAN programming language, 401
Forward buttons, 78
forward software compatibility, 136
four-color printers, 42
fourth-generation computers, 384
fourth-generation programming languages (4GLs), 401
fps (frames per second), 161
frames, 493, 495
Free Software Foundation GNU project, 173
freeware, 135–136
FreewareHome.com, 135
frequently asked questions (FAQs), 133
Friendly Robotics, 12
FROM statements (SQL), 445
front panel (system units), 44–46
FrontPage (Microsoft), 128, 414
FrontRange Solutions Gold Business Contact Manager, 123
FTP (File Transfer Protocol), 66, 77, 529
anonymous FTP archives, 529
functions, 524
how it works, 529
FTP Voyager, 529
full software installation, 137
function (F) keys, 33
functions
function code, 404, 406
in spreadsheet cells, 111
fuzzy logic, 461

G

games
computer, 404
consoles for, 169
controls for, 37
interactive, 95
software for, 131
using grid computing, 269

Gannt charts, 122–123
Gates, Bill, 376, 379
GB (gigabyte), 18
generations, of programming languages, 400–401
geocaching, 328
geographical position systems (GPS), 128
Get Input task (top-down algorithm design), 397
Ghost (Norton), 193
GHz (gigahertz), 215, 355
Gibson Research firewall testing service, 275
gigabyte (GB), 17–18
gigahertz (GHz), 215, 355
Global Positioning Systems (GPS), 123, 543
   embedding in textiles, 320
   how it works, 328
   for PDAs, 326
   uses for, 328
GNOME user interface (Linux), 173
GNU project (Free Software Foundation), 173
Gnutella, 321
GoBack utility (Roxio), 192
GoldMine Business Contact Manager (FrontRange Solutions), 123
Google search engine, 80, 82, 134
.gov domains, 78
GPS (Global Positioning Systems), 123, 543
   CAD (computer-aided design) software with, 128
   embedding in textiles, 320
   for PDAs (personal digital assistants), 326
   how it works, 328
   uses for, 328
Graffiti notation with tablet PCs, 330
Graffiti text system, 322
grammar checking tools, 109
graphic designers, interface designers, 390
graphical data, 434–436
graphical user interface (GUI), 176–177
   desktops, 184–186
   history, 380–381
graphics, graphics file
   computer-aided design software, 128
   players for, 95
   sending to cell phones, 316
   Web page authoring software, 128
graphics software, 123–127
graphs (spreadsheets), 112
grid computing, 269
groupware, 132
GUI (graphical user interface), 176–177, 184–186
   history, 380–381

H

hackers, crackers, 270, 272, 498–499
   computer access approaches, 274–275
   denial of service (DoS) attacks, 273–274
   identity theft, 270–272
   Trojan horses, 272–273
   understanding threats from, 4
handheld computers, 8, 321
handshaking, handshakes, 526
handspring Treo cell phone, 327
hard disks, drives, 18
   data retrieval speed, 217
   data transfer rates, 221–222
   defragmenting, 196–197, 237
   how they work, 221–222
   popularity of, 221
   ROM (read-only memory), 49
   storage capacity, 221, 223
   upgrading, 226–227
hardware *See also* mobile devices
   cable modems, 70
   defined, 17
   firewall devices, 275
   input devices, 18
   for Internet connections, 73
   managing of by operating systems, 179
   modems, 68
   output devices, 18
   storage devices, 18
   system unit, 18
Harvard Mark I, 383
hazards, computer
   data loss, 305–307
   online hazards, 302–304
   physical hazards, 295–297, 311, 314
   theft prevention, 297–299
   unauthorized access, 299–301
head crashes (hard disk drives), 222
health professions, use of computers in, 8
heat, damage from, 295
heat sinks, 217
Help menus, 133

help systems
   accessing using keyboard shortcut, 34
   for software, 133–134
hertz (Hz), 215, 355
hex addressing, 525
hexadecimal number system, 351
hibernation (system units), 45
hierarchical file management structure, 186–188
high-density memory, 325
high-gloss paper, 41
high-speed Internet access, 541
high-speed ports, 237
historical data, 454
historical recreations, 132
History lists, on Web browsers, 78
history, of computers
   Altair 8800, 376
   Apple Computer Company, 377
   application software, 380
   Commodore, Tandy computers, 378
   early computers, 382–384
   electronic switches, vacuum tubes, 348–349
   floppy disk drive, 379
   GUI (graphical user interface), 380–381
   IBM PCs, 378
   Internet, 66
   Microsoft, 376
   operating systems, 379, 381
   Osborne computer, 378
   programming languages, 379
   web browsers/Internet, 381
Hitachi Microdrive, 325
Hoaxbusters (U.S. Department of Energy), 87
hoaxes, e-mail, 87
Hollerith (Herman) Tabulating Machine, 383
home computers, 11–12
home networks, 11–12, 475
   configuring network, 254
   configuring software for, 268
   cost comparisons, 267
   data transfer rate comparisons, 267
   P2P (peer-to-peer) setup for, 266–267
home page, 76
Home Phone line Networking Alliance (HPNA), 258
HomePlug Power line Alliance, 256
Hopper, Grace, 383

INSTALLING SOFTWARE 597

horizontal tiling, 185–186
hosts, 77
"hot spots" for wireless connections, 265
hot-swappable bays, 332
Hotmail e-mail accounts, 86
household appliances, programs for, 416
HPNA (Home Phone line Networking Alliance), 258
H&R Block TaxCut software, 121
.htm/.html files, 190
HTML (Hypertext Markup Language), 66, 413, 531, 535
 embedding scripts in, 533
 functions and limitations, 532
 software for, 414
 tags and elements, 531
 versus XML, 534
 viewing HTML code, 531–532
HTTP (Hypertext Transfer Protocol), 76–77, 524, 531
https:// URLs, 92, 531
hubs (client/server networks), 256, 478, 496
 with Ethernet networks, 261
 expansion, adding to system, 236
 with phone line networks, 258
 with power line networks, 258
 in star networks, 484–485
hyperlinks, 66, 76, 531
 HTML tags for, 531
 hyperlink fields, 434–435
 navigating web sites using, 78
Hypertext Markup Language (HTML), 66, 413, 531, 535
 embedding scripts in, 533
 functions and limitations, 532
 software for, 414
 tags and elements, 531
 versus XML, 534
 viewing HTML code, 531–532
Hypertext Transfer Protocol (HTTP), 76–77, 524, 531
HyperTransport bus specification, 362
Hz (hertz), 215, 355

I

IAB (Internet Architecture Board), 516
IAFIS (Integrated Automated Fingerprint Identification System), 447
IBM PC, 378

ICANN (Internet Corporation for Assigned Names and Numbers, 516, 524
 IP address management, 528
 top-level and second-level domains, 525
Icon view (My Documents), 188
icons
 on graphical desktops, 176
 opening software uisng, 138
 padlock (Secure Sockets Layer), 531
ICQ instant messaging, 539
IDE (Integrated Drive Electronics) hard drives, 223
ideal system, identifying before buying, 213–214
identification chip implants, 14
identity theft, 4, 270–272, 498
identity, protecting in chat rooms, 88
IDEs (integrated development environments), 406–408
IE (Internet Explorer, Microsoft), 76, 86
IEEE (Institute of Electrical and Electronic Engineers)
 FireWire ports, 47, 233
 floating-point standard, 351
IETF (Internet Engineering Task Force), 516
if else keywords (C++), 402
illustration software, 126–127
Illustrator (Adobe), 126
IM (instant messaging), 66
 audio conversations, 88
 how it works, 539–541
 versus chat rooms, 90
iMac (Apple), 44
image quality (digital cameras), 154–155
images
 backgrounds, on desktops, 194
 digitizing, 156
 editing, software for, 124–126, 157–158
 printing, 42
impact printers, 40–41
Import External Data feature (Excel), 113
importing
 data from databases, 448
 digital video, 162–163
incomplete data, problems from, 431
Indeo codec, 165
indexer programs (search engines), 80

indexing database data, 442
Industry Standard Architecture (ISA) buses, 361, 363
.info domains, 78
information
 complex, databases for, 430–431
 defined, 17
 sharing, databases for, 432
 versus data, 16, 430
information systems, 388
 DSS (decision support systems), 460
 functions, 457
 knowledge-based systems, 461
 MIS (management information systems), 459
 model management systems, 460
 OSS (office support systems), 457–391
 SDLC (system development life cycle), 388
 TPS (transaction processing systems), 458–459
information technology careers, 555–556, 558
Infrared Data Association (IrDA) ports, 233–234, 325
infrastructure, of Internet, 516
inheritance (object-oriented algorithms), 398
initial value (loops), 396
injuries, avoiding, 50, 119
ink-jet printers, 41, 43, 159
input, data
 databases, 439
 error-handling approaches, 392–393
input devices, 18
 cellular phones, 314
 for audio, 37–38
 function of, 32
 game controls, 37
 keyboards, 32–35, 37
 for laptop computers, 36–37
 mouse, 35–36
 for PDAs (personal digital assistants), 321
 for tablet PCs, 329–330
Insert key, 34
installation costs, cable networks, 487
installation phase (SDLC), 391
Installation Wizard, 136
installing software, 136–137
 utilities for, 193–194
 versus booting, 182

instant messaging (IM), 66
　audio conversations, 88
　how it works, 539, 541
　versus chat rooms, 90
Institute of Electrical and Electronic Engineers (IEEE)
　floating-point standard, 351
　Wi-Fi standard, 262
instructions, computer (CPU)
　defined, 18, 353
　machine language, 357–359
insurance companies, data warehouses, 454
int (integer) data type (C++), 403
Integrated Automated Fingerprint Identification System (IAFIS), 447
integrated circuits, 348–349, 384
integrated development environments (IDEs), 406–408
Integrated Drive Electronics (IDE) hard drives, 223
Integrated Services Digital network (ISDN), 69, 71
integrated software, pros and cons, 116–117
integrity, data
　maintaining, 433
　enforcing in relational databases, 451
Intel processors, 214, 354, 384
　Centrino, 333
　Pentium M for laptop computers, 332
　Pentium M low-voltage processor, 331
　PRO/Wireless network connections, 333
　3GIO (Third Generation I/O) standard, 362
interactive Web sites, 414–415, 533–534
interface cards (client/server networks), 478, 491, 494
interfaces, tunneling, for VPNs, 540
internal data sources (DSS), 460
internal drive bays, 46
internal management models, 461
International Business Machines (IBM), 383
Internet
　advanced technologies
　　enhanced resource access, 543
　　high-speed access, 541
　　Internet-enabled appliances, 543
　　Internet2, 541

interplanetary communications, 541–542
　Large Scale Networking (LSN), 541
　virtual reality, 542–543
backbones (Tier I), 67, 74, 517–518
careers opportunities, 535
chat rooms, 87–88
clickstream data-capture tools, 455
client/server network structure, 66–67, 520
connections
　broadband options, 68–69
　cable Internet, 70–71
　cell phones, 316
　comparison table, 73
　connection speed, 68
　dial-up, 68–69
　DSL (Digital Subscriber Lines), 69–70
　emerging technologies, 74
　Ethernet networks, 262
　ISDN (Integrated Services Digital Network), 71
　ISPs (Internet service providers), 517–521
　PDAs (personal digital assistants), 326
　routers, 261
　satellite Internet, 72
　sharing using networks, 253
　wireless networks, 264
downloading software from, 134–137
e-commerce, 90–92
e-mail, 83–87, 536
　attachments, 537
　encryption, 537–538
　protocols for, 536
　SafeMessage program, 538–539
　security risks, 537
　servers, 536
ethical issues, 89
freeware and shareware, 135
GPS (Global Positioning System), 543
history, 66, 381
HTML (Hypertext markup language), 66
hyperlinks, 66
IM (instant messaging), 88, 90, 539, 541
infrastructure ownership, 516
interactivity
　applets, 533–534

　CGI (Common Gateway Interface), 532–533
　client-side applications), 533
　embedded scripts, 533
　HTMLs, 535
　XMLs, 534–535
Internet cache, 528
interplanetary Internet, 541
IP addresses, 524–525, 528–529
　linking Excel spreadsheets to, 113
　locating software tutorials, 133
multimedia, 94–95
multiple access, 251
as network, 476, 517
navigating, 76–79
newsgroups, 90
packet-switching methodology, 522–523
peer-to-peer networks, 520–521
protocols, 66, 76–77, 522
　FTP (File Transfer Protocol), 77
　http (Hypertext Transfer Protocol), 76–77
　open systems, 522–523
　PTP (Parcel Transfer Protocol), 541–542
　speed versus accuracy, 527
　SSL (Secure Sockets Layer), 531
　TCP/IP protocol suite, 523–524, 527, 529
pop-up window blockers, 303–304
research using, 10–11
risks from using, 302–304
software help, 133
spyware, adware, 304–305
telephony, 538
temporary files, 86
using, 4, 65–66
web browsers, 75–76, 530
web documents, 531–532
web searches, 79–83
World Wide Web, history of, 530–531
Internet Architecture Board (IAB), 516
Internet Corporation for Assigned Names and Numbers (ICANN), 516, 524
　IP address management, 528
　top-level and second-level domains, 525
Internet Engineering Task Force (IETF), 516
Internet Explorer (Microsoft), 75, 381
　Favorites menu, 79
　removing temporary files, 86

Internet protocol (IP)
  domain names, 516, 524–525, 528–529
  dotted decimal notation, 524
  firewall protection, 276–278
  functions, 523–524
  home computer address, identifying, 525
  IP addresses, 67
  IPv4 versus IPv6, 524–525
  managing allocation of, 516, 524
  static versus dynamic addressing, 525
  versus MAC (media access control) addresses, 495
Internet service providers (ISPs), 21, 73–75, 481
ISPs (Internet service providers), 73–74, 481
  choosing, 75
  hierarchical structure, 517
  NAPs (network access points), 519–520
  POPs (points of presence), 519, 521
  T-lines, 518
  Tier 1 (backbones), 517–518
  Tier 2 (regional, national), 518
  versus OSPs (online service providers), 75
  wireless devices, 75, 316
Internet Society (ISOC), 516
Internet2, 541
interplanetary Internet, 541–542
interpreters versus compilers, 405
interrupts, 177
intranets, 476, 540
Intuit
  Master Builder, 123
  QuickBooks, 122
  Quicken, 121–122
  TurboTax, 121
inventory management, 6
Internet protocol (IP)
  domain names, 516, 524–525, 528–529
  dotted decimal notation, 524
  firewall protection, 276–278
  functions, 523–524
  home computer address, identifying, 525
  IP addresses, 67
  IPv4 versus IPv6, 524–525
  managing allocation of, 516, 524

static versus dynamic addressing, 525
  versus MAC (media access control) addresses, 495
ipconfig command, 525
iPod MP3 player (Apple), 317–318
IrDA (Infrared Data Association) ports, 233–234
  wireless transfers using, 325
iRobot, 12
ISA (Industry Standard Architecture) buses, 361, 363
ISDN (Integrated Services Digital Network), 69, 71
ISOC (The Internet Society), 516
ISPs (Internet service providers), 73–74, 481
  choosing, 75
  hierarchical structure, 517
  NAPs (network access points), 519–520
  POPs (points of presence), 519, 521
  T-lines, 518
  Tier 1 (backbones), 517–518
  Tier 2 (regional, national), 518
  versus OSPs (online service providers), 75
  wireless devices, 75, 316
IT (information technology)
  careerschoosing, criteria for, 558
  job outlook, 555–556
  security professionals, 283
italics formatting, 34

J

jacks, RJ-11 versus RJ-45, 260
Jacquard (Joseph) Loom, 382
jam signals (star networks), 485
Jasc Paint Shop Pro, 124, 158
Java programming language, 412–413
  applets, 412, 533–534
  Java Server Pages (JSP), 414–415
  Java Virtual Machine (VM), 412
  Javabeans, defined, 411
JavaScript programming language, 414, 533
Jobs, Steve, 377
join queries (SQL), 445
JourneyEd.com, 134
joysticks, 37, 131
JPEG file format, 155
JSP (Java Server Pages), 414–415
junk e-mail, 86

K

Kahn, Robert, 66
Kazaa file sharing service, 321, 520–521
KB (kilobyte), 17–18
kernel (operating system), loading during boot process, 182
kernel memory, 218
Kernel Memory Table (Windows Task Manager), 218
key fields, 435
key pairs (public-key encryption), 537
keyboard shortcuts
  accessing Search tool, 34
  accessing Task Manager, 197
  bold formatting, 34
  closing current window, 34
  copying text, 34
  cutting text, 34
  italics formatting, 34
  opening existing files, 34
  opening new documents, 34
  opening Start menu, 34
  opening Windows Help, 34
  pasting text, 34
  printing, 34
  saving documents, 34
  warm boots, 45
keyboards
  Dvorak layout, 32–33
  ergonomic, 50
  keystroke shortcuts, 34
  for laptop computers, 34–35
  maintaining, 33
  navigation keys, 34
  onscreen, 200
  for PDAs (personal digital assistants), 34–35, 37
  QWERTY layout, 32
  standard keys, 33–34
  toggle keys, 34
  wireless, 35
keys (database), 451–452
keys, encryption, 538
keywords, 80, 402
Kilby, Jack, 384
Kildall, Gary, 379
kilobyte (KB), 17–18
KIWE (Kid's Internet World Explorer) browser, 76
Klingonese passwords, 301
Knowledge Base (Microsoft), 134
knowledge-based systems, 461
Kodak DCS Pro 14n, 154
Kyocera Smartphone, 315

600 LABELS

L

labels (spreadsheets), 111
languages
  binary, 17, 348
  HTML (Hypertext Markup Language), 66
  machine versus assembly, 357–358
  markup, 531
    HTML, 413–415, 534–535
    XML, 415, 534–535
  natural, NLP systems for, 461
  scripting, 414, 533
  SQL (structured query language), 443–446
  programming, 400, 409
    C/C++, 412
    choosing for specific projects, 410
    data types, 403
    interactive web pages, 414
    Java, 412–413
    JavaScript, VBScript, 414
    language generations, 400–401
    Visual Basic, 411–412
  keywords, 402
  learning, recommended languages, 410
  operators, 403
  portability, 401
  statements, 402
  syntax, 402
LANs (local area networks), 254, 476, 540
laptop computers, 312, 332
  accessories, 335
  batteries, 334–335
  hardware, 332
  input devices, 34–37
  operating systems, 168, 333
  PC cards, 68
  ports and connections, 333
  processors, 332
  theft prevention, 297–298
  upgrading, 335
  versus desktop computers, 335
  wireless connections, 334
Large Scale Networking (LSN), 541
laser printers, 41, 43
latency (hard disk drives), 222
law enforcement, computer use in, 9–10
layers (OSI networks), 492–493
LCD (liquid crystal display) monitors, 38
  customizing, 193
  how they work, 39
  organic displays (OLEDs and FOLEDs), 52–53
  performance, 39
  upgrading to, 230
  versus CRT (cathode ray tube) monitors, 40, 52
LeakTest (Gibson Research), 278
legal issues, Internet-related
legal reference tools, 130
letter characters, notation systems for, 351–352
Levels 1, 2 and 3 cache memory, 357
Li-ion (lithium-ion) batteries, 335
library automation software, 123
licenses, software, 139
lightening, computer damage from, 296
lighting, ergonomic, 51
Lindows, 174
line charts, 112
linear bus network topology, 481–482
links
  hyperlinks, 66, 531
  between tables in relational databases, 436
Linux operating system, 168, 173
liquid crystal display (LCD) monitors, 38
  customizing, 193
  how they work, 39
  organic displays (OLEDs and FOLEDs), 52–53
  performance, 39
  upgrading to, 230
  versus CRT (cathode ray tube) monitors, 40, 52
Lisa (Apple), 381
List view (My Documents), 188
lists, data in, 430–431
lithium ion batteries, 335
local area networks (LANs), 254, 476, 540
local buses, 361
local ISPs (Tier 3), 74, 517–518
locks
  for desktop computers, 298
  for laptop computers, 335
logic (runtime) errors, testing for, 408
logic bombs, 280
logical ports, 2750276
LoJack transmitters, 299
LookSmart subject directory, 80
loops (algorithms), 395–396, 402
Lotus
  1-2-3, 110, 380
  SmartSuit, 120
LSN (Large Scale Networking), 541
Lycos subject directory, 80
Lynx (RTOS) web browser, 76, 168

M

MAC (media access control) addresses, 495–496
Mac OS, 168
Mac Tools, 195
machine cycle, 215
machine cycle (CPU)
  control unit, 356
  cycles per second (hertz), 215
  decode stage, 355, 357–359
  execute stage, 355, 359
  fetch stage, 355–357
  storage stage, 355, 359
  system clock, 355
  tasks, 355
machine dependent programming languages, 400
machine language, 357–358
Macintosh operating systems, 172
  Apple platform, 20–21, 381
  disk check-and-repair utilities, 197
  file compression utilities for, 195
  file naming conventions, rules, 191
  Mac OS, 172
  microprocessors for, 354
  Motorola processors on, 214
macro viruses, 280
Macromedia
  Dreamweaver, 128
  FlashAnimations, 95, 415
  Shockwave Player, 95
macros, 116
macular degeneration, chip implants for, 14
magnetic card readers, 500
magnifying screen displays, 199
MailWash spam filter, 303
mainframe computers
  operating systems for, 168, 170
  versus supercomputers, 21
maintenance, 4
  CDs and DVDs, 226
  floppy and Zip disks, 225
  following system development, 391
  keyboards, 33
  monitors, 39
  mouse devices, 36

system
  Task Scheduler utility, 199–200
  utilities for, 195–198
managing
  businesses, tools for, 122–123
  computers
    files, 186–188, 191–192
    hardware, devices, software, 179
    memory, 178
    processors, 176–177
  databases, 433, 437–446
  e-mail, 84–87
  personal finances, tools for, 121–122
  projects, 122–123
management information systems (MIS), 459
management models 460
mapping programs, 123
Marketing Plan Pro (Palo Alto Software), 122
markup languages
  features of, 531
  HTML, 413–415, 534–535
  XML, 415, 534–535
marts, data, 456
Massachusetts General Hospital Utility Multiprogramming System (MUMPS), 417
master boot record, 279
Master Builder (Intuit), 123
mathematical operations, 403
Mauchly, John W., 383
MB (megabyte), 17–18
McAfee Oil Change, 239
MCI, 74
.mdb files, 190
media access control (MAC) addresses, 495–496
media players, 128
media, digital
  advantages, 153
  cameras, scanners, 154–157
  versus analog, 150
media, transmission, client/server networks, 478, 486–491
medical applications, 8, 132
Medical Educational Technologies, Inc., 8
medical reference tools, 130
megabits per second, megabytes per second (mps), 221
megabyte (MB), 17–18
megahertz (MHz), 215
megapixels (MP), 154

memory
  cache memory, 356–357
  in cellular phones, 314
  CMOS memory, 181
  configuration of, checking during boot process, 182
  DRAM (dynamic RAM), 360
  management of, 178
  flash memory, 317, 319, 324
  PDAs (personal digital assistants), 321, 324–325
  and printer performance, 42
  RAM (random access memory), 49, 217–220, 356
  ROM (read only memory), 49, 361
  SRAM (static RAM), 361
  tablet PCs, 331
  volatile versus nonvolatile, 217
  VRAM (video RAM), 229
memory bound systems, 219
memory card readers, 228
memory cards/modules, 49
memory chips, capacity of, 212
memory effect, 335
memory modules, 218
memory resident files, 182
Memory Stick (Sony), 46, 319, 324
menu-driven user interface, 176
messages
  e-mail
    attachments, 537
    encrypting, 537–538
    SageMessage, 538–539
    security risks, 537
    sending and receiving, 536
  instant
    sending and receiving, 539, 541
    security risks, 541
meta search engines, 82
metadata, 439
methods (objects), 398, 436
MHz (megahertz), 215
Micro Instrumentation and Telemetry Systems (MITS), 376
microbrowser software, 316
microdisplays, 53
Microdrive (Hitachi), 325
microphones, 37–38
microprocessors
  evaluating
    clock speed, 215
    how it works, 214–215
    processor types, 214
    speed requirements, 215
    viewing CPU usage, 216
  history, 384

  performance improvements, 212
  printable, 358
Microsft Excel
  linking spreadsheets to Internet, 113
Microsoft
  Access, 114, 438, 442, 446
  Disk Operating System (MS-DOS), 170
  Excel, 110, 113, 190, 380, 430
  founding of, 376, 379
  FrontPage, 128, 414
  Internet Explorer, 75, 79, 86, 381
  Knowledge Base, 134
  Money, 121–122
  Movie Maker, 127
  MSN, 74
  .NET program, 134
  Office, 116–120, 457
    creating Web pages using, 128
    Office Assistant, 133
    with tablet PCs, 331
  operating systems, 168–171, 322–323, 381
  Outlook, 84
  Paint, 126
  PictureIt!, 126
  Pocket PC, 322–323
  PowerPoint, 113–114, 190
  Project, 122–123
  Streets & Trips, 123
  Transcriber PDA notation system, 322
  Visio, 126–127, 396
  Visual Basic programming language, 411–412
  Visual C++ programming language
    IDE for, 407
  Windows Explorer, 187
  Windows Media Player, 95
  Windows Messenger, 539
  Windows Update feature, 409
  Word, 108–110, 190, 448
  Works, 117
MIDI (musical instrument digital interface)
  players for, 95
  ports, 235
MIME (Multipurpose Internet Mail Extensions) protocol, 537
.mil domains, 78
mining data, 461–462
Minolta Dimage S414, 154
minus (–) sign, in decimal notation, 351

MIS (management information systems), 459
MITS (Micro Instrumentation and Telemetry Systems), 376
MMCs (Multimedia cards), 319
MMS (Multimedia Message Service), 316
mobile computing devices
    cellular phones, 312–316
    comparison table, 312
    laptop computers, 312, 332–335
    MP3 players, 312, 317, 319–321
    paging devices, 312–313
    PDAs (personal digital assistants), 309, 312, 321–326
    pros and cons, 310–311
    tablet PCs, 312, 329–331
mobile switching centers (cell phones), 315
model management systems, 460
modeling
    CAD (computer-aided design) software for, 128
    using DSS (decision support systems), 460
modems, 68
    cards for, 48
    connection speed, 68
    DSL modems, 70
    ports for, 46–47
modulating digital signals, 68
modules (program code), 404
Money (Microsoft), 121–122
monitors, 18
    bistable displays, 53
    CRT (cathode ray tube), 38–40
    installing, 231
    LCD (liquid crystal display; flat panel), 31, 38–39, 52–53
    microdisplays, wearable screens, 53
    positioning correctly, 50
    refresh rates, 231
    screen resolution, 230–231
    upgrading, 230–231
Moore's Law (Gordon Moore), 212, 363
Mosaic web browser, 381, 530
motherboard, 18, 49
    components on, 353
    CPU (central processing unit), 49, 217
    local buses, 361
    RAM (random access memory), 49, 218
Motorola
    DragonBall PDA processor, 322
    microprocessor chips, 214, 384
    PowerPC G4/G5 microprocessors, 354
    Talkabout T900 pager, 313
    T720 cellular phone, 315
mouse/mice, 35–36, 380
mousepads, 36
.mov files (QuickTime), 164
movement, of dancers, computer analysis of, 7
Movie Maker (Microsoft), 127
moving files, 191
Moving Picture Experts Group (MPEG), 164
MP (megapixels), 154
MPEG4 (.mpg) format, 164–165, 322
MP3 players, 95, 312, 317
    ethical issues, 319–321
    file transfers, 319
    flash memory, 317, 319
    non-music files on, 319
    ports for, 48
    storage capacity, 317
MS-DOS operating system, 168, 170
MSN (Microsoft) subject directory, 75, 80
multimedia
    defined, 94
    electronic encyclopedias, 130
    files, 436multimedia cards (MMCs), 319
    Multimedia Message Service (MMS), 316
    plug-ins and players for, 95
    software, 123–124
    streaming adio and video, 94
    video-editing software, 127–128
multipartite viruses, 281
multiple access (star networks), 485
multiple documents, printing, 177
multiple windows, 185–186
multiprocessor systems, 178
Multipurpose Internet Mail Extensions (MIME) protocol, 537
multitasking, 176
multiuser operating systems, 168, 170
MUMPS (Massachusetts General Hospital Utility Multiprogramming System), 417
.museum domains, 78
museum virtual tours, 10–11
musical instrument digital interface (MIDI) ports, 235–236
My Computer icon, 216
My Documents folder, 188–189
MySQL, 174

N

nack (negative acknowledgement), connectionless protocols, 527
.name domains, 78
naming computers for networking, 268
nanoscience, nanotechnology, 13–14, 336
nanoseconds, 217
nanostructures, 13, 336
NAPs (network access points), 519–520
Napster, 320
Narrator utility, 200
NAS (network attached storage) devices, 481
NASA (National Aeronautics and Space Administration), 516
NAT (network address translation), 277
National Center for Supercomputing Applications (NCSA), 530
national ISPs (Tier 2), 517–518
National Science Foundation (NSF), 516
natural language processing (NLP) systems, 461
natural programming languages, 401
navigating
    navigation arrows, on keyboards, 34
    navigation devices, client/server networks, 256, 478, 494–498
    web sites, 78–79
Navigator (Netscape), 76, 86
NEC Earth Simulator, 365
negative acknowledgment (nack), connectionless protocols, 527
NerdyBooks productivity tips, 116
.net domains, 78
.NET program (Microsoft), 134
netiquette, 90
NetMass online storage, 307
Netscape
    Communicator, 531
    Navigator, 76, 86
network access points (NAPs), 519–520
network adapters, 256, 262, 478, 491, 494
network address translation (NAT), 277
network attached storage (NAS) devices, 481

networking, networks. *See also* Internet
adapters, 256, 262
bandwidth, 255
career opportunities, 474, 499
centralized versus decentralized models, 475
client/server networks, 254, 520, 474–476
  navigation devices, 478, 494–498
  network adapters/interface cards, 478, 491, 494
  network topologies, 478, 481–486
  NOS (network operating system) software, 478, 491
  servers, 478–481
  transmission media, 478, 486–489
components, 255
connections, 68–74, 267–268
data coordination, 503
defined, 252, 474
device intercompatibility, 262
Ethernet networks, 261
extranets, 540
grid computing, 269
home networks, 267–268
Internet access, 251
intranets, 476, 540
LANs (local area networks), 254, 476
LSN (Large Scale Networking), 541
navigation devices, 256
network adapters, interface cards (NICs), 48, 71, 256, 479, 491
  for Ethernet networks, 259, 494
  wireless, 494
network layer (OSI networks), 492
operating systems for, 168, 170, 256, 478, 491
passwords, 182
protocols, 491
P2P (peer-to-peer) networks, 253
  choosing, 266–267
  Ethernet networks, 259–262
  file sharing services, 520–521
  phone line networks, 258
  power line networks, 256–257
  wireless networks, 262, 264–265, 267
SANs (storage area networks), 481
scalable networks, 475
security risks and tools, 251, 498
  authentication and access restrictions, 498–499
  authentication servers, 500

backup procedures, 502
firewalls, 501
hackers, 498
restricting physical access, 500
topologies, 478, 481–486
transmission media, 255
ubiquitous networks, 332, 473
VNC (Virtual Network Computing), 490
VPNs (virtual private networks), 540
WANs (wide area networks), 254, 476–477
wireless, 264, 476
uses for, 251–253, 473–474
ubiquitous networks, 332
Network Solutions domain name registry, 528
New Technology File System (NTFS), 196
newsgroups, 90, 214
newsgroups, online
Next generation Internet, 541
Next keyword (Visual Basic), 402
NeXT operating system, 530
NIC (network interface card), 71
nickel-cadmium (Ni-Cad) batteries, 335
NICs (network interface cards), 256, 478, 491
  device drivers, 494
  for Ethernet networks, 259
  wireless, 494
Nikon Coolpik camera, 154
NLP (natural language processing) systems, 461
nodes, network, 252, 256, 483
Nokia 3650 cellular phone, 315
NOMAD MuVo MP3 player (Creative Labs), 318
Non-system disk or disk error message, 182
nonimpact printers, 40–43
nonprocedural programming languages, 401
nonvolatile memory, 314, 324
nonvolatile storage devices, 49, 217, 220–221
  hard disk drives, 221–223
  portable storage, 224–226
  CD and DVD disk drives, 225–226
  flash memory cards, 224
  floppy and Zip disks, 224–225
normalizing data (relational databases), 449–451

Norton
  CleanSweep, 193
  Disk Doctor, 197
  Ghost, 193, 307
  SystemWorks, 193
  Utilities, 195
NOS (network operating system) software, 256, 478, 491
NOT operator, 82
notation systems, for PDAs, 322
notebook computers, 332
notes view (PowerPoint), 114
Novell Netware, 168, 170
NSF (National Science Foundation), 516
NTFS (New technology File System), 196
Num Lock key, 34
number systems
  base 10, 349
  binary, 350–352
  defined, 349
  hexadecimal, 351
numbers
  in calculations, 434
  computed/computational field, 434
  in databases, 440
  dotted decimal notation, 524–525
  numeric fields, 434
  in spreadsheet cells, 111
  using text data type for, 434
numeric checks (databases), 441
numeric keypad, 33
numeric pagers, 313

O

object-based (visual) programming languages, 401
object-oriented analysis, 398
object-oriented databases, 436
object-relational databases, 436–437
Occupational Safety and Health Administration (OSHA), 50
octets (dotted decimal notation), 524–525
Office (Microsoft), 116–118
  creating Web pages using, 128
  Office Assistant, 133
  Pro extensions, 331
  speech-recognition feature, 118
office setup, ergonomics, 50–51
office support systems (OSS), 457

Office XP
  speech-recognition feature, 118
  XP Pro extensions, 331
Oil Change (McAfee), 239
OLAP (online analytical processing) software, 456
OLEDs (organic light-emitting displays), 52
OLTP (online transaction processing), 458
Olympus P-400 dye-sublimation printer, 158
OMAP (Texas Instruments) PDA processor, 322
on/off controls (system units), 44
online auctions, 91
online data analytical processing (OLAP) software, 456
online hazards, 302–304
online help, 133
online newsgroups, 214
online service providers (OSPs), 74–75
online shopping, transaction processing, 414. 458
online software vendors, 134–135
Onscreen Keyboard, 200
Ontrack System Suite disk cleanup utility, 193
Open Directory Project subject directory, 80
open source software, 121, 173–174
Open Systems Interconnect (OSI) Reference model, 492
opening computer cases, 50
opening files, documents, 34, 191
opening software, 137–138
OpenOffice, 174
open-system protocols, 522–529
Opera web browser, 76
operating systems (OS), 19
  for desktop computers, 170–173
  functions, 168, 175
    file management, 186–188, 190–191
    hardware, device management, 179
    memory management, 178
    processor management, 176–177
    software coordination, 179
    user interface, 176–177
  history, 379–381
  laptop computers, 333
  loading during boot process, 182
  Macintosh versus PC, 20–21
  multiuser systems, 168, 170

NOS (network operating system), 168, 170, 256, 478. 491
OS/2, 168
PDAs (personal digital assistants), 322–323
reinstalling, 239
RTOS (real-time operating systems), 168–169
single-user, multitask operating system, 168–169
single-user, single-task operating system, 168–169
tablet PCs, 331
updating, 282
operators
  Boolean, 82
  in programming languages, 403
opportunity identification (SDLC), 388
optical media, 225–226
optical mouse, 36
OR operator, 82
Oracle
  8i/9i (object-relational database), 436–437
  introduction of SQL, 444
.org domains, 78
organic light-emitting displays (OLEDs), 52
organizing e-mail, 86
OS (operating systems), 19
  for desktop computers, 170–173
  functions, 168, 175
    file management, 186–188, 190–191
    hardware, device management, 179
    memory management, 178
    processor management, 176–177
    software coordination, 179
    user interface, 176–177
  history, 379–381
  laptop computers, 333
  loading during boot process, 182
  Macintosh versus PC, 20–21
  multiuser systems, 168, 170
  NOS (network operating system), 168, 170, 256, 478. 491
  OS/2, 168
  PDAs (personal digital assistants), 322–323
  reinstalling, 239
  RTOS (real-time operating systems), 168–169
  single-user, multitask operating system, 168–169

single-user, single-task operating system, 168–169
  tablet PCs, 331
  updating, 282
Osborne computer, 378
OSHA (Occupational Safety and Health Administration), 50
OSI (Open Systems Interconnect) Reference model, 492
OSPs (online service providers, 74–75
OSS (office support systems), 457
Outlook (Microsoft), 86
output, output devices
  cellular phones, 314
  digital video, 164–165
  MIS (management information systems), 459
  monitors, 38–40, 52–53
  printers, 40–43
  sound/audio, 42
  TPS (transaction processing systems), 459
Output Results task (top-down algorithm design), 397
outputting data (databases), 448
outsourcing, defined, 389

P

packages (program code), 404
packets, 256
  acknowledgment of, in connectionless protocols, 527
  amplification, navigation devices, 494–498
  bus networks, 482
  contents, 522
  filtering, screening (firewalls), 276, 501
  packet sniffers, 272
  packet switching, 522–523
  sending, 522–523
padlock icon (SSL), 531
PageMaker (Adobe), 124
pagers, paging devices, 312–313
pages per minute (ppm), 42
paging, page files
  defined, 178
  excess (thrashing), 178
  storing RAM in, 219
pain control programs, 132
Paint (Microsoft), 126
Paint Shot Pro (Jasc), 124, 158
painting software, 124–127
palm computers, 321

Palm OS, 168–170, 322–323
Palm Tungsten W, 327
Palo Alto Research Center (PARC), 380
Palo Alto Software
 Business/Marketing Plan Pro, 122
PANs (personal area networks), 476
paper, for printers, quality issues, 41
Paradox (Corel), 114
parallel ports, 46, 233–234
parallel-processing, 365
Parcel Transfer Protocol (PTP), 541–542
Pascal programming language, 410
Pascalene mechanical calculator (Blaise Pascal), 382
passive topologies, 482
passive-matrix displays (LCD monitors), 39
passwords, 299–300
 entering during boot process, 182
 good versus bad, 300–301
 protecting against identity theft, 271
pasting text, 34
patches, for software, 239, 409
paths
 for files, 190–191
 in URLs, 78
patient simulators, 8
PatriotGrid, 269
PB (petabyte), 18
PCI (Peripheral Component Interconnect) buses, 362–363
PCMCIA (Personal Computer Memory Card International Association) cards, 68
PCs (personal computers)
 microprocessors for, 354
 operating systems for, 20–21, 168
 PC cards, 68
 tablet, 329–331
PCShowandTell Web site, 133
PDA bomb software, 300
PDA Defense, 301
PDAs (personal digital assistants), 312, 321
 access protection, 300
 accessories, 326
 batteries, 324
 bistable displays, 53
 data transfers, 325
 hardware, 321
 input devices, 34–35, 37, 321
 Internet connections, 326
 memory, 324
 multimedia options, 322

 operating systems, 168–169, 322–323
 ports for, 48
 screen displays, 322
 software, 122, 326
 synchronizing with desktops, 325
 theft prevention, 297
 uses for, 8, 309
 versus tablet PCs, 331
 wireless ISPs, 75
.pdf (Portable Document Format) files, 95, 190
PDLC (program development life cycle), 387, 391
 algorithm development, 393–394
  decision points, 394–396
  flowcharts for, 396
  object-oriented analysis, 398
  pseudocode, 396–397
  top-down design approach, 397–398
 coding tools, 406–408
 compiling, interpreting code, 404–406
 debugging, 408
 documenting and training, 409
 problem statement, 392–393
 programming, 399
  programming languages, 400–403
  reusable components, 404, 406
  writing code, 401, 404–405
 updating, fixing software, 409
 user testing, 409
Peachtree Accounting, 122
peer-to-peer (P2P) networks, 253
 bus topologies for, 482
 in business settings, disadvantages, 475
 choosing, 266–267
 Ethernet networks, 259–262
 file sharing, 320–321, 520–521
 operating system support for, 256, 268
 phone line networks, 258
 power line networks, 256–257
 versus client/server networks, 474–475
pens, digital, 329–330
Pentium (Intel) microprocessors, 331, 332, 354
performance
 AGP (Accelerated Graphics Port) buses, 362
 and upgrade decisions, 217, 237
 bus width/clock speed, 361–362

 cable Internet connections, 71
 computer crashes, troubleshooting, 237
 CPUs (central processing units), 363–365
 data retrieval, 217, 217
 DSL (Digital Subscriber Lines), 70
 expansion buses, 361
 P2P (peer-to-peer) networks, 267
 printers, 42
 RAM and battery life, 334
 supercomputers, 365
 tools for enhancing
  Disk Defragmenter utility, 237
  file maintenance, 237
  software patches and updates, 237–239
  Startup folder maintenance, 237
  system utilities, 195–197
 using virtual memory, 219
Performance tab (Task Manager)
 CPU usage section, 216
 Kernel Memory Table, 218
 PF Usage section, 219–220
 Physical Memory Table, 218
Peripheral Component Interconnect (PCI) buses, 362–363
peripheral devices
 connecting during set up, 31
 port connections, 46–47
 testing during boot process, 181
permissions, access, 299–301, 499
personal area networks (PANs), 476
personal computers. See also computers
Personal Computer Memory Card International Association (PCMCIA) cards, 68
personal finance management software, 121–122
personal firewalls, 275–278, 301
personal ID chips, 14
personal information management (PIM) software, 113, 310
personnel, for systems development, 390
PestPatrol spyware removal program, 305
petabyte (PB), 18, 454
PF Usage section (Task Manager), 219–220
PGP (Pretty Good Privacy) encryption program, 538
phone line networks, 258
phonemes (speech-recognition software), 118

phones, cellular, 312
photo-editing software, 19, 124–126
photographs
   digital versus traditional photography, 159
   editing software, 124–125, 157–158
   printing, 41, 158–159
   uploading, 156–157
Photoshop (Adobe), 124, 158
PhotoSuite (Roxio), 126
physical layer (OSI networks), 492–493
Physical Memory Table (Windows Task Manager), 218
PictureIt! (Microsoft), 126
pie charts, 112
piezoelectric inkjet process, 43
PIM (personal information management) software, 113, 310
Pinnacle Studio DV, 163
pipelining, 363–364
piracy, software, 139
pixels
   and digital camera resolution, 154
   with CRT (cathode ray tube) monitors, 38
plagiarism, 89
planning, strategic, software for, 122
Plastic Logic, 358
platforms
   defined, 170, 173
   platform independent programming, 401, 412
   Macintosh versus PC, 20–21
players, plug-ins, 95–96
plotters, 42
PnP (Plug and Play) devices
   for multimedia files, 95
   versus device drivers, 179
Pocket PC (Microsoft) operating system, 168–170, 322–323
point-of-sale (POS) terminals, 6
points of presence (POPs), 519, 521
policing, role of computers in, 447
polymorphic viruses, 280
pop-up blockers, 303–304
Pop-Up Defender, 304
Pop-up Stopper, 304
POPs (points of presence), 519, 521
portability, of programming languages, 401, 412
Portable Document Format (PDF) files, 95, 190
portable storage devices, 224–226
ports

Bluetooth-compatible, 234–235
connectivity ports, 46
Ethernet ports, 233
on front panel, 47
function of, 233
IrDA ports, 233–234
on laptop computers, 333
MIDI ports, 235
parallel ports, 46, 233
serial ports, 46, 233
USB ports, 46, 233
upgrading, adding, 235–237
POS (point-of-sale) terminals, 6
positive acknowledgment (ack), connectionless protocols, 527
POST (power-on self test), 181
postcardware, 135
posture, when typing, 50
power consumption and performance, 331
power line networks, 256–257
power outages, surges, damage from, 295–297
power supply, 48
power-on self test (POST), 181
PowerLeap.com, 217
powerline connectivity, 74
PowerPoint (Microsoft), 113–114, 190
ppm (pages per minute), 42
.ppt files, 190
Premiere (Adobe), 127, 163, 165
Presentation (Corel), 113
presentation layer (OSI networks), 492
presentation software, 113–114
presentations, multimedia, players for, 85
Pretty Good Privacy (PGP) encryption program, 538
Priceline.com, 15
primary key (databases)
   validating, 451
   creating relationships using, 451–452
   identifying, 435, 450
primary storage, 49
print queues, 177
print servers (client/server networks), 478
printable microprocessors, 358
printers, 18
   buffers, 177
   costs, 42
   dot-matrix, 40–41
   downloading digital images to, 156–157
   impact versus nonimpact, 40

inkjet, 41, 43
laser, 41, 43
paper for, 41
performance, 42
plotters, 42
setting up, 31
spooling, 177
thermal, 42
printing
   keyboard shortcut, 34
   digital images, photographs, 158–159
   multiple documents, 77
privacy issues, risks
   cookies, 92
   GPS (Global Positioning Systems), 328
   hackers, 498
   surveillance cameras, 15–16
   understanding as component of computer fluency, 4
private investigators, computer use by, 9–10
private keyword (C++), 402
private networks (VPNs), 540
private-key encryption, 537–538
privileges, access, 499
.pro domains, 78
PRO/Wireless (Intel) network connections, 333
problem identification (SDLC), 388
problem statement (PDLC), 392–393
Process Data task (top-down algorithm design), 397
processing digital images, photographs, 157–158
processing data (system units), 44–49
processing, batch/real-time (TPS), 458
processors. See also computers; microprocessors
   in cellular phones, 313
   in laptop computers, 332
   managing of by operating systems, 176–177
   multiprocessor systems, 178
   in PDAs (personal digital assistants), 322
   in tablet PCs, 331
productivity software, tools, 108
   databases, 114–115
   macros, 116
   presentation software, 113–114
   software suites, 116–117, 120–121

speech recognition software, 118–119
spreadsheet software, 110–113
standalone versus integrated software, 116–117
suites, 118–119
templates, 115
tips for using, 116
wizards, 115
word processing software, 108–110
program development life cycle (PDLC), 387, 391
 algorithm development, 393–398
 debugging, 408
 documenting and training, 409
 problem statement, 392–393
 programming, coding, 399–408
 updating, fixing software, 409
 user testing, 409
program files, backing up, 306
program specifications, 389
programmers, 390
programming. *See also* languages;
 program development life cycle (PDLC)
 choosing, 410
 code sample, 404–405
 coding tools, 406–409
 compiling, interpreting code, 404–406
 component reuse, 404, 406
 data types, 403
 debugging, 408
 defined developing new software, 387
 documenting and training, 409
 defined, 391
 for household appliances, 416
 for interactivity, 533–534
 keywords, 402
 learning, tips for, 410, 417
 object-oriented analysis, 392–393, 398
 operators, 403
 problem statement, 392–393
 programming languages, 400–403
 source code, 174
 syntax, 402
 updating, fixing software, 409
 visual, 411
 for Web applications, 413–415, 532
 for Windows applications, 410–413
 writing, 401, 404
programs. See also software, 108
Project (Microsoft), 122–123

project management software, 122–123
project managers, 390
PROLOG (PROgramming LOGic) programming language, 401
promotional Web sites, 7
Properties dialog box
 viewing hard drive capacity, 223
 VRAM information, 229
property management software, 123
property, unclaimed, databases for, 453
proprietary software, 123, 174
proprietary-system protocols, 522
protected keyword (C++), 402
protecting computers
 access restrictions
  biometric authentication devices, 301
  passwords, 299–301
  PDA bomb software, 300
  personal firewalls, 301
 data backups, 305–307
 from online hazards, 302–305
 from physical dangers, 295–297
 theft prevention, 297–299
protocols
 client/server networks, 478, 491, 494
 defined, 517, 522
 e-mail
  MIME, 537
  SMTP (Simple Mail Transfer Protocol), 536
 FTP (File Transfer Protocol), 77
 http (Hypertext Transfer Protocol), 76–77
 management, development of, 516
 open systems, 522–523
 proprietary systems, 522
 PTP (Parcel Transfer Protocol), 541–542
 SSL (Secure Sockets Layer), 531
 TCP/IP (Transmission Control Protocol/Internet Protocol) suite, 523–529
 in URLs, 77
 versus topologies, 484
 WAP (Wireless Application Protocol), 326
prototyping, 411–412
proxy servers, 501
pseudocode, 396–397
PTP (Parcel Transfer Protocol), 541–542
P2P (peer-to-peer) networks, 253

bus topologies, 482
 in business settings, disadvantages, 475
 choosing, 266–267
 Ethernet networks, 259–262
 file sharing, 320–321, 520–521
 operating system support for, 256, 268
 phone line networks, 258
 power line networks, 256–257
 versus client/server networks, 474–475
public keyword (C++), 402
public safety, networking tools, 503
public-key encryption, 537–538
publishing, desktop (DTP), 124–125
punch cards, 382–383
purchasing decisions, 4

Q

QDOS (Quick and Dirty Operating System), 379
QNX 4 (RTOS), 168
QuarkXPress, 124
queries (database), 443–446
query languages, 401
queues, printer, 177
.qt files (QuickTime), 164
Quicken (Intuit), 121–122
QuickLaunch toolbar, 138
QuickTime (Apple), 95, 128, 164
quotation marks (" "), in web searches, 83
QWERTY keyboards, 32

R

RAD (rapid application development), 412
radio frequency (RF) devices, 35, 487
 ports for, 235
RAM (random access memory), 18, 49
 access time, 359
 adding RAM, 220
 DRAM (dynamic RAM), 360
  SDRAM (synchronous DRAM), 360
 evaluating, 217
 identifying system needs, 219
 installed, viewing information about, 218
 locating on motherboard, 218
 management of, 178
  adding RAM, 178

memory modules, 218
PDAs (personal digital assistants), 324
RAM hierarchy, 359
SRAM (static RAM), 361
system needs, 218
tablet PCs, 331
types of RAM, 218
use of, by CPU, 356
versus hard drive storage, 217
virtual memory, 219
Rambus Inline Memory Modules (RIMMs), 218
Rand McNally StreetFinder, 123
RAM (random access memory), 18, 49
access time, 359
adding RAM, 220
DRAM (dynamic RAM), 360
evaluating, 217
identifying system needs, 219
installed, viewing information about, 218
locating on motherboard, 218
managing, 178
memory modules, 218
PDAs (personal digital assistants), 324
RAM hierarchy, 359
SRAM (static RAM), 361
system needs, 218
tablet PCs, 331
use of, by CPU, 356
versus hard drive storage, 217
virtual memory, 219
range checks (databases), 440
rapid application development (RAD), 412
read only memory (ROM), 49, 361
in Apple II computer, 377
PDAs (personal digital assistants), 324
read only memory (ROM), 49, 361, 377
read/write heads (hard disk drives), 222
Real Networks Real One Player, 95
real numbers, 403
real-time operating systems (RTOS), 168–169
real-time processing (TPS), 458
RealMedia file format, 164
RealOne (Real Networks)player, 95, 128
recalculations, automatic (spreadsheets), 111

Recording Industry Association of America (RIAA), 228
records (database), 115
defined, 435
fields, primary ke, 435
sorting, 442
viewing, 442
recovering deleted files, 192
Recycle Bin, 192
Red Hat Linux, 174
reference software, 129–130
referential integrity, 451
refresh rates (monitors), 231
refrigerators, Internet-enabled, 543
REFS (Remote Fingerprint Editing Software), 447
regional ISPs (Tier 2), 74, 517–518
registers, 355
registry, checking during boot process, 182
reinstalling operating system, 239
relational algebra (SQL), 444
relational databases, 436
creating relationships, 451–452
normalizing data, 449–451
referential integrity, 451
relationships, 436, 448
uses for, 436
Relationships screen (Access), 452
reliability, of system. See performance
REM keyword (Visual Basic), 404
remote access, 490
Remote Fingerprint Editing Software (REFS), 447
renaming files, 191
repeaters (client server networks), 2160, 496
reports
detailed, 459
database, 448
exception, 459
generators fpr, 401
summary, 459
repositioning tiled windows, 185
required fields, 440–441
research tools. See Internet
resizing tiled windows, 185
resolution
digital cameras, 154–156
monitors, 230–231
printers, 42
scanners, 156
resources, sharing, 251–252, 474–476
restarting computers, 45
RestoreIT! (FarStone), 192–193

restoring systems, utilities for, 136, 193, 198–199
retail, using computers in, 6
retinitis pigmentosa, chip implants for, 14
retrieving deleted files, 192
reusability (OOP), 398, 404, 406
RF (radio frequency) devices, 35
RFI (radio frequency interference), 487
RIAA (Recording Industry Association of America, 228
RIMMs (Rambus Inline Memory Modules), 218
ring network topology, 483, 486
"ripping", 317
Ritchie, Dennis (UNIX), 173
RJ-11 jacks, 260
RJ-45 jacks (Ethernet ports), 233, 260, 494
.rm file form, 164
Roberts, Ed, 376
RoboMower (Friendly Robotics), 12
robotic equipment
AI (artificial intelligence), 15
home uses, 11–12
operating systems for, 169
rollerball mice, 35–36
Rolltronics printable processors, 358
ROM (read only memory), 49, 361
in Apple II computer, 377
PDAs (personal digital assistants), 324
Roomba (iRobot), 12
root directory, 187–188
root domain name servers, 529
routers (client/server entworks), 256, 478, 497
DSL/cable routers, 262
with Ethernet networks, 261–262
function, 256
with phone line networks, 258
with power line networks, 258
wireless DSL/cable, 265
Roxio
GoBack utility, 192
PhotoSuite, 126
RTOS (real-time operating systems), 168–169
rules for validation (databases), 440
run length (cable), 486
runtime (logic) errors, testing for, 408

S

Safe Mode, 183–184
SafeMessage e-mail security program, 538–539
safety concerns
  cellular phones, 314
  e-commerce, 92
Salary.com Web site, 558
sales, tracking, 6
sampling rates, 317
SANs (storage area networks), 481
satellite, Internet, 69, 72
satin paper, 41
Sauvante, Michael, 358
saving
  digital images, 157
  documents, 34
  Save As dialog box (Windows), 190–191
  Save command (Microsoft Word), 110
SC Magazine, 307
ScanDisk (Windows) utility, 197
scanners, 156
ScanSoft Dragon Naturally Speaking, 118
scheduling system-related tasks, 199–200
schema, database, 437
screen displays. *See also* monitors
  on mobile devices, 311
  resolution, 230–231
  screensavers, 193
  tablet PCs, 329
screensavers, 193
script kiddies, 270
script viruses, 280
scripting languages, 414, 533
scrollbars, 185
scrolling, 90
SCSI (Small Computer System Interface) hard drives, 223
SDLC (system development life cycle), 388
  analysis phase (Phase 2), 389
  design phase (Phase 3), 389
  development and documentation (Phase 4), 389
  key personnel, 390
  maintenance and evaluation (Phase 6), 391
  problem, opportunity identification (Phase 1), 388
  testing and installation (Phase 5), 391

SDRAM (synchronous DRAM), 218, 360
SDSL (Symmetrical Digital Subscriber Line), 70
Search (Find in File) tool, 34
searching
  databases (queries), 443–446
  the web
    search engines, 79–81
    search strategies, 82–83
    subject directories, 82
search/replace tools, 109
Seattle Computer Products, 379
second-generation computers, 384
second-generation programming languages (2GLs), 400
second-level domains, 528
sectors (hard disk drives), 222
Secure Digital memory, 319
secure HTTP (s-HTTP), 524
Secure Sockets Layer (SSL), 531
security
  authentication process, 182
security risks
  chat rooms, 87
  computer viruses, 278–281
  cybercrime, cybercriminals, 269
  cyberterrorism, 272
  e-commerce, 91–92
  e-mail, 537–538
  hackers, crackers, 270, 272, 498
    brute force attacks, 499
    computer access approaches, 274–275
    denial of service (DoS) attacks, 273–274
    identity theft, 270–272
    Trojan horses, 272–273, 501
  identity theft, 270–271
  understanding, as component of computer fluency, 4
  viruses, 251, 280–281
security professionals, 283, 499
security tools
  authentication and access restrictions, 299–301, 498–499
  backing up data, 305–307, 502
  browser updates, 530
  encryption, 537–538
  firewalls, 275–278, 501
  for laptop computers, 335
  physical access restrictions, 500
  SC Magazine, 307
  system updates, 282
  utilities
    anti-pop-up software, 303–304

  antivirus software, 281–282, 302
  SafeMessage, 538–539
  spam filters, 302–303
  spyware removal software, 304–305
  VPNs (virtual private networks), 540
Securus Systems Ltd. (Safe Message), 538–539
seek time (hard disk drives), 222
SELECT queries (SQL), 444–445
semi-gloss paper, 41
semiconductor technology, 349
sending
  e-mail, 536–537
  packets, 522–523
Sensory Enhanced Net Experience (SENX) device (TriSenx), 542–543
SEQUEL query language, 444
serial ports, 46, 233–234
servers, 21, 67
  authentication servers, 500
  client/server networks, 474–475, 478–481, 520
  defined, 66, 254
  DNS (domain name servers), 528–529
  e-mail, 536
  FTP file servers, 76
  proxy servers, 501
  server-based networks, 474–475
  server-side programs, 533
  Web, 532
service packs, 409
session layer (OSI networks), 492–493
setup program
SEX (Sensory Enhanced Net Experience) device (TriSenx), 542–543
shareware, 135–136, 529
Shareware.com, 135
sharing
  data, 251–252, 431
  files, 529
  resources, 251–252, 474, 476
  software, 474
Sheats, Jim, 358
shielded twisted pair (STP) cable, 487
ShieldUP (Gibson Research), 278
Shockwave Player (Macromedia), 95
Short Message Service (SMS), 316
shortcuts, keyboard, 34
sign bit, 351
signal transmission methods, 487

signatures, virus, 281
signed integer notation, 351
silicon, 349
SIMMs (Single Inline Memory Modules), 218
Simple Mail Transfer Protocol (SMTP), 524, 536
Simple Query Wizard (Access), 446
simulations
 software for, 129
 use of in medical training, 8
Singapore, traffic management in, 477
Single Inline Memory Modules (SIMMs), 218
single-user, multitask operating systems, 168–169
single-user, single-task operating systems, 168–169
six-color printers, 42
slide shows (presentation software), 113–114
slide sorter view (PowerPoint), 114
small business accounting software, 122
Small Computer System (SCSI) hard drives, 223
smart devices, 12
smart displays, 331
Smart Media cards, 319, 324
SmartSuite (Lotus), 120
SMS (Short Message Service), 316
SMTP (Simple Mail Transfer Protocol), 524, 536
SnagIt (Techsmith), 135
software
 anti-spam, 86
 application software, 19, 108, 168
  digital video import/editing programs, 162–163, 165
  image-editing programs, 157–158
 backup software, 307
 buying, 134–136
 for communications, 132–133
 computer-protection, 303
 coordination of by operating systems, 179
 defined, 19, 108
 DVD authoring, 165
 e-mail management, 84
 educational and reference, 129–130
 entertainment, 131
 file transfer, 76
 financial and business, 121–123
 graphics and multimedia, 123–128

 help systems for, 133–134
 history of, 379–381, 384
 home networks, 268
 installing/uninstalling, 136–137, 193–194
 licenses, 139
 NOS (network operating systems), 256, 478, 491
 open-source, 173–174
 opening, 137–138
 operating systems, 168–170, 175–179, 186–188, 190–191
 for PDAs (personal digital assistants), 326
 piracy, 139
 pop-up blockers, 304
 productivity software, 108–110, 113–121
 security tools, 275, 281–282
 sharing, 474
 spam filters, 302
 spyware removal, 305
 system software, 19–21, 108
 tablet PCs, 331
 updating, 237–239
 utility programs, 168, 192–200
 versions, compatibility isses, 136
software development, 387
 choosing programming language
  Web applications, 413–415
  Windows applications, 410–413
 PDLC (program development life cycle), 391
  algorithm development, 393–398
  coding tools, 406–408
  compiling, interpreting code, 404–406
  debugging, 408
  documenting and training, 409
  problem statement, 392–393
  programming, 399–406
  updating, fixing software, 409
  user testing, 409
software suites
 graphics suites, 127
 office suites, 117–121
 pros and cons, 116–117
Sony Memory Stick, 319, 324
sorting
 database data, 442
 lists, limits of, 431
sound
 input devices, 37–38
 output devices, 42
 sound cards, 48, 232–233

Sound Bytes
 buying a computer, 21
 computer architecture, 362
 computer cleanup activities, 237
 computer protection utilities, 307
 connecting to Internet, 75
 C++ integrated development environment, 407
 CPU performance evaluation, 216
 creating an Access database, 441
 creating macros, 391
 creating a Web page, 531
 creating web-based e-mail accounts, 86
 customizing Windows XP, 186
 desktop system tour, 48
 detecting IP addresses, 496
 enhancing an Access database, 451
 enhancing photographic images, 126
 file compression/decompression, 195
 file management, 187
 healthy computing, 51
 how to use, 22
 installing a CD-RW drive, 226
 installing networks, 256
 installing RAM, 220
 Internet security/privacy protection utilities, 92
 network technician careers, 497
 searching the web, 83
 security tools, 282
 surge protectors, 297
 system maintenance utilities, 199
 tablet PCs and laptop computers, 335
 using PDAs, 326
 using speech-recognition software, 120
 utilities for improving Internet usability, 535
 viewing ports, 235
 virtual computer tour, 18, 50
 web site tours, 78
 wireless network security, 265
 wireless networks, access protection, 302
 working with binary, decimal and hexadecimal numbers, 352
sound effects, 163, 165
source code
 defined, 174, 404
 in HTML documents, viewing, 531–532

limiting access to, 174
stems), 458
spaces in filenames, 191
spam, 85–87
    spam filters, 302–303
SpamButcher filter, 303
speaker notes, adding to presentations, 114
speakers (audio), 18, 42, 231–232
special effects, 163, 165
specialty search engines, 82
speech-recognition software, 38, 118–119, 200, 330
speeds. *See also* performance
    data access, transfer, 68
        Bluetooth technologys, 234–235
        CD and DVD disk drives, 226
        Ethernet ports, 233
        FireWire (IEEE 1394) ports, 233
        hard disk drives, 221–223
        IrDA ports, 233–234
        MIDI ports, 235
        modem ports, 46–47
        parallel and serial ports, 46, 233–234
        USB ports, 46, 233–234
        USB 2.0/FireWire ports, 47
    processing, 349
    versus accuracy, 527
spell checking tools, 109
spider programs (search engines), 80
spoolers, 177
spreadsheet software, 19, 110
    automatic recalculation feature, 111
    cells, 110
    charting, graphic features, 112
    formatting options, 112
    formulas and functions, 111
    history, 380
    linking to Internet, 113
    lists, 430
    uses for, 112
    "what if" analyses, 112
    worksheets, 110
Sprint, 74
spyware, 304–305
SQL (Structured Query Language), 401, 443–444
    join queries, 445
    relational algebra, 444
    SELECT queries, 444–445
    wizards for, 446
SQL View window (Access), 446
SRAM (static RAM), 218, 361
SSL (Secure Sockets Layer), 531

staging, data (data warehouses), 455
stand-alone software, pros and cons, 116–117
Standard VGA displays, 230
standards
    for Internet connectivity, 516
    for networking, 492
standby mode (system units), 45
star network topology, 484–486
Star Office System (Xerox), 381
StarOffice (Sun), 121
Start menu
    accessing, 34
    Control Panel, 136
    opening software from, 137
starting computers
    boot process, 180–184
    warm versus cold boots, 45
Startup folder, cleaning out, 237
statements (programming languages), 402
static addressing, 277, 525
static RAM (SRAM), 218, 361
stealth viruses, 281
steering wheels, 37
storage, storage devices
    of BIOS (basic input/output system), 181
    cache memory, 357
    cellular phones, 314
    data marts, 456
    data warehouses, 452, 454–456
    database size, 454
    digital images, 157
    files, 190–191
    flash memory card, 155, 224
    hard disk drives, 49, 221–223
    laptop computers, 332
    MP3 players, 317
    optical media, 224
    portable storage, 224–226
    tablet PCs, 330
    upgrading, 226–227
    volatile versus nonvolatile, 49, 217
    web-based e-mail, 86
    Zip disks, 224
storage stage (CPU machine cycle), 355, 359
storage subsystems, 220–223
STP (shielded twisted pair) cable, 487
strategic planning software, 122
strategies for web searches, 82–83
streaming audio and video, 95–96
Streaming SIMD (Single Instruction Multiple Data) Extensions, 365

StreetFinder (Rand McNally), 123
Streets & Trips (Microsoft), 123
structured data, storing in relational databases, 436
Structured Query Language (SQL), 401, 443–444
    join queries, 445
    relational algebra, 444
    SELECT queries, 444–445
    wizards for, 446
StuffIt file compression utility, 195
styluses, 34–35, 321
subdirectories, 78
subject directories (World Wide Web), 80–83
subroutines (program code), 404
subsystems
    CPU subsystem, 214–217
    evaluating during system evaluation, 214
    memory (RAM), 217–219
    storage, 220–223
subwoofers (speakers), 232
suites, software, 116–120
summary reports
    databases, 448
    MIS (Management Information Systems), 459
Sun Microsystems
    development of UNIX, 173
    StarOffice software suite, 121
supercomputers, 21, 365
    operating systems, 168, 170
    versus mainframes, 21
supervisor progam, 182
surge protectors, 295–296
surround sound, 232
surrounds, 298
surveillance cameras, 15–16
swap files, 178
switching, switches
    binary, 348–349
    client/server networks, 478, 496–498
    circuit switching, 522
    packet switching, 522–523
Symbian OS, 168, 313
symbols, notation for, 351–352
Symmetrical Digital Subscriber Line (SDSL), 70
symptoms of virus infection, 280
synchronizing PDAs, 325
synchronous DRAM (SDRAM), 218, 360
syntax (programming languages), 402

syntax errors, 407–408
system clock, 355
system development life cycle (SDLC), 388
   analysis phase (Phase 2), 389
   design phase (Phase 3), 389
   development and documentation (Phase 4), 389
   key personnel, 390
   maintenance and evaluation (Phase 6), 391
   problem, opportunity identification (Phase 1), 388
   testing and installation (Phase 5), 391
system evaluation, 211, 214
   CPU subsystem, 214–217
   memory (RAM) subsystem, 217–219
   storage subsystem, 220–223
system files, 182
System Properties dialog box (My Computer)
   clock speed, 215
   installed RAM, 218
system requirements
   for new software, 136
   for RAM, 218
System Restore utility (Windows), 193, 198–199
system software, 19, 108, 168
   accessibility utilities, 199–200
   add/remove programs, 193–194
   backup programs, 198
   defragmenters, 193, 195–197
   desktop display, 192–194
   desktop platforms, 170
   disk cleanup programs, 193, 195–196, 198
   disk scanning, 197
   file compression, 193–195
   file management, 186–188, 190–191
   function, 192
   functions, 168, 175
   hardware, device management, 179
   Linux operating system, 173
   Macintosh operating systems, 20–21, 172
   memory management, 178
   multiuser operating systems, 168, 170
   processor management, 176–177
   RTOS (real-time operating systems), 168–169
   single-user, multitask operating system, 168–169
   single-user, single-task operating system, 168–169
   software coordination, 179
   UNIX operating system, 173
   user interface, 176–177
   utility programs, 168
   Windows operating systems, 170–171
   system restore programs, 193, 198–199
   task schedulers, 199–200
System toolbar, 216
System Tools (Windows), 195, 199
system unit, 18, 44
   binary electronic switches, 348–353
   CPU (central processing unit), 18, 353–365
   data handling technologies, 348
   desktop computers, 44–47
   hard drives, 49
   interior components, 48
   memory, 18
   motherboard, 18, 49
   opening, 50
system use, viewing, 216
systems, defined, 388. *See also* SDLC (system development life cycle)
systems analysts, 390
SystemWorks (Norton), 193

T

tables (database), 115, 442
   avoiding repetition in, 450
   creating relationships among, 451–452
   linking in relational databases, 436
tablet PCs, 312
   data transfer, storage, 330
   docking stations, 330
   hardware, 329, 331
   input devices, 329–330
   memory, 331
   software, 331
   uses for, 329
   versus PDAs (personal digital assistants), 331
Tabulating Machine Company, 383
tags
   HTML, 531
   XML, 534–535
Tandy Radio Shack TRS-80 computer, 378
Task Manager utility (Windows), 197
   CPU usage section, 216
   PF Usage section, 219–220
Task Scheduler utility, 199–200
tax preparation software, 121
TaxCut (H&R Block), 121
TB (terabyte), 18
TCP (Transmission Control Protocol)
   functions, 524
   versus UDP (User Datagram Protocol), 526
TCP/IP (Transmission Control Protocol/Internet Protocol) suite, 476, 523–527
technical diagrams, software for, 126–127
technical writers, 390
technological advances, and upgrade decisions, 212
technologies, wireless, 5
TechSmith SnagIt, 135
TechTV, 541
telephone number fields (databases), 434
telephones
   cellular, 313–316
   dial-up connections, 69
   DSL lines, 70
   sharing, with phone line networks, 258
telephony, Internet, 132, 255, 538
Telnet protocol, 524, 529
templates, 115
   for slide shows, presentations, 113
   in graphics programs, 126
temporary Internet files
   clearing, 237
   removing, 86
temporary storage, 49
Teoma (Ask Jeeves) search engine, 80, 82
terabyte (TB), 18, 454
terminals, point-of-sale (POS), 6
terrorism, cyberterrorism, 272
test condition (loops), 396
test preparation software, 129
testing
   data input plans, 392–393
   firewalls, 275
   programs, 408–409
   systems during systems development, 391

Texas Instruments OMAP PDA processor, 322
text messaging, 316
textbook layout, programs for, 124–125
textiles, electronic, 320
theft
 of computers, 297–299
 by hackers, crackers, 272
 identity theft, 270–272
thermal printers, 42–43
ThickNet/ThinNet coaxial cable, 488
third-generation computers, 384
Thompson, Ken (UNIX), 173
thrashing, 178
three-way handshakes, 526
throughput, 68, 486
Thumbnails view (My Documents), 188–189
tiers, for ISPs, 517–518
TIFF file format, 155
Tile newsgroup site, 214
Tiles view (My Documents), 188
tiling windows, 185–186
Time To Live (TTL), 528
time-variant data, 454
titles, adding to digital video, 163, 165
TLDs (top-level domains), 77–78, 525, 528
T-lines (ISP connections), 518
toggle keys, 34
token-ring network topology, 483
Tomlinson, Roy (e-mail), 536
toner, for laser printers, 43
Tool menu (Microsoft Word), 110
toolbars, 185
tools, productivity, 115
top-down design approach, 397–398
top-level domains (TLDs), 77–78, 525, 528
topologies, network (client/server networks)
 bus (linear bus) topology, 481–482, 486
 ring (token-ring) topology, 483, 486
 star topology, 484–486
 versus protocols, 484
Torvalds, Linus (Linux), 173
touchpads, 36–37
touchscreens, 322
tower configurations, 44
toys, operating systems for, 168–170
TPS (transaction processing systems), 458–459

trackball mice, 35–36
tracking software, for computers, 299
trackpoints, 37
tracks (hard disk drives), 222
traffic control networks, 477
traffic, Internet, effect on routers, 523
training
 computer-based, 8, 129
 resources for, 214
 for using software, 133, 214, 409
 virtual reality programs for, 132
transaction processing systems (TPS), 458–459
transceivers, 262
Transcriber (Microsoft), 322
transistor-based computers, 384
transistors, 348–349
translation tools, 109
Transmeta Crusoe 5800 processori, 331
Transmission Control Protocol (TCP), 523–524, 526
transmission media, processes
 client/server networks, 478, 486, 491
 comparisons among, 489
 network communications, 256
 wired, 255, 486–488
 wireless, 488–489
transport layer (OSI networks), 492–493
Trends in IT feature, 22
Treo (Handspring) cell phone, 327
trial periods (shareware), 135
TriSex sensory enhancement devices, 542–543
Trojan horses, 272–273, 280, 501
troubleshooting, 4
 boot process errors, 182–184
 system damage, 193
TRS-80 (Tandy Radio Shack) computer, 378
true color mode displays, 230
TTL (Time To Live), 528
Tungsten W (Palm) PDA, 327
tunneling technology (VPNs), 540
TurboTax (Intuit), 121
Turing Machine (Alan Turing), 383
tutorials, 129
 for HTML, 413
 locating, 133, 214
 sources for, 214
 for using software programs, 133

twisted pair wiring, 69–70, 255, 486–487
two-way pagers, 313
types, data, 434–435
typing breaks, 51

U

ubiquitous networks, 332
UDP (User Datagram Protocol), 524, 526
Ulead DVD Workshop, 165
unamplified speakers, 231–232
unauthorized access, protecting against, 274–275
 biometric authentication devices, 301
 passwords, 299–301
 PDA bomb software, 300
 personal firewalls, 275–278, 301
unclaimed property databases, 453
Unicode, 352
Uniform Resource Locators (URLs)
 domain names, 77–78
 protocols, 76–77
 secure sites, 92, 531
uninstalling software
 utilities for, 193–194
 versus deleting, 137
uninterruptible power supplies (UPSs), 297
unique fields (databases), 435
United Devices, 269
UNIVAC (Universal Automatic Computer), 384
Universal Serial Bus (USB) ports, 46–47
 downloading using, 156–157
 ports, 46, 233, 236–237
 USB 1.1, 234
 USB 2.0, 47
UNIX operating system, 168, 170, 173
unshielded twisted pair cable (UTP), 260, 487
unstructured data, storing, 436
updating software, 239, 409
 operating system upgrades, 239
 web browsers, 530
upgrading computers, 4
 decisions about
  audio subsystem, 231–233
  CPU subsystem, 214–217
  determining ideal system, 213–214
  determining training needs, 214

memory (RAM) subsystem, 217–220
performance and reliability issues, 237–239
port connectivity, 233–237
storage subsystem, nonvolatile storage, 220–227
system evaluation, 214
technological advances, 212
video subsystem, 228–231
laptop computers, 335
versus buying new system, 238, 240
uploading files, 529
UPSs (uninterruptible power supplies), 297
URLs (Uniform Resource Locators)
domain names, 77–78
protocols, 76–77
secure sites, 92, 531
U.S. Department of Labor Statistics job outlook projections, 556
U.S. Federal Trade Commission identity theft information, 270
US HomeGuard program, 15
USB (Universal Serial Bus)
downloading using, 156–157
ports, 46, 233, 236–237
USB 1.1, 234
USB 2.0, 47
used software, pros and cons, 134
User Datagram Protocol (UDP), 524, 526
user interface
command-driven, 176
defined, 176
for binary data, 353
GUI (graphical user interface), 176–177, 184–186
menu-driven, 176
Linux systems, 173
user testing (beta testing), 409
user training, 409
usernames in chat rooms, 88
users
authentication and access restrictions, 182, 498–499
physical access restrictions, 500
Utility Manager (Windows XP), 199–200
utility programs
accessibility utilities, 199–200
add/remove programs, 193–194
backup, 198
defragmenters, 193, 195–197

desktop customizing programs, 192–194
disk cleanup programs, 193, 195–196, 198
disk scanning programs, 197
file compression/decompression programs, 193–195
function, 168, 192
screensavers, 193
system restore programs, 193, 198–199
task schedulers, 199–200
UTP (unshielded twisted pair) cable, 260, 487

V

vacuum tubes, 348, 383–384
validation rules (databases), 439–441
values
in database fields, 438
in spreadsheet cells, 111
variables, in program code, 401, 404
VBScript programming language, 414
VeriChips, 14, 328
versions, software, 136
vertical tiling, 185–186
video
cameras, 160–162
capture devices, 160
cards, 48, 228–229
clips, storing, 434–436
digital video, 163–165
editing software, 127–128, 163, 165
game consoles, 169
monitors, 230–231
streaming video, 94–95
video RAM (VRAM), 229
videoconferencing, 38, 133
viewing on PDAs, 322
viewing
database data, 442
files, folder in My Documents, 188
slide shows, 114
views (My Documents folder), 188–189
virtual dancers, 7
Virtual Machine (VM, Java), 412
virtual memory, 178, 219–220
Virtual Network Computing (VNC), 490
virtual private networks (VPNs), 540
virtual reality programs, 131
ESRB rating, 131
historical reconstructions, 132

Internet-based, 542–543
medical uses, 132
special equipment for, 131
sports and training applications, 132
virtual tours, 10–11
viruses, computer, 251, 278–279
antivirus software, 302
boot-sector viruses, 279
logic bombs, 280
macro viruses, 280
multipartite, 281
polymorphic, 280
protecting against, 281–282
script viruses, 280
signatures, 281
stealth, 281
symptoms of infection, 280
Trojan horses, 280
worms, 280
VisiCalc spreadsheet program, 380
Visio (Microsoft), 126–127, 396
Visual Basic programming language, 401
For and Next keywords, 402
prototyping feature, 411–412
writing comments, 404
Visual C++ programming language, 407
visual programming, 411
visualization approach (data mining), 462
VNC (Virtual Network Computing), 490
Voice over IP (VoIP), 538
voice pagers, 313
voice-recognition software, 118–119
volatile memory
in cellular phones, 314
PDAs (personal digital assistants), 324
volatile storage, 49, 217
VPNs (virtual private networks), 540
VRAM (video RAM), 229

W

Walker, Jay, 15
WANs (wide area networks), 254, 476–477
WAP (Wireless Application Protocol), 326
war drivers, 302
warehouses, data, 452, 454–455
warm booting, 45, 182

WAV files, 95
waves, analog, 151–152
wearable computers, 53, 320
Web, the
 career opportunities, 535
 Web applications, 413–415
 web browsers, 66
  choosing, 75, 106
  features, 76
  history, 381, 530–531
  Mosaic, 530
  navigating, 76–79
  Netscape, 531
  security tools, 531
  updating, 530
  web documents, 531
 the Web versus the Internet, 530
web clipping, 326
Web pages, personal Web sites
 authoring software, 128
 filename extensions, 190
 HTML (Hypertext Markup Language), 531–532
 interactive, 414–415, 532–535
 home page, 76
 hyperlinks, 76
 for wireless devices, 317
 XML (Extensible Markup Language), 415
Web publishers, 535
web searches, 79–83
Web server administrators, 535
Web servers, 481, 520
 cgi-bin directory, 532
 versus file servers, 76
Web site URLs
 academic software discounts, 134
 AvantGo, 326
 Better Business Bureau, 92
 Business Software Alliance, 139
 Car Talk, 314
 chat room security, 88
 Computer Game WYSIWYG, 404
 computer tracking software, 299
 Dell EducateU training site, 214
 Driver Zone, 179
 DSL line availability, 70
 Dummies productivity tips, 116
 e-mail forwarding services, 87
 Entertainment Software Rating Board, 131
 freeware, 135
 for FTP downloads, 529
 Gamelan (Java) site, 411
 geocaching, 328
 Gibson Research, 275

"Hello Word" program in 204 languages, 410
Hoaxbusters, 87
Home Phone line Networking Alliance, 258
HomePlug Power line Alliance, 256
hot spot directories, 265
HTML tutorials, 413
Internet management organizations, 516
Internet software vendors, 134
IP addresses, 67
ISPs (Internet service providers), 74
KIWE browser, 76
Large Scale Networking program, 541
Librarians' Index to the Internet, 83
Lindows, 174
Lynx web browser, 76
Microsoft, 268
Microsoft anti-piracy program, 139
Microsoft Windows Update, 409
NerdyBooks, 116
Network Solutions, 528
online backups, 307
Opera Web browser, 76
OSHA, 50
PDA Defense, 301
for PDA software, 326
PGP (pretty Good Privacy), 538
personal firewall providers, 301
plug-ins and players, 95
pop-up blockers, 304
popular search engines and subject directories, 80
PowerLeap.com, 217
programming jobs, 417
promotional, 7
QuickTime player, 164
RAM resellers, 218
RealNetworks, 164
Safe Message program, 538–539
Salary.com, 558
SC Magazine, 307
shareware vendors, 135
Smart Home, 12
software tutorials, 133
spam-blocking software, 87
spyware removal software, 305
TechTV, 541
Tile newsgroup site, 214
TriSenx, 542–543

for unclaimed property, 453
United Devices, 269
U.S. Bureau of Labor Statistics, 556
U.S. Federal Trade Commission, 270
Virtual Network Computing, 490
web browser updates, 530
web-based e-mail, 85–87, 536
Web-enabled devices, 310
webcam monitoring, 15–16
WebCT course management software, 129
Webmasters, 535
Webnests.com, 133
"what if" analyses, 112
WHERE statements (SQL), 445
white-hat hackers, 270
whole numbers in binary number system, 350
whole-house surge protection, 296
Wi-Fi (Wireless Fidelity) standard, 262, 334, 489
wide area networks (WANs), 254, 476–477
wildcards in web searches, 83
windows, 185
 in GUIs (graphical user interfaces), 176
 multiple, viewing in desktops, 185–186
 pop-up, 303–304
Windows (Microsoft)
 accessibility utilities, 199–200
 applications, programming languages for, 410–413
 Clear Type feature, 193
 controls, keyboard shortcuts, 34
 disk check-and-repair utilities, 197
 Explorer, 187–188, 191–193
 file compression utilities, 195
 filenaming conventions, rules, 191
 help systems, 34
 home networking, 268
 importing digital video, 163
 Media Player, 95, 128, 164
 Messenger, 539
 microprocessors for, 354
 Movie Maker (Microsoft), 162–163
 operating systems, 168–170, 185
  GUIs (graphical user interfaces), 176
  Mobile 2003 operating system, 168

Tablet PC operating system, 331
Windows 2000, 170
Windows 95 (Microsoft), 381
Windows 98/ME, 268
Windows CE, 168–169
Windows XP, 163, 170–171, 193, 216, 268, 331
Recycle Bin, 192
Save As dialog box, 190–191
Task Manager, 216–218
Update service, 409
WinZip file compression utility, 190, 193, 195
wired data transmission, 486–488
wireless "hot spots", 265
wireless access points, 264–265, 494
Wireless Application Protocol (WAP), 326
wireless data transmission, 476, 488–489
wireless devices, 5
   ISP (Internet service providers) for, 75
   DSL/cable routers, 265
   keyboards, 35
   laptop computers, 334
   mice, 36
   network interfacecards (wireless NICs), 494
wireless networks, 262
   hot spots, 265
   installation problems, 264
   Internet connections, 264
   network adapters, 262
   pros and cons, 267
wiring, preexisting, for P2P networking, 267
wizards, 115, 446
.wmv file format, 164
Word (Microsoft), 108
   filename extensions, 190
   importing database data to, 448
   saving documents and files, 110
   Tool menu, 110
Word for MS-DOS, 380
word processing programs, 19, 108–110, 380, 430
word size (ALU), 359
WordPerfect (Corel), 108, 118, 190, 380
WordStar, 380
Works (Microsoft), 117
worksheets, 110
World Wide Web (WWW), 66. *See also* Internet; Web
World Wide Web Consortium (W3C), 516. 535
worms, 280
Wozniak, Steve (Apple), 377, 379
W3C (World Wide Web Consortium), 516, 535
WWW (World Wide Web), 66. *See also* Internet; Web
WYSIWYG (What You See Is What You Get principle, 380

XYZ

Xeon (Intel) microprocessor, 354
Xerox, 380
XML (Extensible Markup Language), 415, 534–535
.xls files, 190
XP Network Setup Wizard, 268
Yahoo!, 75
   Briefcase, 520
   Messenger, 539
   subject directory, 80, 82
   web-based e-mail accounts, 86
Yepp 55-v MP3 player (Samsung), 318
Zip disks, drives, 46
   adding to system, 227
   caring for, 225
   how they work, 224
   storage capacity, 224
.zip files, 190
zombies, 273
Z1 mechanical calculator, 383
zTraceGold computer tracking software, 299
Zuse, Konrad, 383

CREDITS

page 1 — Piotr Powietrzynski/Index Stock Imagery/PictureQuest

Chapter 1

Chapter Opener	© Mendola/Doug Chezem/CORBIS
Figure 1.1	© Chuck Savage/CORBIS
Figure 1.4	© Index Stock Imagery/Tomas Del Amo
Figure 1.5	© Derek Bencomo
Figure 1.6	COPYRIGHT 2003 GEORGIA TECH, CREDIT: STANLEY LEARY
Figure 1.7	Source: Medical Education Technologies, Inc. © 2003
Figure 1.9a	Wyeknot, inc.
Figure 1.9b	21st Century Forensic Animation
Figure 1.10	National Museum of American History, Behring Center. © Smithsonian Institution.
Figure 1.11a	© Aqua Products, Inc.
Figure 1.11b	Friendly Robotics® and Robomower® are are registered trademarks of F. Robotics Acquisitions Ltd.
Figure 1.11c	© iRobot
Figure 1.12	Courtesy of C. Durkan, University of Cambridge
Figure 1.13	© DR PETER HARRIS / PHOTO RESEARCHERS, INC.
Figure 1.14	© MPI BIOCHEMISTRY / VOLKER STEGER / PHOTO RESEARCHERS, INC.
Figure 1.15	Image of VeriChip courtesy of Applied Digital Solutions
Figure 1.16	DAVID PARKER / PHOTO RESEARCHERS, INC.
Figure 1.20	Open-sided computer photo © BONNIE KAMIN / PHOTO EDIT
Figure 1.21	Microsoft Office box shot reprinted with permission from Microsoft Corporation.
Figure 1.21	Microsoft Project box shot reprinted with permission from Microsoft Corporation.
Figure 1.21	Lotus SmartSuite box shot courtesy of International Buisness Machines Corporation. Unauthorized use not permitted.
Figure 1.21	Microsoft Works Suite box shot reprinted with permission from Microsoft Corporation.
Figure 1.21	Microsoft Flight Simulator box shot reprinted with permission from Microsoft Corporation.
Figure 1.21	CorelDraw Graphics box shot reprinted with permission from Corel Corporation.
Figure 1.21	WordPerfect box shot reprinted with permission from Corel Corporation.
Figure 1.21	Adobe PageMaker box shot reprinted with permission from Adobe Systems Incorporated.
Figure 1.21	Adobe Illustrator box shot reprinted with permission from Adobe Systems Incorporated.
Figure 1.23a	Presley Salaz, LANL
Figure 1.23b	© F64 /Getty Images

Chapter 2

Chapter Opener	© Dave Robertson/Masterfile
Figure 2.4a	© TRBfoto/Getty Images
Figure 2.4b	COURTESY OF DATADESK TECHNOLOGIES INC.
Figure 2.19	Screen shot(s) reprinted by permission from Microsoft Corporation.
Figure 2.21	© BONNIE KAMIN / PHOTO EDIT
Figure 2.23	© BONNIE KAMIN / PHOTO EDIT
Figure 2.26	COURTESY OF DATADESK TECHNOLOGIES INC.
Figure 2.28	Image courtesy of Universal Display Corporation
Figure 2.29	Courtesy of MicroOptical Corporation.

# CREDITS

Chapter 3

Chapter Opener	© Matthew Wiley/Masterfile
Figure 3.2a	MuVo is a registered trademark of Creative Technology Ltd. In the United States and/or other countries. NOMAD is a registered trademark of Aonix and is used by Creative Technology Ltd. and/or its affiliates under license.
Figure 3.3	© IEI 03
Figure 3.8a	Netscape website screenshot © 2003 Netscape Communications Corporation. Used with permission.
Figure 3.8b	Screen shot(s) reprinted by permission from Microsoft Corporation.
Figure 3.11	Reproduced with permission of Yahoo! Inc. © 2003 by Yahoo! Inc. YAHOO! and the YAHOO! logo are trademarks of Yahoo! Inc.
Figure 3.12	Screen shot(s) reprinted by permission from Microsoft Corporation.
Figure 3.14a	© 2003 Google
Figure 3.14b	Reproduced with the permission of Overture Services, Inc. All rights reserved.
Figure 3.18	Reproduced with permission of Yahoo! Inc. © 2003 by Yahoo! Inc. YAHOO! and the YAHOO! logo are trademarks of Yahoo! Inc.
Figure 3.19	Screen shot(s) reprinted by permission from Microsoft Corporation.
Figure 3.20	Courtesy of Childnet International
Figure 3.21	AOL Instant Messenger screenshot © 2003 America Online, Inc. Used with permission.
Figure 3.22a	Copyright 1999–2003 Dell Computer Corporation
Figure 3.22b	© 2003 target.direct. The Bullseye Design and Target are Registered Trademarks of Target Brands, Inc.
Figure 3.22c	Copyright © J. Crew
Figure 3.22d	Copyright 1997–2003 Vivendi Universal Net USA Group, Inc.
Figure 3.23	Copyright © 2002, Harleysville Savings Bank
Figure 3.24	Screen shot(s) reprinted by permission from Microsoft Corporation.

Chapter 4

Chapter Opener	© Sanford/Agliolo/CORBIS
Figure 4.2	Screen shot(s) reprinted by permission from Microsoft Corporation.
Figure 4.3	Screen shot(s) reprinted by permission from Microsoft Corporation.
Figure 4.4a	Screen shot(s) reprinted by permission from Microsoft Corporation.
Figure 4.4b	Screen shot(s) reprinted by permission from Microsoft Corporation.
Figure 4.5	Screen shot(s) reprinted by permission from Microsoft Corporation.
Figure 4.6a	Screen shot(s) reprinted by permission from Microsoft Corporation.
Figure 4.6b	Screen shot(s) reprinted by permission from Microsoft Corporation.
Figure 4.6c	Screen shot(s) reprinted by permission from Microsoft Corporation.
Figure 4.7	Screen shot(s) reprinted by permission from Microsoft Corporation.
Figure 4.8a	Screen shot(s) reprinted by permission from Microsoft Corporation.
Figure 4.8b	Screen shot(s) reprinted by permission from Microsoft Corporation.
Figure 4.9	Microsoft Office box shot reprinted with permission from Microsoft Corporation.
Figure 4.9	Lotus SmartSuite box shot courtesy of International Business Machines Corporation. Unauthorized use not permitted.
Figure 4.9	WordPerfect box shot reprinted with permission from Corel Corporation.
Figure 4.9	Sun StarOffice 7 box shot © Sun Microsystems 2003
Figure 4.10a	© Royalty-Free/CORBIS
Figure 4.10b	Screen shot(s) reprinted by permission from Microsoft Corporation.
Figure 4.12	TurboTax screen shots and product box shot © 2002 Intuit Inc. All rights reserved. Used by permission. TurboTax is a registered trademark of Intuit Inc.
Figure 4.13	Screen shot(s) reprinted by permission from Microsoft Corporation.
Figure 4.14	Screen shot(s) reprinted by permission from Microsoft Corporation.
Figure 4.18	© Chuck Savage/CORBIS
Figure 4.21	Courtesy of Helsinki Polytechnic in cooperation with University of Industrial Arts, Helsinki Figure Photographer, Mr. Esa Kyyrö
Figure 4.22	Copyright © 1997–2003. Blackboard Inc. All rights reserved. Blackboard and the Blackboard logo are registered trademarks of Blackboard Inc.
Figure 4.23	Screen shot(s) reprinted by permission from Microsoft Corporation.
Figure 4.25	Copyright © 1999–2003 ESRB/IDSA
Figure 4.26	Photo by Gretchen Carrougher, U.W., image (on right) by Duff Hendrickson, U.W. Both copyrighted by Hunter Hoffman, U.W.
Figure 4.27	© R.W. Jones/CORBIS
Figure 4.28	Screen shot(s) reprinted by permission from Microsoft Corporation.
Figure 4.29	Screen shot(s) reprinted by permission from Microsoft Corporation.
Figure 4.30	Copyright © 1999–2003 Thraex Software
Figure 4.31a	Screen shot(s) reprinted by permission from Microsoft Corporation.
Figure 4.31b	Screen shot(s) reprinted by permission from Microsoft Corporation.
Figure 4.31c	Screen shot(s) reprinted by permission from Microsoft Corporation.
page 146	Lotus SmartSuite box shot courtesy of International Business Machines Corporation. Unauthorized use not permitted.
page 146	WordPerfect box shot reprinted with permission from Corel Corporation.
page 146	StarOffice box shot © Sun Microsystems 2003.
page 148	Adobe Illustrator box shot reprinted with permission from Adobe Systems Incorporated.
page 148	Adobe PageMaker box shot reprinted with permission from Adobe Systems Incorporated.
page 148	Lotus SmartSuite box shot courtesy of International Buisness Machines Corporation. Unauthorized use not permitted.
page 148	Microsoft Works Suite box shot reprinted with permission from Microsoft Corporation.
page 148	Microsoft Flight Simulator box shot reprinted with permission from Microsoft Corporation.
page 148	CorelDraw Graphics box shot reprinted with permission from Corel Corporation.
page 148	WordPerfect box shot reprinted with permission from Corel Corporation.

Chapter 5

Chapter Opener	© Mendola/Jeff Mangiat/CORBIS
Figure 5.2	(space shuttle) NASA
Figure 5.2	(car) © C Squared Studios/Getty Images
Figure 5.2	(stethoscope) © Digital Vision/Getty Images
Figure 5.2	(phone) © Siede Preis/Getty Images
Figure 5.5	Used with permission from CNET Networks UK Ltd. Copyright 2003. All rights reserved.
Figure 5.6	The Linux penguin, "Tux", was created by Larry Ewing using The GIMP (General Image Manipulation Program).
Figure 5.7	Used with permission from CNET Networks UK Ltd. Copyright 2003. All rights reserved.
Figure 5.10	Screen shot(s) reprinted by permission from Microsoft Corporation.
Figure 5.12	Screen shot(s) reprinted by permission from Microsoft Corporation.
Figure 5.15	Screen shot(s) reprinted by permission from Microsoft Corporation.
Figure 5.16	Screen shot(s) reprinted by permission from Microsoft Corporation.
Figure 5.17	Screen shot(s) reprinted by permission from Microsoft Corporation.
Figure 5.18	Quicken Brokerage screen shot © 2003 Intuit Inc. All rights reserved. Used by permission. Quicken is a registered trademark of Intuit Inc. Microsoft screen shot reprinted by permission from Microsoft Corporation.
Figure 5.19	Screen shot(s) reprinted by permission from Microsoft Corporation.
Figure 5.20	Screen shot(s) reprinted by permission from Microsoft Corporation.
Figure 5.21	Screen shot(s) reprinted by permission from Microsoft Corporation.
Figure 5.23	Screen shot(s) reprinted by permission from Microsoft Corporation.
Figure 5.27	Screen shot(s) reprinted by permission from Microsoft Corporation.
Figure 5.30	Screen shot(s) reprinted by permission from Microsoft Corporation.
Figure 5.31	Screen shot(s) reprinted by permission from Microsoft Corporation.
Figure 5.32	Screen shot(s) reprinted by permission from Microsoft Corporation.

Chapter 6

Chapter Opener	© Colin Anderson/Brand X Pictures/PictureQuest.
Figure 6.5	Screen shot(s) reprinted by permission from Microsoft Corporation.
Figure 6.6	Screen shot(s) reprinted by permission from Microsoft Corporation.
Figure 6.7	© Terra Nova Designs
Figure 6.9	Screen shot(s) reprinted by permission from Microsoft Corporation.
Figure 6.11	Screen shot(s) reprinted by permission from Microsoft Corporation.
Figure 6.12	© Terra Nova Designs
Figure 6.13	© BONNIE KAMIN / PHOTO EDIT
Figure 6.15	Screen shot(s) reprinted by permission from Microsoft Corporation.
Figure 6.24	Screen shot(s) reprinted by permission from Microsoft Corporation.
Figure 6.25	Screen shot(s) reprinted by permission from Microsoft Corporation.
Figure 6.29a	Courtesy of Hack In The Box, www.hackinthebox.org
Figure 6.29b	Courtesy of Hack In The Box, www.hackinthebox.org
Figure 6.29c	Courtesy of Hack In The Box, www.hackinthebox.org
Figure 6.29d	© Terra Nova Designs
Figure 6.29e	Courtesy of Hack In The Box, www.hackinthebox.org
Figure 6.29f	Courtesy of Hack In The Box, www.hackinthebox.org
Figure 6.29g	Courtesy of Hack In The Box, www.hackinthebox.org
Figure 6.29h	Courtesy of Hack In The Box, www.hackinthebox.org
Figure 6.29i	Courtesy of Hack In The Box, www.hackinthebox.org
Figure 6.29j	© Terra Nova Designs
Figure 6.29k	Courtesy of Hack In The Box, www.hackinthebox.org
Figure 6.29l	Courtesy of Hack In The Box, www.hackinthebox.org
Figure 6.30	Courtesy of Hack In The Box, www.hackinthebox.org
page 248	© BONNIE KAMIN / PHOTO EDIT

Chapter 7

Chapter Opener	© H. Prinz/ CORBIS	Figure 7.29	© 1995–2003 Symantec Corporation
Figure 7.2	Screen shot(s) reprinted by permission from Microsoft Corporation.	Figure 7.30	© 1995–2003 Symantec Corporation
Figure 7.21	Screen shot(s) reprinted by permission from Microsoft Corporation.	Figure 7.31	Screen shot(s) reprinted by permission from Microsoft Corporation.
Figure 7.27	Copyright © 2003 Gibson Research Corporation		

Chapter 8

Chapter Opener	© Peter Griffith/Masterfile	Figure 8.9c	Courtesy of Apple. © Apple.
Figure 8.1	© 2002 Wireless 2000, courtesy of Richard S. Anderson, WebMaster (webmaster@phoneswa.com), and Brian Humes, Owner of Complete Wireless Centers DBA Wireless 2000 (brian@phoneswa.com), www.phoneswa.com.	Figure 8.9d	Copyright 2003 Archos, Inc.
		Figure 8.11	Courtesy of Infineon Technologies.
		Figure 8.17	Courtesy of Hitachi Global Storage Technologies.
		Figure 8.20a	Photo courtesy of Handspring, Inc.
Figure 8.4	Reproduced with Permission from Motorola, Inc. © 2003, Motorola, Inc.	Figure 8.20b	Image used with permission of Palm, Inc. Palm is a trademark of Palm, Inc.
Figure 8.6a	Kyocera 7135 Smartphone courtesy of Kyocera Wireless Corp.	Figure 8.23	© Reuters NewMedia Inc./CORBIS
Figure 8.6b	Nokia Inc.	Figure 8.26	© Microsoft
Figure 8.6c	Reproduced with Permission from Motorola, Inc. © 2003, Motorola, Inc.	Figure 8.27	Laboratory for Communication Engineering, Cambridge
Figure 8.9a	MuVo is a registered trademark of Creative Technology Ltd. In the United States and/or other countries. NOMAD is a registered trademark of Aonix and is used by Creative Technology Ltd. and/or its affiliates under license.	Figure 8.30a	BROTHER U.K. LTD.
		Figure 8.30b	Courtesy of Epson America, Inc.
		Figure 8.32	© YURI GOGOTSI/PHOTO RESEARCHERS, INC.

Chapter 9

Chapter Opener	© Denis Scott/CORBIS	page 372	© FERRANTI ELECTRONICS / A. STERNBERG / PHOTO RESEARCHERS, INC.
Figure 9.2c	© FERRANTI ELECTRONICS / A. STERNBERG / PHOTO RESEARCHERS, INC.		

Chapter 10

Chapter Opener	© Paul Cooklin/Brand X Pictures/PictureQuest	Figure 10.19	Screen shot(s) reprinted by permission from Microsoft Corporation.
Figure 10.13	Screen shot(s) reprinted by permission from Microsoft Corporation.	Figure 10.22	© 1995–2003 Macromedia, Inc.
Figure 10.18	Screen shot(s) reprinted by permission from Microsoft Corporation.		

622 CREDITS

Chapter 11

Chapter Opener	© Boden/Ledingham/MasterFile
Figure 11.1	Screen shot(s) reprinted by permission from Microsoft Corporation.
Figure 11.2a	Screen shot(s) reprinted by permission from Microsoft Corporation.
Figure 11.2b	Screen shot(s) reprinted by permission from Microsoft Corporation.
Figure 11.3	(financial aid office) © Lawrence Manning/CORBIS
Figure 11.3	(teacher) © John Henley/CORBIS
Figure 11.3	(student registration) © Jose Luis Pelaez, Inc./CORBIS
Figure 11.3	(parent) © Bill Varie/CORBIS
Figure 11.3	(student housing office) © Jon Feingersh/CORBIS
Figure 11.4	Screen shot(s) reprinted by permission from Microsoft Corporation.
Figure 11.6	Screen shot(s) reprinted by permission from Microsoft Corporation.
Figure 11.7	Copyright © 2003 Oracle Corporation
Figure 11.8a	Screen shot(s) reprinted by permission from Microsoft Corporation.
Figure 11.8b	Screen shot(s) reprinted by permission from Microsoft Corporation.
Figure 11.9	Screen shot(s) reprinted by permission from Microsoft Corporation.
Figure 11.10	Screen shot(s) reprinted by permission from Microsoft Corporation.
Figure 11.11	Copyright © 1997–2003 The Vermont Teddy Bear Company, Inc.
Figure 11.12	Screen shot(s) reprinted by permission from Microsoft Corporation.
Figure 11.13a	Screen shot(s) reprinted by permission from Microsoft Corporation.
Figure 11.13b	Screen shot(s) reprinted by permission from Microsoft Corporation.
Figure 11.14a	Screen shot(s) reprinted by permission from Microsoft Corporation.
Figure 11.14b	Screen shot(s) reprinted by permission from Microsoft Corporation.
Figure 11.14c	Screen shot(s) reprinted by permission from Microsoft Corporation.
Figure 11.15	Screen shot(s) reprinted by permission from Microsoft Corporation.
Figure 11.16a	Screen shot(s) reprinted by permission from Microsoft Corporation.
Figure 11.16b	Screen shot(s) reprinted by permission from Microsoft Corporation.
Figure 11.17	Screen shot(s) reprinted by permission from Microsoft Corporation.
Figure 11.18	Federal Bureau of Investigation, Criminal Justice Information Services Division
Figure 11.19a	Screen shot(s) reprinted by permission from Microsoft Corporation.
Figure 11.19b	Screen shot(s) reprinted by permission from Microsoft Corporation.
Figure 11.20	Screen shot(s) reprinted by permission from Microsoft Corporation.
Figure 11.21	Screen shot(s) reprinted by permission from Microsoft Corporation.
Figure 11.22	Screen shot(s) reprinted by permission from Microsoft Corporation.
Figure 11.23	Screen shot(s) reprinted by permission from Microsoft Corporation.
Figure 11.24	Screen shot(s) reprinted by permission from Microsoft Corporation.
Figure 11.25	Screen shot(s) reprinted by permission from Microsoft Corporation.
Figure 11.26	© 2003 California State Controller's Office
Figure 11.28	(web user) © Reed Kaestner/CORBIS
Figure 11.28	(OLAP query) © LWA-JDC/CORBIS
Figure 11.29	(tele-salesperson) © R.W. Jones/CORBIS
Figure 11.29	(supplier shipments) © R.W. Jones/CORBIS
Figure 11.29	(online shopping) © Paul Barton/CORBIS
Figure 11.29	(online data retrieval) © Tom & Dee Ann McCarthy/CORBIS
Figure 11.30	(step 1) © Norbert Schaefer/CORBIS
Figure 11.30	(step 2) © Royalty-Free/CORBIS
Figure 11.30	(step 4) © Royalty-Free/CORBIS
Figure 11.30	(step 5) © CORBIS
Figure 11.31a	Screen shot(s) reprinted by permission from Microsoft Corporation.
Figure 11.31b	Screen shot(s) reprinted by permission from Microsoft Corporation.
Figure 11.31c	Screen shot(s) reprinted by permission from Microsoft Corporation.

Chapter 12

Chapter Opener   © Chris McElcheran/MasterFile

Figure 12.15   Copyright 1999 AT&T Laboratories Cambridge. Reproduced with permission.

Figure 12.24   U.S. Geological Survey, Earthquake Hazards Program

Chapter 13

Chapter Opener   © Didier Boutet/Brand X Pictures/PictureQuest

Figure 13.10   Screen shot(s) reprinted by permission from Microsoft Corporation.

Figure 13.11a   (left) © Royalty-Free/CORBIS

Figure 13.11b   (right) © Jose Luis Pelaez, Inc./CORBIS

Figure 13.14   Screen shot(s) reprinted by permission from Microsoft Corporation.

Figure 13.15   Copyright © 1999–2001 Futurefiction.com

Figure 13.23   © The Electrolux Group

Technology in Focus: Digital Entertainment

Figure 5a   "CANON", the Canon logo, Powershot and Elura are trademarks of Canon Inc. All rights reserved. Used by Permission.

Figure 5b   Photo © Konica Minolta Photo Imaging U.S.A., Inc.

Figure 5c   Copyright Nikon Inc.

Figure 5d   Image courtesy of Kodak Professional, © Kodak.

Figure 10   Copyright © 1995–2003, Jasc Software, Inc.

Figure 11a   "CANON", the Canon logo, Powershot and Elura are trademarks of Canon Inc. All rights reserved. Used by Permission.

Figure 11b   Olympus America Inc.

Figure 13a   RCA camcorder photo courtesy of Thomson.

Figure 13b   "CANON", the Canon logo, Powershot and Elura are trademarks of Canon Inc. All rights reserved. Used by Permission.

Figure 15a   Screen shot(s) reprinted by permission from Microsoft Corporation.

Figure 15b   Screen shot(s) reprinted by permission from Microsoft Corporation.

Figure 16   Adobe product screen shot(s) reprinted with permission from Adobe Systems Incorporated.

Technology in Focus: Protecting Your Computer and Backing Up Your Data

Figure 3	Image courtesy of American Power Conversion Corporation.	Figure 13	© 1997–2004 Barnesandnoble.com llc
		Figure 14	Copyright © 2002–2003 PestPatrol, Inc.
Figure 8	Screen shot(s) reprinted by permission from Microsoft Corporation.	Figure 15	Screen shot(s) reprinted by permission from Microsoft Corporation.
Figure 12	Reproduced with permission of Yahoo! Inc. © 2003 by Yahoo! Inc. YAHOO! and the YAHOO! logo are trademarks of Yahoo! Inc.		

Technology in Focus: The History of the PC

Figure 1	Courtesy of apple2history.org	Figure 14	Photo courtesy of Apple Computer, Inc.
Figure 2a	Getty Images/Getty Images	Figure 15	Photo courtesy of Apple Computer, Inc.
Figure 2b	© Roger Ressmeyer/CORBIS	Figure 16	© The Computer History Museum
Figure 3	Photo courtesy of Apple Computer, Inc.	Figure 17	© The Computer History Museum
Figure 4	Photo courtesy of Apple Computer, Inc.	Figure 18	Courtesy of Ames Laboratory
Figure 6	© JERRY MASON/PHOTO RESEARCHERS, INC.	Figure 19	Naval Historical Center
Figure 7	© The Computer History Museum	Figure 20	© The Computer History Museum
Figure 8	© The Computer History Museum	page 376	© Boden/Ledingham/Masterfile
Figure 9	TIME MAGAZINE, COPYRIGHT TIME INC.	page 381	©Bill Frymire/Masterfile
Figure 10	© Doug Wilson/CORBIS	page 385	© Piotr Powietrzynski/Index Stock Imagery/PictureQuest
Figure 11	Photo courtesy of Dan Bricklin and Bob Frankston.		
Figure 13	© The Computer History Museum		

Technology in Focus: Careers in IT

Figure 2	Bureau of Labor Statistics	Figure 11	© CareerBuilder
Figure 3	Copyright 2000–2003 Salary.com, Inc.	Figure 13	© Merck & Co., Inc.
Figure 5	(stressed worker) © Jon Feingersh/CORBIS	page 554	© Colin Anderson/ Brand X Pictures / PictureQuest
Figure 5	(marketing department) ©LWA- JDC/CORBIS	page 555	© IT Stock Free/ PictureQuest
Figure 5	(golfer) © Mark A. Johnson/CORBIS	page 557	(left) © DigitalVision/ PictureQuest
Figure 5	(supervisor) © Gabe Palmer/CORBIS	page 557	(right) © IT Stock Free/ PictureQuest
Figure 5	(customer) © Larry Williams/CORBIS	page 558	(top left) © Corbis Images/ PictureQuest
Figure 7	Copyright © 1999–2001 Futurefiction.com	page 558	(bottom left) © IT Stock Free/ PictureQuest
Figure 9	© Royalty-Free/CORBIS	page 564	(background) © DigitalVision/ PictureQuest